Water
Sustainability

Water Sustainability

A Global Perspective

J. A. A. Jones

Routledge
Taylor & Francis Group

LONDON AND NEW YORK

To my wife Jennifer, and to Claire and Mark and their families

First edition published 2010 by
Hodder Education, a member of the Hodder Headline Group

Published 2014 by Routledge
2 Park Square, Milton Park, Abingdon, Oxfordshire OX14 4RN
711 Third Avenue, New York, NY 10017

Routledge is an imprint of the Taylor and Francis Group, an informa business

First issued in hardback 2015

British Library Cataloguing in Publication Data
A catalogue record for this book is available from the British Library

Library of Congress Cataloging-in-Publication Data
A catalog record for this book is available from the Library of Congress

ISBN 978-1-4441-0488-2 (pbk)
ISBN 978-1-138-16794-0 (hbk)

Cover image: © José Luis Gutiérrez/iStockphoto.com

Typeset in 10 on 13pt Sabon by Phoenix Photosetting, Chatham, Kent

Contents

Conclusions

This book has a companion website available at
http://cw.tandf.co.uk/geography/

Preface

"If humanity can avoid nuclear holocaust in the 21st century, then water is going to be the most important issue to deal with."

Professor András Szöllösi-Nagy
Rector, UNESCO-IHE Institute for Water Education
and former Director, Division of Water UNESCO

When the world entered the twenty-first century there were more people without access to safe water and sound sanitation than ever before. A third of the world population were living in countries suffering moderate to severe water stress. The Millennium Development Goals set down by the UN in 2000 included a commitment to begin to rectify the situation, at least to halve the number of people without access to safe water and basic sanitation by 2015. Unfortunately, we are fast approaching 2015 and way short on the numbers. International aid to developing countries has fallen well short of the promises. And the prospects seem even bleaker following the credit crunch of 2008–9.

Many factors are working against a solution, some old, some new. But the greatest of these is population growth. Population growth is solely responsible for the dramatic reduction in per capita water resources in recent decades, and its exponential rise is predicted to reduce resources by another third over the coming half-century. It is in the developing world and emerging economies that the pressure will be felt most. Not only are these mostly the regions with the fastest population growth, they are also countries that suffer from extreme poverty, rapid industrialization and heavy water pollution, and often wasteful water management, sometimes singly but most commonly in combination.

In writing this book I have endeavoured to present a balanced and in-depth view of the problems and their root causes, and to try to foresee the soundest ways forward. There is no 'magic bullet', indeed no ultimate solution at all. Too many interests militate against any solution of the population issue. But we can do things to mitigate the effects of climate change, and even reduce the rate of global warming itself. Many new technologies will help

in this and with water conservation. However, for the vast majority of the needy, simple 'human scale' solutions will go a long way. If not eliminating the problems completely, at least these simple approaches can alleviate the worst – education and aids to self-help are high on the list. Commitment comes first, funding a strong second.

We offer this book as a contribution from the IGU Commission for Water Sustainability to the serious debates now ongoing, and especially to students of water resources in the hope that it will encourage them to delve more deeply into the issues it raises and maybe get involved in some of the work that is so urgently needed. The issues cover a wide range of disciplines and it is also to be hoped that students of economic, politics and law might find many issues of interest: your participation is sorely needed.

The book is also an official contribution to the UN-designated International Year of Planet Earth (2008–9), for which I had the privilege of leading the groundwater team. The IYPE marks 50 years since the International Geophysical Year, which proved the springboard for the development of global collaboration and field observations in the geosciences. In 1957–8, the only water-related activities were in glaciology and the emphasis was purely on the physical science. The world has changed very much since, in many ways it seems more hazardous, and science is focusing more and more on service to humanity. The themes of the IYPE include megacities, climate change, soils, natural resources, hazards, and health and life, as well as deep geology and oceanography.

At the same time, the UN designated three other significant International Years. The International Polar

Year continues the theme of the IGY, but with the social urgency caused by the effects of rapid climate change – as a weather vane for the Earth – rather than the pure physics of glacier movement and mass balance, studied for its own sake. The International Year of Sanitation focuses on one of the major Millennium Development Goals, while the International Year of the Potato sounds rather more esoteric. Daft as it may sound, however, the IYP has a very important message for water resources too. In the world as a whole, rice is the staple carbohydrate and 21 per cent of all water used goes into growing rice. Only 1 per cent is used to grow potatoes. Of course, more rice is grown than potatoes, but just look at the difference in water consumption; it is astounding. On average, 3400 m³ of water is used to produce a tonne of white rice; in India it actually takes 4254 m³/tonne. The world average for a tonne of potatoes is just 250 m³, a saving of more than three million litres per tonne. The drive should be on to get farmers in the Developing World to switch. Unfortunately, culture and tradition militate against success. But with food production consuming over 80 per cent of world water supplies and with rising populations and increasing affluence in the emerging economies demanding more food, agriculture has to be a prime focus for conserving water.

I am indebted to many colleagues and organizations for assistance and financial support in this project. I am especially indebted to the International Geographical Union and the International Council for Science (ICSU), who funded my main research assistants, and in particular to IGU Presidents and Secretary-Generals Bruno Messerli, Eckhart Ehlers, Anne Buttimer and Ron Abler and Vice-President Changming Liu for their unerring faith in the project. I also owe a great deal to my publishers for their support and latterly to Bianca Knights, Liz Wilson and Lyn Ward for their invaluable advice on the final product. The project started life as an atlas – and some of this is still evident in the text. But the issues raised clearly require far more than a conventional atlas, and it was fortuitous that Hodder decided to curtail its atlas series while the work was in progress.

I received enthusiastic support from the managers and heads of a number of international data archives: Dr-Ing Thomas Maurer (Global Runoff Data Centre, Koblenz), Dr Richard Robarts (GEMS/Water, Burlington, Ontario), Dr David Viner and Professor Phil Jones (Climatic Research Unit, University of East Anglia, England), and Professor Dr Wilhelm Struckmeier (WHYMAP Programme, German Federal Institute for Geosciences and Natural Resources (BGR), Hanover). Thanks also to the UN Food and Agriculture Organization for use of the AQUASTAT and FAOSTAT archives, to the management of Eurostat and to the International Union for Conservation of Nature (IUCN). Frank Holsmuller and his colleagues Andrew Parry and Matt Beven at ESRI (Environmental Systems Research Institute, Inc.) gave valuable advice on mapping and Kent Lethcoe at USGS/EROS Data Center, Sioux Falls, guided us through the Interrupted Goode Homolosine map projection. Thanks also to Arjen Hoekstra for tutoring Róisín Murray-Williams in water footprint analysis.

In addition to those colleagues who have made written contributions to the book (listed separately), I would like to thank colleagues who have offered support and advice, especially the Commission's Vice-chairmen Mingo-ko Woo and Kazuki Mori, also John Gash (Centre for Ecology and Hydrology, UK), Des Walling (President of the World Association for Sediment and Erosion Research), Petra Döll (Johann Wolfgang Goethe University, Frankfurt, Germany), Cathy Reidy Liermann (University of Washington, USA), John and Annabel Rodda (formerly president of the International Association of Hydrological Sciences and consultant for the UN Non-Governmental Liaison Service, respectively), Gerry Jones (honorary president, International Commission on Snow and Ice) and his successor John Pomeroy (ICSI Hydrology), David Butler (Professor of Engineering, Exeter University, and former head of the Urban Water Research Group, Imperial College London), Saul Arlosoroff (Mekorot, National Water Corporation, Israel), Stephen Salter, Professor of Engineering Design, Edinburgh University, Arona Soumare (Conservation Director, World Wildlife Fund – West Africa), Ross Brown (Meteorological Service of Canada), Kevin Hiscock (University of East Anglia, UK), Mark Rosenberg (ex-chair, IGU Commission on Health and Environment), Peter Adams, PC (Professor emeritus, Trent University, Canada), Atsumu Ohmura (ETH, Zurich), José Maria García-Ruiz (ex-director, Instituto Pirenaico de Ecología, Zaragoza), Graham Sumner (formerly University of Wales, Lampeter), David Kay (Director, Centre for Research into Environment and Health, Aberystwyth University) and Dick and Duncan Maclean. Nor should I forget my son, Dr Mark Jones, MRCGP, who gave helpful medical advice on Chapter 5.

Without the extensive support from directors and colleagues at the Institute of Geography and Earth Sciences, Aberystwyth, this book would not have been possible.

My thanks go particularly to my Senior Research Assistant, Dr Keren Smith, and to my Senior Cartographer, Ian Gulley, who has done such a fantastic job in transforming our (very) rough drafts into highly presentable maps and diagrams. My thanks also to research assistant Alison Geeves and MSc internship students, Sonia Thomas and Sophie Bonnier, from the Montpellier SupAgro International Centre for Advanced Studies in Agronomic Engineering. Additional cartographic assistance was provided by Antony Smith, Dr Aled Rowlands, Dr Catherine Swain and James Bell. Further support for the research assistants was provided by the Oakdale Trust and by the Walter Idris Bursary (Aberystwyth University).

Anthony Jones
IGU Commission for Water Sustainability
and Aberystwyth University

The author and the publishers would like to thank the following for their permission to reproduce images, data and quotations:

Permission to use data from:
FAO Aquastat (figs. 1.1–1.3, 2.1, 2.3–2.5, 2.7, 2.8) and FAOSTAT (figs. 8.1–8.9, 20.2); Eurostat (figs. 5.3, 5.17); CRU, University of East Anglia (figs. 11.6–11.24); GEMS/Water (figs. 5.1, 5.6, 5.9, 5.21, 6.6, 6.7, 6.15); Global Runoff Data Centre (figs. 11.25–11.30, 18.7); World Health Organization (figs. 5.22–5.26); International Union for Conservation of Nature (figs. 6.11, 6.13); UN Economic and Social Commission for Asia and the Pacific (fig. 3.4); Professor Martina Flörke (figs. 2.2, 10.2); Ebro Hydrographic Department (fig. 7.22); Professor Arjen Hoekstra, Water Footprint Network (figs. 2.6, 8.10–8.14); Ross Brown, Meteorological Service of Canada, Dr David Robinson, Rutgers University, and NOAA (figs. 18.3, 18.4).

Permission for Figures edited and/or redrawn from:
Professor Aondover Tarhule (figs. 3.2, 3.3); Dr C.J.J. Schouten (fig. 5.2); Dr Frank Winde (fig. 5.8); Professor Wang Wuyi, based on original maps published by Tan Jianan, Wang Wuyi, Zhu Wenyu, Li Ribang and Yang Linsheng in *The atlas of endemic diseases and their environment in the People's Republic of China* (1989), Science Press, Beijing (figs. 5.10–5.15); British Geological Survey (fig. 5.16); WHO, based on unpublished WHO document WHO/VBC/89.967 (fig. 5.27); Professors D.E. Walling and B. Webb (fig. 6.17); Emeritus Professor John Lewin (fig. 7.2); Professor A. Hoekstra (figs. 815–17); Dr C. Pilling (fig. 10.1); Dr Thomas Maurer, Global Runoff Data Centre (figs. 11.31, 18.6); Professor Graham Cogley (figs. 12.1, 12.3); Professor Martin Funk, Head of the Division of Glaciology at E.T.H. Zurich, graph based on data from the Swiss Glacier Monitor compiled by Dr Andreas Bauder, Glaciological Commission of the Swiss Academy of Natural Sciences, and supplied by Professor Atsumu Ohmura (figs. 12.2, 12.11); Professors Lydia Espizua and Gabriela Maldonado (figs. 12.8–12.10); Professor Wilhelm Struckmeier, based on *Groundwater Resources of the World 1:25 000 000* (2008 edition), downloadable from www.whymap.org (fig. 12.12); Professors Martina Flörke and Petra Döll (fig. 12.13); Professor D.D. Chiras/Benjamin Cummings (fig. 12.14); Jos Kuipers (fig. 15.1); Dr H. van Hout (fig. 15.3); Professor Jun Xia (figs. 15.7–15.9); Emeritus Professor Stephen Salter (fig. 16.4); Dr Mike Marshall (figs. 19.1, 19.2); Professor Emeritus Tom McMahon and Dr Murray Peel (figs. 19.3–19.5); Water UK (fig. 20.1).

Photographs:
Professor Doracie Zoleta-Nantes (figs. 3.5, 5.4, 5.5); Professor Emeritus Richard Wilson, Harvard University, the Arsenic Foundation Inc., Boston, MA, USA, and Steven Lamm, MD (figs. 5.18–5.20); Dr Simon Benger (figs. 6.3–6.5); Dr Santiago Begueria (fig. 7.20); Società Meteorologica Italiana (Italian Meteorological Society) and Italian Glaciological Committee (figs. 12.4–12.7); Kurobe City Office, Toyama Prefecture, Japan, annotated by Professor Kazuki Mori (fig. 13.1); Professor Jana Olivier (fig. 14.4); Gareth Evans (fig. 15.5); NASA Goddard Space Flight Center and US Geological Survey, Landsat EROS imagery (fig. 15.11); Marcia Brewster and the UN Division for Sustainable Development, Interagency Gender and Water Task Force, UNICEF Photography Unit, Division of Communication, New York, USA – Zimbabwe (UNICEF/HQ02-0352), Rwanda (UNICEF/HQ99-0460) and Iraq (UNICEF/HQ99-0663) photographed by Giacomo Pirozzi; Burma (UNICEF/HQ95-0340) photograph by Franck Charton (figs. 20.3–20.6).

Quotations:
Chapter 4 Public Private Partnerships: Quotation reprinted with permission from: House of Commons Transport Committee, 2008: *The London Underground and the Public-Private Partnership Agreements*. Second Report of Session 2007–08, 25 January, HC45, London, The Stationery Office.

Every effort has been made to obtain necessary permission with reference to copyright material. The publishers apologise if inadvertently any sources remain unacknowledged and will be glad to make the necessary arrangements at the earliest opportunity.

List of Contributors

Dr Abdullah Almisnid, University of Gassim, Unizah, Saudi Arabia.

Dr Simon N. Benger, Geography, Population and Environmental Management, School of the Envrionment, Flinders University, Adelaide, Australia.

Dr David B. Brooks, Director of Research, Friends of the Earth Canada, Ottawa, Ontario, Canada.

Dr Laura Brown, Department of Geography, York University, Toronto, Ontario, Canada.

Marcia Brewster, former Task Manager, Interagency Gender and Water Task Force, United Nations Division for Sustainable Development, Department of Economic and Social Affairs (DESA), United Nations Building, New York, USA. Currently Vice President, UN Association Southern New York Division.

Professor Claudio Cassardo, Department of General Physics, University of Turin, Turin, Italy.

Professor J. Graham Cogley, Department of Geography, Trent University, Peterborough, Ontario, Canada. He is leading the updating of the World Glacier Inventory held at the US National Snow and Ice Data Center (NSIDC), Boulder, Colorado, USA.

Dr L.A. Dam-de Heij, CSO Adviesbureau (Consultancy for Environmental Management and Survey), Bunnik, The Netherlands.

Professor Joseph W. Dellapenna, School of Law, Villanova University, Villanova, Pennsylvania, USA.

Dr Lydia E. Espizua, National Research Council (Consejo Nacional de Investigaciones Cientificas y Tecnicas – CONICET), Argentina.

Professor Dr Mariele Evers, Chair of Sustainable Land Development, Leuphana University of Lüneburg, Lüneburg, Germany.

Dr-Ing Martina Flőrke, Senior Researcher, Center for Environmental Systems Research (CESR), University of Kassel, Kassel, Germany.

Simone Grego, Programme Officer, World Water Assessment Programme, Perugia, Italy.

Dr Glyn Hyett, Managing Director, 3P Technik UK Limited and Celtic Sustainables (Environmental Services), Cardigan, Ceredigion, UK.

Professor Arjen Y. Hoekstra, Scientific Director, Water Footprint Network, University of Twente, Enschede, The Netherlands.

Professor Alain Jigorel, Director, National Institute for Applied Sciences (Institut National des Sciences Appliquées (INSA) de Rennes), Rennes, France.

Dr Peter Khaiter, School of Information Technology, York University, Toronto, Canada.

Professor Nurit Kliot, Department of Geography, University of Haifa, Haifa, Israel.

Academician Professor Changming Liu, Institute of Geographical Sciences and Natural Resources, Chinese Academy of Sciences, Beijing, China. Former Vicepresident, International Geographical Union.

Dr-Ing João-Paulo de C. Lobo-Ferriera, Head of Groundwater Division, National Laboratory for Civil Engineering (LNEC), Lisbon, Portugal.

Emeritus Professor Thomas A. McMahon, Department of Civil and Environmental Engineering, University of Melbourne, Melbourne, Victoria, Australia.

Dr Gabriela I. Maldonado, National Research Council (Consejo Nacional de Investigaciones Cientificas y Tecnicas – CONICET), Argentina.

Dr Michael Marshall, Post-doctoral research assistant, Institute of Geography and Earth Sciences, Aberystwyth University, UK.

Dr-Ing Thomas Maurer, ex-Director, Global Runoff Data Centre, at the German Federal Institute of Hydrology (Bundesanstalt für Gewässerkunde – BfG), Koblenz, Germany.

Róisín Murray-Williams, Scottish Water, Edinburgh; former Water Science graduate, Aberystwyth University, UK.

Dr Rossella Monti, Chief Executive Officer, Hydroaid, Turin, Italy.

Dr Wouter Mosch, Ecopatrimonio environmental consultancy, Murcia, Spain; formerly at CSO Adviesbureau, Bunnik, The Netherlands.

Professor Jana Olivier, Department of Environmental Sciences, University of South Africa, South Africa.

Dr-Ing Luis Oliveira, National Laboratory for Civil Engineering (LNEC), Lisbon, Portugal.

Dr Murray C. Peel, Senior Research Fellow, Department of Civil and Environmental Engineering, University of Melbourne, Melbourne, Victoria, Australia.

Dr Mladen Picer, Institut Ruđer Bošković (IRB), Zagreb, Croatia.

Dr Pierre Pitte, National Research Council (Consejo Nacional de Investigacionas Cieutificas y Tecnicas – CONICET), Argentina.

Professor John C. Rodda, ex-president, International Association of Hydrological Sciences, Wallingford, England. Former Director, Water Resources Division, World Meteorological Organization, Geneva.

Dr C.J.J. Schouten, founding Director, CSO Adviesbureau (Consultancy for Environmental Management and survey), Bunnik, The Netherlands.

Professor Amarendra Sinha, Department of Geology, University of Rajastan, Jaipur, India.

Professor Aondover Tarhule, Chair, Department of Geography, and Consulting Director, Center for Risk and Crisis Management, University of Oklahoma, Norman, Oklahoma, USA.

Sonia Thomas, Agence de l'Eau Rhône Mediterranée et Corse, Marseille. Formerly at Montpellier SupAgro International Centre for Advanced Studies in Agronomic Engineering, Montpellier, France.

Dr H.R.A. van Hout, Director CSO Adviesbureau (Consultancy for Environmental Management and survey), Bunnik, The Netherlands.

Dr Frank Winde, Department of Geography, North-West University, Potchefstroom, North West Province, South Africa.

Professor Emeritus Ming-ko Woo, School of Geography and Earth Sciences, McMaster University, Hamilton, Ontario, Canada. Formerly president, Hydrology Section, Canadian Geophysical Union, coordinator, Global Energy and Water Cycle Experiment (GEWEX) in Canada, and vice-chair, IGU Commission for Water Sustainability.

Professor Wang Wuyi, chair, IGU Commission on Health and Environment. Chinese Academy of Sciences Institute of Geographical Sciences and Natural Resources Research, Beijing, China.

Professor Jun Xia, Director, Center for Water Resources Research, Chinese Academy of Sciences, Beijing, China.

Dr Kathy L. Young, Department of Geography, York University, Toronto, Ontario, Canada.

Professor Doracie Zoleta-Nantes, Chair, Department of Geography, University of the Philippines, Manila, The Philippines. Research Fellow, RMAP-Crawford School, Australian National University.

Richard Zuccolo, Environmental consultant. Formerly at the Italian Agency for Environmental Protection and Technical Services (APAT), Padua, Italy.

# Introduction

1 A looming crisis

While most of the world focuses ever more on climate change and ways and means of controlling the build-up of greenhouse gases, a less celebrated but potentially just as important issue is evolving in the realm of water resources (Figure 1.1). The two issues are not unrelated. Behind both lie the twin factors of a burgeoning world population and rapidly developing economies.

Despite repeated efforts and resolutions by the international community, from G8 summits to the World Water Forum, aimed at improving water supply and sanitation around the world, we are losing ground. There are now more people without safe water and sound sanitation than there were at the turn of the century. The poorer countries are suffering most, and everywhere it is the poor, the young and the old that are most vulnerable. Growing millions of urban poor have no access to safe, centrally supplied water and are forced to use polluted surface or ground water, or else pay over the odds for water from commercial tankers. More than a third of the world population (over two billion people) presently live in countries suffering moderate to severe water stress (less than 1700 m³ per person) and three-quarters of them live in regions classified as 'water scarce', with

less than 1000 m³ per person – that is, below the global average (see Figure 2.1). By 2025, between half and two-thirds of the world population is likely to suffer water stress, perhaps as many as four or five billion out of a world population of eight billion. Nearly 2.5 billion could be in water-scarce regions. Africa is likely to be particularly badly hit as populations grow. By 2025 the number living in water stress or scarcity could increase more than threefold to 700 million, of whom nearly two-thirds will be suffering water scarcity.

Factors behind reduced water resources

The main driving factor behind the reduction in available water resources per head of population is population growth. Climate change is an important factor, but not in terms of total global per capita resources. Actually, it is likely to increase global resources rather than decrease them. Global riverflow, which is the basic yardstick for water resources, is likely to increase by between two and three per cent by 2020, which means

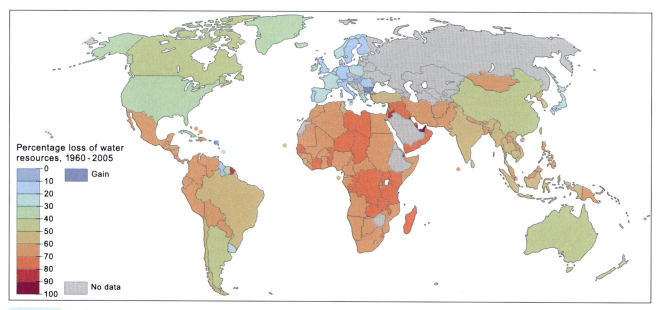

Figure 1.1 Decline in per capita water resources since 1960

Percentage loss of water resources, 1960–2005

Key issues

- 3.6 million people die each year from water-borne disease, of which 2.2 million are in the Least Economically Developed Countries where 90 per cent are children under the age of 5 years.

- A baby dies from water-related disease in the developing world every 20 seconds, some 4500 a day.

- Half the population of developing countries are exposed to polluted water. Many are forced to drink polluted water, buy water expensively from truckers or else drink none at all.

- 40 per cent of the world population, 2.6 billion people, have no access to basic sanitation, rising to nearly 50 per cent in Asia and 70 per cent in Africa.

- Over 1.1 billion people, 17 per cent of the world population, have no access to safe drinking water, including 300 million people in sub-Saharan Africa.

- Population growth and increasing demand is set to reduce global per capita water resources by more than a third in the coming 50 years.

- A third of the world population presently live in countries suffering moderate to severe water stress and this is likely to rise to two-thirds of a much larger population by 2025.

- Poorer countries are the most vulnerable to water stress. Many suffer from harsh climates, unreliable rainfall, poor governance and corruption, and rapid population growth, plus a lack of expertise, technology and finance to overcome problems. The growing urban poor, the old and the very young are worst affected.

- 1 in 10 rivers now run dry for part of the year, some due to climate change, many due to overuse.

- In the Himalayas and Tibetan Plateau, glaciers and ice caps whose seasonal meltwaters are essential to the welfare of 40 per cent of the world population are melting at an accelerated rate.

- Climate change will cause a major redistribution of global water resources – in general regions now short of water will get less, and regions that currently have plenty of water will get more.

- Water wars are in prospect. Shared water resources are a source of conflict: two-thirds of rivers in Africa and Asia cross national frontiers. So do many aquifers.

- Agriculture is the largest water user in the world, consuming some 70 per cent of available resources, much of it in inefficient irrigation. Ten countries use more than 40 per cent of their water resources for irrigation.

- Commercialization, privatization and globalization of water have been detrimental in a number of developing countries, with conflicts between profit motive and service provision, and clashes between multinational companies, national governments and the people. Financial crises are now rapidly transmitted around the world affecting water and sanitation provision.

- World trade is encouraging some countries to use substantial portions of their water resources to produce goods for export. This is not always done in the most efficient manner or without a detrimental impact on domestic supplies.

- Unsustainable groundwater mining – using more than is being replenished – is on the increase. By 2000, Libya and Saudi Arabia were already using considerably more water than their annually renewed resources.

- Rising acts of terrorism now add new fears for water security.

around 800–1200 km³ a year extra resources in a total global riverflow of 34–41,000. Computer models suggest global riverflow could rise by four per cent over the first half of the century and it could reach six or seven per cent by 2100, i.e. 2000–2800 extra cubic kilometres.

The problems with climate change lie elsewhere: in changing the distribution of resources around the world; changing the timing, variability and reliability of rainfall; increasing the occurrence of extremes – floods, droughts and intense storms; melting the glaciers; affecting water quality and acid rain; and through its indirect effects like sea-level rise, the impacts on agricultural crops and changing patterns of disease. All of which are likely to affect the poorer countries more.

This really shows the fallacy of placing too much emphasis on *average* resources. Just as water is not evenly distributed across the land, so uneven growth in population and the uneven effects of climate change are set to make matters worse. Most of the extra resources will be in middle to high latitudes, roughly polewards of 45°. The subtropics and the desert margins will suffer the worst reductions in rainfall. Essentially, the water-rich will get richer, the water-poor will get poorer.

One reason why the CIA took a great interest in computer models from the early days in the late 1980s was the prospect of civil strife and large-scale international migrations being fomented by drought and shortages of food and water. Early models suggested the grain harvest in North America could be reduced by up to 20 per cent, which would drastically reduce the amount available for the food aid on which much of Africa depends. Former British ambassador to the UN, Sir Crispin Tickell (1977, 1991) saw the threats in the Least Economically Developed Countries very clearly. These threats are likely to spill over to developed countries as migrants try to seek asylum. Indeed, the boatloads of migrants entering Europe from North Africa may be the first sign of this. To this may be added flood migrants as tropical storms hit harder.

More subtle changes are already under way in regions that rely on glacier meltwaters for supplies. Virtually all mountain glaciers are melting, as are the polar ice sheets. Melting glaciers provide extra water for a while, until the ice is gone. The glaciers and ice caps in the Himalayas are generally melting more rapidly at present, though their demise is not quite as imminent as the notorious vanishing date of '2035' cited in the IPCC (2007) report, which was exposed as totally unfounded shortly after it was presented to the 2009 Copenhagen climate conference. Even less likely is the suggestion made on America's ABC television in 2008 by Indian glaciologist Syed Hasnain, the originator of the meltdown theory, that the Ganges would run dry by the middle of the century as a result. The ice bodies in the region constitute the largest mass of ice outside the polar regions – some glaciers are several hundreds of metres thick and they lie at altitudes of many kilometres. Professor Julian Dowdeswell, director of the Scott Polar Research Institute, has pointed out that the average Himalayan glacier is around 300 m thick, which means a melt rate of 5 m a year would take 60 years for it to melt completely. It could take hundreds of years at present rates of melting for all the ice to disappear, and there is no guarantee that rates will be maintained; they have fluctuated widely over the past hundred years (Figure 12.2). Nevertheless, many rivers around the world, where the glaciers are thinner and at lower altitudes, are likely to suffer from reduced meltwaters. Glacier meltwaters could run out within decades for some Andean communities.

A question of quality as well as quantity

The effects of global warming on water quality are also likely to be very significant, yet they have been receiving far less attention. Higher water temperatures have important chemical and biological consequences. The water will expel oxygen. Fish and other aquatic creatures can die through lack of oxygen. Higher rates of evaporation and longer, more severe drought periods will reduce river levels, slow down the flow and reduce natural aeration at rapids and waterfalls. The result is less natural self-cleaning in the rivers. At the same time, lower flows mean there is less freshwater to dilute polluted wastewater that is discharged into the rivers. In the early 1990s, during the worst multi-year drought of the twentieth century in south-east England, flow in many rivers was essentially maintained by untreated wastewater and sewage: the National Rivers Authority designated 44 rivers as endangered and suspended permits for discharges into the rivers. Warmer water encourages bacteria to multiply and decompose organic waste, enriching the water with nitrogen and phosphorus. Algae proliferate in the enriched water, killing other vegetation by using up the oxygen and reducing light levels. Toxic algal blooms are an increasing menace on reservoirs. To this may be added increasingly acid rain as it flushes more CO_2 from the air, which has already been blamed for acidifying the ocean.

The quality of water is an increasing problem, even without considering the effects of climate change and warmer rivers and lakes. Some 450 km³ of wastewater are released into the world's rivers and lakes every year without treatment. This is more than one per cent of all the annual riverflow in the world. It takes 6000 km³ of clean water to dilute it to a safe level, but much of it is reused before it reaches that status, and this is the source of many diseases. Even before the water is ingested by humans, the damage to river ecology and wildlife is immense.

Dire predictions made in the 1970s by the Russian hydrologist, M.I. Lvovich, that all the world's rivers would be useless by 2000 because of mounting pollution, have mercifully not materialized. This is in large part due to improved environmental regulation, especially in the developed countries. In Lvovich's Eastern Europe, matters have only progressed since the demise of the Soviet system in 1990, followed by the accession of many of the countries to the European Union making them subject to its environmental legislation. But the process is not yet complete there by any means.

Africa and Southern Asia are now the worst problem areas. Half the population of developing countries are exposed to polluted water. Many are forced to drink polluted water, buy water expensively from truckers or else drink none at all. Under ten per cent of the population in Ethiopia, Eritrea and the Congo have access to proper sanitation facilities (Figure 1.3). Nelson Mandela told a salutary story when he accepted the IGU Planet and Humanity Award in Durban in 2002. He had been asked by a rich American businessman to what he could most effectively donate money to help the young – perhaps school computers? Mandela replied that the greatest need was proper washing and sanitation facilities in schools. Contracting diarrhoea from unwashed hands was the prime cause of lost school days and the main obstacle to education. A report by WaterAid in 2009 called for attention to be given to diarrhoea for another reason. It kills more children than HIV/Aids, TB and malaria put together yet it receives a fraction of the funds.

The emerging powers of India and China are also still well behind the G8 nations. The map highlights two critical problems: most poor nations have inadequate provision and the worst provision is in the tropical regions. The combination of higher temperatures and poor sanitation in tropical regions is a leading threat to public health (see *Water-related infectious diseases* in Chapter 5).

There are well-founded fears that Southern Asia will be beset by rampant river pollution over the coming decades as populations grow, the poor migrate to the cities (often to insanitary shanty town slums) and countries with emerging economies industrialize. India and China are in the forefront. The Yamuna River, which is Delhi's main source of water, is beset with sewage, rubbish and industrial waste. In 2006 it had 100,000 times more faecal coliform than the safe limit. One billion tonnes of untreated wastewater are dumped in the Ganges every day. A typical sample of water from the River Ganges contains 60,000 faecal coliform bacteria in every 100 ml of water. Over 400 million people depend on the river. Many bathe in it in the belief that it has unusual self-cleansing powers: in the largest religious festival in the world, the Pitcher Festival or Kumbh Mela, up to 50 million people attend the bathing ceremonies. Serious action may be about to be taken. The Ganga Sena (Army for the Protection of the Ganges) has over 10,000 student activists. In 2009, hundreds of students staged a demonstration in Varanasi distributing pamphlets and cleaning the ghats – the terraces of funeral pyres beside the river. The World Bank is now making a loan of $1 billion for a five-year clean-up programme on the river beginning in 2010. However, previous attempts, like the 1985 Ganges Action Plan, have come to little: less than 40 per cent of the Action Plan targets were met and there was considerable backsliding afterwards.

Lakes and rivers in China are heavily polluted, particularly from industry. China has the fourth-largest water resources in the world, but over 60 per cent of its 660 cities suffer from water shortages and 110 suffer severe shortages. Pollution plays a large part in this: three-quarters of rivers flowing through urban areas are unfit for drinking or fishing and around a third are unfit for agriculture or industry. Nearly 700 million people drink water contaminated by human or animal waste. During the 2008 Beijing Olympics factories were closed or moved out of town. Even so, the government could only guarantee safe drinking water for the Olympic village, not the whole city. In 2007, China's third largest reservoir at Wuxi City in central China was covered with toxic blue-green algae. In 2006, the Hong Kong based Fountain Set Holdings, one of the largest cotton textile companies in the world, was releasing 22,000 tonnes a day of untreated wastewater contaminated with dye into the local river from its factory in Guangdong, turning the river red. It was fined $1 million and has now installed a treatment plant, but this was not an isolated incident. One way that Chinese companies have kept exports competitively cheap over the last two or three decades has been to dump wastewater into the environment, despite the breach of national environmental legislation: treatment costs an extra 13 cents per tonne. Some companies have been falsifying their record books and officials have colluded with illegal activities. River pollution is said to deprive over 300 million Chinese of clean drinking water.

Water and the natural environment

Protecting the environment is vital for protecting water resources, and *vice versa*. But this requires some entrenched attitudes to change, from governments and multinational companies to peasant farmers. Just as many in developing countries regard protecting the environment as a kind of western fashion that only rich nations can afford – having gained many of their own riches from unfettered exploitation in the past – so some also regard economizing on water use as something of a luxury that they cannot afford. This is especially the case where economies are dominated by the demands of agriculture, like India. Globally, around 86 per cent of all water withdrawals are used in agriculture. As populations grow, so more water will be needed for agriculture. Better informed farmers and more efficient irrigation systems are needed. Overuse of water for irrigation is not just a waste of water, it can destroy soils and poison rivers with salts, pesticides and fertilizers.

The link between land and water, between river basin and rivers, needs wider recognition. Conservation, preservation and restoration are concepts that all humankind needs to embrace, not just to maintain biodiversity and a healthy environment, but also to protect and conserve water resources for our own use and for future generations. This is the essence of 'water sustainability'. Education has a vital role to play here. So too does what is termed 'capacity building' – training local people to solve water problems themselves.

Technological innovation has solved many problems in the past. Dams are the time-honoured solution, but like most technical solutions they have their drawbacks, especially the megadams now being built. But science and technology are rapidly expanding the suite of options. Developed countries may have the resources to overcome technical issues, but what about the developing world and what might mounting hunger, financial and social tensions, migration and conflicts do? Is increasing globalization a help or a hindrance? Is water too fundamental a social need to be in private hands? Is there, indeed, such a thing as a human right to water?

Over-pumping of groundwater and pollution of groundwater, especially by fertilizers and pesticides from agriculture, are growing problems around the world. Much of the western USA, including the grainlands of the Great Plains, is suffering from falling water tables. If this leads to falling crop yields in the coming decades, this will have serious ramifications well beyond the USA, as this is one of the last remaining sources of food aid for the world. Large tracts of both China and India are in a similar situation. It is a leading reason behind grand plans for diverting rivers from one side of the country to the other in both countries. Many Middle Eastern countries are exploiting groundwater that has lain there for thousands of years. It is not being replenished in the present climate. They are using more than 100 per cent of their natural water resources, and it is ultimately unsustainable. Jordan and Yemen are pumping 30 per cent more water from their groundwater than is being recharged by rainfall and Israel is exceeding its renewable supplies by 15 per cent. This led Israel to sign a 20-year agreement with Turkey for the supply of 50 million m^3 of water from the River Manavgat in return for armaments in 2004. Libya has completed the grandest of diversion plans to bring 'fossil' groundwater from the central Sahara to the Mediterranean coastlands. Algeria has similar plans. But how long will such schemes and overdrafts last and what environmental damage will ensue?

From the tropical rainforests to the desert margins, destruction of land resources and wild habitats by agriculture, poor husbandry or pure commercial greed is also destroying water resources and aggravating flooding. The world's wetlands have been decimated over the last half century. Most have been drained for agriculture, and some in the real or imagined belief that they are the source of disease. The result is less storage on floodplains for stormwater, so floods are worse downstream and more of the water that the wetlands might have released after the storm to bolster river levels has passed by uselessly as floodwaters. Reappraisal of flood controls after the disastrous floods on the Mississippi in 1993 has led to a programme to re-establish riparian wetlands in the headwaters. The Netherlands is now actively recreating wetlands for the same reason. Similar arguments surround deforestation. By and large, forests decrease flood hazard, although they do also reduce water resources by lowering riverflow across the board. Deforestation and poor agricultural husbandry also exacerbate soil erosion, which is reaching critical levels in many parts of the world. Soil erosion is not simply a loss for agriculture. The International Rivers Network estimates that severe soil erosion affects 80 per cent of the land surrounding the Three Gorges reservoir on the Yangtze River, much of it due to tree-felling for fuel and building materials.

This is causing 530 million tonnes of silt a year to enter the reservoir, reducing its water-holding capacity and affecting water quality for the towns along its banks.

The authorities in northern China are battling against environmental degradation caused by uncontrolled grazing, over-zealous irrigation, deforestation and over-pumping of groundwater. Desertification is spreading through northern and north-western China largely as a result. Water tables are falling and the desert is expanding by over 2500 km² a year in northern China.

The Millennium Development Goals

The Millennium Development Goals set by the UN Millennium Declaration in 2000 include halving the number of people without access to safe water and sanitation by 2015 (Figures 1.2 and 1.3). Unfortunately, at the time of writing we are already two-thirds of the way there in time and way short on the numbers. International aid to developing countries has fallen well short of the promises. And

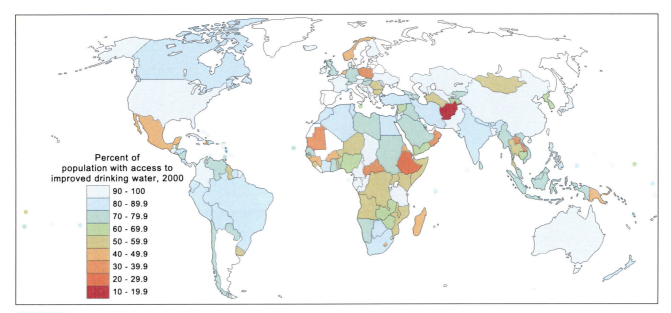

Figure 1.2 Access to safe drinking water when the UN Millennium Goals were set

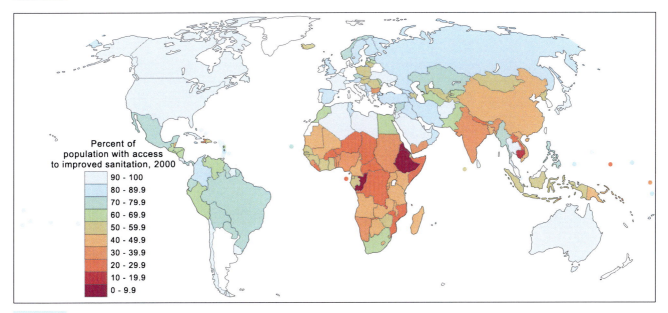

Figure 1.3 Access to safe sanitation when the UN Millennium Goals were set

the prospects seem bleaker following the credit crunch of 2008–9. Delivery of the Millennium Goals is frustrated by the twin forces of reduced international aid and growing populations. India is a clear example of the latter. In 2004–5, safe water and sanitation were provided for 325,000 more people, but over the last 20 years the population has been growing at up to 1.75m a year, four to five times the rate of growth in the provision of water services.

It is also regrettable that the MDG aims with respect to water were only sub-themes of the eight main themes: poverty and hunger, primary education, gender equality, child mortality, maternal health, diseases, environmental sustainability, and developing a global partnership for development. Water is most prominent under the environmental sustainability theme.

Water underpins up to six of the Millennium themes, yet water deserves higher political priority in its own right. One of the difficulties in coordinating international efforts to solve water problems is that so many different bodies have a legitimate interest in water. The World Water Assessment Programme has to deal with some 30 or so UN bodies alone. This is not just an obstacle created by bureaucracy. It is in the very nature of water. It is the most fundamental of resources and is involved in so many issues.

Poverty is a major cause of lack of access to safe water supplies and sanitation. It undermines health, increases exposure to disease, hinders access to education and reduces the ability to work, and so poverty breeds poverty. A key to improving water supply and sanitation for most of the population of the world is to reduce poverty. Yet the gap is widening between the 'haves' and 'have-nots'. The 'Make Poverty History' campaign championed by Prime Minister Tony Blair at the Gleneagles G8 summit in 2005 had little effect. There are now more than a billion people in poverty in the world. The rate of population growth in sub-Saharan Africa means that economies would need to grow by several per cent a year just to maintain present levels of poverty, and this is not happening. Apart from the lack of piped water and sanitation, the urban poor pay between five and ten times as much per litre as the rich in the same city buying from water truckers. Reducing poverty would also be good for the environment.

Water, politics and economics

Despite limited progress with the Millennium Development Goals, water has moved up the political agenda over the past decade or so. The World Water Assessment Programme, established by the UN in 2001, and the five meetings of the World Water Forum since 1997 have played a major role in informing politicians and the public of an impending crisis in water resources and its causes. The UN World Water Development Reports that the WWAP presents at the Forums are extremely valuable and highly influential, most notably in leading to the launch of the UN Water for Life decade 2005–2015, which aims to facilitate the Millennium Goals. The 2002 World Summit on Sustainable Development (WSSD) in Johannesburg also produced a number of useful initiatives. The WASH campaign launched at the WSSD by the Water Supply and Sanitation Collaborative Council aims to raise public awareness and increase the political profile of sanitation and safe water supplies through the media, fact sheets and other publications. Water now appears on the agendas of the World Economic Forum in Davos and of the World Political Forum.

The present and impending problems are not simply questions of supply and demand. There are significant issues in governance, finance and legal frameworks. Rivers and groundwater resources that cross international boundaries are a source of conflict that is still inadequately covered by international law, even though it is now two decades since the idea of 'water wars' was first broached in the run-up to the 1992 Rio Earth Summit. The WSSD rightly identified lack of 'good governance' as a major factor in the problems: the need for governmental institutions to plan for and supply adequate and affordable facilities for all their citizens, as well as to inform them, educate them in water issues and consult them. In 2009, the 5th World Water Forum identified corruption as a major obstacle to good governance. It called for an international tribunal to address violations of anti-corruption rules and appealed for safeguards against corruption to be incorporated in the design of all new water projects.

The question of money is paramount. Money is needed not only for the basic provisions of water and sanitation, but also for the scientific and technical backup in Water Resources Assessment, real-time monitoring, warning and forecasting systems, and predicting the effects of foreseeable changes in the environment, both physical and socioeconomic, so that sound planning can take place. The whole question of funding for water resources projects and the long-held policies of the World Bank and IMF to encourage privatization and public-private partnerships need careful scrutiny in light of recent cases of conflict of interest between profit-making and service provision.

At the G8 summit in May 2003, appropriately held at Evian, the richest nations made a number of potentially valuable resolutions on both governance and finance in their Water Action Plan (see *The G8 Action Plan*). The subsequent appearance of water on the agendas of the World Economic and Political Forums is helping to implement some of these issues.

The World Economic Forum is particularly valuable because it engages large private companies as well as governments. At the 2008 Davos meeting, the UN Secretary-General shared the chair of a debate on the growing portion of world population living in regions of water stress up to 2025 with the heads of Nestlé and Coca-Cola. The same meeting addressed two other growing concerns: security from industrial espionage, especially by cyber attacks, and the impact of buy-outs by hedge funds and foreign sovereign wealth funds whose prime interest is profit. No less a person than the director of the FBI examined the vulnerability of governments, companies and individuals to cyber attack by electronic spies. And the chair of the US Securities and Exchange Commission scrutinised the growing anxieties over the

powerful role of hedge funds in global markets, accompanied by a number of major industrialists. These issues are of real concern to all countries with privatized water industries.

The perspective was widened at the 2009 Davos meeting to water in the context of fuel and food. A session on an integrated approach to energy, food and water security included a call for world leaders to establish a Natural Resource Security Council to cover food and water that can match the UN Security Council for War and Peace in its authority. A world authority such as this is needed to coordinate strategy to improve the efficiency of food supply chains: some 40 per cent of food production is lost or spoiled between the farm and the consumer. The chairman of Nestlé said at the time that he was 'convinced that we will run out of water long before we run out of fuel'. He said that open markets with no subsidies, especially not on biofuel, lead to more efficient water use, and that it is imperative to try to arrest the trend towards structural overuse of freshwater.

The connections between food, energy and water are becoming critical. Irrigated agriculture currently

The G8 Water Action Plan, Evian 2003

1 Promoting good governance:

Assisting countries that commit to prioritize water and sanitation as part of their sustainability strategy, especially as part of the eradication of poverty – helping with integrated management and developing institutional, legal and regulatory frameworks.

2 Utilising all financial resources:

Giving high priority in Official Development Aid to sound proposals as catalyst to mobilise other monetary sources, direct encouragement of International Financial Institutions, and providing technical assistance to domestic financial provision and promoting public-private partnerships (PPPs).

3 Building infrastructure by empowering local authorities and communities:

Help build systems, PPPs, community-based approaches and improved technologies at household level.

4 Strengthening monitoring, assessment and research:

Encourage sharing of information from UN and other systems, including the websites established by the 3rd World Water Forum Ministerial Conference, and supporting research and collaboration on aspects of the water cycle.

5 Reinforcing the engagement of international organizations:

Improving coordination within the UN Organization, and between institutions, like the World Bank, the Bretton Woods institutions and regional development banks. Proposals by the World Panel on Financing Water Infrastructure include funding and insuring risk mitigation schemes, and increasing flexibility in the rules covering loans.

produces nearly half of the food in the world by value. It is the main source of over-pumping of groundwater and falling water tables. Feeding a world population that is 50 per cent larger than now by mid-century will require at least 50 per cent more water unless there is a major shift in technology. It could take even more as people in the emerging economies, like China and India, change their diets and consume more meat, and invest in washing machines. To this may be added biofuel. It is estimated that just supplying five per cent of current energy from biofuel will double water consumption. But this is assuming the current approach and technology, which is changing: the British government for one has revised its plans for expanding biofuel use. Of more concern is biofuel crops taking up land that is supplying food crops.

One issue that caused considerable controversy at the World Economic Forum in 2009 and which has been hotly debated for some time among water policymakers is whether access to water should be a basic human right or simply a 'need'. These two small words hold the key to the difference between public and private in the water industry. If it is a human right, then governments should shoulder the responsibility of securing and subsidizing supplies. The majority of stakeholders at the 5th World Water Forum were strongly pro the right to water, despite two decades of World Bank support for privatization. At Davos 2009, it was argued that governments should provide people with up to 25 litres a day for drinking and hygiene, and that anything above this amount should be chargeable. Some argued that pricing water is necessary to restrain use. Only by attaching a price to water can a system of trading permits be established like those being implemented to constrain carbon emissions. Farmers could then sell unused permits to other farmers and so hopefully reduce overall water consumption. However, the issue of pricing remains problematic: it may be necessary economically and to achieve sustainability, but it can also be politically risky in many countries. There is a compromise move developing among water providers towards pricing for 'sustainable recovery' of costs rather than full recovery, which was identified at the 2009 World Water Forum in Istanbul.

Davos 2010 had yet broader issues in mind following the 'great recession'. The theme was no less than: 'Improve the state of the world: rethink, redesign, rebuild.' The primary aim was to seek ways of creating more financial stability in the world to prevent a recurrence of the near meltdown of international finance that marred 2009 and has led to severe and continuing impacts on the water industry, including deferred refurbishment of old infrastructure, job losses and problems raising loans. Davos alone can only produce ideas. The G8, or more effectively the G20, which includes India and China, is the only place where there is any hope of an effective redesign of global financial systems ever being achieved.

Conclusion

The threats to water sustainability are huge. Political awareness of the issues is increasing. Plans are proliferating. Technologies are advancing. But obstacles are also persistent and new ones are arising. The greatest threats are manmade. Will the fine words and good intentions prevail?

Further reading

World Water Assessment Programme, 2009. *Water in a changing world*. 3rd edition, UN World Water Development Report (WWDR–3). Available online at: www.unesco.org/water/wwap/wwdr/wwdr3/

And, earlier WWDR editions.

Shiklomanov, I.A. and Rodda, J.C. (eds) 2003. *World Water Resources at the Beginning of the 21st Century*. International Hydrology Series, UNESCO, Cambridge University Press, Cambridge. pp. 414–416.

Sir Crispin Tickell 1977. *Climate change and world affairs*. Cambridge, Mass., Harvard Studies in International Affairs No. 37, 76pp.

And 1991. The human species: a suicidal success? *Geographical Journal*, **159**(2), 219–26.

Part 1
Status and challenges

Rising demand and dwindling per capita resources

2

Demand continues to grow, driven by rising population, the drift to the cities, the need for food and the increasing consumption of manufactured goods.

Diminishing resources

In broad outline, the present-day geography of water resources very much follows the pattern predetermined by climate. Even when the size of the population is considered, the pattern remains broadly the same: at least up to this point in time, the countries with the lowest per capita water resources are generally those where nature provides the least.

Economic development and rising affluence are beginning to change this pattern. Per capita resources have fallen by between 40 and 50 per cent over much of Southern Asia during the last half century, mainly through population growth, but they are set to decline even more as economies expand. India as a whole is already suffering water stress and China is moving in the same direction. Provided these two emerging economies avoid civil unrest, they should be able to find the economic and technical means to meet the challenges.

Africa is different. Per capita resources have fallen between 60 and 80 per cent since the mid-twentieth century. Most of subtropical Africa now suffers from water stress, even scarcity. Modelling suggests that a combination of population growth, economic development and climate change will cause demand to double or even quadruple in most of sub-Saharan Africa during the first quarter of this century. Most countries with the exception of South Africa are unlikely to have either the economic or technical apparatus needed to meet this challenge on their own.

By contrast, the huge leap in water demand in western countries last century is unlikely to continue. There has been a drastic reduction in demand in Germany over the last 20 years as a result of water-saving technology and citizen awareness. The demise of heavy water-consuming industries and the effective export of many manufacturing industries to countries where production costs are cheaper have added a widespread downward trend in demand from the industrial sector in the West. Demand is therefore also a product of stage of economic development, rising in the early stages, levelling off and maybe even decreasing later.

Current status

Of the top 20 best-resourced nations, nine have significant ice- and snow-melt resources, the rest have humid tropical climates, some have both (Figure 2.1). Unsurprisingly, Greenland tops the list with nearly 10.5 million m^3 per head. Most of this is inaccessible, although climate change does appear to be increasing meltwaters around the southern margins. Greenland's resources are more than 15 times that of the next, French Guiana (at 679,599 m^3 cap^{-1}), followed by Iceland and Guyana at 569,650 and 326,088 m^3 cap^{-1} respectively. Looking down the list at the top 50 countries reveals that just over a quarter have significant frozen resources, but the proportion of humid tropical nations dependant solely on high rainfall receipts rises to nearly 60 per cent, and they account for over 80 per cent if the top 20 are excluded.

There are, however, some major exceptions to this climatic imperative that are mainly due to the size of population, either too many or too few. Australia appears in the top 50, at number 36, more because of its low population than any inherent climatic advantage – indeed, it is the world's driest inhabited continent and adverse climatic trends are currently hitting it hard and likely to get worse (see Chapter 10 The threat of global warming). India is relegated to number 133 out of the 176 countries reporting to the UN Aquastat database, by virtue of its huge population of over one billion and rising. The other huge nation, China, with over 1.3 billion, comes in at just ten places above India. As the two nations most likely to develop leading roles in the world economy in coming decades, both will have to face serious water resources issues (see Chapter 7 Dams and diversions).

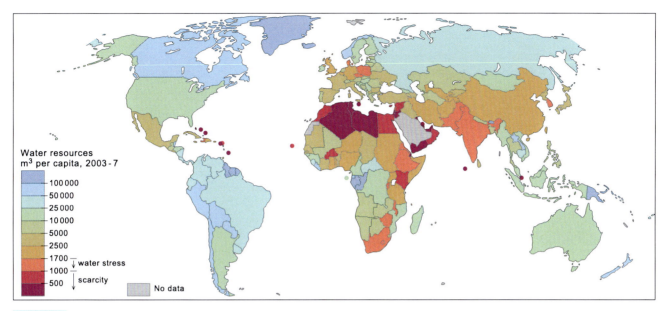

Figure 2.1 Current per capita water resources

Turning attention to the bottom of the list, of the 20 nations on the really critical list with the lowest per capita resources 13 are in the Middle East-North Africa (MENA) region and seven are islands, including Singapore. If this is expanded to the bottom 50, these two categories continue to loom large, but they are joined by a major contingent from the drier regions of sub-Saharan Africa. Over 40 per cent are from the MENA region, a quarter are islands and another sixth are sub-Saharan. But the latter class rises to just over one quarter in the group immediately above the bottom 20, equalling the MENA group there. Considering Africa as a whole, African nations account for one third of all those in the bottom 50 and over 40 per cent, excluding the bottom 20.

All but the top two of the bottom 50 nations, Uzbekistan and Germany, fall into the official range of water stressed (1700–1000 m³ cap⁻¹) and the bottom 27 are officially water scarce (with less than 1000 m³ cap⁻¹). This means that 60 per cent of nations suffering water scarcity are in the MENA region, a third are islands and just two are in sub-Saharan Africa: Kenya and Burkino Faso. That Kenya falls into this group and that East Africa has also been plagued by persistent drought in recent years puts its role as the principal supplier of irrigated flowers to the EU in a critical light (see Chapter 8 Trading water – real and virtual).

Recent trends

More than half of the 20 nations experiencing the greatest reduction in resources over the last five decades are also in regions of water scarcity (see Figure 1.1). This bodes ill if trends continue, as they likely will. And every one of these is in the MENA region with the exception of Kenya. Most of these nations have experienced between 75 and nearly 100 per cent reductions in resources over the last half century. Population growth and overexploitation are clearly hitting some of the worst endowed nations hard.

The breakdown of the top 20 is very familiar: 50 per cent are from MENA and a third from sub-Saharan Africa. When it comes to the top 50 countries with the greatest reduction in resources, however, sub-Saharan Africa accounts for nearly half of all and MENA just over a quarter. In other words, the most extreme reductions over the last 50 years have been in the MENA region, overwhelmingly from the Middle East proper, closely followed by sub-Saharan Africa. Of these, a third are currently suffering water stress (four nations) or water scarcity (12 nations).

At the bottom of this list, the trends are much less severe: indeed, Bulgaria and St Kitts and Nevis are in the rare but enviable situation of slightly increased per capita resources. 85 per cent of the bottom 20 are from the European Union, showing between just 10 and 25 per cent reductions, and Europe continues to dominate the bottom 30, accounting for 80 per cent of those that have lost less than 33 per cent of their per capita resources over the period.

Only five members of the EU have severe water shortages: Poland, the Czech Republic and Denmark (in stress), and Cyprus and Malta (in scarcity). Although the latter two have also shown some of the lowest declines over the past 50 years, at 32 per cent and 23 per cent respectively, it is unsurprising that Malta has turned to desalination. It is also noteworthy that while Europe as a whole, including the Russian Federation, is among the better-endowed regions outside the humid tropics, it is also one of the greatest consumers of virtual water from the rest of the world.

The effect of dams

The discussion thus far has been based on the official UN statistics as presented in Aquastat, and these are based on dividing the average annual net water balance by population size. This is the official assessment. While it provides a valid basis for comparison in itself, there are a number of other aspects that should be considered, which might give a somewhat different view. Not all the water that flows into the rivers is usable, either because it comes in spasmodic floods that are not or cannot be harnessed, or because there are other demands upon that water, be it fisheries and the aquatic environment, or because there are countries downstream that have an established demand, perhaps by treaty. Occasionally, it gets too polluted.

In the 1970s, the veteran Russian hydrologist M.I. Lvovich made his own calculations based on the obvious fact that the main objective of building dams is to harness floodwaters and transform a potential hazard into a potential resource (Lvovich, 1970, 1977). His calculations were very instructive, and they have been extended into the twenty-first century in Table 2.1. The column covering stable per capita runoff without extra capacity shows the effects population growth would have had on per capita resources if no dams had been built.

The view is imperfect because we do not know the true purpose of every dam or the proportion of stored water that is destined for water supply and what may be used purely to control floods or to generate hydropower. The data also purport to cover only large dams, of which Gleick (2003) counted 48,000 dams, though Chao et al (2008) counted only around 30,000. Nevertheless, extending Lvovich's approach over the period since the beginning of the 1970s reveals some very interesting divergences in water resources engineering between regions.

The table suggests that North America has nearly doubled its per capita resources by building dams.

In South America, the huge dam building programme during the 1980s raised per capita resources to above 1970 levels, but dam building has not kept pace with population growth since then. Australasia is the third best endowed region, but while dam building doubled per capita resources between 1970 and 1992, rising population and a cutback in dam building have severely eaten into this, resulting in a 35 per cent deterioration – hence Queensland's problems (see box 'Brisbane's water crisis: a tale of politics, protests and complacency' in Chapter 7). Europe has remained stable, balancing population growth and dam building, while Asia has managed to increase resources and turn around from the decline experienced at the beginning of the 1990s. Lastly, by this measure Africa has lost nearly half of the resources it had in 1970, although there has been little change since the early 1990s.

Future water demand

Predicting future water demand is not easy and prone to error. Individual countries are increasingly modelling demand as a necessary aid to forward planning. Perhaps the earliest national forecasts were undertaken by Canada in 1978 and the USA in 1981. Both predicted water shortages in specific parts of their countries by 2000. Predictions of future demand in the UK by 2025 were produced by the National Rivers Authority (1994), which revealed the parsimonious position of SE England. The UK water industry has since been instructed to update these predictions to include climate change and to adopt as industry standard the so-called 'medium-high emissions' estimate of global warming, which is the scenario produced by the UK Hadley Centre's greenhouse-gases-only run of its General Circulation Model (Hulme and Jenkins, 1998; CRU, 2000).

The addition of climate change compounds the difficulties and increases the range of outcomes. The industry now speaks of a range of 'scenarios' rather than a simple projection of trends. These scenarios consist of different assumptions about demographic and economic trends and greenhouse gas emissions. Errors in one can be multiplied by errors in another. The physical simulation models themselves also contain assumptions and approximations about climatic processes (see Chapter 10 The threat of global warming). The result is a range of possibles that is whittled down to a few probables.

This is no argument for not trying, but it is an important caveat. Forward plans must be laid, especially given the

Table 2.1 The effect of large dams on stable regional water resources

Continent	Natural riverflow km³/year			Regulated runoff km³/year	Total stable runoff km³/year	Percentage increase in stable runoff	Per capita runoff m³/year			
	Total	Stable baseflow		1970 (1992) 2003	1970 (1992) 2003	1970 (1992) 2003	Total	1970	Stable	
		Total	Per cent				1970 (1992) 2003		2003 (1992) Theoretical without extra capacity	2003 (1992) Actual with extra capacity
Europe	3100	1125	36	200 (312) 384	1325 (1437) 1509	18 (28) 34	4850 (4334) 4270	2100	1550 (1852)	2079 (2009)
Asia	13190	3440	26	560 (1198) 8399	4000 (4638) 11839	16 (35) 244	6466 (4210) 3450	1960	900 (1276)	3097 (1481)
Africa	4225	1500	36	400 (564) 1121	1900 (2064) 2621	27 (38) 75	12250 (6536) 4965	5500	1763 (2939)	3080 (3193)
N. America	5950	1900	32	500 (1115) 1922	2400 (3015) 3822	26 (59) 101	19100 (14280) 18252	7640	5828 (5760)	11724 (7236)
S. America	10380	3740	36	160 (4135) 7062	3900 (7875) 10802	4 (110) 189	56100 (35791) 19116	21100	6888 (13447)	19893 (27154)
Australasia	1965	465	24	35 (273) 118	500 (738) 583	7.5 (59) 25	109000 (74274) 61406	27800	14531 (18899)	18219 (27895)
TOTAL	38830	12170	31	1855 (7597) 19006	14025 (19767) 31176	15 (62) 156	10965 (7442) 6163	3955	1931 (2688)	2049 (3789)

Estimating future water withdrawals

Martina Flőrke

The global water use model WaterGAP has been used to estimate the changes in annual water withdrawals up to 2025 (Alcamo *et al*, 2003; Döll *et al*, 2003; Alcamo *et al*, 2000). WaterGAP provides a framework for taking into account the impact of key demographic, economic, technological, climatic and other driving forces on the future of the world's water uses. The model consists of a series of submodels designed to compute water withdrawals and water consumption for the domestic, industrial and agricultural sectors in more than 11,000 river basins. The map in Figure 2.2 shows the calculated changes in annual water withdrawal in 2025 compared to 1995. Total water withdrawals are the total volume of water extracted from surface or subsurface sources for various uses.

The effect of different driving forces on future water use will obviously depend on their assumed future trends. In this study we base these assumptions on the input data used to compute the A2 scenario of greenhouse gas emissions (IMAGE Modeling Team, 2001; IPCC, 2000). Specifically, we evaluate the impact of driving forces such as population and economic growth as well as climate change. The A2 scenario assumes that trends in population and economy are consistent with an economically-oriented world but with a lower level of integration (e.g. world trade and cross-border diffusion of technology) than in an ideal 'globalized' world. Compared to the other IPCC scenarios, population growth is relatively low and economic growth is low to moderate. In this analysis we consider climate data from the HadCM3 model of the Hadley Centre in Great Britain (Gordon *et al*, 2000; Pope *et al*, 2000). This model is used to compute climate conditions under changed levels of greenhouse gas emissions under the A2 scenario for 2025.

The result of the model simulation is a worldwide increase in water withdrawals from approximately 3594 km³/a in 1995 to 4177 km³/a in 2025 under the A2 scenario. Between 1995 and 2025, water withdrawals increase by more than five per cent in about 46 per cent of the river basins under the A2 scenario. Changes in water withdrawals have a very irregular spatial pattern between 1995 and 2025 arising from varying sectoral and national trends and the uneven pattern of irrigated and settlement areas. Withdrawals tend to stabilize or

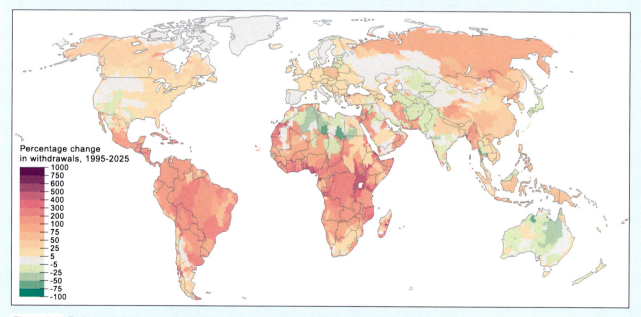

Percentage change in withdrawals, 1995–2025

- 1000
- 750
- 600
- 500
- 400
- 300
- 200
- 100
- 75
- 50
- 25
- 5
- -5
- -25
- -50
- -75
- -100

Figure 2.2 Estimated change in water withdrawals by 2025 according to the WaterGAP model

decrease (changes less than five per cent) in many industrialized countries, because of the saturation of per capita water use, stabilizing population and continuing technological improvements. This is in contrast to the large increases in developing countries where large, currently unfulfilled water demands are being met by 2025. Only about six per cent of all river basins have declining water withdrawals under the A2 scenario in 2025.

timeframe for developing major infrastructure works: Jones (1997) cites 25 years for one central England reservoir. In her article 'Estimating future water withdrawals', above, Dr Martina Flörke of Kassel University, Germany, describes the results of using an elaborate computer model fed with a medium-high climatic scenario.

Rising population and growing cities

The growing world population is the largest single cause of the water crisis. It is being exacerbated by the drift to the cities and by increasing demand for water and products as economies grow.

Rising population

World population is due to double or even quadruple before stabilizing by the end of the century. The best UN estimates suggest a 50 per cent increase in the first half of this century, from six billion in 2000 to nine billion in 2050. However, in 2009 the UN warned that if rates of increase continue at current levels the world population could double its present level of 6.7 billion by 2050. If, in the most unlikely scenario, there were to be a rapid reduction in fertility, this would clip little more than a billion off the best estimate for mid-century.

Worst of all from the water resources point of view, the geographical distribution of this population increase will not only be uneven, but largely concentrated in the Developing World and especially in regions where water resources are already stretched, notably in sub-Saharan Africa, the Middle East and parts of the Indian subcontinent. More than a dozen countries in these regions are expected to show population growths in excess of 50 per cent between 2000 and 2030, and most others will experience increases of more than 30 per cent (Figure 2.3).

The change in the geographical distribution of population since the mid-twentieth century is illustrated by the growth in the Arab world. In 1950, the US population stood at 160 million, while the Arab region had a total population of just 60 million, almost equivalent to the population of the UK in 2010. By 2000, the US stood at 284 million, but the Arab world had increased to over 240 million. From 2010 onwards, the Arab world is predicted to overtake America, possibly rising to 650

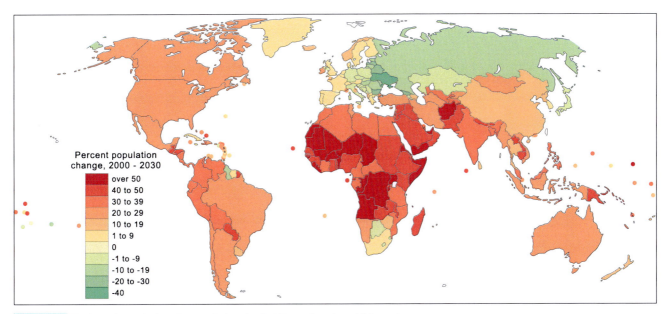

Figure 2.3 Projected population change during the first three decades of this century

million by 2050 against America's 400 million. Over the next 25 years, the population in the Middle East is projected to double – the fastest growth rate outside Africa. There is already a critical scarcity of water resources in the Middle East and North Africa, where 11 of the 20 countries in the region have been using more than 50 per cent of their renewable resources for over a decade. Libya and most of the Arabian Peninsula already use over 100 per cent of their traditionally assessed resources with the support of desalination and fossil groundwater, some of which is 10,000 years old and has not received significant recharge since the Pluvial Period at the end of the last glacial period.

Most of the current expansion in world population is occurring in the Least Economically Developed Countries. In 1950 the LEDCs contained 67 per cent of world population, in 1990 76 per cent, but by 2025 this could rise to 84 per cent. The population in sub-Saharan Africa is expanding rapidly despite rampant disease (Figure 2.3). In the hundred years between 1950 and 2050 the population is likely to have increased more than sevenfold, from barely two-thirds that of the current EU countries in 1950 to twice the EU now and accelerating to five times the EU by 2050.

Another significant change in global demography is the divergence in age structures between Developed and Developing countries. While average age is increasing in most Developed countries – between 1950 and 2000 it rose from 30 to 35 in the USA and from 30 to 38 in Europe – the reverse is happening in most of the Developing World. In many of these countries average age is now in the teens. The greater number of young people is an important factor accelerating overall fertility rates. The International Labour Organization estimates that youth unemployment in the Middle East is already around 25 per cent. There and elsewhere, growth of an impoverished youth class is likely to reduce ability to pay for water services, but worse, it might lead to social unrest, more extremism and terrorism.

The implications for per capita water resources are immense. World resources per head of population are likely to fall by a third or more between 2010 and 2050 solely as a result of population growth. Meanwhile, the increase in global precipitation is likely to reduce this loss by a mere three to four per cent (see Chapter 10 The threat of global warming). However, the differences in regional population growth shown in Figure 2.2 mean that the situation is far worse for most of the

countries already suffering water stress (see Figures 1.1 and 1.2). By mid-century, the number of people living in water-stressed countries is likely to increase by up to 300 per cent to six billion, of whom half will be suffering severe stress.

Climate change will add to regional problems. Although at the global scale climate change will increase resources, in the regions that are currently stressed it will only make matters worse, because of the shifts in rainfall patterns (see Chapter 10 The threat of global warming). Arnell (1998) estimated that climate change will add an extra 66 million people to those suffering moderate stress and 170 million to those suffering severe stress by 2050. These estimates suggest that climate change will add only one per cent as many people as population growth to the numbers living under moderate stress and just five per cent to those with severe water stress. Population growth is clearly the principal problem.

What these estimates do not cover is increasing per capita demand as economies grow and people become richer. The so-called 'multipliers' of the population explosion include improvements in the quality of life from centralized water supply to more washing machines, dishwashers and bottled water, the expansion of cities and centralized sewerage, and the demand for more industrial, commercial and agricultural products. China is going to follow this route over the next few decades. In 1980 per capita consumption in China was under 100 litres a day. It is now around 280 litres. In the UK, households use 150 litres per capita. In America, per capita consumption is 400 litres. An indication of the potential importance of multipliers is given by research in America which concluded that just over half of the increase in national demand during the 1970s could be attributed to population growth, a little over a third was due to increased demands from the pre-existing population as affluence and expectations increased, and the remaining ten per cent was due to rising demands from the new additions to the population.

While current trends suggest that the greatest population growth is going to be in the LEDCs, one of the great unknowns is the degree to which this may be accompanied by increasing prosperity. Prospects do not look good at present, in which case the LEDCs will experience less of a multiplier effect and growth in demand will be more in line with population size. Water consumption in poor rural areas of Africa, Asia and Latin America, is currently around 20 to 30 litres per capita per day. If most of the population expansion were in rural communities, then the extra strain on water resources would not be so great. But increasingly people are drifting to the cities.

The drift to the cities

We have already passed the point where over half of the world's population live in cities. By 2050 it could reach three-quarters (Figures 2.4 and 2.5).

The drift to the cities is occurring everywhere in the world and it has been happening since the Industrial Revolution: 40 per cent of the French countryside now has a lower population than it had 150 years ago, although the national population has doubled. The drift has accelerated during the last two decades and the focus has shifted to the developing world. Within ten years, more than 70 per cent of the population of the developing world will be urban and a quarter will live in cities with over one million people: by 2025 this could reach 80 per cent.

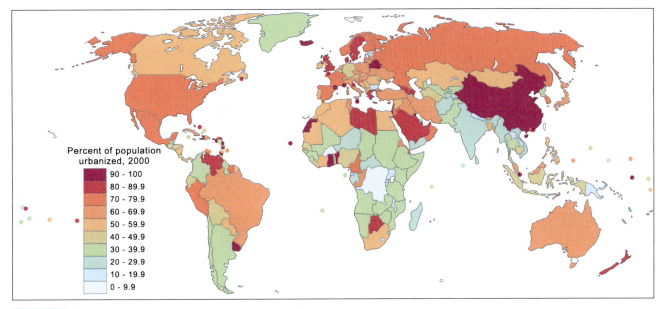

Figure 2.4 Percentage of population living in urban areas in 2000

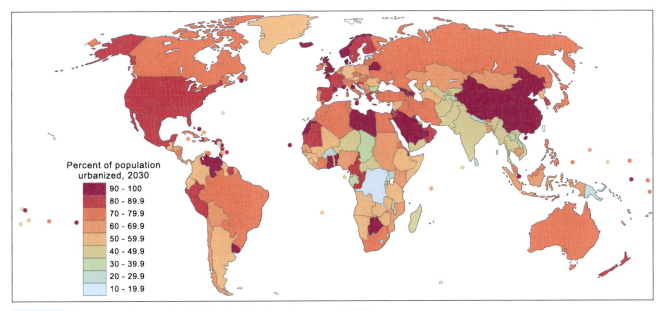

Figure 2.5 Percentage of population likely to be living in urban areas in 2030

Urban dwellers typically consume between 300 and 600 litres a day – between ten and 20 times that of peasant farmers in the Least Economically Developed Countries (LEDCs). Major cities now consume water at rates equivalent to the flow in some of the larger rivers in the world. Indeed, they are often a major factor, along with irrigated agriculture, causing rivers to dry up (Pearce, 2007). The Nile, Colorado and Hwangho (Yellow River) often fail to reach the sea. London consumes over half again as much as currently flows down the Thames, yet a major slice of the government's planned 160,000 new homes by 2016 is set to be in the Thames basin, an area that is predicted to exceed its local water resources in the 2020s (Rodda, 2006). The Environment Agency complained of lack of consultation over the plans and requested that local water resources should be formally considered in planning applications.

China already has some 46 cities of two million or more, of which the fortieth is larger than Paris. Twenty million people a year are migrating to China's cities, bar a temporary reversal during the credit crunch: at least 120 million have already done so in recent years. Sixty per cent of Chinese are still rural, but this is expected to fall to 40 per cent by 2030. In China this drift is likely to be accompanied by improved living standards and therefore more washing machines and more water consumption. It is estimated that China's economic emergence has already raised 400 million people out of poverty. Small wonder that over half of China's cities are experiencing water shortages. 'Instant cities' like Shenzen, which has grown from nothing to five million in just a few years, are further stretching resources in China. China plans to relocate a further 400 million people to new urban centres between 2000 and 2030.

The link between city dwelling and water consumption is strong, but it is not immutable. There are signs that some US cities are curbing consumption. Residential use in New York has fallen by a third in 15 years, from around 760 litres per capita per day to 530. Seattle and Albuquerque have also cut per capita consumption, and total water withdrawals have also fallen in California despite rising population according to Palaniappan (2008) – although the latest revelations suggest that official statistics may not be too accurate (see *Satellites* in Chapter 18). The link is also somewhat decoupled in the LEDCs, which account for most of the new increase in urban population globally. A considerable proportion of these new urbanites live in makeshift housing without mains services (see Chapter 3 Water and poverty).

Agriculture and industry: the big water users

'Agriculture is the greatest user of water worldwide, accounting for an estimated 70 per cent of potable water use.'

Lester Brown
World Resources Institute

Agriculture

Global water use is dominated by agriculture. This is leading to severe water shortages and competition and disputes over supplies in an increasing number of areas. The quote from Lester Brown relates only to 'potable' water, i.e. of drinking grade. But agriculture also uses large quantities of lower grade water, mainly in irrigation. The UNESCO-IHE Institute for Water Education estimates that the growth and processing of agricultural products account for nearly 86 per cent of global water use (Hoekstra and Chapagain, 2008). Agriculture dominates even more so in many of the regions suffering water scarcity. While this may be a question of necessity, it is also true that water for agriculture is priced far below its real cost in most countries of the world and that some of the drive to grow more comes from export-led monetary returns.

Of the remaining water, industry now consumes about ten per cent and the rest is down to domestic users (Figure 2.6). It is important to bear this very uneven breakdown in mind when considering where water consumption might best be cut (see Chapter 13 Cutting demand). Cutting domestic and industrial consumption may be the prime target in Europe and North America, but for the rest of the world it is agriculture.

Most of the water is used for products and services consumed within the same country. But a substantial and generally increasing amount is used to produce products that are traded internationally. This is termed 'virtual water', an interesting concept that is studied in detail in Chapter 8 Trading water – real and virtual. About 13 per cent of all the water used globally goes into international trade in food products (the external water footprint). This amounts to 15 per cent of all the water used in agriculture.

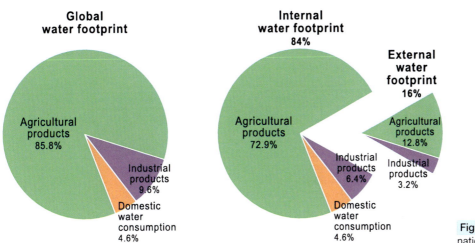

Global water footprint

Agricultural products 85.8%

Industrial products 9.6%

Domestic water consumption 4.6%

Internal water footprint 84%

Agricultural products 72.9%

Industrial products 6.4%

Domestic water consumption 4.6%

External water footprint 16%

Agricultural products 12.8%

Industrial products 3.2%

Figure 2.6 The water footprint of nations by sector

'It is normal for the world to be short of food. The brief era of cheap eating has been a disaster for the world. People in the rich world have eaten too much food … To make food cheap, we have poisoned farmland with chemicals, squandered precious water, abused marginal terrain and undermined biodiversity.'

Felipe Fernández-Armesto
Professor of Modern History, Oxford University, 2008

The global pattern plotted in Figures 2.6 to 2.8 shows the almost total hegemony of agricultural demands for water outside Europe and North America, most notably in the MENA region and the tropics. The overwhelming bulk of this consumption is for irrigation. Only in major beef producing countries like Lesotho and Brazil are substantial amounts used for livestock, and even in Brazil the recent rise of biofuels is diverting increasing amounts of water to irrigated agriculture. Indeed, some critics believe that water will become a major limiting factor in growing biomass for energy (Rijssenbeek, 2007). Bio-oil production can consume up to twice as much water in a semi-arid climate.

Irrigation can easily double crop yields and may be the only way to obtain any worthwhile crop at all. Just 18 per cent of cropland is irrigated, but it produces 40 per cent of the world's food and around 50 per cent by value. Moreover, with world population set to increase by three billion in the next half century and climate change likely to bring less rainfall in many of the developing countries experiencing rapid population growth, irrigated

agriculture is going to be major tool in the struggle to feed the world, irrespective of what plant technology may achieve. FAO projections suggest that by 2030, irrigation in 93 developing countries will account for over 70 per cent of the estimated necessary increase in food production since the millennium (UNESCO, 2006).

Some three-quarters of all irrigated land worldwide is currently in Asia. Around a third of all Asian agriculture is irrigated. India has the largest irrigation system in the world, covering more than 70 million hectares. India also has plans to divert waters from the Himalayas to drier parts of the country (see Chapter 7 Dams and diversions). Much of the water used is glacial meltwater and concerns are being raised about its sustainability with climate change (see Chapter 10 The threat of global warming and Chapter 12 Shrinking freshwater stores). Snowmelt is also a major contributor in North America, which has the second largest area under irrigation.

Irrigated agriculture is expanding faster than other sectors. Global demand for irrigation expanded tenfold over the twentieth century, while the total demand for water increased just fivefold. The irrigated area doubled in the last 30 years of the century. The peak rates of expansion may be over. In the 1970s, almost half a million hectares were being added each year, but the big expansion planned by the communist states of Eastern Europe was curtailed by political changes from the mid-1980s onwards. By the late 1990s expansion had fallen to nearer 100,000 ha a year. Nevertheless, both the amount and area of irrigation are clearly set to continue expanding worldwide. Although the current irrigated area stands at 277 million ha, it is estimated that 470 million ha are suitable by virtue of soil fertility, topography and altitude. The Food and Agricultural

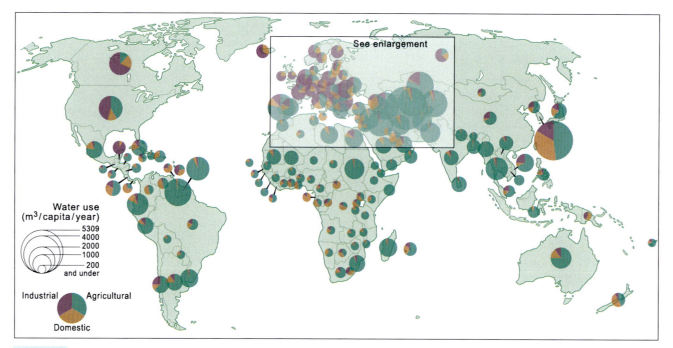

Figure 2.7 The world pattern of water use by sector (see Figure 2.8 for enlarged area)

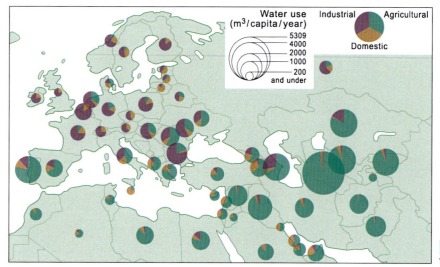

Figure 2.8 Water use by sector in Europe, the Middle East and North Africa

Organization of the United Nations (FAO) estimates that the irrigated area in 93 developing countries will increase by 20 per cent or 40 million ha between 1998 and 2030, even though this is less than half the 100 million ha expansion in the previous 30 years. Worldwide, the FAO calculates that the irrigated area will expand by 34 per cent and that this will increase the freshwater demand by 14 per cent up to 2420 km³ in the first three decades of this century. Substantial though this is – at about seven per cent of global river resources – it is modest compared with the a 55 per cent increase in food production, which could result from improvements in crops and management.

Although most of the increase will be in developing countries – and the World Bank is focusing on support for sub-Saharan Africa – a recent report by the European Environment Agency (2009) describes irrigated agriculture in Europe as a heavy and growing burden. Nearly a quarter of water abstracted in Europe is for irrigation. This may be little more than half the 44 per cent that is used for cooling in power plants, but whereas most cooling water is returned to the environment, only a third of irrigation water is. The situation is worse in southern Europe where 80 per cent of water abstracted goes to irrigation. Most of this is abstracted in summer when there is less in the rivers anyway, thus maximizing the detrimental effects and providing less dilution for pollution. Golf courses are a key bane. A typical golf course in southern Spain consumes 500,000 m³ a year – and the Júcar river basin has 19 courses and 55 more planned (see *Saltwater irrigation of leisure landscapes* in Chapter 16). These also attract many tourists who add to water demands (see Chapter 7 Dams and diversions).

There are, however, a number of problems with current approaches to irrigation that need addressing: wasteful practices, dominant crops that are extremely thirsty, destruction of the soil, especially through salinization, and pollution of surface and groundwater by fertilizers, herbicides and pesticides. Globally, irrigation is extremely inefficient. Of the 2300 km³ of freshwater withdrawn each year the crops effectively consume only 900 km³. Cotton and rice top the list of thirsty crops (see Chapter 8 Trading water – real and virtual). And monocultures like cotton require large quantities of fertilizers and pesticides (see *The Aral Sea: lost and gained?* in Chapter 15). The Green Revolution of the 1960s and 1970s may have fed the world but it also killed wildlife (see Chapter 5 Pollution and water-related disease). The problem is still with us. In the run-up to the Copenhagen climate summit in 2009, agri-business companies producing fertilizers and pesticides lobbied to make agriculture exempt from gas emissions targets. (Farming and food account for 30 per cent of greenhouse gas emissions.)

There is a wide range in the water efficiency of irrigation systems around the world due to differences in technology, culture and climate. The key figures are the net consumptive use, the 'specific water use' and the return rate rather than the gross intake. Specific water use is the intake per hectare per year, and the return rate is the amount that is returned to the rivers. The net consumptive use in the USA is one of the lowest. Specific water use in Israel is less than half that in Iraq. The highest rates of return are in North America and the Commonwealth of Independent States (CIS, the former members of the Soviet Union), which at around 40 per cent are twice as high as the world average or that in India. However, for the CIS this return efficiency is outweighed by the high level of initial, gross intake. As a result, the net consumptive use in the CIS averages more than ten times that of India. The extreme inefficiency of the old Soviet irrigated agriculture contributed to the approval of plans in 1982 to undertake a 50-year programme to divert the arctic rivers to the south (see *US and Soviet diversion plans* in Chapter 7).

It is estimated that over the millennia more than 25 million hectares have been lost to agriculture through salinization and waterlogging, most notably in the Middle East, including substantial areas in Egypt and Iraq as recently as the last quarter of the twentieth century. In part, this is an inevitable consequence of high evaporation rates, raising and depositing salts in the surface soils, but it is also in part due to mismanagement: over-irrigation and lack of adequate drainage. Flood irrigation in paddies or furrows is still the predominant method, yet it is also the most wasteful in evaporation losses. Distribution channels are predominantly leaky and uncovered, yet covering and lining channels could save up to 60 per cent. Converting to spray or drip irrigation could save up to 75 per cent. It is also a question of knowledge and perception – when to irrigate and by how much. In his piece below on irrigation in Saudi Arabia, Abdullah Almisnid shows how wasteful traditional farmers are compared with modern, high-tech commercial farms.

Even non-irrigated or 'rainfed' agriculture affects water resources. Forest clearance is likely to decrease losses through evaporation and transpiration and can aggravate flood problems. Some crops may increase evapo-transpiration and reduce riverflows or groundwater recharge (see Chapter 6 Water, land and wildlife).

Industry and commerce

Compared with agriculture, industry is a minor player on the world scale, accounting for barely a tenth of total global water use (Figure 2.6). Only in Europe and North America is industry a larger user per head of population (Figures 2.7 and 2.8). In most developed countries, industrial water use has also been declining for decades. In England and Wales, whereas industry and power took about two-thirds as much as domestic and municipal sources in the 1960s, by the 1990s it took under a third. Industrial use in England and Wales fell by 50 per cent in the 1970s alone, partly because of the shock caused by water cut-offs during the 1976 drought (perhaps a one in 400 year extreme event) and it has not recovered. In contrast, domestic and municipal demand increased 12 per cent over the decade.

The decline in industrial use continues. In part, this has been because of the contraction in what used to be heavy water users, like iron and steel production, but it is also due to the need to cut costs, increase efficiency and to revised assessments of vulnerability to outside supplies. It used to take 200 tonnes of water to make a tonne of

Irrigation methods, declining water tables and the potential impact of climate change in Gassim, Central Saudi Arabia

Abdullah Almisnid

The aquifers around Gassim are being depleted by abstractions for irrigation, yet the economic activity in the region is almost entirely dependent on the groundwater. The local aquifers are barely being replenished by the low rainfall (92 mm per annum). In the only well with a long record, groundwater level fell by 71 metres between 1979 and 2001. Four other wells show falls of between five and 12 metres from 1997 to 2002. In the Al-Hasa area, south east of Gassim, the recent decline has been as much as four metres per year (Omran, 2004). If this pressure on the water supply continues, even without climate change, the area will face severe water storages with dire consequences for agricultural production.

Both water use and productivity differ significantly between traditional farms, using open furrow irrigation, and modern commercial farms, using sprinkler irrigation. Recent observations show that traditional farms use between 12,663 and 18,874 m³/ha/season to produce only 1.6 to 2 tonnes/ha of wheat (Almisnid, 2005). In contrast, the modern commercial farms applied only 7100 to 9341 m³/ha/season and achieved the much higher productivity of 4.2 to 6.3 tonnes/ha. Farm interviews showed little awareness of the need to conserve water amongst the traditional farmers and a high degree of overuse.

The possible impact of climate change on water availability and use

Calculations based on the output from three GCMs (HadCM3, CGCM2 and ECHAM4) with two emissions scenarios (A2 and B2) show significant increases in annual temperature of 0.4° to 1.6°C by the 2020s for Saudi Arabia as a whole, and 2° to 4.8°C by the 2080s. There is no clear trend in rainfall, however, with results ranging between −20 and +30 mm/year relative to the baseline period of 1971–2000.

Incorporating the consequent changes in evapotranspiration losses, crop water requirements are estimated to increase by three per cent with a 1.3°C increase in average annual temperature by the 2020s. A 4.1°C rise by the 2080s could increase crop water requirements by 12 per cent under scenario A2, and nine per cent under scenario B2. Without any change in agricultural practices, Gassim would therefore have significantly higher irrigation needs, placing the area under even greater water availability pressures. Every 1°C increase in temperature could lead to an additional water requirement of 103 m³/ha/season for winter wheat.

steel in the 1930s, but it now takes only 2 tonnes, and 20 years ago semiconductor production took 17.6 litres per cm², but it now takes only 7.6 (Palaniappan and Gleick, 2009).

Declining industrial consumption is also partly due to environmental legislation, the need to establish environmental credentials with shareholders and some genuine concern for the environment. The situation in most developing countries is the opposite. As industrialization spreads, so water use rises, especially when much of the industry is using old, water-intensive processes. China began to rise as a steel producer in the 1970s and is now the largest producer in the world, producing two and a half times that of the EU. India now produces more than half that of the USA.

To a large extent, developing countries and especially the emerging economies are now following what happened in the West last century. North America showed a twenty-fold increase in industrial water demand during the twentieth century and by the end of the century North America and the CIS accounted for half of all industrial withdrawals.

We normally focus on the impacts of industry *on* water, but a report by the banking group J.P. Morgan entitled

'Watching Water: a guide to evaluating corporate risks in a thirsty world' focuses on the impacts of water on industry. It points out that industry and commerce generally underestimate their sensitivity to water issues (World Resources Institute, 2008). Exposure to water scarcity and pollution is not limited to onsite production, but may have a greater effect via the supply chain. The industries most exposed to such risk are: power generation, mining, semiconductor manufacturing, and food and beverages. The authors conclude that disclosure of exposure to water-related risks is 'seriously inadequate' and is typically included only in environmental statements prepared for public relations rather than in the regulatory filings that investors use. They recommend that investors include vulnerability to water shortages and pollution in their assessments.

Power industry

There is increasing evidence of the impact of the power industry on water. Cooling power plants run on fossil and nuclear fuels consumes large amounts of water. The large 2000 MW Didcot A coal-fired station in England took in 24 million m^3 from the River Thames in 2007 and returned just 17 million. Its sister plant, the 1360 MW gas-fired Didcot B, took in 11 million m^3 and returned just 5.5 million. Didcot also used over a million cubic metres of mains water taken from the Thames and treated to drinking water standard to top up the boiler circuits. Coal-fired plants also use substantial amounts of water in the process of scrubbing sulphur dioxide out of the gas emissions – a technology that became widely used following the fears associated with acid rain deposition in the 1980s. Power stations also return warm water to rivers and lakes with likely effects on aquatic life. At Wales's now defunct Trawsffynydd nuclear plant, fish reared in the lake grew faster and larger and were exported to stock other lakes for fishing. However, excessive water temperatures reduce the oxygen content of the water and could asphyxiate fish.

Power generation in the United States uses 514 million m^3 of freshwater a year and the thermoelectric power sector accounts for 39 per cent of all freshwater withdrawals in the country (World Resources Institute, 2008). Coal plants use 2800 litres to produce one MWh of electricity, and gas plants use 2300. The 882 MW gas-fired Riverside generator in Oklahoma uses 62,000 m^3 a day – nearly 25 million m^3 a year – which is about as much as a city of 180,000 in America. Nuclear plants need even more cooling: 3100 litres per megawatt hour.

Britain's Dungeness station takes in 1892 million m^3 a year. This is seawater, but France's stations on the Loire use freshwater, as do many in the USA like the Cowan's Ford plant in North Carolina.

When current debates on reducing greenhouse gas emissions are being used by governments to revive the nuclear industry as a supposedly 'green' source of energy, it is important to note its impact on water resources and on the natural water environment. This is rarely if ever considered in the debate, yet if it were it could substantially alter the verdict.

The power industry is also vulnerable itself to changes in the quantity and quality of water available. Events in France are a good illustration. In 2003 and again in 2009, exceptionally high water temperatures and low river discharges caused the power company EDF to close down many of its nuclear power stations. EDF lost €300 million in 2003 closing down a quarter of its nuclear plants and importing electricity from abroad. The Brown's Ferry nuclear plant that serves Memphis was forced to close in the summer of 2007 because the temperature of the intake water was too high for cooling the plant, just as city demand was peaking for air conditioning.

This is discussed further in the section on hydropower in Chapter 7.

Peak water?

We hear a lot about 'peak oil', but is there such a thing as 'peak water'? The term peak water was invented by the American media. The idea may have some value, but it is not true in the same way as for oil. Oil is a finite and limited resource. Water is essentially finite, but it is unlimited, because it is being continually renewed by the water cycle (see Chapter 11 The restless water cycle). When we consume it, we are merely transferring water from one place or one medium to another. There can be no global peak water, but there can be local peak water, as local water resources diminish. This is more obvious in some parts of the water cycle than others. Exploiting groundwater that is not being recharged under the current climate is a case where potential exhaustion is possible (see *Mining groundwater* in Chapter 12). And cases where the rate of extraction exceeds the rate of recharge will fall back to a maximum yield that is dictated by the slower rate of recharge.

'It signals we are at the end of the age of cheap, easy water. In the same way that Peak Oil has meant the end of cheap, easy to access sources of petroleum, Peak Water means we are going to have to go further, spend more, and expect less in the realm of freshwater.'

Meena Palaniappan in Keynote address to
American Planning Association
Salt Lake City, 2008

The point made by Meena Palaniappan in the quotation from her address to the American Planning Association is that water is very expensive to transport over large distances compared to its low value. Either we reduce demand to the levels of locally renewable water or we move to more expensive options like desalination or bulk transportation. As an engineer, she is very familiar with the costs of engineering water (the hard path), but she sees the future in 'soft path' options, like controlling demand, as the best way forward – decentralized and in the hands of planners and communities (see box 'Water soft paths: the route to sustainable water security' in Chapter 13).

One valuable new concept that arises from this is 'peak ecological water'. This is discussed in Chapter 15.

Conclusions

The rise in global demand for water shows no sign of abating. Agriculture is the prime user and it is bound to absorb more water as world population continues to grow. Urbanization is demanding more and climate change will add to demand in all sectors.

Economies of use are badly needed in every sector. Industrial demand is in check in most developed countries, but for reasons more of cost-cutting, changes in products and importing goods from abroad than for saving national resources. In many parts of the developing world, however, industry is a growing user. Important elements of both agriculture and industry in the developing world are expanding to supply goods to developed nations. In effect, developed countries are exporting their water demand to the developing world, where many countries are less able to support it through water shortages of their own. Again, providing clean water and sanitation in the developing world is bound to increase their domestic consumption.

Discussion points

- Analyse the validity and implications of the concept of peak water.
- Will increasing urbanization inevitably lead to more water crises?
- Consider the links between growing population, food demands and the use of water in agriculture.

Further reading

Science journalist Fred Pearce's book is very readable and informative:

Pearce, F. 2007. *When the Rivers Run Dry – what happens when our water runs out?* Transworld Publishers, London.

A more complete scientific assessment is available in:

Shiklomanov, I.A. and Rodda, J.C. (eds) 2003. *World Water Resources at the beginning of the 21st century.* International Hydrology Series, UNESCO, Cambridge University Press, Cambridge.

The following paper and book by Lvovich are worth reading as historical records:

Lvovich, M.I. 1970. Water resources of the world and their future. *International Association of Scientific Hydrology Pub.* No. 92: 317–22.

Lvovich, M.I. 1979. *World water resources and their future.* Washington, D.C., American Geophysical Union.

Jones provides a critical summary of Lvovich's views in:

Jones, J.A.A. 1997. *Global Hydrology: processes, resources and environmental management.* Addison Wesley Longman, Harlow.

The concept of 'peak water' is engagingly presented by Meena Palaniappan in her address to the American Planning Association, available at the website of the Pacific Institute:

Palaniappan, M. 2008. 'Peak Water'. Pacific Institute, 11pp. At: www.pacinst.org.

and

Palaniappan, M. and Gleick, P.H. 2009. Chapter 1 Peak Water, in *The world's water 2008–2009*, Pacific Institute, 1–16.

3 Water and poverty

'We should all be more alarmed that with the halfway mark to 2015, it is clear that most of these targets will not be met. The cause is not a lack of resources, but a lack of global political will.'

Dr Rowan Williams, Archbishop of Canterbury,
Head of the worldwide Anglican Communion

'Poverty must be eradicated and if we all work together for change, poverty will be eradicated.'

Gordon Brown, former UK Prime Minister
(Both at the Lambeth Conference, 2008.)

Poverty and lack of adequate drinking water and sanitation go hand in hand. The WHO Commission on Social Determinants of Health (2008) reported that social inequalities are killing people on a grand scale: a girl born today may have an average life expectancy of 80 years in the USA compared with just 45 years in the developing world. However, the Commission was optimistic that equality could be achieved within a generation if the will is there.

The 1.1 billion people who live in what the World Bank classifies as 'extreme poverty' have little or no chance of lifting themselves out of their predicament on their own (Chen and Ravallion, 2004). Living on less than $1 per day in equivalent purchasing power, they are incapable of saving and investing, of paying for health care and schooling, of buying new farm implements, or of buying equipment for water treatment or even water storage tanks. Their life is dominated by the daily fight to eat and live. Many exist in declining circumstances, in a negative feedback cycle of failing crops and failing health. Workforces are weakened by malnutrition, and debilitated and decimated by diseases like malaria and HIV/AIDS. In many parts of rural Africa, whole villages have lost their able-bodied men and women to AIDS, ageing grandparents are left to raise orphaned children, often up to half a dozen a time, or else the eldest child, perhaps barely in their teens, is forced to care for their younger siblings. The young have not yet learned the farming knowhow of their parents and the old are incapable of putting it into practice. Education is curtailed, schools close and the prospects for advancement elsewhere are cut off. Worse, inheritance rules often divide the farmland between heirs, thus reducing the area of cropland available to sustain each new family. Most countries where more than a quarter of the population are extremely poor lie in the Indian subcontinent and in most of the countries in sub-Saharan Africa. Many of the countries in Latin America and eastern Asia are also in 'moderate poverty', where more than a quarter of the population earn between $1 and $2 a day in equivalent buying power. These too are regions where both existing water resources are hard-pressed and demand is projected to increase markedly in the near future (see Figure 2.2).

'The daily conditions in which people live have a strong influence on health equity. Access to quality housing and shelter and clean water and sanitation are human rights and basic needs for healthy living.'

Commission on Social Determinants of Health,
World Health Organization 2008

Environmental change is also playing its part. Haiti is a tragic case, where denudation of the hillsides caused by desperate attempts to get food and fuel has left the country even more vulnerable to devastation by hurricanes. Climate change has increased the frequency and persistence of drought in parts of Africa, especially in the Sahel. Deforestation is widespread and expanding as people seek out fuel and the lucky few clear more land for farming. And climate change and deforestation can interact, with disastrous consequences. As the land is laid bare, the soil is more vulnerable: rainwater runs off the surface of the sun-baked earth, enters desiccation cracks and washes away the topsoil, and with it the nutrients and organic matter needed to maintain the fertility of the soil. Crop yields dwindle. More intense droughts and the occasional heavier cloudbursts, fuelled by more convective activity over warmer land, intensify soil loss. With more rapid runoff and less water stored in the soil or evaporated by vegetation, floods increase (Figure 3.1). The floods are more of a hazard than a potential resource, as they tend to exceed the capacity of village dams and even wash the simple earth dams away.

The eminent American climatologist Reid Bryson has even suggested that deforestation and agricultural activity in the semi-arid Sahel may change the climate: that fine soil particles may be blown high into the air, increase the reflectivity of the air to incoming sunlight and so cool the air aloft. The cooled air sinks, gradually heating up under the increasing weight of the more and more atmosphere above, and eventually forms an 'inversion', a zone of air that is warmer than most convective air cells rising from the ground. The inversion inhibits updrafts and so suppresses rain-forming clouds (Figure 3.1). Bryson suggests that this process may be extending the natural subtropical inversion that is responsible for the Sahara Desert southwards and holding back the seasonal northward advance of the West African monsoon. See Aondover Tarhule's article 'Drought in the Sahel', below.

Urban slums and shanty towns

As subsistence farming becomes less viable, migrants head for the cities. The more able and better educated may manage to raise their standard of living to the next level, the 'moderately poor' on $1–2 per day or occasionally better still, but many are simply increasing the numbers in extreme poverty in the cities. The ever-expanding shanty towns on the edge of cities in Africa, southern Asia and Latin America are foci for disease and desperation. Still, as in the villages they left, lacking centrally supplied water and sewage, they are now prey to the extra hazards born of close living, jerry-built housing, lack of the old community support and exploitation by ruthless city 'entrepreneurs'. The vast difference in water prices between piped public water supplies and water purchased from mobile tankers, which is all even the more affluent poor may have access to in these unplanned settlements, is illustrated in Figure 3.4. These new urban poor are also confined to live in the worst topographic locations, on hillsides prone to soil erosion

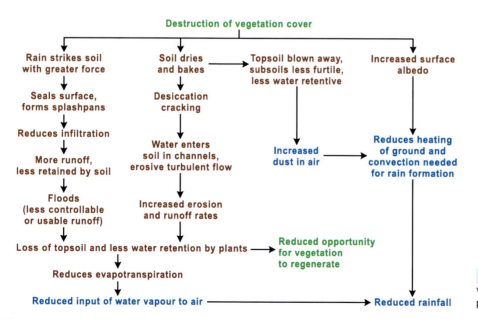

Figure 3.1 The effects of destroying vegetation cover on hydrological processes in semi-arid regions

Drought in the Sahel

Aondover Tarhule

The semi-arid Sahel savannah zone in Africa is experiencing the longest drought recorded anywhere on Earth in modern times. A narrow bioclimatic zone, the Sahel (an Arabic term meaning 'desert fringe') measures only about 500 km on average from North to South but stretches across the entire southern margin of the Sahara Desert from the Atlantic Ocean on the west to the Red Sea on the east (Figure 3.2). Beginning in the late 1960s, the total seasonal rainfall throughout the zone declined abruptly by between 20 and 40 per cent (Figure 3.3). This was due, principally, to a reduced number of high intensity rain events. The onset of this drought, especially the period between 1968 and 1973, was devastating. Famine induced by the drought is generally blamed for the deaths of between 100,000 and 250,000 people along with 12 million head of livestock in the six worst affected Sahelian countries (Senegal, Mauritania, Mali, Burkina Faso, Niger and Chad). In the early to mid-1980s, drought again ravaged the Sahel. This time, the hardest hit areas were in Sudan and Ethiopia. Approximately one million people starved to death in Ethiopia alone, and nearly 30 million were forced to abandon their homes in search of food and water. Since the late 1980s the total seasonal rainfall in the Sahel has recovered somewhat but the pre-drought levels have yet to be reached and the threat of drought remains. Indeed, in 2005, poor rains, combined with a locust outbreak, caused major famines that threatened nearly three million of Niger's 12 million inhabitants.

The link between drought and famines in Africa is frequently confused in public discussion. Drought is a meteorological condition that denotes a deficiency of water with respect to a place, time or activity for which the available water is normally adequate. Famine, on the other hand, refers to the widespread shortage of food and/or mass starvation of people as a result of a complex web of interacting variables including drought or other climatic phenomena (e.g. floods, locust outbreaks), conflicts or wars, poor transportation infrastructure, and macroeconomic policies among many others. Thus, in parts of the USA, Australia, or Israel with climates like that of the

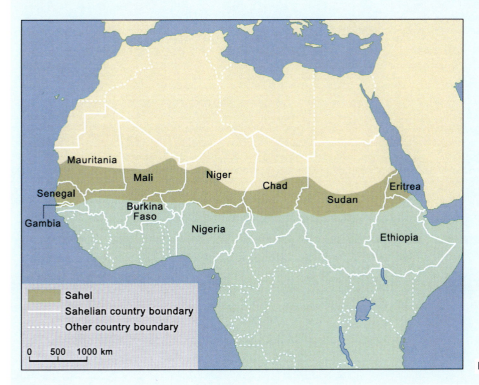

Figure 3.2 The African Sahel Zone

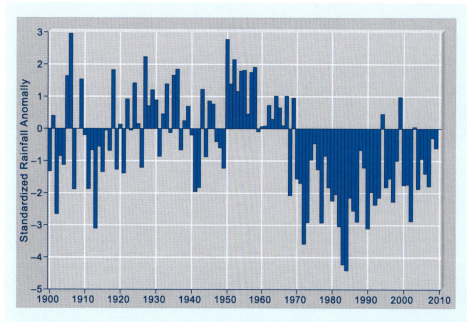

Figure 3.3 Rainfall variability in the Sahel during the twentieth century – departures from long-term average

Sahel, drought might lead to a slight increase in the price of food products but not result in famine. In sub-Saharan Africa, however, droughts and famine are strongly linked primarily because agricultural production is at a near subsistence level and the margin between normal food supply and starvation is extremely small. At the same time, owing to high poverty rates, many households that do not grow enough food to eat either have very low-paying jobs or no jobs at all, and cannot earn enough to buy food even when it is physically available.

Scientists have pondered the question of whether multidecadal droughts, such as the one currently affecting the Sahel, have previously occurred in the region during the last few centuries or whether the current drought sequence represents a new phenomenon in the climate dynamics of the region. Unfortunately, the instrumental climate records are relatively short (most only go back to the 1930s) and available proxy records, primarily from lake-bottom sediments, have low stratigraphic or chronological resolution (Street-Perrott et al, 2000). However, qualitative historical information can be retrieved from reports referring to the landscape, such as the extent of vegetation and sand dunes, river discharge, population settlements, the size and depth of Lake Chad, droughts and famines. This information is available in various forms including travellers' accounts and diaries, folklore, oral histories, colonial records and palace chronicles from the many empires and kingdoms that have occupied the Sahel zone over the centuries. These sources indicate that the Sahel as a whole was wetter than present prior

to the sixteenth century. They also show that widespread and severe droughts occurred at least once during each of the past four centuries, with a few between 10 and 20 years in duration.

The cause of Sahel drought has been the subject of much research since the early 1970s. From the plethora of studies have emerged two leading but sharply differing hypotheses. The first emphasizes the role of anthropogenic impacts, while the second focuses on changes in sea surface temperatures (SSTs) and their links to Sahel rainfall. The premise of the first hypothesis is that rapid and large-scale human-induced changes in vegetation cover throughout the Sahel created a positive albedo-rainfall feedback loop in which reduced vegetal cover increases surface albedo, leading to desiccation of the land surface and diminished rainfall, which causes further decrease in vegetation cover and still further decrease in rainfall (Glantz, 1994; Nicholson, 2000). Thus, once initiated, the process is self-reinforcing, creating persistent drought (Figure 3.1).

Both hypotheses appear correct. During the past half century, for example, the Sahel has supported one of the highest rates of population growth in the world, doubling approximately every 25–30 years. The major urban centres have grown even faster, with growth rates of between two to three times those of the national population. To feed the burgeoning population, agriculture and livestock grazing expanded correspondingly, often extending into marginal areas ecologically unsuited to such activities. The expansion in cultivated land was especially large because crop yields remained relatively low as the result of very low applications of fertilizers and pesticides. Despite the increased area in cultivation therefore production tonnage remained static. Demand for firewood in the urban centres contributed further to massive deforestation in the surrounding areas. Around major centres like Niamey, tree cover has been so reduced that from the air, the city itself appears as by far the most heavily wooded area within a radius of between 30 and 40 km.

These dynamics therefore support the theory of a possible land surface feedback mechanism. This has been replicated to varying degrees in many computer models. The problem is that the amount of change in the land surface and the albedo required to simulate the observed rainfall change are much larger than those observed in the real world.

Scientists have recognized since the late 1970s that variations in SSTs at key locations have some influence on Sahel rainfall (Lamb and Peppler, 1992; Giannini et al, 2003). Recent modelling indicates that changes in global sea surface temperatures between 1930 and 2000 can account for between 25 and 35 per cent of the observed rainfall change in the Sahel. These results are important because they confirm what scientists had already suspected, but they also reveal that SST by themselves cannot explain all of the Sahel's rainfall variability and drought.

A consensus is beginning to emerge that models incorporating the effects of both vegetation change and SST variations would explain a larger proportion of observed rainfall patterns and provide a basis for predicting future rainfall and perhaps drought. In line with this thinking, scientists from more than 20 countries are collaborating on a new multidisciplinary initiative called AMMA (African Monsoon Multidisciplinary Analysis). The goal of AMMA is to assess the state of knowledge regarding rainfall variability and predictability for West Africa as a whole and to develop new strategies for monitoring the causative land, oceanic, and atmospheric processes (Lebel et al, 2003). Such international collaboration is essential for reducing the real human suffering and economic hardships caused by drought in the Sahel, one of the poorest regions of our world.

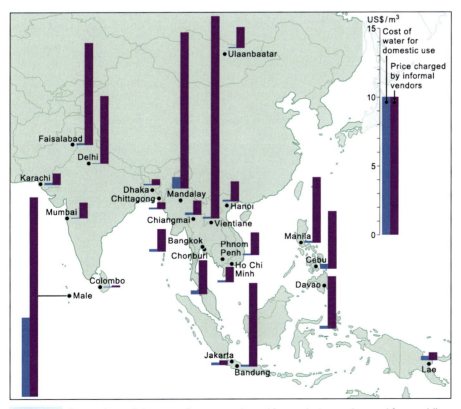

Figure 3.4 Comparison of the cost of water purchased from private vendors and from public supplies in SE Asia

and landslides or floodplains prone to inundation, typically disowned and unsupported by city authorities (see 'Floods and poverty in Metro Manila' by Doracie Zoleta-Nantes, later in this chapter).

The lack of mains water supplies for many of the new urban dwellers in the LEDCs may reduce per capita water consumption, but it also increases pollution in urban watercourses and health risks from water-related diseases.

The World Health Organization estimates that between 2004 and 2015 the number of people living in cities without sanitation will rise from 611 million to 692 million. Numbers living in slums could reach two billion by 2030. Over half the population of Mumbai, seven million people, live in slums and three million of those are forced to defecate in the open. Sewage rots in the creeks and disfigures the shore. Some 60 per cent of the population of Nairobi live in slums like Kibera, where plastic bags are used as 'flying toilets'. Two-thirds of the population of Johannesburg live in Soweto. Around one million people, a quarter of the population of Rio de Janeiro, live in its 946 'favelas'. Rocinha *favela*, the largest slum in South America, where sewers overflow in the street and rats infest the homes, houses 160,000 people. The *favelas* are creeping up the surrounding hills, destroying the forests. The situation is repeated around the world as people migrate to the city margins attracted by jobs or forced off the land by drought, falling yields, agribusiness or dams.

Inevitably, the urban poor are forced to live in the worst locations. This often means they are more exposed to natural disasters like floods and landslides. When typhoon Ketsana struck Manila in September 2009, it was the poor living in shanties in the low-lying areas of city who suffered worst, as shown in Doracie Zolenta-Nantes' article. Landslides hit the inhabitants of Rio de Janeiro shanty towns after a heavy rainstorm in April 2010.

Floods and poverty in Metro Manila

Doracie Zoleta-Nantes

Christina is 13 years old and is the eldest of seven siblings. She does not go to school and lives with her unemployed and sickly mother. She begs for money with her youngest sister on the street pavements of Manila and earns roughly one US dollar a day. Christina often experiences floods as high as her waist on street pavements while she begs, as well as inside her family's shanty that is made of rusty sheets of galvanized iron and worn out scraps of wood. The family's shanty is built on top of one dried-up creek in the district of Sampaloc. Annual floods always destroy her family's shack. Her mother and her siblings rebuild their huts together, with the help of their neighbours, using whatever materials they are able to salvage from the floods. They need food, shelter, dry clothing and medicine for respiratory illnesses that they often experience during flooded days. Floods inconvenience everyone in her family, except perhaps for her father who is currently in jail for theft.

Christina Lungcay and her family are among millions of Filipinos who are affected by the annual flooding in Metro Manila. About 4500 hectares of the 65,000 hectares of land comprising Metro Manila are prone to annual flooding. Metro Manila has a tropical marine climate with two distinct seasons: rainy (mid-June to October) and dry (mid-November to early May). It experiences a mean annual precipitation of 2069 mm, but it can reach 6800 mm some years. The country is hit by about 20 storms and typhoons a year. When a typhoon event coincides with torrential monsoon rains, caused by warm moist south-westerly winds from the Indian Ocean between June and October, the resulting flooding events can be devastating. During this time, the water levels in most reservoirs rise up to 30 m above normal. This necessitates an emergency release of water at a time when the rivers have already swelled considerably. This and the combined effects of high tides and storm surges in the adjoining waters of Manila Bay contribute to extensive flooding in the region.

The flooding is made worse by deforestation in the catchment areas of major rivers such as the Pasig River, and rampant expansion of housing on sloping site to meet the demands of the population, which is growing rapidly mainly due to immigration from rural areas. The lack of an adequate solid waste management system and weak enforcement of anti-dumping and anti-littering ordinances and other environmental laws compound the problems. An estimated 14 million people and hundreds of industries indiscriminately dump plastic wastes, hazardous chemical wastes and other solid refuse into the metropolis's waterways and drainage systems. The encroachment of the squatter population in search of living space and the building of structures by the government, commerce and industry on most of the waterways that dry up during the summer months has already led to the loss of 60 km of *esteros* (drainage arteries). The lack of political will on behalf of the government to build sufficient drainage and sewage systems or to dredge most channels of the antiquated drainage systems is further compounded by the rapid conversion of former swamplands and river banks into housing estates and industrial sites. Moreover, the piecemeal allocation of flood countermeasures at both the national and local levels and the lack of public

Figure 3.5 All *esteros* or drainage canals in Greater Manila Area are a convenient waste dumping site for millions of Metro Manila residents and squatters

cooperation to systematically solve the problems of flooding and the half-hearted support of the political leaders who live in places not affected by floodwaters further add to the complexity of addressing the root causes of flood disasters in Metro Manila. Villages that were built on former rice and swamplands are submerged in floodwaters to depths of between 1 and 6 metres, causing some people to regularly move up to the rooftops of their houses for safety during the rainy season. Thousands of commuters get stranded on flooded streets and pedestrians wade through, facing the risk of plunging into open manholes or public infrastructure excavations and the danger of coming in contact with floating and maybe disease-infected human wastes. Occasionally, they may bump into swimming rodents and snakes. Floodwaters contaminate some sources of food and drinking water and cause upsurges in the incidence of diarrhoea, cholera, leptospirosis and bronchitis, influenza and pneumonia. People drown and the shanties and informal housing structures of most of the urban poor are destroyed and carried away by the ravaging floodwaters; floods further marginalize poor families and render them homeless.

Flood mitigation measures have been undertaken in the past and are continuously being proposed for the future. One is squatter relocation and dismantling the shanties and other illegal constructions on the banks of waterways together with the strict observance of the 3-metre easement rule, which prohibits the construction of any type of structure within 3 metres of any waterway. Clearing, desilting and de-clogging drainage canals and waterways, and the strict enforcement of anti-littering and anti-dumping ordinances are badly needed, as well as the additional construction and installation of flood inlets and pumping stations in low-lying areas. The provision of more public assistance, free transportation and other community health services during the flood season is also salient.

Local and international programmes to 'Make poverty history' are entirely necessary to eliminate urban slums and shanty towns, but they will inevitably put great pressure on local water resources and require far more careful planning than is presently the norm. Indeed, at present very little of the money that cities spend on improving water supply and sanitation reaches the urban slums. Mumbai's slums benefited from only ten per cent of the World Bank money that funded flush toilets, treatment plants and provided a long sea outfall for nearly half the city's population in 2002. It was estimated that Mumbai

slums needed 65,000 new toilets, but by the beginning of the second sewage scheme (2008–2011) only a tenth of these had been built and authorities were committed to provide only 4700 more.

Sadly, the problems are not all confined to the LEDCs as the EU and the USA receive more immigrants from deprived regions of Africa, Asia and Latin America. Lisbon's Cova da Moura is probably the worst urban slum in Europe, a belated legacy from the country's colonial past as 'reverse colonization' brings immigrants from

Brazil and Angola, some illegally, to live in a Latin American style *favela*.

One of the great tragedies for the children of slum dwellers is the loss of school days caused by sanitation-related disease (see *No plumbing disease* in Chapter 5).

The political economy of poverty: how the vast differences have developed

In the Millennium Development Goals the UN aims to halve the number of people living on less than a dollar a day by 2015. At the G8 summits in 2005 and 2006, the leaders of the wealthiest nations pledged their support to actually 'end poverty'. These are ambitious aims and ones that, given the limited success of many previous grand international political aims, like the International Drinking Water Supply and Sanitation Decade (1981–90), have a high chance of not being realized. Many politicians and political activists alike see international debt relief as a major route to this goal, if not the sole panacea. This debt is certainly a critical factor that has developed as a result of 'aid' from the developed world over the last half century, aid that has often been given specifically to develop irrigation projects and hydropower from dams constructed by western companies (see Chapter 7 Dams and diversions). The burden of debt has frequently been worsened by being linked to fluctuating interest rates, which subsequently reached new heights, especially in the 1970s. Widespread moves to cancel debt for the poorer countries in the mid-2000s appear to be a step forward. However, wiping out debt may discourage financial probity and reward spendthrift or corrupt governments.

Contrary to the impression given by many well-publicized campaigns, notably ones led by certain entertainment personalities, debt is not the only cause of national poverty. Financial burdens have also been exacerbated by corrupt regimes that have diverted funds, by civil war and by the direct or indirect costs of schemes that have failed. Emigration of the intelligentsia and qualified technicians – doctors, nurses, school teachers and engineers – has also taken its toll. Education too often leads to the realization that a better lifestyle is achievable abroad: a recent World Bank report claims that emigration has been 'disastrous'.

Parallel calls to increase western aid, while well-meaning and potentially of value, similarly need careful and critical evaluation. At the Gleneagles G8 summit lead-ers promised to double aid to Africa by 2010. This has not materialized, but the aim is still there to increase aid to 0.7 per cent of GDP. In 2006, the UK pledged to give £8 billion to Africa over the following ten years as part of a new aid package designed to eradicate poverty. But clearly, the first lesson from the recent past is not to repeat the same sort of aid packages that have intensified debt, fed corruption and often been inappropriate for a particular physical and cultural environment. Improved criteria are needed for giving aid, funding schemes that have been better designed in terms of the social and physical outcomes that are required in each country and, most importantly, including tighter controls on the use of the money. Both the World Bank and the IMF have been rather lax in the latter area (Sachs, 2005). So have many western governments. In 2010, Britain's Department for International Development pledged to improve after admitting that the £300 million aid given to Malawi to alleviate poverty had just been handed to the government with no requirement to account for how it was spent.

Aid might also be given in smaller tranches for 'pump-priming' schemes, rather than supporting grandiose projects. Aid need not always be financial: encouraging self-help through the transfer of knowledge and skills, so-called 'capacity-building', is increasingly recognized as an important solution. The ultimate aim must be to enable the poor countries to help themselves. In the words of travel writer Paul Theroux (2006): 'the impression that Africa is fatally troubled and can be saved only by outside help … is a destructive and extremely misleading conceit.' In his view, Africa is more resilient and self-sufficient than often portrayed, but these countries should force people like medics and teachers to stay where they were trained.

The causes of poverty vary around the world, with different mixes of geographical context and historical development. Landlocked countries like Bolivia or Nepal are denied the advantages of ocean trade routes that have boosted the economies of most of Western Europe. Although both Nepal and Bolivia are blessed with high per capita water resources by virtue of their mountain location, their economies have failed to develop sufficiently to support extensive networks of sanitation and freshwater supply. Many developing countries also suffer from adverse climates with higher exposure to extreme climatic events and endemic diseases that put a strain on weak economies. And a lack of natural resources, like coal and oil, can be a stranglehold on economic development. Indeed, the economic differences between nations were proportionately far less before the Industrial Revolution.

Jeffrey Sachs, Special Advisor to the UN secretary-general on the Millennium Development Goals, pointed to the need for authorities to recognize these differences and constraints in proposing solutions, and to analyse each case afresh (Sachs, 2005). He notes that the development policies of the World Bank and the IMF over the past 20 years, dominated by the notion of 'structural adjustment' as a universal panacea, have been largely detrimental. Their thesis has been that the sole causes of poverty are: too much state ownership, excessive government spending and intervention in the markets and generally poor governance. The solutions they imposed have been 'belt tightening', liberalization of markets, privatization, and improved governance. Many are reminiscent of similar IMF prescriptions for more developed nations and reflect the premise that: 'Poverty is your own fault. Be like us … enjoy the riches of private-sector-led economic development' (Sachs, 2005) (see *Privatizing water* in Chapter 4). Sachs points to the disastrous effects on countries that hardly had an economic 'belt' to tighten, including the destabilization of regimes. Nevertheless, closed markets, poor governance and corruption – often encouraged by loose controls on the use of aid monies – have undoubtedly played a part (see *Corruption and poor governance – Africa's burden* in Chapter 4).

In analysing the reasons why the economies of some poorer countries have advanced over the previous quarter-century, while others have not, Sachs (2005) suggests that food productivity is a major factor. Of the 62 low-income countries with per capita incomes below $3000 in 1980, the economies of 37 have grown and 25 have declined. Sixteen of the 25 are in Africa. Sixteen of the 37 successes are in Asia. While East Asia had higher per capita food productivity in 1980 (2016 kg/ha) and has increased production, sub-Saharan Africa started with low productivity (927 kg/ha), which has since declined. Sachs links this with higher population densities, better roads to transport fertilizer and grain, greater use of irrigation, and more donor aid supporting higher yield varieties in Asia. In 1980, the percentage of cropland under irrigation in East Asia was 37 per cent compared with only four per cent in sub-Saharan Africa. The 'poverty trap' in rural areas, especially in Africa, is a reinforcing cycle: low food productivity, lack of surplus food and infrastructure for trade, and so lack of money, lack of investment in irrigation, fertilizer and Green Revolution crops, thus less food, more illness and a diminished workforce for growing food. In contrast, Sachs sees higher food productivity as the 'platform for Asia's extraordinary growth'.

There are, however, other factors. Higher levels of education and literacy, and lower fertility rates and infant mortality have given East Asia an advantage. According to Sachs (2005), poor countries that have higher populations also tend to do better, perhaps because they develop more infrastructure and larger markets. Once again, Asia scores over Africa, with India and China prime examples, although there are still areas of extreme poverty within these advancing nations.

Conclusions

The eradication of poverty is a vital part of the current international political agenda, but it is making little headway. Poverty and water poverty go hand in hand. There are areas in the western world where this exists, but the biggest challenge lies in the developing world. The poorest countries also tend to lie in regions that suffer climatic disadvantages – irregular and declining rainfall, and frequent extreme weather events like floods, droughts and hurricanes. These countries also have fragile environments, where human destruction of the environment – over-grazing, clearing native vegetation for agriculture and the destruction of trees for firewood or building – is exploited by harsh weather and climate change. The drift to the cities is creating a population of urban poor, which is now expanding at an unprecedented pace. The urban poor often suffer worse, crammed in high density shanty towns with no clean water or sanitation and occupying the areas most prone to floods and landslides.

Discussion points

- Study the links between poverty and harsh climatic environments.
- Analyse the historical causes of poverty in one region of the world.
- Consider the prospects for alleviating poverty and water poverty.

Further reading

Jeremy Sachs has written a readable and thought provoking book:

Sachs, J. 2005. *The End of Poverty: how we can make it happen in our lifetime.* Penguin, London.

4 Governance and finance

Inadequate governance has played a large role in the water crisis and is set to continue to be a major contributor. Fragmented responsibilities, ineffective management, overlapping interests and inter-institutional rivalry, private profit-making, lack of coordination between upstream and downstream users, inappropriate political interference, poor legal structures and lack of enforcement mechanisms, historical water rights inhibiting fairer water distribution, and lack of public or stakeholder involvement are all responsible in different degrees in different places. Communist countries have suffered from over-centralization, underfunding, lack of performance incentives and jobs for life for Party members. At the other extreme, countries with more laissez-faire capitalist politics have often suffered from too much fragmentation among institutions and an excessive profit motive. Many countries of all shades of the political spectrum also continue to suffer from corruption. Corruption continues to be a particular problem for developing countries, but it is by no means exclusive to them.

Finance is an integral part of governance. Until recently, finance has been mainly a question for national governments, but the rise of UN agencies in recent decades has added another and largely welcome dimension, especially by introducing an ethical element into the granting of funds. This theme will be taken up more fully in Chapters 20 and 21.

Corruption and poor governance – Africa's burden

In no part of the world more manifestly than in sub-Saharan Africa, poor governance and institutional corruption have added to the woes of the poor. In his history of modern Africa, Guy Arnold (2006) points to 'disappointed hopes and catastrophic failures' in terms of per capita income, life expectancy and infant mortality. Most have developed since colonial independence. Widespread poverty is becoming increasingly restricted to sub-Saharan Africa as Asia is rapidly assuming the world economic dominance it had prior to 1750. Much of the problem can be put down to international monetary aid that has been misdirected by corrupt, inept or unstable governments. Between the early 1970s and 2010, Africa received over £260 billion in aid, yet very few countries outside South Africa have shown any sustained economic 'takeoff'.

The Democratic Republic of Congo, Ethiopia and Sudan are prime examples. Soon after receiving the money collected by Live Aid, the first global appeal for famine relief in 1985, Ethiopia waged war on Eritrea. Billions of dollars in aid is alleged to have been siphoned into private accounts in Nigeria and Malawi. In March 2010 a former general in the Ethiopian rebels revealed that rebels posing as merchants diverted most of the funds, mostly to buy weapons, some to corrupt politicians. He claims that only five per cent of the $10m aid received in 1984–5 reached the starving people. Oxfam and Bob Geldof disagree and say it was properly accounted for, but there must be strong doubts. Accounting in such situations can be extremely difficult and loose accounting is a big contributor to corruption.

Even in relatively successful Mozambique, according to Professor Joseph Hanlon of the Open University, there was little corruption before western aid arrived in quantity. The aid also encouraged dependency and discouraged entrepreneurship. Donors bear much of the responsibility through their natural propensity to publicize success rather than failure, ignoring waste and not insisting on proper auditing procedures. Commonwealth Secretary-General Don McKinnon has said: 'too much is left to the privacy of the corporate-client relationship.'

The Commission for Africa is now committed to eradicating corruption and is pressurizing western banks to repatriate the money. As with international debt relief and aid focused specifically on eliminating poverty, how successful this will be remains to be seen. At the very least, it is an encouraging new mindset.

Privatizing water

'There is little doubt that the headlong rush to private markets has failed to address some of the most critical issues and concerns about water. Our assessment shows that rigorous, independent reviews of water privatization efforts are necessary to protect the public. Water is far too important to human health and the health of our natural world to be placed entirely in the private sector.'

Dr Peter H. Gleick
Director of the Pacific Institute, California

Many water enterprises began as private initiatives, especially in the USA. The twentieth century tended to see more governmental involvement, both locally and nationally. But the theories expounded by economist Milton Friedman which were embraced by the Reagan and Thatcher governments in the 1980s led to a new phase that has been widely copied: return of water administrations into private hands. The creation of the World Trade Organization (WTO) in 1995 and its decision to treat water as a commercial good has proven a major encouragement. The foundations for this commercialization of water were laid beforehand by the international meetings of GATT (the General Agreement on Tariffs and Trade) in the years between its formation in 1948 and the establishment of its permanent successor, the WTO. GATT strove to establish the principle of free trade. In the 1970s, it was supported by the Trilateral Commission in which the USA, Europe and Japan, supported by major international banks, stated their support for free trade, fewer tariff barriers and less 'excess of democracy'.

The free trade philosophy led to the 1994 North American Free Trade Agreement (NAFTA), which has become a source of disagreement between the USA and Canada over the extent to which this covers water. NAFTA sees water as a commercial good, a service and an investment. It overrules national law like Canada's Water Resources Protection Act (1999) in all cases except where exploitation can be proven to be environmentally damaging. It supports foreign investment in water goods and services and gives foreign companies the right to sue national and provincial governments that block these in favour of their own companies. The first such case was in 1998 when the Sun Belt Water Company of California sued the

British Columbian government for a reported $10 billion for stopping it exporting water from BC to California. The arguments are not all one way. The Nova Group in Canada wanted to use the NAFTA legislation to export Lake Superior water to Asia – 600 million litres by 2002, but was stopped by public outcries on both shores. And there has been a longstanding fight between the Newfoundland government and the national government over Newfoundland's desire to use NAFTA to export water to the USA. Within America itself, the Supreme Court has ruled that water is an article of interstate commerce.

The legislation continues to become more and more invasive. Further economic integration is being introduced under the provisions of NAFTA's new Security and Prosperity Partnership, which could force public funding of water pipelines from Canada to the USA. Under WTO legislation, any government that wishes to stop foreign takeover of water services can be challenged under international trade rules with enforcement procedures. The WTO's new GATS (General Agreement on Trade in Services) contains further threats to water sovereignty strengthening judicial authority over domestic laws – as does the Free Trade Area of the Americas, still in the planning. The Council of Canadians, an influential NGO chaired by Maude Barlow, is implacably opposed to water privatization and exports and to these trade agreements (Barlow and Clarke, 2003). The NGOs meeting at the 2nd World Water Forum in the Hague, 2000, recorded their clear-cut opposition: 'Water is the common heritage of mankind and should not be negotiated as a commodity' (Morely, 2000). There is also strong opposition in certain quarters of the UN.

'The Commission views certain goods and services as basic human and societal needs – access to clean water, for example, and health care. Such goods and services must be made available universally regardless of ability to pay. In such instances, therefore, it is the public sector rather than the marketplace that underwrites adequate supply and access.'

Commission on Social Determinants of Health
World Health Organization 2008

The principal argument in favour of privatization is that private companies can go to the stock market and banks for funding and overcome the problems of under-investment that had led to ageing and inefficient infrastructure. They could also afford to hire managers with strong career experience as professional managers rather than engineers with less experience of the marketplace. And competition should be good for the consumer and reduce prices.

The model does not fit water as well as most other commodities, in so far as it tends to be a monopoly industry – competition is limited by its very nature as a heavy, bulky resource largely constrained by the pattern of river basins and aquifers. Nevertheless, the World Bank claims that the other advantages of privatization have led to marked improvements in the provision of water and sanitation around the world. The World Bank also argues that privatization will help developing countries pay off their international debts. The World Bank, the International Monetary Fund and the Asian Development Bank have all advocated privatization and have often made it a condition of receiving international aid as part of 'structural adjustment programmes' since the early 1990s. In 2007, the Bank reported that 84 per cent of all the systems that were privatized in the 1990s remain so. However, 97 per cent of all administrations in developing countries have not been privatized, and a significant number of countries – 24 in all – have reclaimed their water from private hands. In many cases this has been achieved with some acrimony.

Britain and America have been in the forefront of privatization. Water Authorities in England and Wales were privatized in 1989. America is more varied with many local community-owned operations remaining public and others opting to try privatization, such as: Atlanta, Georgia; New Orleans; Stockton, California; Chattanooga, Tennessee; Jersey City; Peoria, Illinois; and Lee County, Florida.

Types of privatization

There are numerous forms of privatization. In England and Wales privatization involved the complete sell-off of assets. Under a concession agreement as practised in France, the private company is given licence to run the entire system for a fixed number of years. Under a leasing contract, the company leases the assets but investment is controlled by the public sector. Finally, under a management only contract, the company runs the business and is responsible for investment, but does not own the assets or keep the profits – it is merely paid to do a job. And there is another way that was established in Wales in 2000 when Welsh Water, under its new owning company Glas Cymru became a non-profit organization, reinvesting all profits back into the company and its customers. This could be a blueprint for a fairer solution – no shareholders to show a profit for, concentration on the primary public services but benefiting from business acumen and the ability to raise money from commercial sources. But it was fortuitous. It came about as a management solution to financial difficulties in 2000, when it was nearly taken over by a Japanese company. And ironically, Welsh Water, under the operating company Hyder, had been one of the more acquisitive companies following privatization, investing in electricity and hotels, which proved a step too far and were subsequently sold off before Hyder's demise.

The growth of the transnationals

A number of water companies have come to dominate international investment in water services around the world as a result of the freeing up of restrictions. Private sector water companies in France have competed for local authority contracts for over 50 years, and this has given them the experience that explains why France has so many global players: Vivendi (now, Véolia Environnement), Suez (now Suez Environnement) and Saur. In 2000, Vivendi and Suez were respectively 91st and 118th in the list of top corporations, operating in over 130 countries. Both companies have now spun off their core water interests, but Véolia Environnement still stood at number 135 in the list in 2009.

In terms of global water companies alone, Suez tops the list serving 125 million customers, Véolia is second with 110 million and Saur fourth with 29 million. The German company RWE ranks third. Between them, the French water companies serve a massive 40 per cent of the world's privatized drinking water. Other substantial international players include Britain's Thames Water and United Utilities. In 2000, Suez, Vivendi and Thames supplied 80 per cent of the world's privatized drinking water. Now just five companies still supply around 50 per cent of the privatized public water supplies in the world, despite the entrance of a few more competitors (STWR, 2010).

These companies have a tremendous global reach, often incorporating engineering wings selling water and waste treatment equipment as well as supplying water directly. They also commonly incorporate an energy wing. Vivendi and Suez began buying up water rights in India in the 1990s. Saur operates in Poland and Spain. United

Utilities operates in Bulgaria, Estonia, Poland and the Philippines, serves half a million people in South Australia, and hopes to build a seawater desalination plant for Adelaide, although the recession caused it to consider divesting some of its foreign interests to raise £250,000. Véolia began operating Sydney's desalination plant in February 2010.

The other constant feature of the industry over the last 20 years has been its fluidity. Mergers and acquisitions have burgeoned this century, and the global recession has added to the turmoil. RWE bought Thames Water in 2001 and sold it in 2006 to a consortium, Kemble Water, which includes the investment company Macquarie of Australia. RWE bought the largest private water company in the USA, the American Water Works Company, in 2002 and sold what is now American Water in 2009. Domestically in the UK, United Utilities bought Southern Water and is currently contracted to manage Scottish Water's capital investment programme. Ofwat objects to company mergers that reduce competition and diminish Ofwat's ability to regulate them. It argues that a certain number of companies are needed in order to have comparators, but if the market is allowed to work unfettered, the most efficient or most capital-rich companies will absorb the others and consolidation is the result.

The triumph of the multinationals carries at least four major concerns for water. Is it good for a handful of companies, mostly from one country, to control such a large proportion of the water supply systems in the world? Is foreign ownership a potential threat to local democratic control of vital resources and utilities? Is it good for the same companies to have a joint stranglehold on electricity *and* water? And is the inconstancy of ownership a sound basis for long-term management and for developments that are sympathetic to the local needs? One noticeable feature of the opposition that Ofwat received from water companies when it published its plans to force them to reduce water prices over the five-year period to 2015, is that the only really serious threats came from foreign-owned companies. Thames, owned by Australia's Macquarie threatened legal action but withdrew just before the deadline for acceptance. Bristol Water, owned by Aguas de Barcelona, did challenge it and published plans to increase prices by a third.

Many takeovers and sales seem to suggest that interests are purely financial. The growing interest from private equity firms in utility companies, generally only concerned with profits – even with a reputation for asset stripping – is also a matter for concern.

Successes and failures

It is a truism that the media report failures rather than successes. There have been many successes. United Utilities for one has worked with its partner in the Philippines, Manila Water, to raise clean water supplies from 26 per cent in 1997 to 92 per cent of the population in the area served. Water quality in England has improved significantly since privatization due to a £70 billion investment over the past 20 years, double that spent before privatization, but this is also due to new EU directives and the threat of fines from the Environment Agency forcing change.

In the UK, during the first eight years of private companies water prices rose around 50 per cent in real terms and company profits increased by 147 per cent, i.e. more than doubled. The UK water industry makes over £1 billion in annual profits. Some companies invested more, but others did not, and many, like SevernTrent, paid substantial amounts of the profits back to shareholders. Investment by some companies, like Welsh Water, has had a marked effect on sewage treatment and generally enabled a year-on-year improvement in the cleanliness of beaches and the number of blue flag beaches (see Chapter 5 Pollution and water-related diseases). But many companies have also received substantial fines for polluting rivers, especially with raw sewage. The fight to stop deliberate release of raw sewage into rivers is still ongoing more than a decade after the policing authorities were set up (see Chapter 5). And in social terms, the industry shed 30,000 jobs in the first two years. A very small, but technically important number of jobs were lost by hydrometric officers (see Chapter 18 Improving monitoring and data management).

The original privatization plan produced by the Thatcher government in the 1980s had the private companies policing themselves in England and Wales. Swift objections by environmentalists and the threat of taking the issue to European authorities resulted in backtracking. The National Rivers Authority and subsequently the Environment Agency were created to cover environmental issues like pollution and floods, while Ofwat was established to police commercial aspects. The industry in Scotland has remained in the public sector, but with the Water Industry Commission for Scotland setting price limits and the Scottish Environmental Protection Agency policing the environment. However, doubt is developing. The Centre for Public Policy for the Regions (2010) suggests privatization is the only way to update the infrastructure and cure the 40 per cent leakage losses.

Meanwhile, Ofwat and the Environment Agency are looking at ways of stopping water companies blocking plans to move water from the north to SE England, and of introducing competition in order to cut water bills. Ofwat suggests the principle of 'common carriage' could be used to create competition. Modelled on the privatized electricity industry, this would nominally allow different companies to use the same pipe network: it becomes a more complex accounting exercise. The regulators also accuse companies of choosing to build unnecessary reservoirs and desalination plants because these boost profits.

In the wider world, during the first ten years of privatized water in South Africa, ten million people had their water supplies cut off. A number of forays by western companies into developing countries have resulted in disaster. Tanzania famously terminated the contract of the British company Biwater in 2005 to regain control, just two years into its ten-year lifespan, after marked increases in tariffs and lack of progress in extending the water supply system. In Bolivia, organized public opposition, widespread riots, and a number of deaths in 2000 caused the government to terminate concessions to Aguas del Tunari in Cochabamba, Bolivia's third city, and to Aguas de Illimani in the capital La Paz. Both companies were established after the World Bank refused to renew its $25 million loan to Bolivia unless it changed the law to allow privatization (Barlow, 2001). Aguas del Tunari was formed by a consortium from the USA, UK, Spain and Italy, but the US company Bechtel received most of the opprobrium. It had a 40-year concession to manage water and sewage, and to extend the network to the poorer parts of the city, and was guaranteed a 15 per cent return per annum. The company immediately raised tariffs – to levels equal to between 20 and 30 per cent of household incomes – on the grounds that it needed to build a new dam, the Misicuni. It began cutting off non-payers, and was given the power to seize their homes and outlaw the use of well water. It claimed to have extended water pipes to the poor areas of Cochabamba, but although it had laid pipes down the roads most inhabitants did not get connected to them because of the high costs. Following water supply cutoffs, some 300,000 contracted cholera between 2000 and 2002, and at least 300 died from it in Cochabamba. La Coordinadora de Defense del Agua y la Vida organ-

ized a strong opposition movement and helped advise citizens. Aguas de Illimani, a subsidiary of Suez, was similarly relieved of its contract in La Paz following the protests in the suburb of El Alto.

Despite the victory for local democracy, very little has changed since. The public companies are still cash-strapped, the infrastructure inadequate for the current population sizes and the poor pay ten times as much for their water as the richer citizens with piped supplies. Access to water supplies has improved, although water is available only five days a week in these areas.

Other countries have not been so proactive. This is not surprising as the economic power of some of these companies can exceed that of a small state. Legal actions have proliferated, but they can be very expensive. Both Biwater and Bechtel subsequently sued the national governments, as is their right under WTO rules, although Aguas del Tunari signed a contract with the Bolivian government in 2006, which committed both sides to drop any financial claims. Conversely, the giant French companies, Vivendi and Suez, were charged with corruption in Argentina and Indonesia, and in 2002 the US engineering company Halliburton was convicted of bribing a Lesotho government official in order to win the contract to build a dam funded by $8 million from the World Bank to enhance the water supply for Johannesburg.

The moral of these stories must be that if privatization is forced upon developing countries in debt, it should be done with more regard for the local situation and the local communities. This is the argument behind the newer version of privatization: public-private partnerships (PPPs).

Public-Private Partnerships

In its *Water Action Plan*, the 2003 G8 summit in Evian gave its support for a compromise solution, public-private partnerships (PPPs), to utilize private sources of funds and commercial management expertise, while theoretically retaining governmental control. The big question is 'can such partnerships work?' One successful scheme was Aquafin, which was established in 1990 by the regional government of Flanders with 51 per cent owned by UK's SevernTrent Water. It was described as an 'effective operator', but in 2005 it was returned to state control by order of the European Commission, which ruled that the original contract broke EU competition rules. Two advantages resulted: a significant reduction in the cost of raising loans, and this was passed on in price reductions to the consumers. Yet in 2008, the government began to consider returning to a PPP (Hall, 2008).

Three-quarters of France's water services are in PPPs and a study in 2004 showed that there were similar higher water prices under them – 16.6 per cent higher than in services provided by municipalities. Chile has a particularly successful record of PPPs, which has raised access to piped water from 27 per cent in the 1970s to 99 per cent in 2005. The European Investment Bank is one of the biggest funders of PPPs, including the water sector. Hungary and the Czech Republic have established PPPs in water services. Italy has invested more than €2.6 billion in water PPPs. The UK established PPPs in the non-privatized water services in Scotland and Northern Ireland during the 1990s. Lack of complaints seems to suggest that many of these have been successful, such as United Utilities' contract with Scottish Water. Welsh Water is another example. Though not strictly in public hands, it has contracted work out to United Utilities and Kelda Water and reports that this enabled it to reduce costs more than any other privatized water company in the UK. However, the regulator has required the company to reduce water prices over the five-year period between 2010 and 2015. It also decided to take the United Utilities work back in-house in 2010 and to reconsider the Kelda contract in order to help meet the necessary 20 per cent reduction in costs.

The outright failure in 2007 of the PPP with Metronet, the operator of London Underground part owned by Thames Water, has, however, led to serious doubts about such schemes in Britain. This is the conclusion of the parliamentary committee:

'The Government should not enter into any further PPP agreements without a comprehensive and accurate assessment of the level of risk transfer to the private sector and a firm idea of what would constitute an appropriate price for taking on such a level of risk. If it is not possible in reality to transfer a significant proportion of the risk away from the public purse, a simpler – and potentially cheaper – public sector management model should seriously be considered ... we are inclined to the view that the model itself was flawed and probably inferior to traditional public-sector management. We can be more confident in this conclusion now that the potential for inefficiency and failure in the private sector has been so clearly demonstrated. In comparison, whatever the potential inefficiencies of the public sector, proper public scrutiny and the opportunity of meaningful control is likely to provide superior value for money. Crucially, it also offers protection from catastrophic failure. It is worth remembering that when private companies fail to deliver on large public projects they can walk away—the taxpayer is inevitably forced to pick up the pieces.'

House of Commons Transport Committee, January 2008.

A special issue of the journal *Public Money and Management* concurred. The editor, Joop Koppenjan of Delft University, maintains that profit is the primary motive for company involvement, that PPPs frequently do not offer the best of both worlds, and that involving the private sector in national infrastructure has made it harder for governments to uphold public sector values (Koppenjan *et al*, 2008). All too often the public sector 'merely provides cover for the pursuit of private interest incompatible with the public interest'. Jones (2008) concludes that the best that can be aimed for is a 'workable (imperfect) mix'. Profits can come from the increased value of shares as well as sales: Hall (2008) reports that shares in PPP companies rose 250 per cent between 2001 and 2007, while the global average was only 100 per cent.

Given that the effectiveness of PPPs has been questioned in the UK, the prospects for success in developing countries must be open to serious doubt. The revelation in Bernard Marr's book *Managing and Delivering Performance* (2008) that nearly 70 per cent of managers massage data in order to meet performance targets, even in developed countries, adds to the concerns. Bechtel's claims that they had met water provision targets in Cochabamba are an illustration. In Britain, Ofwat has fined private water companies more than £60 million for misrepresenting customer service levels and/or misreporting leakage data: SevernTrent £35.8 million, Southern Water £20 million and Thames Water £9 million. And there is always the grave danger that the private sector takes any profits, but the public purse shoulders any losses and the job is not even satisfactorily completed (as in the case of London Underground). Worse is the potential loss of sovereignty over national resources.

Even the World Bank's own 2006 review of private participation in infrastructure (PPI) over the period 1983 to 2004 concluded that 'PPI has disappointed' (Annez, 2006).

The changing geography of wealth and influence

The management, improvement and sustainability of water resources around the world, from the local level to the international, are highly dependent on both politics and economics. The main international institutions that currently govern the world, the United Nations Organization, the International Monetary Fund and the World Bank, were set up to provide peace and security at the end of the Second World War, and the dollar was established as the international reserve currency. The stability provided by this economic system survived the decoupling of the dollar from the gold standard in 1973 and the floating of currency exchange rates. Two books written by historians embody the certainty of the western world view: J.M. Roberts's *The Triumph of the West* (1985) and Francis Fukuyama's *The End of History* (1992), written following the rapid downfall of Soviet communism after 1989. It was assumed that the West would continue to dominate the world, both politically and economically, and that the 'western' model of democracy would continue to spread throughout the world without serious competition.

Those certainties began to crumble with the turn of the millennium. In 2002, Jim O'Neill of Goldman Sachs coined the term the 'Bric' economies to identify the rapidly advancing economies of Brazil, Russia, India and China. In terms of importance, the acronym should really be read in reverse. Initial doubts as to the likely power of these economies now seem unfounded. The Bric countries contain 40 per cent of world population and already produce 15 per cent of global GDP, half that of the USA. O'Neill (2009) predicts that China's economy could equal the USA by 2027 and make the total GDP of the Bric economies larger than that of the G7. By 2050, China's GDP may be twice the size of the USA.

Two significant events occurred in 2009. China called for more representation on world governing bodies, especially the IMF and World Bank, and for a new world economic system founded on an alternative currency to the dollar that is not linked to the viability of the economy of one single country (a reaction to the credit crunch). The second event was the first meeting of heads of Bric countries in June 2009, marking the first overt recognition that they may have shared interests and might even represent a new force in international politics as a united group. Unity was not apparent at the meeting: the group is too diverse. But the challenge to the status quo was established and Bric countries promised a 'multipolar world order'. With the growth of globalization, markets opening up and increased labour mobility, Bric countries, together with some Gulf states such as Abu Dhabi and Qatar, are doing exclusive deals with Africa to get resources, be it food or minerals. In return, some, especially China, are supporting dam-building and water-supply systems.

The world appears to be in the process of the greatest change in the geography of wealth and influence since the end of the Second World War. The dollar seems unlikely to be replaced in the foreseeable future, given that half the official monetary reserves of countries are held in US dollars. But it has received a significant challenge, especially from China. More importantly, the political challenge to the way the world is governed is bound to be addressed. The credit crunch of 2008–9 seems to have made that certain. Despite suffering from the economic recession hitting its trade with Europe and America, China's economy is set to grow further, perhaps as much as 10 per cent p.a. (O'Neill, 2009), and the loss of exports has forced China to foster its domestic market more and so put its economy on a sounder footing.

China and India are clear leaders in the change. But while India is the world's largest democracy and has thus far shown relatively little interest in seeking resources and political influence in other developing countries (as opposed to manufacturing enterprises in developed countries), China is the world's largest dictatorship. And it is China that currently has the economic ascendancy, is most politically active internationally and is energetically buying into strategic resources around the world. Whereas India's companies are independent companies, most of China's have links with the state, and commercial and political interests are inevitably intertwined. China forestalled UN resolutions directed against the repressive Burmese administration, and China has major designs on hydropower resources in Burma. China was the first to call for the International Criminal Court to lift the indictment on President al-Bashir of Sudan for the war crimes that necessitated withdrawal of water, food and medical aid from numerous international agencies (see *The role of charities* in Chapter 21), and the Chinese state petroleum company has control of Sudan's valuable oil reserves. China's normal defence is that it does not interfere in the internal affairs of states it invests in, but defending these countries against international criticism and even UN sanctions goes beyond that.

China is offering to build infrastructure rather than emulate the western approach of offering loans. It is building the Tekeze hydropower scheme on a tributary of the Nile in Ethiopia, and the 40,000 MW scheme at the Grand Inga Dam in the Congo. China's state company Sino Hydro is also helping Ghana overcome dry season problems at the Volta Dam, which restrict electricity production and caused major power cuts in 2009, by taking over development of the 400 MW Bui Dam on the Black Volta. The previous proposal by a consortium under the troubled US company Halliburton failed, partly due to a critical environmental impact statement from a Canadian firm. On completion in 2011, the dam will flood over a quarter of the Bui National Park and threaten the existence of the black hippopotamus. China plans to build 216 large dams abroad in 49 countries, mainly in Africa and SE Asia in the coming decade, including more than ten megadams in the Sudan, Zambia, Gabon, Equatorial Guinea, Mozambique and Angola, as well as in Ethiopia, Nigeria, Ghana and the Republic of Congo, generally as part of a package deal in return for energy and minerals. In 2010, China agreed a loan to Ethiopia to build the £1 billion Gibe III megadam on the Omo River, which the World Bank had refused on environmental and social grounds.

China's foray into Africa is regarded by many as undermining attempts by the World Bank and IMF to improve transparency, reduce corruption and limit environmental impacts. There is little or no public consultation and people are forcibly displaced, like the 50,000 removed from the site of the Merowe Dam in Sudan. The 1350 MW Mphanda Nkuwa Hydropower Dam in Mozambique will undermine attempts to restore the Zambezi Delta, which is a Ramsar site. The World Bank and Western interests shunned the project on social and environmental grounds.

China's investment in Africa is being followed up by Russian investment, with not dissimilar aims. Namibia's 4366 tonnes of uranium oxide, ten per cent of the world resources, must be attractive to both (see *Uranium pollution* in Chapter 5). Russia is currently seeking more involvement in Nigerian oil and gas. Oil pollution is already a problem in some Nigerian rivers, reportedly caused mainly by attacks on pipelines from nationalists

unhappy with the returns that Nigeria gets from foreign companies.

Martin Jacques's book *When China Rules the World* (2009) charts the trend and gainsays the certainty of Roberts and Fukuyama. Jacques challenges the assumption that China is only interested in economic gain. The recession has actually made it easier for Chinese companies to pursue an aggressive programme of acquisition of strategic resources, just as the West did in the nineteenth century, in order to lay the foundations for a new hegemony. Jacques concludes that the process will be 'highly disorientating and disconcerting'.

The changes afoot do not stop with the Brics. According to O'Neill (2009) the 'Next 11' (including Indonesia, Iran, Nigeria, Turkey and Mexico) will soon follow, though their GDPs are considerably lower than those of the Brics. Their rising influence was recognized in the first meeting of the G20 in London in 2009. Changing demography will add to economic muscle. Whereas in 1900 Europeans accounted for a quarter of world population, this was halved by 2000 and is projected to fall to 7.5 per cent by 2050, by which time Africans will outnumber Europeans five to one. In 1950 there were 60 million people in Arab countries; by 2000 there were 240 million, compared with 284 million in the USA. In time, different political theories, religions and economic viewpoints, including Muslim views on debt, are likely to alter the principles of international aid that have been guiding world institutions. One principle that seems particularly vulnerable in the water resources context is the insistence on privatization or public-private partnerships as prerequisites for funding from the World Bank for water and sanitation projects.

Debt and transfiguration

Debt is the *modus operandi* of modern business, but there are occasions when it gets out of hand. Credit ratings are the essential guidelines for banks and investors, and both companies and governments rely on good ratings to raise money and finance projects. Lower credit ratings mean higher interest rates and make it more expensive, or *in extremis* impossible, to raise funds.

Two events in recent decades have demonstrated the often complex links between finance and the water industry. The first was in the 1980s, when loans for water projects were a substantial factor in an international banking crisis, which led to the movement to rescind Third

World debt that rippled on for decades. In the 1980s, Brazil began its vast expansion of hydropower dams in order to kickstart the economy with 'cheap' electricity. This was financed by loans from the World Bank and other international sources to the state-owned company Eletrobás. It played a significant part in Brazil becoming the largest debtor in the world with $111 million of debt by 1988, at least 25 per cent of which has been said to be unnecessary. At its peak, $8 out of every $10 spent on developments was derived from international loans. And a third of the revenue of national utilities went to servicing the debt. Brazil defaulted on repayment deadlines and the global banking system was rocked by fears that other developing countries would follow.

In the second event, water has been the victim rather than part of the cause: the worst economic recession since the 1930s.

Recession 2009

'Blind faith in progress led to euphoric overconfidence in the financial markets.'

On the Great Depression in:
Euphoria and Depression: the US between the wars
1917–1945, National Geographic *Visual History of the World*, 2005

The credit crunch that began in 2007 with widespread defaults in mortgage repayments among poor, 'sub-prime' households in the USA, became a global recession in 2009. In the 1990s, successive US governments had encouraged financial institutions to help families with poor credit ratings to take out mortgages to increase home-ownership. The failure of numerous, time-honoured financial institutions, and crucially the collapse of Lehman Brothers in September 2008, exposed the foolishness of the novel financial instruments they developed called 'credit default swaps', in which sub-prime mortgages were bundled together in complex mixes that were thought to virtually eliminate the risk from defaults and which banks traded with each other recklessly.

Concerns were raised for the financing of the water industry in many developed countries. In 2009, the IMF calculated that economies would contract by an average of two per cent during the year and the International Labour Organization estimated that an extra 50 million people worldwide would lose their jobs as a result of the recession. Japan suffered badly. In September 2009, Japan's Infrastructure Minister cancelled bidding

for construction contracts on the partially completed Yamba Dam in Gunma Prefecture as part of the government 'war on waste'. He subsequently announced that no funding would be requested for Yamba in the 2010 fiscal budget, even though cost-benefit analysis showed that it was among the highest rated of the 134 dams planned or under construction, with benefits estimated at 3.4 times the cost. The total cost of all the dam projects in hand would have been ¥8 trillion. The 100 of these dams that were still in the planning stage were to be considered for termination, 82 of which scored lower than Yamba in cost-benefit terms.

A key problem that emerged was the reliance of international and national banking organizations on credit-rating agencies, like Standard & Poor, as a guide to sound borrowers. As the agencies down-rated companies and countries, these found it more difficult to raise loans for new projects or even to continue operating. The IMF, the EU and other international organizations stepped in to help countries, but companies were largely on their own. Traditionally in a recession, stock market investors gravitate to the 'utilities', which sell products that everyone needs, as safer investments, and some did benefit from this trend. Nevertheless, many UK companies lost more than a quarter of their value in a year and some faced down-grading.

The steep fall in share prices hit water companies in a number of other ways. Anglian, South East and Southern Waters were all bought in debt-fuelled deals during the buyout boom of 2006. The fall in share values put them at risk of breaching the terms of the loans based on the value of their assets. The private equity firm 3i, which was part of the consortium that bought Anglian Water, was particularly affected.

One company's misfortune, however, can be another, bigger company's advantage. The fall in share prices was ideal for takeovers by companies that could still raise the capital, provided government objections to monopolies can be overcome. Hence, by mid-2009 Thames Water was seeking relaxation of merger restrictions in order to buy up over a dozen small water supply companies in the Thames region. The government was sympathetic to the expected reduction in water tariffs in the areas served by the smaller companies and to consolidation of water supply and sewage treatment: consolidation would add 5 million new water supply customers that currently only received sewage services from Thames, creating a total of 13 million combined sewage and water supply customers. The changes were incorporated into the Flood and Water Management bill, which was coincidentally being formulated at the time.

Early in the downturn, a number of other events in the UK during January 2009 illustrate some of the sensitivity in the water industry as revenues began to fall. Thames Water brought forward 300 redundancies as the number of households unable to pay their bills increased above expectation. SevernTrent Water, announced that it expected revenues to fall by about £25 million due to lower consumption among metered customers. Northumbrian Water and Essex & Suffolk Water warned that revenues would be hit as industrial customers scaled back usage. Anglian Water even put a proposal to government to install 'drip-meters' in households that would not, as opposed to could not, pay their water bills. Ofwat was sympathetic to the idea of 'trickle valves' which would make showers, baths and garden hoses impractical, allowing flows of one or two litres a minute, as against the normal ten litres. The critical question was how to distinguish between inability to pay and unwillingness to pay: real poverty or playing the system, perhaps many cases fit both. Thames Water reported a 16 per cent rise in bad debt from customers, amounting to £45 million in 2008–9. By October 2009, UK water companies were owed £1.25 billion in unpaid domestic bills. Ofwat reported that 20 per cent of customers – five million – failed to pay.

In the event, British companies like Thames, Northumbrian and South West Water managed to make an operating profit for the financial year 2008–9 by raising tariffs, despite falling demand and bad debts. They were helped by falling interest charges on their debts. Some companies also circumvented the law prohibiting cutting supplies to non-payers by fitting pre-payment meters. Even so, water companies pressed the regulator Ofwat to change its policy of reviewing water tariffs only at five-yearly intervals, and Thames Water proposed a further tariff rise of 17 per cent in order to support planned improvements.

The next event was quite different. As a consequence of the UK Treasury's need to raise money, it proposed to sell some of the British Waterways' £400 million worth of canal properties and halted a plan to reopen another canal in the Cotswolds. This would result in a significant loss of income from rents etc. for British Waterways, which depends on its commercial activities for a third of its income and already had a £30 million deficit. It

could affect its future roles in flood relief and water supplies (see *Don't forget the canals* in Chapter 15). British Waterways subsequently proposed going independent because of the lack of available funds from central government.

Many of the problems experienced by the UK water companies were prompted by the British law that prevents companies turning the water off in households that fail to pay their bills, which is rightly designed to protect the poor and vulnerable. As a result, Northumbrian Water raised its provision for bad debt from domestic customers from £2 million to £30 million and along with other companies asked the government's review of water charges to consider making landlords liable for tenants' unpaid water bills. With government support, the regulator, Ofwat, began working on 'social tariffs' for vulnerable households.

The situation in Britain contrasts with that in Detroit, where the city water company was turning domestic water supplies off in large numbers years before the car industry's near collapse, which dramatically raised unemployment in 2008.

SevernTrent's problem with metered customers brings to mind the classic case of water conservation measures in San Francisco during the water crisis of 1976–8. The city administration requested a 25 per cent reduction in consumption, but 40 per cent was actually achieved. The Public Utilities Commission had granted the Water Department a 43 per cent rate increase in order to maintain its income, which probably contributed to the excessive conservation. To cover the extra loss in revenue, the Water Department requested an additional 22 per cent levy and then, at the height of the drought, encouraged customers to consume more to prevent rates going higher!

The speed and the way in which the credit crunch developed in 2008–9 seem to have a number of elements in common with the way events unfolded in that San Francisco water crisis: the lack of prediction and foresight, the unforeseen interlinkages within the system, well-intentioned decisions that made matters worse, and the lack of overall coordination of response.

What began as a crisis in the banking industry hit the 'real' economy as soon as banks failed, especially after Lehmans: investments were lost, administrators called in loans and banks refused new loans – credit dried up. And this is where its true relevance to water supply comes in: the potential to halt investment in new water-related infrastructure, to cripple investment by foreign companies in developing countries, to reduce government aid to developing nations, and to deprive NGOs and charities of vital operating funds from hard-pressed company sponsors and the public. In autumn 2008, the UK Water Industry announced plans to spend £27 billion on improving infrastructure from 2010 to 2015: how much of this it will be able to raise is currently a moot point. Some may come from the four per cent increase in water bills agreed for 2009 by the regulator Ofwat back in 2004, but most of the funding for the new infrastructure works has to be raised from the financial market. More generally, the downward spiral is aggravated when credit rating agencies downgrade companies, making them less able to raise funds.

What is worse, whole countries were forced to seek bailout funds from the IMF: Iceland, Hungary, Ukraine, Latvia, Belarus, Pakistan, Serbia and Greece. And EU banks suffered further heavy losses on large loans they had made to the transition economies of east and central Europe since the demise of the communist regimes. Even China, once seen as a potential savour of foreign institutions, suffered from the downturn in international trade and was forced to inject $700 billion into the economy. The fate of environmental protection in Latvia is probably an indication of what may have happened in many east European countries. The Latvian Environment, Geology and Meteorology Agency suffered crippling staff cuts.

Impacts on developing countries

Although a number of emerging economies seem to have been less affected by the global recession, most notably China, the effects have been transmitted to many developing nations. India weathered the crunch mainly because of the large population of rural consumers with little exposure to outside finance. Nevertheless, the charity ActionAid estimates the recession will cost developing countries more than $400 billion between 2008 and 2011. Compare this to the $180 billion that would be enough to supply everyone in the world with safe water. The IMF warned that sub-Saharan Africa could be badly affected and calculated that the 22 poorest countries would need as extra $25 billion, or even $140 billion, in 2009 alone to cope with the recession.

A number of factors transmitted the recession to Africa: multinational companies had less money to invest in

projects, international aid programmes were cut back, reduced consumer spending in developed countries reduced demand for goods, bankrupt companies could no longer give aid, western banks were less able to fund development loans and interest rates on any loans that were made would be higher, and the ability of the countries to raise loans for themselves at an acceptable price was severely reduced.

It has hit the poorest hardest, with a likely fall of $46 in annual per capita income among the poor in sub-Saharan Africa (Tutu, 2008; Seager, 2009; Parry, 2009). Lehmans itself was a major donor, particularly to shanty towns and the WHO (2009) says many more people will have to continue to live in areas that are flooded each year with a high risk of contracting leptospirosis and other diseases. Italy cut aid by 50 per cent. The Interaction Coalition predicted that US donations would fall by $1 billion in 2009, down nearly ten per cent.

The recession set back anti-poverty campaigns and an extra 90 million people are likely to fall into poverty (DfID, 2009) – the World Bank estimates that 64 million of them will be relegated to extreme poverty. Some of the effect is due to lower remittances from migrant workers abroad: half the population of Tajikistan live below the poverty line and prior to the recession, these remittances were equal to 45 per cent of national GDP.

In response, the G20 proposed a rapid response fund and a Global Poverty Alert System, and the IMF suggested that such a fund would comprise $1 trillion. However, renewed promises of aid tied to GDPs are also not so generous when GDP is falling. The EU pledged to give 0.56 per cent of GDP a year in 2010, but falling GDPs mean an actual reduction of $4.6 billion (Seager, 2009).

Impacts on water charities

Reports from UK charities reveal substantial cutbacks as a result of pressure on a number of fronts: losses in Icelandic banks, reduced donations and the fall in the value of the pound. British charities had invested nearly £120 million in Icelandic banks before their collapse and the UK government refused compensation. This was partly alleviated when the UK government decided to channel renewed aid to Zimbabwe through international charities. Nevertheless, charities' incomes fell by more than 50 per cent in 2009 while demands rose by over 70 per cent in the wake of natural disasters.

Transfiguration – a new world order?

In 2009, world leaders were united in condemning the banking practices that led to the recession, referring to the activities as 'shameful' (US President Barack Obama), 'irresponsible' (UK Prime Minister Gordon Brown), and 'greedy' (Chinese Premier Wen Jiabao). The Deputy Governor of the Bank of England spoke of the need for 'profound' changes in international financial systems and methods, and many at the World Economic Forum in Davos 2009 called for more socially responsible banking. With governments owning large parts of the banking industry, albeit by default, improvements certainly should be possible. In response to apparent backtracking by banks in 2010, President Obama initiated a renewed attempt to institute domestic and international reform.

> 'Before I die I hope someone will clarify quantum physics for me. After I die, I hope God will explain turbulence to me.'
>
> Albert Einstein, quoted by
> Alan Greenspan, former chairman of the US Federal
> Reserve in *The Age of Turbulence* (2007)

Economists at the IMF spoke of 'uncharted waters' now (Stavrev *et al*, 2009), and the challenge is twofold: to improve regulation of the financial system, which may involve new institutions, and to improve economic forecasting. Will it be any more feasible to develop a global financial system that is proof to sudden shocks than it is to guard against all possible floods? The principal weakness in both economic and hydrological models is that they rely on past data and experience, yet future events are rarely if ever the same. The details will be dealt with more fully in Chapter 19 Improving prediction and risk assessment.

> 'The economic world is driven primarily by random jumps. Yet the common tools of finance were designed for random walks in which the market always moves in baby steps …The inapplicability of the bell curve has long been established, yet close to 100,000 MBA students a year in the US alone are taught to use it to understand financial markets … In bell-curve finance, the chance of big drops is vanishingly small and is thus ignored. The 1987 stock market crash was, according to such models, something that could happen only once in several billion billion years. In power-law finance, big drops – while certainly less likely than small ones – remain a real and calculable possibility.'
>
> From: *How the financial gurus get risk all wrong*
> Mandelbrot and Taleb, 2005

Writing in *The Times* in 2009, resident economist Anatole Kaletsky called for new economic models to replace the current ones, which are based on the assumptions that the market is always correct and 'efficient' and that investors always behave 'rationally'. He extols the virtues of the models for predicting extreme events based on fractal, power-law mathematics developed by Benoît Mandelbrot. In their book *The (Mis)behavior of Markets: a fractal view of financial turbulence* (2004), Mandelbrot and Hudson extend Mandelbrot's theories into the realm of human behaviour by considering the psychological bases which generate over-confident investment in the boom years and panic withdrawals when the bust appears. So are we doomed to continue to suffer the cycles of boom and bust? Are they inherent in the capitalist financial system as we know it? Former chairman of the US Federal Reserve, Alan Greenspan, thinks they will continue to occur, but that each one will be different from the last (Hasell, 2009).

Conclusions

Significant changes are happening in the balance of influence and wealth around the world. The rise of the emerging countries, the Brics and others soon to join them, is wresting control of international finance from the old western sources, and this is bound to affect water resources developments. The role of China is already making its mark, with less concern for social and environmental issues than the World Bank and IMF have come to adopt. Corruption is a major problem, especially but by no means exclusively in Africa. Privatization is still being promoted by most funding agencies, although its record has been a mixed success, especially in the developing world. It raises important questions about sovereignty of resources and there is little doubt that domination of the world market by a few transnational companies is a dangerous state of affairs. Ofwat's policy of limiting company mergers in the UK is based on sound principles that would be a good blueprint for the world if there were any comparable authority to enforce it.

Perhaps the most important issue of all is the stability of the world financial system, as demonstrated by the recent recession. We can only hope that the more dire forecasts of a multiyear global depression and of new national debts blighting a generation to come, are just bad predictions. At least, we might reasonably hope that human resilience will result in new and fairer international financial systems. But the primary need is to prepare new systems that can cope with extreme shocks.

Discussion points

- Research the successes and failures of privatization.
- Explore the role of corruption in water resources contracts.
- Follow the developing news on redesigning world financial systems.

Further reading

The UN World Water Development Reports provide detailed discussions on governance and regular updates.

The book *Blue Gold: the fight to stop corporate theft of the world's water* by Maude Barlow and Tony Clarke (Earthscan, 2002/2003) gives a very detailed analysis of developments in privatization.

A number of books try to predict the future economic geography of the world. One authoritative account is:

King, Stephen D. 2010. *Losing Control: the emerging threats to western prosperity*. Yale.

5 Pollution and water-related disease

Water pollution escalated during the twentieth century as populations grew, industry expanded and agriculture took to artificial fertilizers and pesticides. While population grew three-fold, water use increased six-fold and more and more waste products were washed into the environment as water treatment failed to keep pace. Controls like the EU Water Framework Directive are stemming the tide in most developed countries, but there is still a substantial legacy lying in the landscape, in soils and river deposits, waiting to be washed out. Moreover, it is now the emerging and developing economies that are becoming the big polluters.

The Global International Waters Assessment Report found that pollution is the main water problem in a fifth of the 97 UNEP regions (UNEP, 2006). This is exceeded only by the 23 per cent of regions where shortage of water is the prime concern. Among the most significantly polluted drainage basins are: the Mississippi basin, the Atlantic seaboard basins of Brazil, parts of West Africa, the Congo and Lake Victoria basins, the Murray-Darling and drainage entering the Great Australian Bight, and the Black and Baltic Sea drainages. Many islands are also suffering, especially in the Caribbean, Indian Ocean, Coral Sea and Sea of Japan.

Defining and determining pollution

Most established water quality standards have been based on the available evidence of their medical effects. They generally relate to human drinking water supplies rather than to water in the environment. There are two major problems with this. First, medical evidence is changing; as new links are found, so standards are revised and generally the recommended limits are revised downwards (for example, see the sections headed 'Arsenic in groundwater' and 'Uranium pollution in South Africa' later in this chapter). Second, our knowledge of the effects on wildlife is generally much poorer than our medical knowledge. This is despite the fact that links were mooted and established between water pollution and pathological effects in fish, mammals and birds half a century ago and widely publicized in Rachel Carson's classic 1962 book *Silent Spring*. Furthermore, individuals' resistances vary, as do those of different wildlife species. For wildlife populations, sustainability is not only a matter of individuals' resistance to toxins, but may also be a question of behavioural reactions, the availability of environmental refuges and the effects on predators and food stocks.

For humans as well as wildlife, the practicalities of water testing also mean that generally only spot checks are made or (somewhat better) a continuous monitoring devise samples too infrequently to pick up a pollution event, or retains only a statistical average. In such cases, critical peak levels of pollution may be missed. It is also possible to deliberately arrange to sample the water when the risk of pollution is judged to be low. This issue has been raised in criticism of sampling undertaken in some Mediterranean countries to meet the requirements of the EU Bathing Waters Directive. Discharges from sewage plants are often regularly timed and predictable. Worse, faecal pollution is commonly higher in rivers after a rainstorm has washed manure off agricultural fields, so a false impression of the safety of the water could be deliberately created by avoiding sampling on such days. This is a tricky question for the design of legislation.

Where possible, in the following sections we have used data from the international water quality data bank, GEMS/Water, to map the global distribution of pollutants. The maps show long-term mean concentrations. World coverage is still patchy. In general, the best coverage is in Europe and Southeast Asia. However, the database currently provides the best overview of the state of surface waters worldwide and will hopefully develop into a major resource as more stations are added.

For convenience, we have divided the chapter into two broad halves: the first half deals with pollutants that are poisonous, the second deals with infectious diseases. Again for ease of presentation, the first half is broadly divided into agricultural sources and urban-industrial sources. However, as will become clear, not everything fits neatly into this categorization.

Part 1 Toxic substances

Pollution from agriculture

Agriculture is taking over from industry as the greatest source of water pollution in developed countries. This is largely because legislation and environmental protection agencies have focused more on point discharges from industry but find pollution from agriculture, especially from farmland, generally more difficult to pinpoint. The Green Revolution of the 1960s and 1970s also contributed to widespread pollution in developing countries due to reliance of artificial fertilizers, pesticides and herbicides. Intensive livestock rearing, with its slurry ponds and silage liquor, is now adding to river pollution, mainly through accidental releases and storm runoff. The infectious diseases associated with this will be discussed in Part 2.

Nitrates

Nitrates derived from agricultural fertilizers have been an increasing threat to water quality over the last half century. The EU Common Agricultural Policy encouraged large-scale application of fertilizer, sometimes as high as 540 kg/ha/year in the Netherlands, by guaranteeing a minimum price for agricultural products. Use of inorganic nitrate fertilizers increased three-fold in Britain between the 1960s and 1980s, outstripping phosphate or potash fertilizers (DoE, 1986). Urea from animal excreta also increased, providing an additional source.

There has been a parallel increase in nitrate concentrations in British river water over the last 50 years. Nitrate levels are highest in the agricultural areas of lowland England, where levels were generally in excess of 5.5 mg/L of nitrogen in nitrate during the 1990s (Betton *et al*, 1991). In the Great Ouse River in East Anglia, levels of nitrogen as nitrate rose dramatically from around 2 mg/L at the start of the 1960s to near the WHO recommended limit of 10 mg/L during the 1980s (Roberts and Marsh, 1987). The EU sets a limit at 50 mg/L of nitrate, which was adopted in the 1980 *Directive on the Quality of Water for Human Consumption*, while the limit finally adopted by the WHO in 1984 is equivalent to 45 mg of nitrate per litre.

The 1991 EU *Directive Concerning the Protection of Waters Against Pollution Caused by Nitrates from Agricultural Sources* provides protection of aquifers used for drinking water. But despite strict controls on the application levels for nitrogenous fertilizers and farm manures, the designation of nitrate sensitive areas and groundwater protection zones within the EU, the high levels of pollution accumulated during the years of prolific application are going to affect groundwater quality in some areas for decades to come.

The GEMS/Water archive demonstrates relatively high concentrations in many European rivers (Figure 5.1), and there is no reason to suppose that countries not represented, like France and Italy, are any different. Relatively high levels are also found in parts of the Indian subcontinent, notably Gujarat, West Bengal and Bangladesh, as well as in south west USA.

> 'Science is simply not always able to provide neat and clean answers, and in order to protect the public, expensive policy decisions must sometimes be made ... We seem to be forced to the conclusion that an exceedingly rare toxic condition, methaemoglobinemia in infants, is linked to two episodes of exposure to endogenous nitrite, but generated by two entirely different mechanisms.'
>
> Roger P. Smith,
> Professor Emeritus of Pharmacology and Toxicology,
> Dartmouth College, 2009

The reasons for concern

A scientific paper by Henry Comly published in 1945 raised the issue of a link between nitrates in well-water and 'blue baby syndrome' or methaemoglobinemia. The paper became a classic and was reprinted in the *Journal of the American Medical Association*'s landmark series in 1989. As a result, nursing mothers are advised to avoid drinking water contaminated with high levels of nitrate for fear of methaemoglobinemia. The main symptom is cyanosis, a bluish coloration of the skin due to lack of oxygen in the blood, but Comly's work was based on just two cases. A subsequent survey covering all 50 US states in 1951 reported 278 cases and 39 deaths (Smith,

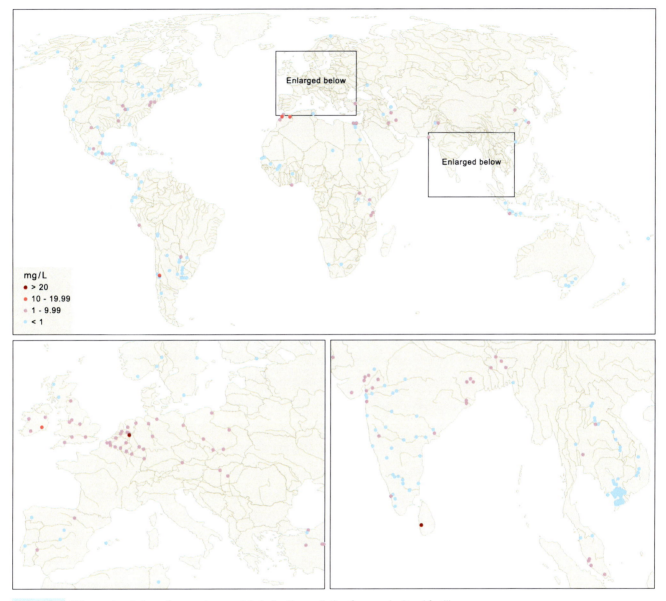

Figure 5.1 Nitrogen levels in surface waters, mainly indicating pollution from agricultural fertilizers

2009). The survey found no cases with levels in drinking water below 10 ppm of nitrogen in nitrate, and just two per cent up to 20 ppm, but above that the severity of the symptoms rose in parallel to the increase in nitrates. Due to this early work, the 10 ppm limit became the US, WHO and EU standard and was reaffirmed in America in 1995 by a National Research Council report. However, the number of cases fell steadily during the 1950s without any preventative action. Professor Smith (2009) speculates: was this due to improved public water supplies in rural areas, or to public awareness of the dangers, or to the popular trend to breastfeeding, because

high levels of nitrogen had been found in formula milk? He notes that no cases have been reported in America since 2000.

Although blue baby syndrome has received the most publicity, there is evidence of other medical effects. Barrett *et al* (1998) report statistically significant links between nitrate levels in drinking water in Yorkshire and cancer of the brain and central nervous system, although not of the stomach or oesophagus as has been supposed. Chronic health problems, like cancer and heart disease, can increase sensitivity to nitrates. The Wisconsin

Department of Natural Resources recommends that pregnant women avoid high levels of nitrate as it may reduce oxygen supply to the foetus.

Nitrate in the gut

The very young are most vulnerable because the common microflora of the gut of infants under six months convert nitrate to nitrite, which is then absorbed into the blood. In adults, the acid stomach and duodenum kill nitrate-reducing microflora, but the high pH in the gut in early infant allows them to enter the small intestine and reduce more nitrate.

Another outbreak of infantile methaemoglobinemia occurred in America during the 1980s and 1990s. In 1982, 11 babies under three months old were affected even though nitrate levels in the drinking water were well below the recommended limit. Between 1983 and 1996 there were more than 90 cases, then nothing. Smith (2009) asks: was this outbreak caused by 'endogenously produced' nitric oxide or exogenous drinking water? Some viral infections can increase nitric oxide in the colon and the 1980s outbreak was associated with diarrhoea and gastroenteritis. He believes well-water could have been the cause in the first mini-outbreak studied by Comly, but not in the 1980s outbreak.

More generally, however, nitrosamines, created in the human gut in the presence of high levels of nitrogen, have been identified as carcinogens.

Nitrate in the landscape

The removal of nitrate from public water supplies is expensive. Conventional treatment is ineffective and the only method currently used is an expensive ion exchange process. Hence, prevention is far better than cure, which means avoiding nitrates reaching the rivers, groundwater and water abstraction sites. Whereas acidification and organic pollution are largely problems for surface waters, as they tend to be neutralized or filtered respectively by passage through rocks, nitrates affect both surface and subsurface waters and are a prime source of 'nonpoint' pollution of aquifers beneath agricultural land.

Up to 50 per cent of inorganic nitrates are leached out of the soil and Dudley (1990) estimated that 80 per cent of nitrates entering British seas come from agriculture and factory farms. Some excess nitrogen is assimilated by organic compounds in the soil, but a large proportion is washed out, especially after droughts like the '400-year event' in 1976 in England. Land drains on floodplains speed the drainage of nitrate fertilizers into the rivers. They also dry the soils out, which encourages cracks and quicker drainage. Technically, they encourage 'bypass flow' and reduce 'residence times' for nitrate within the soil. In contrast, if the floodplains remain wet and undrained, natural 'denitrification' processes decrease the amount of nitrate reaching the river. Denitrifying bacteria reduce the nitrate to nitrite, and thence nitrogen gas, which is released into the atmosphere. Some may also be released by volatilization as nitrous oxide gas. A study on the River Windrush in England found a reduction of 82 per cent in nitrate concentrations as water drained across the floodplain, which was attributed mainly to denitrification (70 per cent) with the remainder due to dilution (Burt and Haycock, 1992). Hence one proposed solution is to leave the riparian zone untouched to allow natural removal of nitrates.

Temperatures and the timing of fertilizer application affect the amount of water pollution that occurs. Biochemical processes tend to operate more rapidly in higher temperatures. During the growing season plant uptake tends to absorb nitrate fertilizers, but applications on bare soil in autumn in preparation for winter crops can cause excessive nitrate leaching, especially when assisted by autumn rains.

Important biochemical processes also operate within the watercourses. Fertilizer leachates and animal excreta contribute to eutrophication, in which algae take over water bodies over-rich in nutrients. We consider its effects on wildlife in Chapter 6.

From the point of view of human water resources, the pollution of groundwater by nitrates is a major concern. Because the rate of nitrate movement into and through the aquifers tends to be very slow, many water companies are only now experiencing problems created by nitrogen fertilizer applied many years, even many decades, ago. In some cases, peak levels of pollution may still lie within the aquifer system and are yet to appear in the springs that feed the rivers. Equally, preventative measures taken now, like nitrate protection zones that limit fertilizer applications in the vicinity of public abstraction points, can take years to show results. Figure 5.2 illustrates the problem in the southern Netherlands.

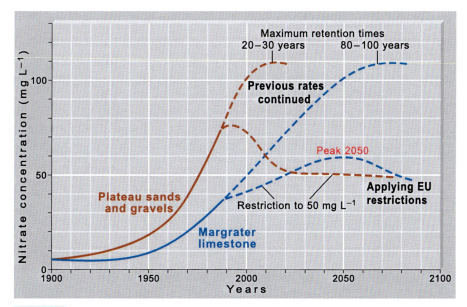

Figure 5.2 Nitrate in groundwater in Limburg will take many years to clear (after Rang and Schouten, 1988)

Pesticides and herbicides

Pesticides and herbicides were a key element of the Green Revolution in the 1960s and their use has continued to expand. In the current Genetic Modification Revolution, considerable effort is being invested by companies like Monsanto into producing crops that are more resistant to herbicides and pesticides so that more may be applied. These become a potential hazard for water resources when they are washed off the land or into aquifers. Until the 1970s it was believed that soil naturally filters them out. Pesticides bind to soil particles. Some, like paraquat and glyphosate (the principal ingredient in Monsanto's best-selling Roundup) bind tightly, but others less so. However, soil erosion can still carry them into the rivers. Roundup is not approved for the aquatic environment and should not be applied in wet conditions. Herbicides are adsorbed less on soils when the soil is wet, allowing more to drain away. New research at the University of Caen's Comité de Recherche et d'Information Indépendantes sur le génie Génétique (CRIIGEN) suggests that Roundup is more toxic to human cells than pure glyphosate because of other ingredients (Gasnier *et al*, 2009).

It is now also clear that pesticides and herbicides are reaching groundwater. The problem was not immediately apparent because they take a long time to filter down. The USGS suggest they could take ten years to penetrate down to the level of the average groundwater monitoring well and up to 50 years to reach the level of a deep drinking water well. Along the way they may be broken down by microbes or chemical reactions, which may make them safe or may produce other toxic compounds.

Two key aspects of environmental safety are the rate of decay of the pollutant and its breakdown products. Organophosphorus and carbamate pesticides and herbicides, such as malathion and parathion, which have been introduced progressively since the 1950s, breakdown far more rapidly in the environment than organohalides such as DDT and dieldrin (Figure 5.3). DDT has a half-life of eight years and dieldrin three years (Alloway and Ayres, 1993). However, this advantage may be offset by soil organisms that increase their solubility, so that atrazine and its sister herbicide simazine can get washed into rivers well before they have decayed. Rates of decay also vary very much according to environmental conditions. Dieldrin and aldrin are among a dozen chemicals banned under the 2001 UNEP Convention on Persistent Organic Pollutants (POPs), which finally came into effect in 2004.

Highly toxic dioxins have been used in herbicides, notably in Agent Orange. Many other dioxins are created by accident as chemical by-products or in the incineration of commercial polymers. Although they are relatively insoluble, they may still cause water pollution

53

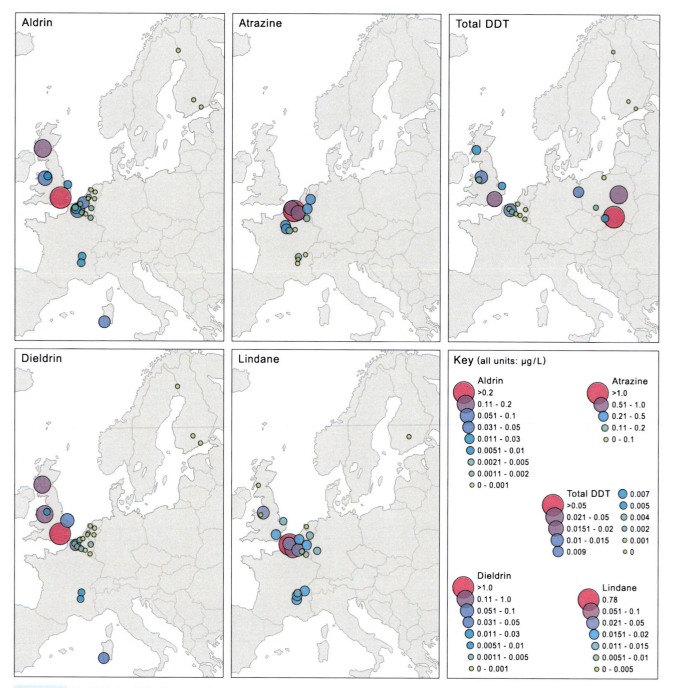

Figure 5.3 Pesticide levels in European rivers

as the fine soil particles they adhere to are washed out. An iconic incident happened in 1982 in a city dumpsite near Niagara Falls, NY, when the partially dug Love Canal exposed buried dioxins that were washed out and emerged in sediment in the storm sewers (Kamrin and Rodgers, 1985).

There has been a welcome trend for more recently introduced products to be designed to breakdown into harmless products, like the selective weedkiller Verdone, which breaks down when it hits the soil. Moreover, some of the older pesticides, like DDT, have now had national or international bans imposed, but their use continues in many places, either illicitly or with no locally imposed bans. The USA banned DDT in 1973 as a result of Rachel Carson's book *Silent Spring* (1962), and many other countries followed suit. However, DDT is still widely used in the developing world. The UNEP Convention on Persistent Organic Pollutants fell short of an outright ban, principally because of its value in combating malaria, although the UN has expressed an aim to rid the world of DDT by 2020 and called for alternatives to be found. But there are other interests campaigning against a ban. DDT is still widely seen as the most effective weapon against the tsetse fly and DDT levels are dangerously high in many African rivers, like the Zambezi, that flow through cattle-rearing country. When it does decay, DDT produces DDE, which is an endocrine disrupter (see section headed *Endocrine disrupters – a sea of oestrogens* later in this chapter). DDT is still used in China. Kondratjeva and Fhisher (2009) report levels of DDT and the pesticide hexachlorocyclohexane in fish in the Amur River on the Chinese-Russian border that are comparable to levels in British fish before restrictions were introduced in 1972. They cite Chinese agriculture as the principal source.

Pesticides and herbicides are poorly monitored globally. Despite detailed recording by the EPA in the USA and national bodies in a few other countries, the only usable regional coverage held in the GEMS/Water archive relates to Europe. European Union drinking water standards have a limit of 0.0001mg/L for total concentration of pesticides (98/83/EC). WHO drinking water standards cite limits of 0.03µg/L for aldrin and dieldrin and 2µg/L for the herbicides lindane and atrazine. Atrazine is on the British Environment Agency's 'red list' of dangerous substances. The Environment Agency has been carefully monitoring atrazine in England since the early 1990s, when it caused pollution of some water supply sources and its use was restricted to maize fields: satellite surveillance has been used to track maize fields and farm visits organized to advise farmers on safer use. Some of the worst levels of dieldrin, exceeding the WHO standards, are recorded in British rivers. Levels of aldrin exceed WHO standards in many rivers in the Benelux countries and in Finland. Levels of atrazine are high in parts of Belgium, the Netherlands and the River Seine basin, though generally within WHO limits for drinking water (Figure 5.3). The same applies to lindane in southern Belgium.

Levels of DDT, though generally within the WHO guideline, are still remarkably high in some British rivers, despite the ban in 1984. Concentrations of DDT in Polish rivers are part of the legacy of the lack of environmental protection under the old communist regime. In this respect, Polish rivers are probably representative of the situation covering wide areas of Eastern Europe where data are not readily available.

Sheep dip is another hazard. Organophosphate dips were widely used in Britain to combat ectoparasites in the skin and wool until the 1990s, when they were withdrawn following evidence of neurological effects on farmers. They were replaced by cypermethrin dips. These proved devastating for aquatic life and could continue to be washed off the sheep by the rain for a month after treatment. In 2005, the Environment Agency successfully prosecuted ten farms for polluting rivers with sheep dip. One of the worst events was on the River Mint in Cumbria, which killed over 5,000 white-clawed crayfish, a protected species. Sales were suspended in 2006 and the Environment Agency finally banned their use in April 2010. A 70-year record of numbers of aquatic species in the River Teifi in Wales shows a 50 per cent fall in biodiversity over the period, attributed in part to pollution from agriculture, in part perhaps to acid rain. Organophosphate dips are now back in use, but with much stricter controls and instructions.

Bio-accumulation

A particularly worrying feature of many products released into the environment is bio-accumulation. They become more and more concentrated up the food chain as animal eats animal. This has been noted especially in the case of endocrine disrupters, but may equally well be problematic with any reasonably stable pollutant or one that decays into other toxic substances. Lead and mercury are cases in point. Mercury poisoning has affected fishermen from Japan to Canada. The well-publicized cases in Japan, at Minimata in 1956 and Niigata in 1965, were linked to methyl mercury in industrial effluent ingested by the fish. The result became known as 'Minimata disease', causing ataxia, weakening of the muscles and the senses. In the Canadian and Russian Arctic, elevated levels of mercury in rivers and lakes have been caused by the slow anaerobic decay of forest vegetation flooded by reservoirs in the cold climate. The Cree Indians stopped eating fish from the reservoirs of the James Bay hydropower scheme as a result.

Bio-accumulation is a major problem with organochlorines and some aromatic hydrocarbons. Early evidence of accumulation in the food chain came from Clear Lake, California's largest freshwater lake, in the 1950s. The lake was sprayed with DDD to control mosquitoes and the lake itself appeared relatively unpolluted with DDD concentrations of just 20 parts per billion

(Moriarty, 1988). However, birds that eat fish from the lake showed levels of 1600 ppb in their fatty tissues.

Endocrine disrupters – a sea of oestrogens

Research over the last two decades has revealed the dangers of many man-made substances now in common use that act as 'endocrine disputers', causing the body's endocrine or hormone release system to malfunction. These compounds affect the sexual development of animals and humans and they tend to bio-accumulate in the body. Endocrine disrupters may either act like female sex hormones, mimicking oestrogens; they may block the normal receptors for the sex hormones, effectively turning off the hormonal activity; or they might activate the receptors in a different way, perhaps stimulating some activity.

Many everyday products act as endocrine disrupters, either in their original state or in the secondary substances they produce as they degrade. They are by no means restricted to agricultural use, although it was pesticides that first alerted biologists to the dangers. The problem chemicals include many chlorinated hydrocarbons, DDT and its breakdown product DDE, DES (diethyl stilbestrol), some PCBs, parabens used in cosmetics, dioxins, several pesticides such as atrazine, and fungicides. Many of these substances are resistant to biodegradation. But some substances that are non-toxic or have low toxicity may nevertheless break down into other substances in the environment that are unstable and more toxic. Members of one major class of plasticizer, called phthalates, used for example to make plastics flexible and to improve the flow of paints, are known to be endocrine disrupters in animals.

Although research in the early 1990s first alerted medical and environmental science to the dangers, manufacturers have mounted strong opposition to legislative controls. Only now are governments beginning to introduce controls, more than a decade since expert scientists presented their findings to the US Congress and called for legislation. Around 200 million tonnes of plasticizers are produced in North America annually. The American Chemistry Council – which represents the industry and is not as might appear a scientific organization – claims that there is no sound evidence that these chemicals affect the endocrine system and points out that they play many valuable roles, including producing flexible tubing for medicine and coatings for pills.

The International Programme on Chemical Safety (IPCS), run by the WHO, the International Labour Organization and UNEP, was asked to investigate the claims and present evidence to the contrary (Damstra et al, 2002). There is very strong evidence for effects on wildlife, even though the exact mechanisms are often poorly understood. The evidence for effects on humans is less clear, partly because obtaining experimental evidence with humans is more difficult than with laboratory-controlled experiments on animals. Even so, strong statistical evidence was produced in the early 1990s for effects on humans as well as animals. Sharpe and Skakkebaek (1993) and Dibb (1995) stated that at least 37 different chemicals had been identified that disrupt hormone operations and appear to be linked to testicular cancer and other sexual abnormalities in humans and animals. The IPCS report concluded that although the evidence for effects on humans in 'weak', this only means that more rigorous studies are urgently needed. Recently, Swan et al (2005) published meticulous research that establishes a link between genital abnormalities in boys and their prenatal exposure to phthalates as measured in the mother's urine.

'Despite an overall lack of knowledge of mechanisms of action of EDCs [endocrine-disrupting chemicals], there are several examples where the mechanism of action is clearly related to direct perturbations of endocrine function … The evidence that high-level exposure may impact both humans and wildlife indicates that this potential mechanism of toxicity warrants our attention.'

WHO/ILO/UNEP International Programme on Chemical Safety, Damstra et al (2002)

Research in the Great Lakes in the 1990s revealed major problems with the development of fish in Lake Erie (exploding thyroids and trans-sexual development) even though the lakes have been largely cleaned up. Scientists recently identified the 'Great Lakes Embryo Mortality, Edema (or oedema) and Deformity Syndrome' among fish-eating birds, which has been directly related to PCBs. At the 2009 Annual Conference on Great Lakes Research held in Toledo, Ohio, Parrott and colleagues from Environment Canada presented evidence of a link between the synthetic oestrogen used in birth control pills, ethinyl estradiol (or ethynylestrodiol), and reduced male characteristics in fish and decreased ability to fertilize eggs. Concentrations as low as one nanogram per litre had an effect and these levels have been found in some municipal wastewater effluents. Experiments exposing male fish to municipal wastewater effluents from Ontario decreased egg production by over 50 per cent. Intersex male perch in the Lakes have also been linked to ethinyl estradiol in the water. And male fathead minnows captured from a creek in Saskatchewan that receives municipal wastewater effluents have fewer sex characteristics.

Similar evidence emerged in the 1990s from Lake Apopka in Florida. There was a major spillage of the pesticide kalthane around the lake in the early 1980s, but neither this nor its decay products can now be found in the water. The lake water appears clean, yet there is an abnormally high incidence of alligators with trans-sexual characteristics, no Florida panthers have descended testicles, and many birds' eggs are infertile or have extremely fragile shells. In fact, the breakdown products of the pesticide, notably DDE, are there, but they are stored in animal fat and they are being recirculated through the predatory cycle. In Canada and elsewhere, fish downstream of pulp and paper mills show similar sex changes. Males become feminized and females masculinized.

Experiments in England by Professor John Sumpter and colleagues at Brunel University in the 1990s showed that male fish kept in cages below sewage outfalls rapidly begin to produce female hormones and develop female characteristics. When the results were presented to the water industry, the official response was that there were no plans to test for these substances in sewage outfalls. However, recent studies on 39 UK rivers confirm the general tenor of the earlier work. They show that both the incidence and severity of intersex conditions in wild roach (*Rutilus rutilus*) are significantly correlated with the estimated concentrations of the natural estrogens 17β-estrodiol and estrone (E1 and E2) from humans and the synthetic contraceptive pill constituent oestrogen (EE2), as predicted by computer models of water quality in sewage outfalls (Jopling *et al*, 2006). Similar feminization of roach was found in the River Lee in Hertfordshire in the 1970s and led to the banning in 1987 of the antifouling compound TBT (tributyl tin) used on boats, but the dog whelks have still not recovered from the feminization. TBT is still used as a fungicide in paints. On the River Aire in Yorkshire, male fish have become feminized downstream of a textile factory using an alkylphenol-based detergent. In America, the US Geological Survey has reported that male smallmouth bass in the Potomac River have started to produce eggs.

Of particular concern is the fact that standard water treatment does not remove these hormones. Estrone has been recorded in tap water in Tulane at levels of 45 parts per trillion – low but potentially a matter for concern.

Governments are now responding. The US Congress imposed severe restrictions on the content of children's toys. The EU passed similar legislation and the Australian government introduced restrictions in 2010. Under new legislation, the American EPA issued the first test orders in late 2009/early 2010 to manufacturers and importers of 67 pesticides, including five phthalates, requiring them to prove the substances are safe. Many phthalates are already being phased out of many products in North America and Europe because of health concerns. However, the American Chemistry Council points out that if forced by law to test specific phthalates, manufacturers will simply withdraw them and choose another. This is yet another case where companies may dodge the law and put the burden of proof on the State. The EPA could be testing up to 15,000 chemicals for endocrinal effects.

Urban and industrial pollutants

As we have already seen in relation to endocrine disrupters, modern chemistry has blurred the distinction between agricultural pollutants and those from urban and industrial sources. The same applies to drugs reaching the rivers and returning in tap water – some are from medical sources, some from veterinary sources and others from addicts.

The earliest industrial pollutants, dating back millennia, were heavy metals – copper, tin, silver, lead and gold – to which mercury, cadmium, zinc, aluminium and uranium were added more recently. Until the mid-twentieth century, the main non-metallic pollutants were coal and oil, but a vast array of synthetic products have been added over the last century: polychlorobiphenyls (PCBs), nonyl phenols, plasticizers etc. The decline of the European otter and illness or death among the seal and dolphin populations of the North Atlantic have all been ascribed in part or all to PCBs (Alloway and Ayres, 1993).

The GEMS/Water archive, though very limited at present, does show that parts of southern Belgium continue to rank high in long-term average levels of PCBs, phenols and benzene despite recent moves to conform to EU standards. Levels of phenols also exceed WHO standards in parts of the Rhône-Saône basin and the Netherlands. WHO drinking water standards for various chlorophenols are in the range of 0.1 to 0.3 µg/L.

Drugs in the water – an emerging threat

Until very recently, the focus of concern has been on pesticides and non-pharmaceuticals which pose greater threats to the aquatic environment than drugs. Since 2006, however, the European Medicines Evaluation Agency has required all new pharmaceuticals to be tested for environmental impact, though the legislation is not retrospective. Both the US EPA and the UK Drinking Water Inspectorate have been criticized for complacency and lack of monitoring (Watts, 2008). Clearly, drugs have been passed into rivers ever since they were invented, but two things have changed: increasing drug usage and increasing awareness of the potential dangers. In many ways, because the threat is 'new', there is a dearth of scientific knowledge, both on the effects of individual breakdown products and on their combined or cumulative effects on health. In an issue of the *Philosophical Transactions of the Royal Society A* in 2009 devoted to the theme 'Emerging chemical contaminants in water and wastewater', Zuccato and Castilioni report substantial amounts of morphine, metamorphine and ecstasy entering sewage plants, and suggest that their presence in surface waters as complex mixtures of breakdown products may be toxic to aquatic life.

The EPA drastically expanded testing following a Senate hearing in April 2008, which included evidence that the drinking water of over 41 million people in 24 metropolitan areas across America contains traces of many drugs like antibiotics, mood stabilizers, like Prozac, anticonvulsants and sex hormones. Some of this water passes into irrigation systems and cattle. US manufacturers legally release over 123,000 tonnes of pharmaceuticals into waterways annually, including lithium to treat bipolar disorder and nitroglycerine for heart treatment (and explosives). Antibiotics were prominent in Tucson's drinking water in 2008 and Dallas tap contained anti-epileptic and anti-anxiety medication.

Particular concern focuses on cytotoxic or cancer drugs like fluorouracil developed for rapid action in hospital,

'We know we are being exposed to other people's drugs through our drinking water, and that can't be good.'

Dr. David Carpenter, Director,
Institute for Health and the Environment,
State University of New York at Albany, 2009

but which patients, once they are home, can continue to excrete into public sewage systems that are not designed to remove them. Many surplus drugs are simply flushed down the toilet, despite labelling advising otherwise. Possible solutions including more pre-treatment of hospital wastewater, an extra day or two for patients in hospital and better water monitoring have been suggested. And more research on toxicity is clearly needed.

Veterinary medicines are also found in tap water and they are rising. Steroids were found to increase fourfold downstream of a feeding lot in Nebraska and male flathead minnows showed reduced levels of testosterone and had smaller heads.

There are now new concerns over the possible health effects of 'cocktails' of drugs in drinking water, even when the concentrations of the individual drugs may be measured in just nanograms or even picograms per litre and well within recommended limits. There is already evidence that drug cocktails can inhibit the growth of embryonic cells (Watts, 2008): young foetuses may be especially vulnerable.

In Canada, the Environment Canada's National Water Research Institute found nine drugs in water near 20 drinking water treatment plants in Ontario, including ibroprofen, neproxin, gemfibrzil and cholesterol-reducing medication, all in parts per trillion. Levels were worst downstream of sewage treatment plants. The anticonvulsant caramazepine has been found in tap water across Canada, and the NWRI reports that the St Lawrence River contains a cocktail of drugs. A national survey of the US by the US Geological Survey found active pharmaceutical ingredients in 80 per cent of streams, especially steroids, the insect repellant DEET, caffeine and the disinfectant triclosan. The USGS reports bass in the Potomac River containing nicotine-related chemicals and caffeine.

Illicit or classified drugs, like cocaine, amphetamines, opiates and cannabis derivatives, may also be an increasing problem. Metabolites of cocaine have already been reported in the River Po in Turin near public water abstraction points. Illicit drugs have been found together with endocrine disrupters and personal care products in rivers in south Wales (Kasprzyk-Hordern *et al*, 2008). Zuccato *et al* (2008) report metabolites of illicit drugs in rivers near Bristol coming from treated sewage. They calculate that the loading of drugs in rivers tested in Italy and Britain ranged from tenths of a gram to hundreds of grams a day and that most residues still have potent pharmacological effects.

In England, 25 per cent of public water supply comes from lowland rivers that are among the most vulnerable in Europe because of the low levels of dilution afforded to wastewater effluents from major cities by rivers with low discharges, especially in summer. As ever, dilution is the key: during the multi-year drought in south east

England in the early 1990s, the National Rivers Authority banned water abstractions from 40 rivers it designated as 'endangered', many of them nearly 100 per cent sewage effluent in the summer (see *Sewage overflows* in Part 2 of this chapter).

A problem common to most of these pollutants is that there is no regular sampling of drinking water sources for these substances, indeed, mostly no testing at all. Authorities generally have no idea of the levels of contamination. In America, there is also an unfortunate division of responsibility between the EPA covering industrial chemicals and the Food and Drug Administration responsible for the safety of pharmaceuticals. The call for 'sustainable pharmacy' is only just beginning.

The degradation of Manila Bay

Doracie Zoleta-Nantes

Manila Bay is one of the finest harbours in Asia. The economic activities that the Bay facilitates contribute almost half of the Philippines' GDP. However, the rapid urbanization of Metro Manila and its environs has contributed greatly to the decrease in the productivity of the shallow estuaries of the Bay. The remaining pockets of its mangrove forests have been significantly reduced due to the large-scale reclamation of its inter-tidal areas for commercial and entertainment activities, housing projects and roads.

Tonnes of rubbish generated daily by the 17 cities and municipalities of Metro Manila find their way into the coastal and estuarine waters of Manila Bay. Other rapidly urbanizing communities located along the major river systems that drain into the Bay also contribute to the immensity of solid waste that pollutes the Bay.

A century ago, the solid wastes were mostly biodegradable. Banana leaves were used as food wrappers, while *pandan* and *nipa* leaves were used for small bags. Many Filipinos continue to see water bodies as boundless sinks for human waste, but plastic bags and other non-biodegradable materials have long since replaced the biodegradable food wraps and bags. The Pasig River carries a regular load of plastic garbage, wood by-

products, animal carcasses and human faeces disposed of by the residents of Metro Manila. The river also delivers chemical compounds and industrial effluents from the 95 industries that are located along its banks (Table 5.1). All the major waterways in Metro Manila are used by hundreds of industries as convenient dumping sites for their effluents. The waste either floats as unsightly scum on the shallow waters or settles on the river banks and Bay coast

Table 5.1 Waste generated by industries that dump their waste into the Pasig River

Waste generated by industries on the Pasig River	Quantity
Non-hazardous solid wastes	33,682.1 tonnes
Non-hazardous liquid wastes	7,732,932.59 cubic metres
Hazardous wastes stored on site	132,170.33 tonnes
Hazardous wastes removed from site	721,146.87 tonnes

Source: *The Study on Solid Waste Management for Metro Manila in the Republic of the Philippines. Final Report Supporting Report*, PCCIKKCL. Japan International Cooperation Agency and Metropolitan Manila Development Agency.

as foul-smelling sludge deposits. They contribute to the increased biological oxygen demand and the decrease in the amount of the dissolved oxygen level on the Bay.

Most local government units do not allocate much time and resources to cleaning up the rivers. The nutrient-laden runoff contributes to red tides (algal blooms) in the estuaries of Manila Bay. The red tide outbreak in 1992 cut off 38,500 fisherfolk from their source of livelihood and they suffered an economic loss of about 3.4 billion Philippine pesos ($770,000 or £500,000).

Ships add to the pollution, spilling and dumping solid waste, oil and sewage, even though it is unlawful. According to the records of the Philippine Coast Guard, there were a total of 227 oil spill incidents between 1975 and 2004, 88 of which were land based, while 139 occurred in various other water bodies. However, the Philippine Coast Guard has allowed the dumping of wastes in eight designated areas within Philippine marine waters.

If this pollution continues unabated, the productivity of the shallow waters of the Bay will be greatly reduced, and sea level rise will compound the problems. The Philippines government and other interested parties urgently need to initiate a number of antipollution campaigns and environmental measures to protect the shallow waters of Manila Bay against further degradation. Undertaking this initiative, however, will not be easy. Despite the large number of comprehensive environmental laws enacted by the government to protect its marine and aquatic waters over the past eight decades, these laws are frequently ignored by the offending parties. In some cases, the laws are not properly implemented or enforced.

Figure 5.4 Rubbish along Manila Bay. Roxas Boulevard with the US Embassy in the background

Figure 5.5 The Pasig River and its tributaries are categorized as biologically dead with zero level of dissolved oxygen from its mouth up to 6 km upstream

Heavy metal pollution

Some heavy metals are necessary for human health in 'trace' amounts, but they soon become toxic as concentrations rise. Lead is particularly toxic. High levels in the blood can cause liver, kidney and brain damage, and there is increasing evidence that even quite low levels can affect the intelligence and development of children, hence its removal from petrol. Replacement of lead waterpipes has largely made 'plumbism' a thing of the past.

Heavy metals present a long-term hazard as they do not degrade like organic compounds. Many river floodplains contain heavy metals that date back to past periods of mining and industrial activity. These can be mobilized when floods erode the riverbanks. In the floodplains of

the Maas (Meuse) in the Netherlands peak levels of contamination are buried 3 metres down in the deposits laid down a century and a quarter ago when the industrial revolution began. Prior to the recent cleanup in the Maaswerken Project, the River Maas reworked 500 tonnes of these deposits each year, supplying the river with over 1500 kg of zinc, 330 kg of lead, 40 kg of copper and 6 kg of cadmium annually (Rang *et al*, 1986). No matter how much current wastewater treatment is improved, this historical legacy would remain, hence the focus of the largest floodplain cleanup operation ever undertaken (see box 'Cleaning up contaminated sediments on the Meuse floodplain' in Chapter 15).

Abandoned spoil tips can also be a source of pollution when heavy rains wash material into the rivers. Acid drainage waters from abandoned mines can add to this. Spoil tips in mid Wales abandoned nearly a century ago still carry high levels of zinc and cadmium into rivers. On the River Rheidol, zinc concentrations can reach 0.88 mg/L and up to 13 tonnes can be shifted in a single month. Some is dissolved, some carried as particles of ore, and some as coatings on fine sediments that may be ingested by fish. Floodwaters mobilize most sediments and also dissolve more heavy metals, because the water is commonly more acidic due to increased yields from the mine adits (drainage tunnels) and from the peaty mountain moorlands. Concentrations of dissolved metals can also be high during low flows in summer, but there is less sediment movement.

Figure 5.6 shows the distribution of lead levels in rivers according to data from the GEMS/Water archive. It shows high concentrations in many mining areas of coastal Brazil, Kenya, Thailand and Indonesia. But the fact that Europe is largely pollution-free is only partially true and reveals some of the limitations of water sampling and of reportage. Many rivers have been cleaned up over the last 50 years. Salmon and sea trout (sewin) have returned to Welsh rivers like the Rheidol. Regular sampling tends to record the most common situation, but it is likely to miss the occasional storm that carries high metal concentrations and would raise the average. Table 5.2 lists some of the medical effects of heavy metals.

mg/L
- 0.2 – 0.5
- 0.1 – 0.199
- 0.05 – 0.099
- 0.011 – 0.049
- ≤ 0.01

Figure 5.6 Lead pollution in rivers

Table 5.2 Selected metal pollutants and known medical effects

Metal	Principal sources	Health impacts
Cadmium (Cd)	Phosphate fertilizers	Carcinogen
Cobalt (Co)	Sewage treatment plants, fertilizers, manure	Carcinogen
Copper (Cu)	Sewage treatment plants, fertilizers, manure, water pipes	Liver damage
Lead (Pb)	Phosphate fertilizers, sewage, sludge	Neurological and behavioural problems, migraine, paralysis, liver and kidney damage
Nickel (Ni)	Industrial wastewater, sedimentary rocks, soils contaminated by fossil fuel burning and industry, phosphate fertilizers	Toxic in high concentration but may be valuable in traces. Insoluble forms worse and potentially carcinogenic. Inhibits growth in wildlife invertebrates if food supply limited
Paladium (Pd), platinum (Pt) and rhodium (Rh)	Vehicle catalytic converters, also mined for jewellery	Asthma, sensitizing airways, breathing difficulties
Uranium (U)	Mining and weapons	Carcinogen
Zinc (Zn)	Pesticides, mining and manufacturing	Intestinal irritant, diarrhoea, jaundice, seizures, long-term exposure in drinking water can be deadly (but trace amounts essential for health)

Sources: Alloway and Ayres (1993), Alloway (1995), Olajire and Ayodele (1997), Nabulo *et al* (2006) and Colombo *et al* (2008).

The search for fuel

The two great pillars of modern energy production – oil and nuclear power – are both capable of creating serious water pollution during both mining and use. Hydropower is dealt with in Chapter 7, but it is worth noting at this stage that it, too, is not entirely free from causing pollution.

Oil and water don't mix

The extraction and transportation of crude oil is the source of numerous environmental disasters. Accidents at sea, like the burst at the BP oil rig in the Gulf of Mexico in 2010 and the grounding of the Exxon Valdez in Alaska in 1989, attract widespread media coverage; indeed, recent evidence indicates that the Alaskan coast has still not fully recovered from that event. On land, oil spills around pipelines in Nigeria, some blamed on terrorist activity, some on poor maintenance, are also well publicized. Rapidly rusting pipelines are also a problem in Siberia, where the Tomsk wetlands are 'cursed by regular spills over large areas' (Kotikov, 2003): 11 million tonnes of crude oil are exported from the area each year, accounting for 40 per cent of the regional government's budget and the environment is not high on the agenda. Georgia is particularly concerned to protect its rivers from any possible leakage from the new Baku-Tbilisi-Ceyhan oil pipeline, because it crosses a seismically active region. Their scheme involves constructing five permanent, water-tight tanks in the riverbeds and blocking valves on six rivers (Devidze, 2009).

The price of oil is now encouraging the exploitation of ever more difficult sources, which not only have great potential to pollute the environment but also use huge volumes of water in the extraction processes. This is the case with extracting petroleum from oil shales. Oil shales are already being exploited in Alberta and there are plans to develop major reserves of oil and natural gas in the USA, China and Europe. Shale gas could add 30 per cent to US gas reserves.

Water is a vital ingredient in the extraction method called 'hydro-fracturing' or 'fracking'. A New York City report expressed grave concerns in 2009 for the damage that expansion of hydro-fracturing in the Catskill Mountains and Upper Delaware basin could cause to the city's water supply, particularly contamination of groundwater supplies by the toxic chemicals used to free the oil. The technique is being used to extract oil from the Marcellus Shale, which is the largest of the American oil shale formations, underlying parts of New York State, Pennsylvania and West Virginia. It involves injecting vast quantities of water down wells to fracture the shale and force the oil and gas out: 11–30m m^3 per well of which only a half to a third is recovered, all heavily polluted with oil and the chemicals used in the process. New York politicians urged the government to ban fracking and the Federal Government passed a FRAC Act in 2009 putting the EPA in charge of policing the practice. It remains to be seen how effective this is.

In the Paris basin, Toreador of Texas has licences to explore over 750,000 km² between St Dizier and Montargis where they believe the organic-rich sedimentary rocks could yield 50,000 barrels a day within a few years. The total yield from the Paris basin could reach 65 billion barrels, making it nearly double Nigeria's 36 billion. Poland may hold 50 per cent of Europe's gas reserves and is actively developing them to reduce its dependence on Russia's Gazprom. Germany, Hungary and Sweden could follow suit, with huge implications for European water resources.

Most publicity has been given to Canada's tar sands. Much of Canada's prosperity over the last decade has been due to Alberta's fuel resources and even though there was a slump in investment when oil prices fell during the recession, extraction is bound to expand in the coming years. The pollution is appalling (see the article about the Mackenzie Basin, below). It is also tragic that in the 1980s there were grand plans for inter-province cooperation to protect the whole Mackenzie basin that could have been a flagship for Canada and the world. The pollution has already affected wildlife and there are complaints that it is affecting human health.

But this is not the only mining operation to depart from the plans and aspirations of the Mackenzie River Basin Committee (1981). Uranium City tells another tale (see the section headed *Uranium mining*, later in this chapter).

The Mackenzie Basin – gained and lost?

Extraction of oil from Canada's tar sands in Alberta has created the world's largest area of open cast mines, 520 km² of mines and toxic lakes. The extraction process uses prolific amounts of water, mainly from the Athabasca River, and returns highly toxic water to storage ponds, many of which are thought to be leaking into the environment and into the Athabasca River itself. An oil company official admitted on CBC television in 2009 that water will eventually be purposefully released from the ponds into the Athabasca River and claimed some water is already clean enough. This was news to the Dene people, an Indian First Nation that live downstream of Fort Chippewyan on the Athabasca River, who report higher rates of cancer since the tar sands project began. In April 2008, 1600 ducks died landing on an oil slick lake and Syncrude subsequently erected orange scarecrows, dubbed 'bit-u-men'.

Yet in the 1980s the Mackenzie basin was developing as a pioneering model of environmental protection. The Mackenzie River Basin Committee (MRBC) was set up in 1977, not long after Canada rejected American proposals to include the basin in a continent-wide water grid (see *The North America Water and Power Alliance* in Chapter 7), to lay the foundations for coordinated development and environmental protection. Although one major dam

had already been built, the WAC Bennett in British Columbia, the basin had not suffered the wholescale river diversions that were underway in northern Quebec (see *James Bay Hydro – redesigning a landscape* in Chapter 7). There was also little economic development in the larger portion of the basin north of the Lesser Slave Lake in the Athabasca basin (Figure 5.7). The Athabasca catchment now threatened by oil extraction is one of the main headwaters, responsible for about a sixth of the total discharge of the Mackenzie River.

The tar sands, or 'oil sands' as the oil companies prefer to call them, cover an area of 141,000 km², and are estimated to contain at least 174 million barrels. This would mean that Canada holds at least a third of the world's oil reserves, second only to Saudi Arabia; some estimates put the volume even higher at 270 billion m³ of bitumen. Production dipped as oil prices fell during the peak of the credit crunch in 2008, but the long-term view is for expansion. Shell and BP are extracting 1.3m barrels a day and plan to double this by 2015, perhaps reaching 6m by 2030. Further improvements in technology might even raise the yield to 1.7 trillion barrels. The project was delayed for decades because of the high cost of extraction. The sands contain an admixture of oil,

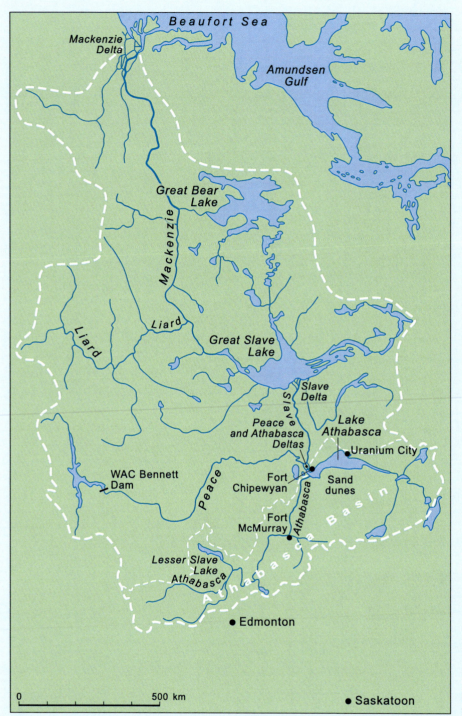

Figure 5.7 The Mackenzie basin, where oil and uranium have trumped environmental protection

bitumen and clay. Extraction is particularly expensive and energy intensive as two tonnes of sand are needed to provide just one barrel of crude oil. The energy needed to heat the sands is equivalent to two barrels of oil for every three produced. But rising oil prices have led to a 'gold rush' since 1997.

Much of the oil is exported by pipeline to the American mid-West for refining. It now provides petrol for half of Canada's road vehicles and supplies the USA with a fifth of its needs, more than Saudi Arabia. Between 2000 and 2008, the Canadian Government earned C\$430 million from the project and so far only two per cent of the oil has been extracted. A quarter of a million jobs now depend on it.

In a ground-breaking book which won the Rachel Carson Environment Book Award from the American Society of Environmental Journalists, Andrew Nikiforuk (2010) blames the scheme for destroying rivers and forests in the Athabasca region and describes it as an 'out-of-control megaproject'. Apart from the social ills associated with the largest homeless population in Canada, attracted by boom towns like Fort McMurray, the scheme is said to be poisoning the drinking water and draining the Athabasca River. It also threatens to affect the delicate ecology of the Athabasca Sand Dunes, which was singled out for special attention in the 1981 Environmental Protection Plan.

Opposition is mounting. In 2000 the Northwest Territories Association of Communities, representing 33 communities, passed a resolution calling on the governments of Alberta and Canada to stop approvals for new oil sands projects until they had negotiated enforceable transboundary water agreements to ensure that riverflows are 'clean, uncontaminated and unimpeded'. In practice, the resolution was all but ignored even though it was only reiterating what the 1981 agreements aimed for. Nearly ten years later, Greenpeace protesters occupied a processing plant, and the First Nation movement now estimates that the financial compensation they need could exceed the total royalties received by the Alberta Government. The website Tar Sands Watch documents the pollution and by 2009 the Canadian media were carrying frequent news items on the controversial project. In apparent response, Syncrude recently handed back the first piece of reclaimed land to the government, but it amounts to less than 1.5 km² on the edge of toxic lakes.

The 1981 Environmental Protection Plan

The ten-volume prospectus produced by the Mackenzie River Basin Committee in 1981 was world-leading in its range and depth. The scheme covered the whole of Canada's largest river basin, an area of 1,787,000 km², 6095 km of channel and 65 550 km² of lakes (Figure 5.7). The MRBC Report included: agreements on transboundary water management, minimum acceptable flows, water quality and river regulation between four Provincial governments, one Territorial government and the Federal government; a permanent, integrated network of sta-

tions for hydrometric and meteorological observations, including snowcover, sediment and water quality; a suite of computer models to predict responses in all parts of the water budget, including a daily routing model for flow in the main river network; a series of special basin-wide and local area studies to cover key environments and processes, including the most important annual event, spring ice breakup sediment transport, and the ecology of alluvial areas and the Athabasca Sand Dunes. The Report recommended that every new project should be judged according to an Environmental Impact Assessment, and that all information should be readily shared through the national water document database, WATDOC.

This should have been a blueprint for the world. Although the hydropower potential within the basin was estimated at 14,000 MW, development has been limited, partly by the realization that large reservoirs would reduce the spring melt freshet, damage alluvial ecosystems and reduce the stability of ice cover on rivers and lakes. But perhaps more importantly, the basin was saved by the excessive capacity developed in eastern Canada – until the price of oil made the tar sands project profitable.

The future

Canada is facing increasing domestic and international criticism, though more for its carbon footprint than for effects on the water environment. Canada dropped its commitment to the Kyoto agreement in 2006 and tar sands have now raised CO_2 emissions to 26 per cent above the benchmark 1990 levels. Oil extraction releases three times as much CO_2 as conventional oil and the mines are now the largest industrial source of CO_2 emissions in North America. Shell plans to introduce carbon storage and capture (CCS) by 2015, but this is unlikely to reduce emissions by more than a fifth. Several US states have banned the product because of its carbon footprint, but neither this nor any other foreseeable obstacle is likely to halt expansion. What is needed is tighter regulatory control to minimize the environmental impact.

Some work is in-hand. The Oil Sands Regional Aquatics Monitoring Program (RAMP) was begun in 1997 in order to establish baseline data and predict the impacts of development. The programme covers water quality, climate and hydrology, sediment quality, wetlands, benthic invertebrates, fish populations, aquatic vegetation and acid sensitive lakes. But experience from the recent past suggests that such evidence will have a hard time against strong economic interests. (See Notes.)

One potential area for concern that does not appear to have received attention is the possible effect of changing riverflows on sea ice in the Beaufort Sea. The Mackenzie-Peace is the eleventh longest river in the world and discharges 8000 m³/s into the Arctic Ocean – 250 km³ a year. The discharge plays a major role in reducing the salinity of the ocean surface. Reducing the river discharge will reduce sea ice formation and increase the frequency of an open water Northwest Passage, as first experienced in 2008. Disappearing sea ice is a key factor in the disproportionate warming of the Arctic.

The impact of the tar sands project on global warming might not be confined solely to carbon emissions.

Radioactive pollution

'Increased incidence of childhood leukaemia has now been verified near most of the main sources of radioisotopic pollution in Europe … A reassessment of the hazard to health of such exposure should be the subject of urgent research effort since the problem of risk from such pollution carries important human health policy decision implications.'

Chris Busby and Molly Scott Cato,
Energy and Environment, 2000

As more countries return to nuclear power as a means of reducing their carbon footprint and to service burgeoning energy demands, water pollution by radioactive substances is an increasing concern. Radioactive pollution of the environment by the nuclear power industry has been an issue since the 1950s, strongly censured by environmental pressure groups like Greenpeace and equally strongly denied, and sometimes covered up, by governments. The pollution can have many causes: accidents, unsafe disposal of spent fuels and other solid waste, or the regular release of low-level wastewater into water bodies. It is not known whether the extra fear of terrorist activity has ever materialized.

Before the international Nuclear Test Ban Treaty in 1963, most of the globe was covered by fallout from nuclear bombs. Caesium-137 was the predominant caesium isotope, with very little if any caesium-134. Caesium-137 has a half-life of around 30 years. In contrast, leakages and fallout from nuclear power stations characteristically contain high levels of caesium-134, which has a half-life of just two years, as well as caesium-137.

Exposure to radionuclides, especially by ingestion through food and drink, has been associated with genetic damage and various human cancers, notably leukemia. There are reports of increases in human mini-satellite mutation rates in children living in areas affected by Chernobyl fallout in former Soviet Union countries. Busby and Cato (2000) maintain that proven clusters of leukaemia near sources of radioisotopic pollution call into question the models used to assess risk based on external exposure at Hiroshima rather than ingestion. They claim the calculated risk factors can be in error by orders of magnitude: a 100-fold underestimate for Chernobyl, 15-fold near La Hague, France, 10-fold near Sellafield and 20-fold near the UK Atomic Weapons Establishment. They suggest this indicates that a more cautious approach is needed in policy decisions about nuclear power.

Release of low-level radioactive wastewater from the Sellafield complex in Cumbria into the Irish Sea formed part of an official complaint from the Irish government and has been blamed for pollution on the coast of North Wales. The local campaign group, Cumbrians Opposed to a Radioactive Environment (CORE), claims that huge amounts of waste are discharged into the local air and water. The site has a long history of accidents and has been called the most contaminated industrial site in western Europe: it houses two-thirds of Britain's nuclear waste, some from the 1950s. Early accidents in the 1950s, when it was called Windscale, were glossed over in government reports to the media; one cleanup activity involved pouring contaminated milk into drains. Understanding of the environmental and health hazards has improved since those days. Even so, after the Nuclear Decommissioning Authority passed control of the plant to a private consortium in 2008, a long-standing leak was discovered and the consortium's chief executive reported a 'lack of urgency' culture built on decades of mismanagement. Some 83,000 litres leaked from a cracked pipe in 2005.

Safe disposal of nuclear waste remains largely unsolved. Only a tiny minority of countries have built any permanent storages. Before it was disbanded, the British government's Nirex spent decades searching unsuccessfully for a secure deep burial site where there is no risk of groundwater ever being contaminated. Britain is now exporting waste to Gorleben in Lower Saxony, where most of the German radioactive waste is stored. Fears emerged in 2010 that some storage sites in Germany may not be as secure as they should be. Nuclear waste buried in a 750 m deep disused mine in Asse, Lower Saxony, may be leaking into the groundwater. Some 12,000 litres of water are thought to be leaking into the mine, which is said to hold 100 tonnes of uranium, 87 tonnes of thorium and 25 tonnes of plutonium. Plans have been laid to transfer canisters to another mine at Schacht Konrad. Another suggestion is to export waste to eastern Siberia. Much of the nuclear waste generated between 1967 and 1978 was buried in a depopulated zone near the East German border when safety standards were less strict and political considerations meant that inventories were kept vague. Although the German government yielded to Green activists with plans to be nuclear free, it is now trying to extend the life of the remaining nuclear plants, as are Britain and the USA. Germany still consumes 400 tonnes of nuclear fuel a year, as well as importing over 17,000 tonnes of waste from Britain and France. The EU is now considering a collective high-level waste storage facility in one of the new member states in Eastern Europe, perhaps with monetary compensation.

The worst ever nuclear disaster, at Chernobyl in the USSR in 1986, polluted rainfall in Scandinavia and the UK as well as local rivers in the Ukraine. The Soviet government tried to cover it up. The alarm was first raised by scientists at a Swedish facility. If the wind had been the normal westerly wind and had carried the toxic cloud over the Soviet Union rather than through Europe, the event might have gone largely unnoticed. Cleanup activities in Kiev involved hosing down the streets and washing the pollution into road drains, and thence into the rivers. In Britain, the pollution was picked up by a soil scientist, not by any professional surveillance system. As the winds carrying the radioactive dust passed over the UK and encountered areas of heavy rain, the rainfall created radioactive hot spots. Over four million ha in the mountains of North Wales were heavily polluted. Even though the caesium-134 isotope that was a signature of Chernobyl has a very short half-life, monitoring in the area around the Trawsfynnedd nuclear plant by the

University of Wales Medical School in the early 1990s revealed continuing high levels of the radioactivity. Was it being topped up from the local Magnox plant or was the radioactivity from the remaining caesium-137? The local plant was beyond its designed lifespan and was subsequently decommissioned in 1991. The experience led to the formation of RIMNET in 1988, a network of 50 government establishments now managed by the Met Office, which continuously monitors radiation levels and issues automatic alerts. Even so, the effects of the single event at Chernobyl persist in upland Britain: in 2004 there were still 53,000 ha of Welsh farmland affected by restrictions on sheep movement and revised predictions indicated that high levels of radioactivity might last a further 50 years. The area has the highest cancer rate in Wales.

The only good news is that the radioactive elements bind to the soil and are not readily washed out into surface waters unless the soil itself is washed into the rivers. This happens, however. Walling *et al* (1992) report levels of caesium-137 increased by two orders of magnitude in suspended sediments in the headwaters of the River Severn in mid-Wales immediately after the Chernobyl disaster, reaching up to 1450 mBq/g, and it remained an order of magnitude higher than before until monitoring stopped in 1988. They note that radiocaesium is carried downstream and deposited in river channels and floodplains.

Some rain also falls directly onto surface waters and some drains from the land by fast routes as 'bypass flow' through macropores and natural soil pipes (Jones, 2010). Soon after the Chernobyl event, research by the Ministry of Agriculture Fisheries and Food in Britain reported that Chernobyl fallout was readily detected in all sectors of the aquatic environment (Mitchell *et al*, 1986).

There is abundant evidence of poor management of nuclear material in Russia. One of the most remarkable incidents resulted from the Soviet plan to reverse the arctic rivers (see *Reversing Russia's arctic rivers* in Chapter 7). Work to reverse the Pechora and Karna rivers so they drain into the Volga to serve the grainlands of the Ukraine began in 1973, immediately following the severe drought that led to the 'Great Grain Robbery' the previous year and America's subsequent limit on the USSR's purchases of American grain. 'Controlled' nuclear explosions were used to blast away the basin divides. This produced a toxic lake of radioactive water – and the scheme was halted (Zherelina, 2003). In Siberia, Kotikov (2003)

67

reports that liquid radioactive waste used to be poured directly into the Tom River, a tributary of the Ob. This has left the riverbed 'horribly polluted'. Despite closure of the reactors, measurements in the early 2000s show caesium, strontium-90 and phosphorus-32 in the river-bed – and the phosphorus has a half life of just 14 days, indicating recent topping up from some source. Fish in Tomsk market were 20 times above the radioactive limit.

After decades in abeyance, both Britain and America are planning new nuclear stations. They aim for safer technology, but an additional criterion since 9/11 is ability to survive a direct impact by aircraft and one of the designs evaluated in the UK may fail the test. Even if the power plants themselves are more secure, however, the burgeoning demand for uranium – with China alone building 21 new nuclear stations and many countries extending the life of old plants – there is an additional hazard in areas surrounding the sources of the uranium.

Uranium mining

Far less publicity has been given to pollution from uranium mining. There is now belated public concern over its possible effects on the health of local communities in South Africa. Pollution has existed for some time, not just in relation to high-grade uranium ores but also because uranium is commonly associated with gold deposits and has been an inadvertent and unwanted byproduct of the gold mining industry particularly in South Africa. With the expansion of nuclear power around the world, people and environments could suffer for other nations' attempts to create climate-neutral energy, as mothballed mines in South Africa are reopened and prospecting extends throughout southern Africa.

Much of the pollution occurs through leakage and infiltration from the waste material held in tailings deposits and slimes dams, and it enters streams directly or via groundwater. Some of this streamwater is used untreated as drinking water in poor townships near the mines. Streams are also polluted by dewatering the mines, pumping out water that has been in contact with uraniferous ore bodies. Now, as the price of gold and uranium rises, many old tailings are being reworked to salvage metal, with the risk of remobilizing the metals.

It is an 'inconvenient truth' that has been fairly successfully hidden by commercial interests since the 1960s. Now, however, South African media are at last exposing the health risks, due in no small part to the work of Dr Frank Winde of North-West University (see his article below, 'Uranium pollution in South Africa'). The media attention is valuable, but quantifying the risk scientifically faces three common obstacles: gaps in medical understanding of the dangers for human health; lack of freely available data on pollution levels; and lack of effective and permanent monitoring programmes for both water and people. The long-term impact of radiation pollution on health is, fortunately, somewhat reduced in the mining communities by their more rapid turnover of population. This reduces individuals' exposure to radiation, but it also means that people whose health may have been damaged have often left the area and are no longer traceable for medical research.

In Canada, the land around Uranium City on the north shore of Lake Athabasca in the Mackenzie basin once hosted 52 pits and 12 open cast mines extracting gold and uranium (Figure 5.7). They were largely closed down in 1983, leaving the north shore heavily contaminated. Buried and ponded mine tailings have polluted the waters affecting fish and fishing communities among the Dene and Cree people (MacKenzie, undated). Revival of international interest in uranium led to renewed activity from 1995 around McClean Lake. Deposits at McArthur River contain the world's largest and highest grade uranium. The renewed operations now have stricter environmental controls and since 2000 the McClean Lake operation has had an ISO 14001 environmental management certificate; activities are intensely monitored and tailings are stored in lined pits. But in 1998 one mining company was convicted of contaminating the environment and not reporting it.

In common with many other toxic pollutants, the recommended 'safe' levels of uranium pollution have been reduced over the years, but the scientific basis for these is still inadequate. The question remains: is any concentration safe if ingested over a long period?

Other effects of geology on human health

Oil and radioactive minerals are clear examples of pollution from geological sources. But there are other, less high-profile geological sources that affect the healthiness of drinking water. There is still debate as to whether the hardness of drinking water – the amount of dissolved calcium and magnesium salts – affects health. There is

Uranium pollution in South Africa: 'Death in the water'

Frank Winde

In company with an increasing number of countries, South Africa recently embarked on a nuclear expansion programme. Since South Africa holds the world's fourth largest deposits of uranium, the expansion of mining activity and uranium exports, which began with the government's 2007 declaration of uranium as a 'strategic mineral', is going to outstrip the requirements of its own nuclear power industry, with important implications for protecting water resources. The Nuclear Fuels Corporation of South Africa (NUFCOR) is the largest continuous producer of uranium concentrate worldwide, with a total of 240,000 tonnes of uranium oxide (U_3O_8) sold, and it is estimated that at least 25 new uranium mines will be needed by 2020. Already, tailings deposits left by past mining in the goldfields of the West Rand and the Far West Rand are estimated to contain well over 100,000 tonnes of uranium (U), constituting a large reservoir for ongoing future water pollution. Other sources of uranium pollution include: runoff from mining rock dumps, water decanting into streams from old boreholes and mine shafts, water pumped out of operating mines, and uranium-rich river deposits.

A 2006 report by the Water Research Commission of South Africa, the so-called WRC 1214, on mining-related uranium pollution and associated health risks, attracted widespread media attention to radioactive water pollution in the Wonderfonteinspruit (WFS), Potchefstroom, North West Province (Coetzee *et al*, 2006). The National Nuclear Regulator (NNR) was initially sceptical of the WRC findings, but the independent report it commissioned from the German consultancy Brenk Systemplanung, which had overseen the rehabilitation of the Wismut uranium mining area in East Germany during the 1990s, confirmed significant risks in 2007. These were mainly due to food crops and animal fodder grown by irrigation.

In July 2007, *The Sowetan* newspaper ran the front page headline 'Death in the water'. Throughout 2007 and 2008, over 50 articles appeared in national and local newspapers on the topic. Under headlines such as 'Toxic shock' (*Potchefstroom Herald*, 8 February 2008), 'Lives at risk as mines coin in' (*The Sowetan*, 27 July 2007),

'Living in fear of a toxic tsunami' and 'Far West Rand residents claim poisoning' (*Saturday Star*, 12 April 2008) some of these articles, often with a certain degree of sensationalism, linked water pollution to a number of serious health effects, unsettling the general public. The topic was even raised in radio and television broadcasts in South Africa and on Aljazeera, the UN Integrated Regional Information Network (IRIN) and numerous websites with headings like 'SA radioactive stream – 400,000 at high risk'. Environmental activists prepared special submissions to parliamentary portfolio committees and Members of Parliament. Mining-related pollution was also at the heart of several legal actions taken or threatened by local municipalities, land owners and environmental pressure groups against various gold mines operating in the WFS catchment, as well as against government authorities for neglecting law enforcement. One environmental pressure group, the Public Environmental Arbiters (PEA), approached the Human Rights Commission of South Africa on the issue.

The first signs of progress are beginning to appear. In 2006, South Africa signed a five-year agreement with the International Atomic Energy Authority (IAEA) aimed, among other things, at water resources development and integrated pollution control. In a generally antagonistic situation between public perceptions of the dangers and government's nuclear expansion programme, this may provide the first steps toward finding common ground. And in 2007, the Department of Water Affairs and Forestry (DWAF) and the NNR launched a joint initiative to remediate contaminated sites in the WFS, based on a methodology developed by Winde (2008), which has identified 36 sites along the course of the river for priority intervention.

Uranium toxicity and drinking water guidelines: increasing awareness of the dangers

Evidence is mounting that the radioactive heavy metal uranium may pose a more severe health risk than previously thought, even at comparably low concentration. Uranium is the heaviest, naturally occurring element on earth, with a global background concentration in the earth's crust of c.2 mg/kg (or ppm). It is approximately 1000 times more common than gold and ten times more abundant than other toxic heavy metals like cadmium (0.3 ppm) or mercury (0.4 ppm). Granite and shales display elevated levels (typically 3.4 and 3.7 ppm, respectively). But there are over 200 uranium-bearing minerals and the ability of uranium to form soluble complexes with a large range of ions explains its exceptionally high geochemical mobility in aquatic environments.

Like all heavy metals, uranium is non-biodegradable and tends to accumulate in the biosphere. Winde (2009) reports several instances where secondary accumulation of waterborne uranium in the environment reached much higher levels than in the original source of pollution. Elevated uranium-levels in several municipal water supply systems in Germany, especially in agricultural areas, have been linked to long-term, large-scale applications of uraniferous, phosphate-based fertilizers (Foodwatch, 2008; Schnug et al, 2005). Surveys of bottled mineral waters from a number of countries by Krachler and Shotyk (2008) found uranium levels as high as 27.5 µg/L, mainly due to uranium-rich geological settings. Other sources report up to 72 µg/L in Swedish mineral water, and close to 15,000 µg/L or even 40,000 µg/L in Finnish groundwater, generally associated with uraniferous granites (Winde, 2010a and b).

Although most natural uranium (^{238}U) decays extremely slowly and is therefore relatively harmless, the isotopes ^{234}U and ^{235}U, despite making up less than one per cent of naturally occurring uranium, account for roughly half of the total radioactivity. All the isotopes only emit alpha particles that are absorbed within a few centimetres of the air and barely penetrate the skin, but it is a very different matter when they are ingested. Deposits in organs such as kidneys, lungs, brain and bone marrow directly attack surrounding tissues. Shorter-lived, highly radioactive decay products such as radium (^{226}Ra) and radon (^{222}Rn) add to the radiotoxicity.

Research at Stellenbosch University identified high levels of leukaemia and related blood anomalies in 1996 in a remote arid area in the Northern Cape where borehole drinking water is high in uranium. The link was confirmed by a subsequent WRC report. Other work has pointed to possible links between long-term consumption of contaminated mineral water and kidney cancer, and with DNA and brain damage. Longstanding rumours in the gold mining town of Carletonville link uranium-polluted drinking water to an abnormally high number of children with learning difficulties in a mining community that relied on groundwater pumped from the mine. Recently, Raymond-Wish et al (2007) added uranium to the long list of known endocrine disruptive compounds that might increase the risk of fertility problems and reproductive cancers at levels so far regarded as safe in drinking water even by the US EPA.

A major problem for authorities trying to set safe limits for uranium concentrations in drinking water remains the lack of comprehensive medical studies on human beings exposed to relatively low levels of pollution over long periods of time. So far, most limits have been based on short-term experiments on animals, especially on just one experiment exposing rats and rabbits to uranium for 91 days. Others have used evidence from nuclear explosions, which were intense and caused chromosomal damage but which were instantaneous and did not involve ingestion. Retrospective studies of former mine workers from the Wismut uranium mines in Germany found the risk of contracting liver cancer from occupational exposure to uranium was 20–70 times higher than indicated by extrapolating data from the study of atomic bomb survivors.

The large variation in official drinking water limits in Table 5.3 (3500 per cent) may be partly explained by different dietary assumptions about the amount of water

Table 5.3 Comparison of recommended safe limits for uranium concentrations in water

Organization	Date	Limit µg/L
WHO	1998	2
	2003	9
	2004	15
DWAF (S Africa)	1985	1000/4000* (drinking)
	1996	70 (drinking)
		10 (irrigation)
EPA (USA)	2006 (latest)**	30
Germany	existing	2 (mineral water and table water)
Germany/EU	proposed	10 (for life-long exposure)

Key:
* maximum limit for 'insignificant risk'; the limit for 'no risk' is 1000; Kempster, P.L. and Smith, R., 1985: Proposed aesthetic/physical and inorganic drinking water criteria for the Republic of South Africa. *Research Report no. 628. National Institute for Water Research, CSIR, South Africa.*
** US Environmental Protection Agency, 2006: *Drinking Water Standards and Health Advisories.* US Environmental Protection Agency, Office of Water, Washington D.C.

consumed. It may also reflect differing political and economic considerations of the legal enforceability of the limits. A more worrying observation is the 1959 agreement between the WHO and the IAEA which stated that 'whenever either organisation proposes to initiate a programme or activity on a subject in which the organisation has or may have a substantial interest, the first party shall consult the other with a view to adjusting the matter by mutual consent' (WHO, 1999). Bertell (1999) pointed out that this constitutes a serious conflict of interest and that since 1959 the promotional goals of the IAEA frequently took priority over regulatory aspects. This may explain why the limits set by the WHO were raised twice within a comparably short period when more uranium was needed to satisfy a growing global demand (Winde, 2010).

Table 5.4 shows the estimated differences between DWAF standards in 1996 and 2002. When the two guidelines are converted into the same units of measurement, this reveals considerable differences even within the same organization in South Africa over a short period of time.

In contrast to heavy metals such as iron and zinc, which are needed for human metabolism and only toxic above certain threshold levels, uranium has no health benefits at any level. Ideally, uranium and other non-essential metals and half metals such as mercury, arsenic and cadmium should not be in drinking water at all (Figure 5.8). However, owing to their natural abundance in the environment, the costs of completely removing all potentially dangerous metals would render tap water unaffordable for many.

Table 5.4 Comparison of limits for radioactivity in water in different DWAF guidelines based on ^{238}U-concentration (DWAF, 1996a) and annual equivalent dose through drinking water (DWAF, 2002) with associated risks of uranium-related cancer

Water quality class	DWAF (1996a)*			DWAF (2002)		
	^{238}U-concentration (upper limit)	Dose (calculated upper limit)	Associated annual fatal cancer risk	Dose (upper limit)	^{238}U-concentration (calculated upper limit)	Associated annual fatal cancer risk
	µg/L	mSv/a	1 per number	mSv/a	µg/L	1 per number
1 (ideal/ TWQR)	70	0.11	4,000,000	0.1	62	1,000,000
2 (good)	284	0.39	1,000,000	1	754	100,000
3 (marginal)	1420	1.87	200,000	10	7677	10,000
4 (poor)	>1420	>1.87	>200,000	100	76,908	1000
5 (unacceptable)				>100	769,215	<1000

* **Bold columns** are values given in source, *italics* show calculated values. For methods see Winde (2010a and b).
† 435,000 tonnes of uranium or approximately eight per cent of the world's total of 'reasonably assured resources', plus inferred uranium resources recoverable for under US$ 130/kg, January 2007 (World Nuclear Association, 2009).

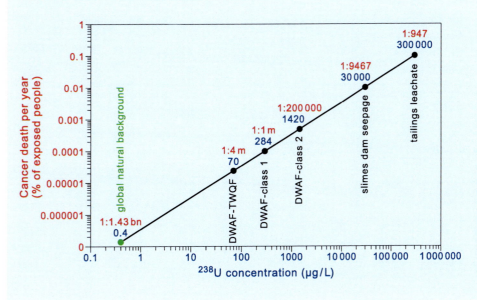

Figure 5.8 Relationship between uranium-238 concentration in drinking water and annual risk of fatal cancer according to DWAF (1996a) compared with the global natural background and extrapolated to accommodate uranium levels observed in the Wonderfonteinspruit, North West Province, South Africa

a long-standing belief that 'soft' water, low in dissolved salts, increases the risk of cardiovascular diseases. The WHO (2003) has decided that the evidence remains statistical and that it does not prove a causal relationship, except that soft water with less than 100 mg/L of solutes can increase the solution of heavy metals in the waterpipes – cadmium, lead, zinc and copper – though acidity is also an important factor. A survey by Dr Margaret Crawford of the London School of Hygiene and Tropical Medicine concluded that there is a stronger correlation between soft water and deaths for cardiac arrest and to a lesser extent bronchitis, in the UK than elsewhere, but notes there could be many confounding influences (Crawford, 1972). Perhaps the best piece of evidence comes from a Canadian review that while cardiac arrhythmia is more prevalent in soft water areas, adding magnesium can rectify it (Marier, 1978).

Similarly, a deficiency of iodine in drinking water has also been linked to health disorders. At the other end of the spectrum, high levels of fluoride and arsenic have clear impacts on health. While fluoride may be added deliberately to drinking water for health reasons, arsenic is a poison that appears to have become more prevalent in recent years in parts of the world where more groundwater is being exploited, often because surface waters have become more polluted.

Iodine, goitre and cretinism

Goitre is a persistent enlargement of the thyroid gland. The thyroid is an endocrine gland that produces the iodine hormone thyroxin, which increases oxidation and affects the body's metabolism. Goitre is especially prevalent in areas with a deficiency of iodine in the soil or water: Derbyshire in England and parts of Switzerland have traditionally had high rates of goitre. Iodine deficiency in the local water is becoming a relatively more important factor compared with food grown in the local soil, as food provision becomes more globalized and imported.

Cretinism is most commonly caused by severe underfunctioning of the thyroid, associated with extreme and persistent iodine deficiency. Unless treated from the very earliest years of life, it results in mental retardation, stunted growth and a prematurely aged appearance.

In his article 'The health effects of iodine and fluorine in drinking water', Professor Wang Wuyi looks at the distribution of goitre and cretinism in China in relation to water quality (Figures 5.10 and 5.11).

Fluorine and fluorosis

Fluorine is used in the human body to build bones and teeth. It is present in all natural waters in varying concentrations, commonly between 0.1 ppm and 6 ppm, usually in fluorides of potassium or sodium leached from weathering minerals. Only relatively small amounts are needed for human health, less than 1 ppm (or 1 mg/L), and, as with many other substances required in small doses, when concentrations get high it becomes toxic. Toxicity may also be induced by a combination of drinking water with other sources, like excessive intake of foodstuffs high in fluorine, such as tea or beer, or by breathing polluted air.

The GEMS/Water archive shows high levels of fluoride in the southern Low Countries and East Africa, together with moderate levels in the western British Isles (Figure 5.9). Some of the British and Irish cases may be related to granitic rocks, but as Professor Wang Wuyi shows, there are many possible sources, and mining and industry may be important locally.

Artificial fluoridation of public water supplies has been adopted in many countries and by individual water companies and authorities over the last 60 years, especially in North America and Europe, in response to evidence that it increases the resistance of teeth to dental caries (decay) in young people. The incidence of caries nearly halves as concentrations are increased from 0.6 to 1 ppm.

This is, however, a very controversial subject. Despite overwhelming, though not complete, support from the medical professions, there is the very slight probability of toxicity being induced by ingestion from other sources adding to the fluorine in the water. There is also the problem that some of the fluorine accumulates in the body over a period of time, so that people who drink more tap water may be more at risk. More importantly, there is the ethical argument against mass medication, because it removes personal choice and it is unnecessary for mature adults. Some have argued, with little foundation to date, that fluoridation could be the precursor to further mass medication. Those in favour argue that it is much cheaper than curing caries, and painless, and that there is no evidence that it has caused harm.

Toxicity manifests itself in fluorosis. Dental fluorosis is a mottling of the enamel and is associated with concentrations above 5 ppm. More severe fluorosis can appear as deformed bones. Fluorosis caused by drinking water that is naturally high in fluorine is found throughout the

Figure 5.9 Fluoride concentrations in surface waters

world. Its prevalence in China is illustrated in Figures 5.10 to 5.12 and analysed by Professor Wang Wuyi in his article.

Arsenic in groundwater

Arsenic is a matter for increasing concern in many parts of the world, especially in groundwater, where it is released by oxidation of pyrite as groundwater levels are lowered by over-exploitation or climate change. It is a major problem in the Indian subcontinent (see Amarendra K. Sinha's article below, 'Large-scale human exposure to arsenic in groundwater'): some 75 million people live in affected areas of Bangladesh and thousands suffer from chronic arsenicosis. Significant problems have also been reported elsewhere, notably in China, Argentina,

Chile and parts of Eastern Europe (Figure 5.16). More minor cases have been reported from Minnesota, Utah, Ontario, Nova Scotia, New Zealand and Japan. The medical effects continue to become more widespread and apparent and, as with so many pollutants, the 'safe' limits are being adjusted downwards: the EU reduced limits in drinking water from 50 to 10 µg/L (0.01 mg/L) in 2004, the US EPA followed suit in 2006, and New Jersey even went down to 5 µg/L the same year. This is also the WHO's currently recommended limit. In addition to drinking water, vegetables and rice irrigated with water polluted by arsenic create a further source of arsenic poisoning. The global coverage of GEMS/ Water data is sparse, but for Europe it does suggest that only five per cent of rivers exceed the EU drinking water standard (Figure 5.17).

The health effects of iodine and fluorine in drinking water in China

Wang Wuyi

Endemic goitre and endemic cretinism

Endemic goitre is caused by inhabitants drinking deep groundwater with a low iodine content. It occurs in some counties along the Bohai Bay in Hebei and Shandong provinces, and both deep and shallow groundwater in several low-lying interior areas of Xinjiang, Shanxi and Hebei (Figure 5.10). It may also be caused by eating kelp, kelp salt and its pickled products in some coastal areas.

Endemic goitre is widespread (Figure 5.11). It is highest in the mountain areas, including the Himalayas, eastern Pamirs and the Da Xinggan and the Changbai mountainous areas in Northeast China. It is relatively high on hilly areas like Nanling and the western Guangxi highland in South China, but generally low in the plains, and lowest on the coast. However, it also occurs in some ancient river beds on the plains, in peaty and swampy areas, as well as along parts of the coastal sandy belts. It is also prevalent in Shanghai.

Endemic cretinism occurs in most of the same areas as serious goitre and there appears to be a link (Figure 5.12). It is prevalent in all regions except Shanghai and Jiangsu Province. No cretinism is found in coastal areas even where goitre occurs. Clinically, endemic cretinism in most areas of China is of the nerve or mixture type. More patients suffering from myxoedema endemic cretinism are found in southern Xinjiang, eastern Qinghai and Liangcheng county of Nei Monggol.

Endemic fluorosis

Endemic fluorosis is a chronic disease affecting the whole body, including the teeth and the skeleton, which is related to high fluoride concentration in drinking water (Figures 5.13 to 5.15).

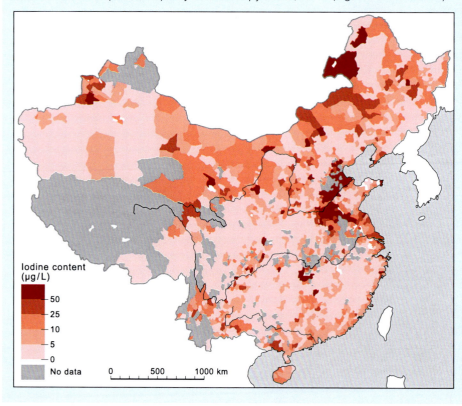

Iodine content
(µg/L)

- 50
- 25
- 10
- 5
- 0
- No data

0   500   1000 km

Figure 5.10 Iodine in water in China (based on *The atlas of endemic diseases and their environment in the People's Republic of China*, by Tan Jianan et al, 1989)

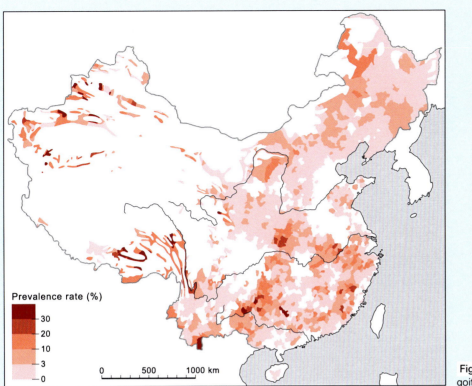

Figure 5.11 The incidence of goitre in China

Prevalence rate (%)

30
20
10
3
0

0   500   1000 km

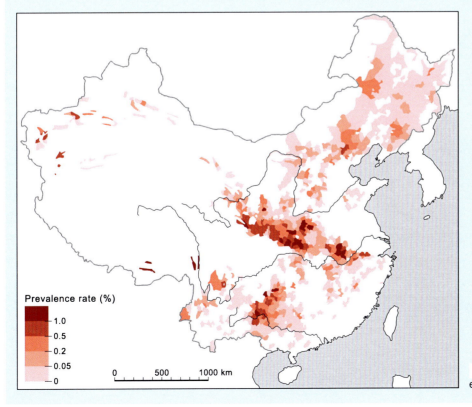

Figure 5.12 Prevalence of endemic cretinism in China

Prevalence rate (%)

1.0
0.5
0.2
0.05
0

0   500   1000 km

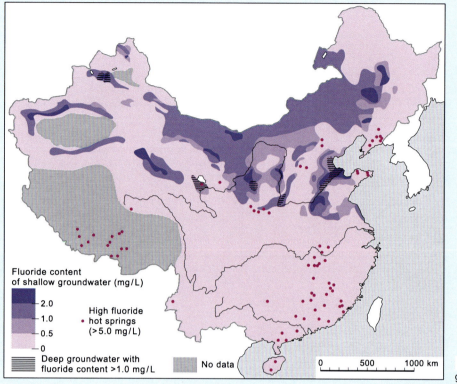

Fluoride content
of shallow groundwater (mg/L)

2.0
1.0
0.5
0

High fluoride
hot springs
(>5.0 mg/L)

Deep groundwater with
fluoride content >1.0 mg/L

No data

0   500   1000 km

Figure 5.13 Fluoride in shallow groundwater in China

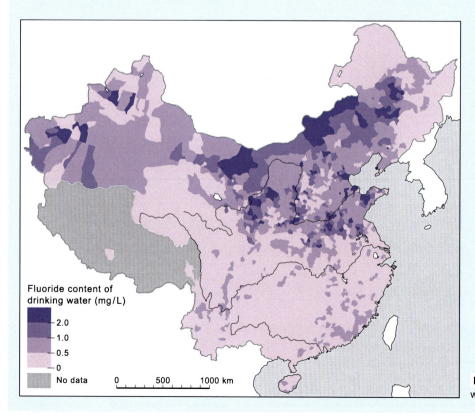

Fluoride content of
drinking water (mg/L)

2.0
1.0
0.5
0

No data

0   500   1000 km

Figure 5.14 Fluoride in drinking water in China

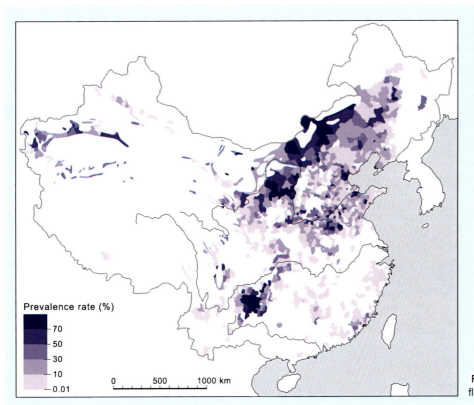

Figure 5.15 The prevalence of fluorosis in China

Endemic fluorosis is widely distributed throughout the world, covering over 40 countries. It is also extensively distributed in China. It is relatively serious in North China, ranging from Heilongjiang in the northeast to Xinjiang in the northwest. There is an uneven spread in the disease-affected areas. It is also prevalent in the vicinity of high-fluoride hot springs and areas where there is fluoride-rich coal and rocks, but the extent of such areas is rather limited.

The higher the fluoride concentration and the longer the exposure, the more serious is the disease. The teeth of children are affected readily up to the age of seven or eight years, but once the enamel is formed, dental fluorosis does not develop further. Skeletal fluorosis is more prevalent in women and the rate increases above the age of 20. People not native to a high-fluoride locality are more readily affected.

The fluorosis-affected areas in China can be roughly divided into five types of environment on the basis of geographical characteristics and sources of fluoride:

1 Shallow high fluoride groundwater type

This type affects the largest area in China and is mainly in the vast arid and semi-arid areas in North China. Here the groundwater derives fluoride from fluoride-rich rock formations, such as volcanic lava and granite, and the soils have relatively poor drainage, and suffer from fairly serious salinization and alkalinization. The climate is dry with a high evaporation rate, causing a further concentration of fluoride ions in the surface water and shallow groundwater, which may reach up to 32 mg/L.

2 Deep groundwater type

This is mainly found along the coastal plains, as well as in the southern part of the Junggar Basin in Xinjiang. The groundwater bearing strata are alternating marine and freshwater sedimentaries. Influenced by the palaeo-geographical environment, the fluoride concentration in the deep groundwater is very high, generally as high as 7 mg/L and occasionally in excess of 20 mg/L. Although the fluoride concentration in the shallow groundwater is not high, the water is undrinkable because it is bitter to taste due to the high salt content.

3 High fluoride hot spring type

This is mainly found on the periphery of continental plates and fracture zones. Fluoride concentrations in drinking water are high, owing to hot springs seeping into the surrounding areas.

4 Abundant fluoride rock formation type

This is mainly found in exposed sites of fluorite and apatite mines, where fluoride concentration is high.

5 High fluoride coal type

This occurs mainly in the mountainous areas of south-west China. These areas are characterized by low fluoride concentrations in the drinking water (all below 1.0 mg/L), but high content in food and indoor air caused by burn-ing the local coal. Since the climate is cold and wet, the local high-fluoride coal is used as the major fuel for cooking, heating and drying grain. As chimneys have not been built, the indoor fluoride concentration rises, and the fluoride content in the dried maize and chilli increases up to several hundredfold. Some researchers, however, maintain that the high fluoride content of the grain originates in the soil surrounding the local coal formation.

In some areas, fluorosis is also caused by eating salt with a high fluoride content.

Figure 5.16 Documented cases of arsenic problems in groundwater related to natural contamination (source: British Geological Survey, 2001, with additions)

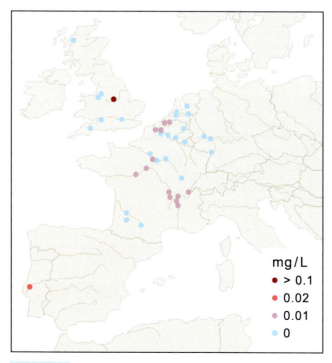

Figure 5.17 Arsenic in European rivers

A variety of hypotheses have been put forward to explain the problem. Most have some validity and in truth there are a number of causes, ranging from purely natural to totally man-made. In most cases, it appears to be the product of specific combinations of geological environment and water regime. Hydrothermal waters permeating through fractured granites in mountain ranges, like the Himalayas, and the weathering of volcanic rocks can release arsenic, which may then be carried by rivers and deposited in concentrated form in floodplain sediments. Post-glacial deltaic sediments are a major source, especially in the Bengal basin. In New Jersey, the black shales are considered a source. Alkaline waters generally favour arsenic release. However, human modification of the water regime is almost universally the trigger for enhanced groundwater pollution.

A key medical observation is that millions of people in some of the areas now badly affected drank well-water for thousands of years but with little evidence of toxic effects until the late twentieth century. This appears to imply a major effect on the environment due to increased human influence, most likely related to modern water management. In the Indian subcontinent, the problem was exacerbated by three factors which led to a dramatic increase in the use of groundwater as a supposedly safer source of drinking water and a major resource for expanding irrigated agriculture: 1) the worsening biological contamination of surface waters in the mid-twentieth century; 2) the Green Revolution in agriculture; and 3) the severe cholera outbreak in south Bengal in the 1960s. The World Bank was a principal promoter of the switch to groundwater and funded thousands of tubewells for irrigation from the 1960s onwards.

Arsenic may be released by oxidation of pyrite, arsenopyrite and clay minerals, which have arsenic adsorbed on their surface. Oxygen enters aquifers when the water table is lowered; the arsenic is then free to pollute the groundwater when the rocks are reflooded. Arsenic pollution may thus result from groundwater overdraft.

It has even been suggested that reducing surface water discharges can have a similar effect: diversion of the Ganges, the Tisha and 28 other transboundary rivers in India may have contributed to the problem in Bangladesh by reducing groundwater recharge. Even though millions of tubewells were installed in Bangladesh before the 1970s, medical records only show a significant rise in pathological symptoms during the 1980s in West Bengal and 1990s in Bangladesh. Some observers have linked this with the completion of the Farraka barrage in 1975, which diverted Ganges water to Calcutta and reduced the discharge entering Bangladesh.

A second hypothesis focuses on the opposite situation: excess water. In organic-rich clayey sediments, saturated, anaerobic conditions create a chemically reducing environment, which leads to the dissolution of iron oxyhydroxides containing sorbed arsenic. Iron-reducing bacteria may also play a part. This situation is typical of natural wetlands and could also be a dangerous by-product of over-irrigation.

Yet a third hypothesis blames pollution from human sources, especially from agriculture. Nitrate from fertilizer can oxidize sulphate minerals and release sorbed arsenic. And arsenical pesticides were widely used in America for 100 years until banned in the late twentieth century, although their low solubility suggests that they are not a major source of water pollution. Current research in Botswana's Okavango delta is trying to determine whether the calamitous arsenic pollution that has affected the town of Maun is due to excessive drawdown of the water table or human sewage (Huntsman-Mapila *et al*, 2006, in press).

Solutions

In recent years, some communities in the Bengal basin have begun to revert to using surface waters as the enormity of the problem has been realized. Yet, ironically, low-cost technology is already available for purifying well-water in the domestic environment. One method is to mix the water with ferric chloride and potassium permanganate to oxidize it in a plastic bucket and strain it through a cloth into a second bucket filled with sand to clarify it. Results reported to the United Nations University show arsenic concentrations reduced to less than 20 µg/L, even with water containing up to 500 µg/L (Ali *et al*, 2001). Many other sorbents can be used, especially laterite, a clay soil rich in iron and aluminium, providentially common in tropical regions. Education and government help are the keys to encouraging adoption.

'Reversion to surface water-based supply … will involve enormous financial outlay and would nullify the investment of around Rs 10,000 crores [*100 billion rupees or around $2 billion*] already made in West Bengal alone over the last 30 years. It is necessary that such major policy shift should not be *ad hoc*, but a science-driven decision.'

S.K. Acharyya, former Director General in the Geological Survey of India, 2002

Better still would be expansion of the public water supply systems in which filtration and coagulation methods are applied centrally. However, the best solution is to identify the specific causes responsible in each area and either to manage water usage appropriately or else to clean up at source. Considerable research effort is currently being devoted to this end, especially in India and Bangladesh. One promising approach is bio-remediation within the aquifer, introducing microbes to precipitate iron sulphide that will sorb arsenic.

Large-scale human exposure to arsenic in groundwater

Amarendra K. Sinha

Over the last 15 years or so, the widespread nature of human exposure to arsenic in drinking water has become apparent in many countries (Figure 5.16), especially in the Indian subcontinent. Groundwater high in arsenic has been reported from Bangladesh, India, Nepal and Pakistan, but the contamination in the Ganges delta region of Bangladesh and West Bengal represents the largest poisoning of a population in history, with millions of people exposed. In Bangladesh, 79.9 million people in an area of 92,106 km^2 are suffering from arsenic levels above the World Health Organization's maximum permissible limit of 50 µg/L, and a further 42.7 million in West Bengal, India, in an area of 38,865 km^2. Numerous sources of contamination have been cited: 1) natural – weathering of arsenic bearing mineral ores, arsenic rich geothermal waters, and concentration in accumulated sediments (in the Ganges delta and Bengal plain), or 2) human-induced – past mining activities, combustion of fossil fuels, chemical weapons dump sites and metallurgical processes.

Health implications

Drinking arsenic contaminated water has been found to increase the risk of disorders of the skin, cardiovascular and nervous systems, and especially the risk of cancer (Table 5.5).

The early stages of arsenic poisoning are characterized by dermatitis, keratosis, conjunctivitis, bronchitis and gastroenteritis. This is followed by peripheral neuropathies, hepatopathy, melanosis, depigmentation, hyperkeratosis. The most advanced stage is marked by gangrene in the limbs and malignant neoplasm (cancers).

Even when the concentration in drinking water is reduced to the World Health Organization's recommendation of 10 µg/L (WHO, 1993), potential cancer risks remain high (NRC, 2001). The effects are more acute in countries where people are poor, malnourished and drink groundwater.

Conclusion

There is no technological 'magic bullet' that will solve the arsenic problem quickly and easily. Although our understanding of the problem has improved in recent years, many uncertainties and wide information gaps still exist. Numerous studies have shown that arsenic may be successfully removed from solution by adsorption onto clays, co-precipitation on laterites or metal ion precipita-

Table 5.5 Pathological effects associated with arsenic poisoning

Organ system	Problems
Skin	Dermatitis, symmetric hyperkeratosis of palms and soles, melanosis, depigmentation, Bowen's disease, basal cell and squamous cell carcinoma
Liver	Enlargement, jaundice, cirrhosis, non-cirrhotic portal hypertension, hepatopathy, cancers of liver and gall bladder
Nervous system	Peripheral neuropathy, hearing loss, conjunctivitis
Cardiovascular system	Acrocyanosis and Raynaud's disease, hypertension
Hemopoietic system	Megaloblastosis
Digestive system	Gastroenteritis, vomiting, diarrhoea
Respiratory ystem	Cancers of lung and nasal passage, bronchitis
Endocrine system	Diabetes mellitus and goitre
Urinary system	Cancers of prostate and bladder
Limbs	Gangrene, muscular cramps, collapse

tion. However, we still need a much better understanding of arsenic migration processes, accurate and low-cost field water-testing methods, a good understanding of the health effects and dose-response patterns of chronic and low-level arsenic exposure, possible clinical treatment methods, and efficient and practical arsenic removal technologies. Accelerated research efforts are urgently needed. Yet the immediate challenge is how to take effective action without delay amid the many unknowns.

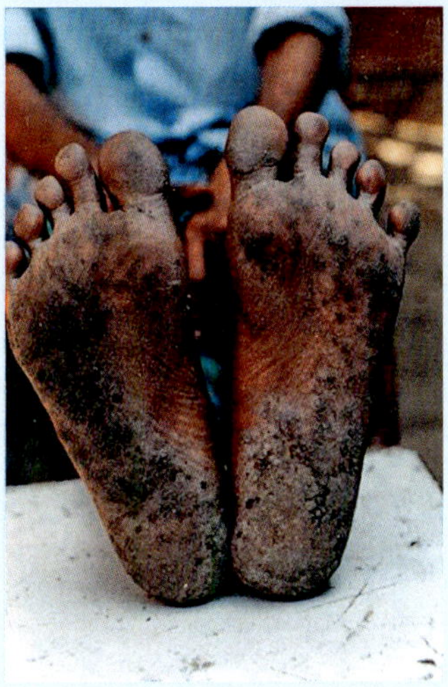

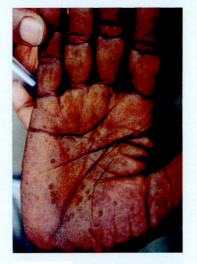

Figure 5.18 Keratosis on the feet (Courtesy: Arsenic Foundation Inc.)

Figure 5.19 Melanosis of the hand (Courtesy: Arsenic Foundation Inc.)

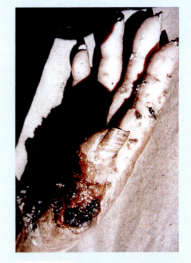

Figure 5.20 Carcinoma of the hand (Courtesy: Arsenic Foundation Inc.)

Part 2 Infectious diseases

Water-related infectious diseases are mostly spread via two routes: from the bites of animal 'hosts' or vectors carrying the disease, or via the 'faecal-oral' route. Appropriate water and wastewater management can play a vital role in reducing the incidence of both. In the second half of the twentieth century, spraying water bodies with pesticides became the preferred option for controlling vectors, but avoiding low water levels, which promote warmer water, and slack flows and stagnant water, which favour mosquitoes, or waterfalls that favour black fly, are more eco-friendly approaches. Installing adequate sewage treatment systems, along with hygiene education, is the main route to preventing faecal-oral disease. Poor sanitation can also increase the number of vector animals. Many vectors, like mosquitoes, only become infectious once they have bitten a human carrying a disease. Growing population densities increase the chances of infection in tropical and subtropical countries, but the low level of baseline infection mitigates against this in more northerly countries. Other vectors, like the water snails which carry the flukes that cause schistosomiasis, acquire the agent of disease without human contact.

'The most practical investment we can make in global public health is plumbing.'

John Sauer, Communications Director, Water Advocates, Washington, DC, 2009

No-plumbing disease

Almost four children die every minute from poor sanitation and water supply. The World Bank estimates that two million people die of sanitation-related disease every year, 90 per cent of them children. John Sauer coined the term 'no-plumbing disease' in an internet article he published in January 2009. Sauer is Communications Director for Water Advocates of Washington, DC, a non-profit organization devoted to publicizing water problems. He points out that clean water and sanitation receive far less attention and funding than malaria or Aids, but they kill more children than Aids/HIV, malaria, measles and tuberculosis combined. In total, some two million deaths are linked each year to unsafe water and poor sanitation and hygiene, equivalent to ten 2004 Indian Ocean tsunamis and 1000 Hurricane Katrinas (Table 5.6). Sauer argues that if all the diseases associated with them were lumped together as one disease, then they might be taken more seriously. 'No-plumbing disease' comprises more than 25 separate diseases, including cholera and typhoid, campylobacter enteritis, giardia, amoebic dysentery, bacillary dysentery (shigellosis), Weil's disease, hepatitis A and *Escherichia coli* diarrhoea. Poor sanitation and unsafe water fill half the hospital beds in the developing world and account for 10–15 per cent of the entire global disease burden.

In a poll conducted by the *British Medical Journal*, basic sanitation, good hygiene and access to safe drinking water were rated the most important advances of the twentieth century. But not everywhere is benefiting from those advances. In 2008, the WHO report *Safer water, better health* calculated that providing safe water, sanitation and hygiene education could save healthcare agencies $7 billion pa, and gain 320 million productive working days and 272 million extra school days a year. The overall payback would be $84 billion from the $11.3 billion annual investment needed to meet international drinking water and sanitation targets. WaterAid estimates that in Madagascar alone a million working days and 3.5 million school days a year are lost through ill health caused by unsafe water and sanitation.

Diarrhoea has many causes, mostly related to water and food contaminated with a range of diseases stemming from poor hygiene and sanitation. The same applies to gastroenteritis, which is inflammation of the gut producing vomiting and diarrhoea. Infections like cholera and dysentery are common causes, but some causes are not related to water at all.

Amoebic dysentery is a tropical disease caused by ingesting the single cell *Entamoeba histolytica*, derived from excrement. Far more common, but less serious, is bacillary dysentery, which occurs in all climates and is caused by *Shigella* bacteria.

Infective hepatitis A is commonly caused by bathing in faecally-contaminated water. It causes inflamed liver, jaundice and debilitation, and may be fatal to old people. Weil's disease or infectious leptospirosis, commonly known as swamp fever, can also be contracted by bathing in contaminated water, especially stagnant water. It is caused by a spirocaete, *Leptospira icterohaemorrhagiae*, spread from animal urine entering water bodies, usually from rodents but occasionally cats and dogs. There is a 20 per cent mortality rate through damage to the liver and kidneys.

Giardiasis is yet another disease that originates from exposure to faecal material. It is rare in developed countries and where it is endemic it can be relatively mild, but it can cause diarrhoea and intestinal gas. It is caused by the single-celled *Giardia lamblia*.

Faecal coliform

Concentrations of faecal coliform are the best indicators of organic pollution of rivers by human and animal faeces. For drinking water, the WHO, the US EPA and the EU all set a standard of zero faecal coliform per 100 ml. Faecal material can carry not only bacteria like the ubiq-

Table 5.6 Deaths due to water-related diseases and poor water management. Percentages of total deaths according to WHO (2008)

	WHO region						
	Africa	Americas	Europe	SE Asia	Western Pacific	Eastern Mediterranean	All regions
*Due to unsafe water, sanitation and hygiene**							
Diarrhoeal diseases	55	55	44	60	39	62	56
All diseases	72	66	53	78	45	85	73
Due to poor water management							
Water-related diseases†	21	3	0	3	4	4	12
Drowning‡	3	14	35	7	41	6	8

Data relate to 2002.
Key:
* No-plumbing disease.
† Diseases carried by vectors: mainly malaria, dengue, onchocerciasis and Japanese encephalitis.
‡ Most but not all due to floods.

uitous *Escherichia coli* (*E.coli*) and *Salmonella*, but also viruses like *Rotovirus*, parasitic worms like the tapeworm *Taenia saginata*, and the protozoan *Cryptosporidium*. All are prime causes of gastroenteritis and diarrhoea.

Cryptosporidium oocysts live in the intestines of livestock and can be washed from animal faeces into rivers and reservoirs. Cryptosporidiosis causes acute diarrhoea, vomiting and fever for a week or two. It has caused a number of public health emergencies in Britain and America in recent years, breeding in the warm water of reservoirs in summer. The 1989 incident at Farmoor Reservoir, Oxfordshire, which infected over 400 people, originated from runoff from fields spread with farm slurry. There were 575 cases in Torbay in August 1995. Part of that problem may have been the water company's decision to recycle water through the treatment plant to save water during the drought, and not passing it through all the treatment processes. In Milwaukie in 1993, 100 people died and half the city population of 800,000 fell ill. Both cryptosporidiosis and giardiasis are essentially diseases of surface water: well water is generally safer.

Over the last decade or so, there has been a better understanding of the biology of infections and monitoring using genetic fingerprinting is helping to control the risks. New regulations were introduced in England in 1999, requiring risk assessments at all treatment plants. Since 2001, companies are required to limit the average number of oocysts to below one per ten litres and health authorities must alert water companies after 28 cases have been reported. Noncompliance is a criminal offence. Welsh Water was fined £60,000 for incidents in North Wales in 2005–6. But the problem often returns in warm summers, as in East Anglia in 2008 and in Cornwall and North Wales in 2009. It is resistant to chlorination and most oxidizing agents. The water must be boiled for a number of minutes or else passed through 1 μm filters. Cleaner pipes, better monitoring and warning alarms are helping. South West Water is modernizing 7500 km of mains and 1500 km of pipes and fitting turbidity alarms.

The map in Figure 5.21 shows that barely three per cent of the surface waters recorded in the GEMS/Water global archive typically have less than one faecal coliform per 100 ml and only around ten per cent have under 100 per 100 ml. The most extreme pollution, in excess of half a million faecal coliform per 100 ml, is found in just a few locations in Ecuador, mainland Malaysia, Java and Belgium. Sadly, data for Africa are totally inadequate and it is likely that many more sites there and elsewhere

are beset by high levels of faecal pollution, especially in cities and shanty towns.

The map (Figure 5.21) shows that high levels of pollution are by no means a preserve of developing countries. Despite some of the most rigorous environmental controls in the world, the European Union still has many rivers with high mean concentrations. Only six per cent of sites in the pre-2002 Union average less than 100/100 ml and around a quarter show in excess of 10,000.

Cholera

Cholera is an acute infection of the gastrointestinal tract contracted through the faecal-oral route from contaminated drinking water or food. It is caused by the *Vibrio cholerae* bacterium. Death can result from chronic loss of water and electrolytes as a result of intense diarrhoea and possible kidney failure.

Cholera killed over 36,000 Londoners in four epidemics during the nineteenth century. The cholera outbreak in London's Soho in 1854 was a landmark event in medical history. Nearly 11,000 died from cholera after drawing water from a public well near the surgery of Dr John Snow. Snow identified the pump and persuaded the council that it was polluted by sewage and should be disabled. He also noted that the workers at the nearby brewery, which had its own water supply, were unaffected. He had already published a paper 'On the Mode of Communication of Cholera' five years earlier, in which he also argued that the practice of flushing sewage from cesspools and 'water-closets' (flush toilets) into rivers made cholera epidemics worse. After another epidemic, which killed over 5000 in Whitechapel in 1866, caused by sewage contamination in the East London Water Company's reservoir, the government finally accepted his findings and Joseph Bazalgette, chief engineer to the Metropolitan Board of Works, proposed extending his groundbreaking network of mains drainage throughout the city.

The events had interesting impacts on the 'private or public' debate, both for and against. In his report on the Whitechapel epidemic, William Farr, statistician to the Registrar General, railed against the public blame placed on air pollution for the outbreaks. 'As the air of London', he said, 'is not supplied like water to its inhabitants by companies the air has the worst of it … while Father Thames and the water gods of London have been loudly proclaimed immaculate and innocent.' Yet it was not till the passage of the Metropolis Water Act in 1902 that London's water supplies were removed from private

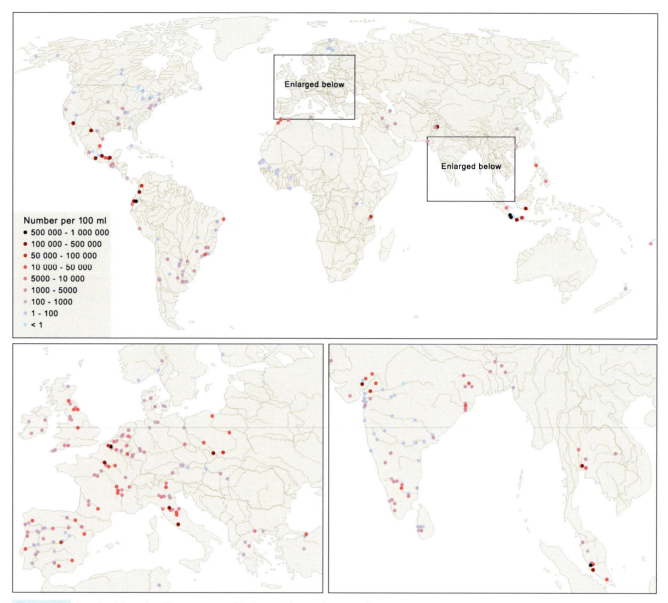

Figure 5.21 Levels of faecal coliform, a general indicator of water-borne pathogens

companies, partly perhaps because Bazalgette's sewer system had been so effective in controlling cholera, partly because of strong opposition from the companies (Halliday, 2004).

'If the general use of water-closets is to increase, it will be desirable to have two supplies of water in large towns, one for the water-closets and another, of soft spring or well water from a distance, to be used by meter like the gas.'

Perspicacious observations by Dr John Snow in the *Medical Times and Gazette*, 1858, after the cholera epidemic in Soho, London

The disease is associated with brackish water and river estuaries, as well as poor management of water and sanitation. It appears to have originated in the Ganges delta. Cholera is often associated with algal blooms, which are a common indicator of high levels of plant nutrients in the water caused by agricultural fertilizers or faecal pollution.

Cholera is yet another disease that is now most prevalent in sub-Saharan Africa and India (Figure 5.22). South Africa is worst, but Mozambique and the Democratic Republic of Congo are close behind. It tends to occur in discrete pandemics, of which seven major ones have been identified since the nineteenth century. The WHO Global Task Force on Cholera Control has

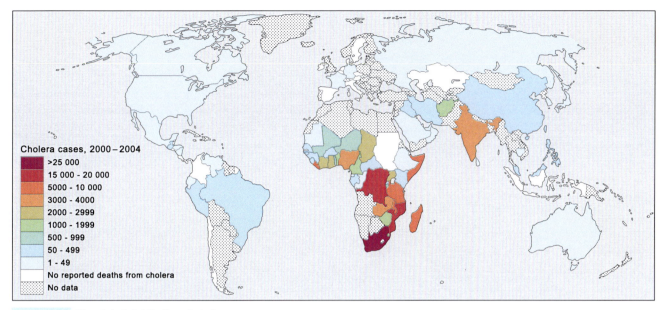

Figure 5.22 The global distribution of cholera

made great strides in cholera control through hygiene education since it was launched in 1992, matched by improvements in the provision of sanitation around the world. However, all the campaigns have not kept pace with the rise in the number of people without access to basic sanitation caused by burgeoning populations and by wars and natural disasters that disrupt water supplies and sanitation and cause thousands of refugees to be confined in overcrowded conditions. Cholera has re-emerged as a major problem since 2005. The WHO reports a 79 per cent increase in one year alone between 2005 and 2006, bringing the number of cases back to levels last seen in the late 1990s. The WHO recorded nearly a quarter of a million cases and more than 6000 deaths in 52 countries in 2006. Major outbreaks also occurred in Thailand, Iran, Iraq, India, eastern China and Zimbabwe between 2007 and 2009. In Zimbabwe, the unreliability of public water supplies has led many to dig shallow wells, which are prone to faecal contamination. As a result, a cholera epidemic broke out in August 2008 infecting over 100,000 and killing more than 4000. Although officially declared over the following summer, another 150 lives were lost by early 2010 and there is every reason to suppose that this will be a long-running issue.

Unfortunately, the WHO estimates that it only receives information on perhaps ten per cent of actual cases. Global data are therefore highly unreliable and in all probability a gross underestimate of the real situation.

Typhoid

Typhoid fever is caused by ingesting the bacterium *Shigella typhi* through drinking water or food. It is not to be confused with typhus, which is spread human-to-human by body lice hosting a rickettsial bacterium. Despite the availability since the mid-twentieth century of two effective vaccines, the WHO still regards it as a serious public health problem and it is a factor in its recommendations for local Water Safety Plans, to manage drinking water 'from catchment to consumer' (WHO, 2005). It estimates that on average 21 million people are infected annually and over 200,000 die. With good antibiotic treatment barely one per cent should die, but without it up to 20 per cent could die. Poverty and poor water supplies mean that the disease is most fatal in the Least Economically Developed Countries. Recent outbreaks have occurred in the Democratic Republic of Congo 2004–5, infecting 42,564 and killing 214, and in 2010 in Chandigarh, India, on Fiji and in Harare, Zimbabwe.

For the same reasons as cholera, it is particularly common in disaster areas and there were fears of an outbreak of typhoid and cholera in Haiti as the rainy season arrived after the devastating earthquake early in 2010.

Poliomyelitis

Polio is a curious case of a disease that is seen as due both to lack of plumbing and too much plumbing! The

Drink beer, not water

Many breweries were established in the east end of London during the nineteenth century by Christian Protestant Nonconformists, in apparent contradiction of contemporary puritanical views on alcohol consumption. The reason was simple: by killing most pathogens, the fermentation process and the acidity of the beer made it safer to drink than the water from the city's rivers and wells. It is still a sound principle to follow in many parts of the world today.

The practice is time-honoured, and appears to have acquired a religious following at least 6000 years ago with a recipe for beer-making appearing in the Sumerian 'Hymn to Ninkasi', a goddess thought to reside in Mount Sabu, literally 'the mountain of the tavern-keepers' (Hart-

Davies, 2004). Washing with beer and hot water was recommended for disinfecting wounds.

Beer might even be a by-product of ancient irrigation practice. A Sumerian farming handbook *The Farmer's Instructions*, dated at around 1500 BC, contains the oldest known reference to solving the 'inevitable' consequence of irrigation in a warm climate, salinization: flushing and leaching the soil with large quantities of freshwater. But the technique was no more successful then than Saddam Hussein's flushing of the Euphrates Marshes in the 1990s. The solution was to plant barley rather than wheat and other grains that are less salt-tolerant. And barley is the basic ingredient of beer.

virus is passed through faeces and faecal contamination of water and causes permanent paralysis of the limbs. But the prevalence of polio in developed countries during the first half of the last century has been blamed, ironically, on *improvements* in public water supplies. Repeated exposure to the virus appears to offer some immunity, so that travellers to areas where polio is endemic tend to be at greater risk than the inhabitants. Loss of immunity through drinking cleaner water, it has been argued, increases susceptibility. The incidence peaked in the 1950s, when public swimming pools were often closed when risks were judged to be high – the incidence was very seasonal, rising significantly during the warm, summer months.

Polio has all but disappeared in most of the world. The development of the Salk vaccine inoculation and subsequently the Sabin oral vaccine were milestones in its eradication, helped by campaigns like the 'Stop Polio' campaign pioneered by Save the Children. However, this also poses a medical problem, because little research has been conducted since the 1950s. It means that medical knowledge is very limited compared with most other diseases. Worse, there has been a marked resurgence in India due to contamination of drinking water by faeces and urine, caused by open defecation in the forest and faeces in open ditches beside the road: 660 million people defecate in the open. There was a 26 per cent increase in the number of cases in India between 2008

and 2009, mainly in Uttar Pradesh and Bihar. In parts of Uttar Pradesh less than two per cent of households have a toilet.

There is now a determined move to improve sanitation in India. In the State of Haryana, north west of Delhi, a 'No toilet, no bride' scheme was launched in 2005 to encourage men to install a toilet before taking a bride. The State has built over 1.4 million toilets. Dr Bindeshwar Pathak, who founded the Sulabh International Sanitation and Social Reform Movement, which has built toilets for tens of millions in India, received the 2009 Stockholm Water Prize for developing eco-friendly and cheap toilets. The Indian government also advises people not to drink surface water or even very shallow groundwater that might be polluted by faecal material: wells should be at least 70 feet (21 m) deep to avoid contamination.

India is one of only four countries that have a serious current problem; the others are Nigeria, Sudan and Yemen. In total, 300,000 are estimated to be infected, but one billion people could be at risk.

Guinea worm (Dracunculiasis)

This is a thin parasitic worm that is a serious threat in parts of Africa. The larvae of *Dracunculus medinensis* live in copepod crustaceans, minute water fleas that lack a carapace. The fleas are extremely common in freshwater and they can be swallowed in untreated drinking water. Adult worms live in the leg muscles and can grow to 100 cm long. They cause ulceration and can incapacitate the sufferer. It is a disease of poverty and lack of safe drinking water. Through incapacitating adults it also contributes to poverty and can put extra burden on children to do adults' work.

The Global 2000 Programme supported by UNICEF and the WHO has made great strides in containing the disease. Larvicides are effective in the environment but there is no cure, short of physical or surgical removal for the adult worms. Education is a major weapon. Filtering the water with a cloth before drinking will remove the fleas. Drinking only deep well water rather than surface water is also advisable, as is preventing infected persons washing and relieving their leg pain in public pools, because the worms can release a milky liquid containing millions of larvae into the water.

In 1986, 3.5 million people suffered from the disease, but by 2007 there were less than 10,000 reported cases. The maps (Figures 5.23 and 5.24) show that it used to be problematic in the Indian subcontinent, but receded

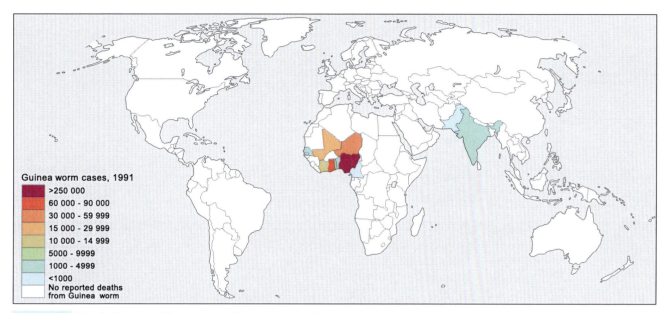

Figure 5.23 The distribution of Guinea worm disease at the beginning of the 1990s

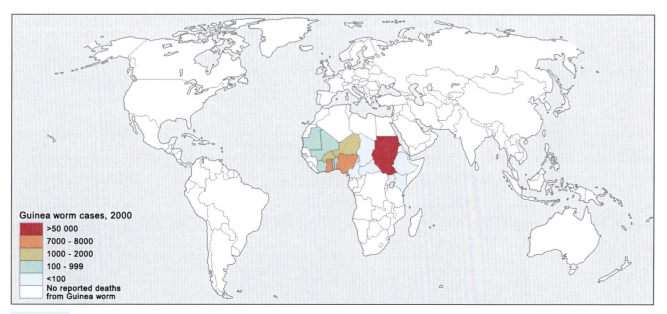

Figure 5.24 The distribution of Guinea worm disease at the beginning of the twenty-first century, showing a marked retreat

87

during the 1990s. Asia is now essentially free of the disease. The WHO certified Pakistan free in 1996, India in 2000 and the Yemen in 2004. It is also less prevalent in West Africa, but became a serious problem in the Sudan around the turn of the century. Sudan now accounts for over 60 per cent of all cases. Protracted civil war and a lack of provision from central government have been factors there. It is now largely confined to the Sahel. Mali, Ghana, Niger and Nigeria are the only other remaining countries where dracunculiasis is a problem. Senegal was certified free in 2004, Cameroon and the Central African Republic in 2007.

Vector-borne diseases

Malaria

Malaria is one of the most prevalent water-related diseases on Earth, second only to diarrhoea and dysentery. Every year between 350 million and 500 million new cases are recorded, between one and three million people die from it, and 40 per cent of the world population are judged to be at risk. A baby dies from malaria every 30 seconds and nearly three-quarters of a million children under five years old die from it every year. Malaria is still predominantly a disease of tropical and subtropical Africa, despite significant improvement over the last two decades (Figures 5.25 and 5.26). Around 90 per cent of all malaria cases are in sub-Saharan Africa, where 3000 children die each day: it is a major contributor to a child mortality rate of 30 per cent. More Africans die daily from malaria than from any other cause and it is currently estimated to account for 40 per cent of the public health spend in sub-Saharan Africa and to cost Africa's economy £6.8 billion pa in lost GDP. Yet it is in retreat at the global scale. It was finally eliminated in Europe in the latter part of the twentieth century. It has not been common in Britain since the 1880s, but it was not totally eradicated from Kent and Essex until after the First World War. The maps show a significant contraction in the area at risk even since 1990. And even in countries where it is still prevalent, the maps show that the number of cases has fallen since the 1990s. However, although the total number of malaria cases peaked in the 1990s, the most severe form, falciparum malaria, did not peak till 2000.

The retreat is due to a number of factors: improved standards of living, improved water and sewage management, improved medical cover and drugs, and the spread of bed nets.

However, there are also many reasons for being less optimistic: developing resistance to antimalarial drugs, the spread of poverty and failed states, the continued spread of other diseases, like Aids and malnutrition, that weaken the immune system, and climate change and natural disasters.

Thus, the worst death rates are not in the hottest or the wettest regions, but appear to be partly related to levels of malnutrition, care and antimosquito measures. The western Sahel is especially badly affected, with the highest rates of all in Mali and Niger, closely followed by Sierra Leone, Liberia, Guinea, Upper Volta and Chad. In southern Africa, Angola, the Congo and Mozambique show the highest rates, but these are roughly half as severe as in West Africa. In contrast, there have been marked reductions in the number of cases in Thailand, and China, Argentina, Paraguay and Indonesia are now largely free.

A prime example of the effect of a failed state occurred in Zimbabwe in 2009 when the government health service collapsed through lack of funds and the free supply of mosquito nets stopped, as did chemical sprays in public places. There were real fears of a major malaria outbreak.

The ecology of malaria

The disease is caused by one of four species of *Plasmodium* parasite in the red blood cells transmitted in the bite of a female *Anopheles* mosquito that has bitten a person carrying the gametocytes – immature parasites. The parasites multiply in the liver and blood. Each species has its own characteristic symptoms, with periodic or more continuous fevers, but *Plasmodium falciparum* is the most serious and can infect the brain and kidneys and cause death. As they reproduce in the body, each new generation of parasite causes a fever attack.

The mosquitoes breed in stagnant or slow-moving water. While the females drink blood, the males only drink plant juices.

Developing strategies for elimination

There are two standard approaches: to eliminate the host mosquitoes or to treat the infection. Probably the most cost-effective control is to eliminate the shallow, stagnant water where the carrier mosquitoes breed. This was well understood as long ago as the ancient Greeks and Romans. The Emperor Trajan began drainage of the Pontine Marshes in Central Italy around 100 AD in

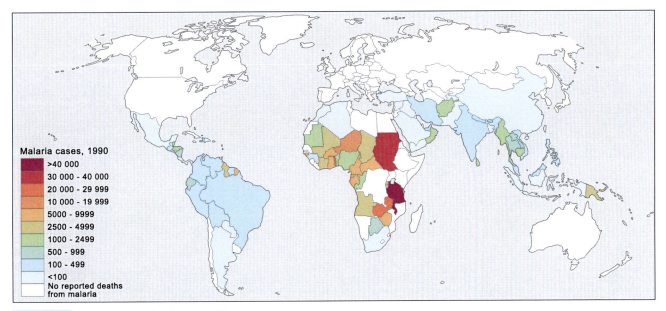

Figure 5.25 The incidence of malaria in 1990

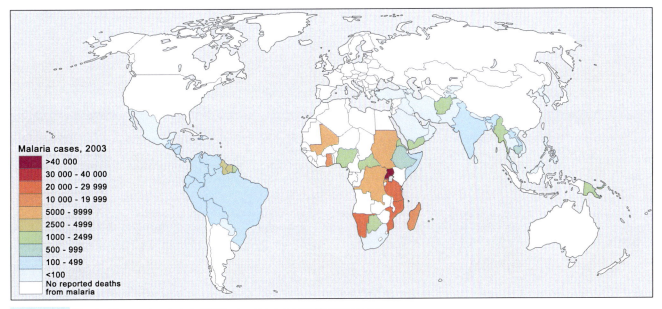

Figure 5.26 The incidence of malaria in 2003, showing a marked improvement

order to eliminate the malaria that had caused depopulation of the area – a scheme only finally completed by Mussolini in 1926. The English Fenlands were drained progressively from Roman times partly to eliminate malaria. But the view that wetlands are inherently refuges for pestilence has also led to the demise of many of the world's wetlands, to the detriment of flood control and wildlife (see Chapter 6 Water, land and wildlife).

Since the invention of insecticides in the twentieth century, the preferred approach in many parts of the world has been to spray water bodies with insecticide from the air at the beginning of the breeding season. This is still the most widespread solution, despite polluting the watercourses and destroying water resources. A wide range of insecticides is employed including organophosphates like malathion and naled, natural pyrethrin, and synthetic pyrethroid-based permethrin and resmethrin. The juvenile growth hormone methoprene, which is non-toxic to humans, is also used in drinking water cisterns to regulate mosquito development. Spraying has increased in the USA since the arrival of West Nile fever in 1999.

Weston *et al* (2006) report tests of the impact on other aquatic life of aerial spraying of pyrethrin and the synergist piperonyl butoxide around Sacramento, California. Although the effects were limited, the authors noted that the sprays doubled the toxicity of the river sediments for the amphipod *Hyalella azteca*, a shrimp-like animal often used as an indicator of pollution. These 'Mexican freshwater shrimps' or scuds live in the bottom sediments, which were found to be storing pesticides from previous applications in the urban area.

Modern medical approaches face two challenges: increasing resistance to drugs and lack of health care workers. *Plasmodium falciparum* is the first malarial parasite to develop resistance to chloroquine, the main antimalarial drug since quinine was phased out. Fortunately, to date *Plasmodium vivax*, common in North Africa, has yet to develop resistance. A WHO World Health Report in 2006 noted a 'serious lack of health workers' in 36 of the 57 countries in sub-Saharan Africa and estimated that four million extra workers are required.

The drug manufacturer GlaxoSmithKline is planning a totally new approach with the first antimalarial vaccine, Mosquirix, due to start trials in East Africa in 2011–2015.

A more radical approach is using genetic modification. Water Aid is championing genetically modified mosquitoes. Initial field trials are being run in India. Research at Imperial College London focuses on producing sterile males to reduce reproduction, while at Johns Hopkins researchers have produced mosquitoes that produce a protein that blocks the malarial infection. Another approach is to genetically modify the mosquitoes' food. Research at the University of Florida has focused on producing genetically engineered chlorella algae, a green slime that is a favourite food of mosquitoes, to replace DDT. The algae have been modified to produce a hormone found in mosquitoes' digestive systems to make them feel full. They die of malnutrition. The hormone is natural and enters the food chain naturally when the mosquito is eaten by higher animals. The algae do not remain indefinitely in the environment, but disappears naturally when eaten by other water life or killed by viruses that attack it.

A more ecological solution comes from the Indian Ministry of Environment and Forests that proposed using frogs to control mosquitoes. The scheme gets farmers in Haryana and Uttar Pradesh to catch frogs and put them into over 500 rivers and lakes around Delhi, especially the Yamuna River. Over the last 20 years that river has been highly polluted from sewage, pesticides and industrial effluent, which has killed the frogs and increased the risk of malaria in the city. River pollution has also killed other predators – spiders, fish and birds – and the year 2006–7 saw a 45 per cent increase in the mosquito population. Other approaches in India include traps that lure females of the *Aedes* mosquito to deposit their eggs in them, and introducing mosquito-eating fish like the guppy and *Gambusia affinis*. The same approach to mosquitoes is being taken in California where they are stocking newly constructed wetlands with *Gambusia* fish in preference to other chemical or biological solutions.

A very simple solution is provision of latrines. Christian Aid has been collaborating with a local community movement to build latrines in Nicaragua to combat malaria and cholera, with dramatic effect.

As a more high tech solution, the EU has funded the Africa Monsoon Multidisciplinary Analysis (AMMA), which aims to improve prediction of the West African monsoon. The monsoon brings the heavy rains that produce the spawning grounds for the mosquitoes and it has been difficult to predict recently. Satellites can also be used to identify areas of recent rainfall to help direct spraying operations. The satellite approach was first successfully employed in the 1970s to track likely sources of locust infestation in north east Africa.

Moves to reverse the loss of wetland habitats and to reduce the amount of pesticides sprayed into water bodies, and even to recreate wetlands to improve biodiversity and serve as natural flood control measures are all acting against the established approaches to controlling mosquito populations. Since these recent trends all have strong merit (see Chapter 6 Water, land and wildlife), the development of alternative biological and medical approaches is to be welcomed. In the last resort, this includes new methods for quicker diagnoses and treatment, such as the infrared sensor system developed at McGill University by Professor Paul Wiseman, which makes secretions from the parasite glow blue in the blood cells, as well as the design of robots to undertake repetitive pathological analyses produced by the robotics team under Professor Ross King of Aberystwyth University (King *et al*, 2009).

Other diseases transmitted by mosquitoes

Mosquitoes are vectors for a wide range of diseases well beyond the geographical confines of malaria. In southern

Sweden, where they are reconstructing wetlands near settlements, there is concern about possible mosquito and midge-borne Sindbis or Sindbis-like virus, which causes Ockelbo disease. The disease was identified in the 1960s in Ockelbo, Sweden, and spread to Finland where it is known as Pogosta or Karelian disease. It causes fever and lesions on the joints.

Of more widespread concern, however, are dengue fever, yellow fever, Rift Valley fever, chikungunya virus, filariasis and West Nile fever.

Dengue fever killed 60 people and infected 3000 after a plague of mosquitoes hit Delhi in 2006. The frog-stocking scheme (mentioned earlier) was introduced to protect against dengue as well as malaria The virus is currently spreading outside the tropics, following the expansion of mosquito habitats. Classic dengue fever resolves itself in a few weeks, but the haemorrhaging variety is frequently fatal.

Yellow fever is caused by a virus that is endemic in monkeys in Africa and South America and is transmitted to humans by the bite of the female *Aedes Egypti* mosquito, which breeds in shallow pools and swamps. It may be mild in people normally resident in these areas, but it can also be fatal.

Rift Valley fever mainly infects livestock, but is also spread to humans by mosquito bites. It is most common in years of heavy rainfall and floods in sub-Saharan Africa, especially in the East African Rift Valley, and Madagascar. It can cause hundreds of deaths through inflammation of the brain. Its appearance in the 1970s in Egypt was blamed on the Aswan High Dam, and similar outbreaks have been reported around the Diama Dam in Senegal and Volta Lake in Ghana (McSweegan, 1996).

Chikungunya virus is carried by mosquitoes of the *Aedes* species, *Aedes Egypti* and *Aedes albopictus*, and is prevalent around the Indian Ocean from Mozambique and Tanzania to India and as far as Thailand. The virus causes a highly debilitating fever, headaches and joint pain. A severe outbreak in 2006 affected a third of the population of Réunion, with 5000 new cases a week and 155 deaths. On Mauritius, the government regularly operates a programme of insecticide spraying, along with a public education programme. Over a million people were infected by the virus in mainland India in 2006 and there have been recent outbreaks in Madagascar and the Seychelles.

Filariasis is a common disease of tropical climates caused by tiny nematode worms (filaria). The larvae are transferred by a variety of mosquito species to humans, where they mature and settle in the lymph nodes. These block the lymph vessels and cause oedema or elephantiasis (gross swelling of the legs or arms).

West Nile fever was first identified in Uganda in 1937 and remained in the 'old world' – Africa, the Middle East and western Asia – till 1999 when it jumped to New York City. It has now spread throughout the conterminous USA into Mexico and parts of the Caribbean and into seven provinces of Canada. It is caused by a flavivirus that is carried by birds and transmitted by a number of different species of mosquito. It produces encephalitis and occasionally fatal inflammation of the brain. There were 720 cases in the USA in 2009. The US Geological Survey recommends eliminating standing water and cleaning out gutters. The US Air Force now has a special unit for aerial spraying in emergencies, with particular regard to West Nile fever.

Even when not transmitting a disease, mosquitoes can be problematic. In 2007, more than a thousand residents took legal action against Thames Water over their management of the large sewage plant at Mogden, which serves 13.6 million customers in West London. The litigants claimed it was responsible for a plague of *Culex pipiens* that had caused some people to be hospitalized. Numbers of mosquitoes increased dramatically that year over a wide area of southern Britain following warm, wet weather, and the National Health Service reported nearly a third more calls over mosquito bites in August.

Tick-borne diseases

Although ticks do not breed in water, in some parts of the world changing rainfall patterns are making it easier for pathogens to survive and spread, and changes in water availability are bringing farm animals into more frequent contact with wildlife that carry the ticks. Tick bites can spread cholera, Lyme's disease and babesiosis, a malaria-like disease spread by *Babesia* protozoa.

River Blindness (Onchocerciasis)

Like malaria, river blindness is a disease that is carried by vectors that breed in water. Unlike malaria, however, it is associated with waterfalls and fast-flowing rivers, where oxygen levels are high, rather than swamps. It still affects around 37 million people worldwide, and 300,000 have been permanently blinded, despite one of the most successful eradication programmes in history.

No one appears to have died from the disease directly, but it shortens life by a decade or more because of its effects on the immune system.

The infectious agent is a parasitic worm or nematode, *Onchocerca volvulus*, carried by blackfly of the genus *Simulium*, which lay their eggs in the water. The disease is spread from person to person by fly bites. Female flies ingest worm larvae by biting infected people and release macrofilarial worms into the flesh of the next person they bite. These adult worms then release minute microfilariae. It causes nodules to develop predominantly in subcutaneous tissue, accompanied by itching and depigmentation of the skin. If the microfilarial worms reach the eyes, they cause impaired eyesight and eventual blindness. It is the world's second largest infectious cause of blindness. The disease has been especially problematic in West Africa, especially Nigeria, where whole settlements have been abandoned in the river valleys and re-established on higher ground (Cliff *et al*, 2004). The high incidence of the disease, with its debilitating effect on the mainly agricultural workforce, and the move to less fertile land away from the rivers have impaired agricultural productivity in the region. In the worst affected areas of West Africa, before the WHO Control Programme began in the 1970s, ten per cent of the population were blinded and 30 per cent had severe visual impairment and 250,000 km² of good agricultural land was abandoned.

The map (Figure 5.27) shows that river blindness is predominantly an African disease, where it affects 36 countries. It probably originated in Africa. The limited areas in Central America and the Yemen where the disease has been reported may have originated due to slave trading, but the American cases are also circumscribed by the distribution of suitable inter-human vectors, i.e. flies of the *Simulium* genus. Some species like *S. gonzalezi*, *S. soubrense* and *S. damnosum* are zoophilic, preying on cattle rather than humans. These predominate in southern Africa. Muro and Raybould (1990) report an interesting case of declining incidence in Tanzania, which was associated with destruction of the breeding sites for the *S. woodi* blackfly in small shaded streams due to deforestation. The removal of shade is reported to have reduced the populations of *S. neavei* group flies in other parts of East Africa, without any deliberate control programme. This may be one of the very few good consequences of deforestation.

Eradication measures

The WHO began the Onchocerciasis Control Programme (OCP) in West Africa in 1974, with support from the UNDP, FAO, the World Bank and international aid, as well as the private sector including free drugs from Merck. The campaign was supported by 11 countries covering West Africa from Senegal, Guinea, Guinea Bissau and Sierra Leone through Mali, Côte d'Ivoire, Ghana, Togo, Benin and Burkino Faso to Niger. The OCP ended officially in 2002. The WHO estimates that 34 million people have been protected against the disease, over 400,000 cases of onchocerciasis have been prevented, 1.5 million people have been cured and more than 11 million children born during the eradication programme have been free from risk.

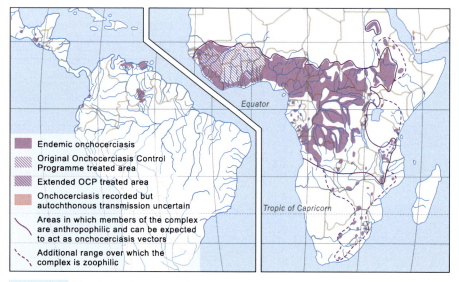

Figure 5.27 The global distribution of river blindness

The programme focused on both the environmental sources and the human reservoir of worms. Helicopters sprayed more than 50,000 km of rivers with larvicides to kill the blackfly larvae over an area of 1.3 million km², and people have been given medication to kill the nematodes. The human reservoir of macrofilarial worms dies out after their cycle of development has been interrupted for about 15 successive years. Larviciding has now ceased in the area that has been treated since the beginning of OCP. There the programme now runs mainly as a surveillance scheme to monitor any return of the disease, integrated with surveillance of the other main endemic diseases. In the remaining 40 per cent of the OCP area, where larvicide treatment began ten years later, it is still continuing, together with a human medication programme. Most of the work is now run locally with teams trained down to the district level.

In addition to the direct benefits to human health, elimination of the disease has allowed recolonization of the richer agricultural land in the river valleys, opening up 25 million hectares for cultivation, which is sufficient to feed 17 million people and so reduce their susceptibility to other debilitating diseases.

River blindness remains a significant problem in African countries outside the OCP, with 6.5 million suffering from dermatitis and 27,000 blinded. In 1992, the Onchocerciasis Elimination Programme in the Americas (OEPA) was launched. Areas of endemic river blindness in the northern part of Chiapas province in Mexico and the San Antonio region of Colombia were no longer active even before the launch of OEPA (WHO, 1989). The OEPA has since virtually eliminated the problem in the Americas.

Schistosomiasis (Bilharzia)

Schistosomiasis, bilharziasis or 'snail fever' is spread by parasitic flukes or flat worms that live in tropical waters and irrigated fields. Over a billion people are thought to be at risk and 300 million are commonly infected. It is endemic in Africa and Latin America and parts of the Far East. It is commonly contracted through bathing in rivers or lakes. It is the primary cause of death for Egyptians aged 20 to 44: its re-emergence in Egypt has been partly blamed on the Aswan High Dam.

The eggs of the flukes (or trematodes) hatch in water and the swimming miracidia larvae infest water snails. There they develop into a further larval stage (cercariae) before emerging into the water again and potentially infecting humans or animals. The flukes can burrow into the skin and live in the veins for years. If ingested in drinking water, the flukes cause irritation in the intestine or bladder, resulting in bleeding or dysentery before being ejected in urine or faeces, whence they may return to watercourses. The worst cases involve severe damage to the kidneys, liver or bladder, possibly leading to failure, others suffer from flu-like symptoms, but some infected people show no symptoms at all. Some research also suggests that schistosomiasis infection is linked to a higher incidence of bladder cancer.

The parasites belong to seven different species of the flatworm genus *Schistosoma*, each with its own geographical distribution, preferred genus of snail host, preferred mammalian host and specific medical impact. *Schistosoma japonicum* is passed from water snails of the genus *Oncomelania* to rodents. Rats occasionally pass on the disease to humans.

The drug praziquantel is an effective cure, but proper sanitation and hygiene are effective preventions. Among the measures recommended by the WHO to combat the disease are provision of safe drinking water, health education, especially for farmers in irrigated fields and freshwater fishermen, and control of snail populations, especially before dam construction begins.

The effects of floods and climate change

Floods and disease

Floods and standing water can be the breeding grounds for disease. In addition, floods can reduce the population's resistance to infection by destroying housing, water and sanitation systems and crops, with sewage overflows, hypothermia, malnutrition and even petroleum pollution in their wake. Floods are also a common cause of drowning.

After every flood disaster in developing countries, there is the fear of disease epidemics. In the autumn of 2009, rains finally returned to East Africa where 23 million people were starving because of the long drought, only to cause floods, landslides, cholera, malaria and hypothermia. The hundreds of thousand of refugees from the civil war in Somalia were especially at risk crammed in refugee camps in Kenya.

An estimated 140,000 people died in the rain and floods of May 2008 in the Irrawaddy delta caused by Cyclone Nargis, the worst cyclone in Burma's history. Some died as a direct result of the flooding, many others due to the typhoid, cholera and malaria that followed. One of the longer-term problems was caused by saltwater driven inland by the winds in storm surges 3.7 to 7.5 metres deep. As freshwater supplies were contaminated or cut off, people dug new wells but the groundwater was salinized. The livelihoods of 2.4 million people were devastated as the floods washed the salt and rice factories away, and millions of tonnes of saltwater flooded the rice paddies, making them unfit for planting for some time. With no employment, large numbers left the area. The buffalo used to plant the crops died drinking the muddy salty water and the farmers were wary of using the mechanical tillers provided by aid agencies. Planting of the new rice crop was delayed for months and yields are likely to be lower for years. The cyclone pushed up world rice prices, with a threefold rise in early 2008.

Sewage overflows

Outflows of raw sewage into rivers and onto beaches have been a scandal in the UK for more than two decades. Considerable improvements have been made by the privatized water companies investing some £80 billion in general system improvements, and their organization, WaterUK, says that £1 billion is earmarked for further improvements in the period 2010–2015. The Environment Agency reported in 2010 that over 8000 overflows and other storm sewage outlets on rivers and coasts have been improved and plans are in place to fit monitoring equipment to another 400. Nevertheless, the European Commission is investigating Britain's use of these overflow systems and has already taken the UK to the European Court over discharges at Whitburn, South Tyneside. This is another case of the taxpayer being held responsible for the activities of private companies. More stringent EU bathing water legislation is due to become active in 2015, at which point a large proportion of the current arrangements will be inadequate.

The problem is mainly caused by the old Victorian system of combined rainfall and wastewater sewers feeding combined sewer overflows. During heavy rains, more water enters the sewer system and to avoid overloading treatment plants or the provision of larger sewage treatment facilities to contain storm runoff, the drains were designed to allow rainwater to flush some of the sewage into the rivers and sea. But there are many other outlets under the control of the water companies, where sewage can be released when the system comes under pressure.

The old systems were not designed to be monitored, so there is a considerable lack of information. Quite apart from the fact that the private water companies claim the Freedom of Information Act does not apply to them, in many cases they do not have much information to give. Current legislation also leaves a loophole allowing companies to exceed the limit in extreme circumstances. Despite EU regulations limiting releases on beaches to three times a season, Southern Water admits some overflow pipes have been discharging more than 100 times in the summer.

Storm runoff from agriculture

There is one element, however, that is even more ubiquitous and difficult to control: storm runoff from agricultural land. Nitrates may be controlled by limiting the application of artificial fertilizers. Slurry and silage releases are obvious point sources that can be pinpointed. But storm runoff carrying urine and excretal material from pastures and upland grazing is much more diffuse and difficult to control. When the City of Birmingham Corporation set up the Elan Valley reservoir system in mid-Wales at the close of the nineteenth century to put an end to typhoid plagues in the city, it bought the whole catchment area and placed limits, still operating today, on the intensity of stocking with livestock. But this is a rare case and it was designed to protect the reservoirs, not the rivers.

The effects on rivers went long unnoticed, more or less until water quality legislation began to be introduced in the 1970s. Like sewage overflows, it is worst in storms and floods. It is often cited as the main reason why some beaches in the UK fail to get European Blue Flag status for cleanliness in wet years, yet achieve it in drier years. There was a threefold increase in the number of UK beaches that failed EU mandatory standards in 2008 as a result of heavy rain the previous summer.

Research by Professor David Kay's Centre for Research into Environment and Health at Aberystwyth University has documented numerous cases of high coliform levels in runoff from agricultural land (Crowther *et al*, 2002a and b; Fewtrell and Kay, 2008; Kay *et al*, 2010). Their findings show that faecal coliform can survive for an average of eight weeks on the land, with longer survivals in winter (a tenfold reduction, taking 13.4 days against 3.3 in summer). There is also normally more runoff from the land in winter increasing the risk of pollution events – and this goes unnoticed when most sampling is undertaken only in summer to meet the requirements of the bathing water legislation.

Kay and colleagues make interesting comparisons between the amount of coliform produced by animals and those of humans (Table 5.7). When it is realized that Wales has 33 million sheep and just three million people, the potential significance for riverwater quality becomes clear. Moreover, human sewage is largely treated, animal faeces are not. While there is potential for controlling pollution from farm animals – and Defra produce guidelines for farmers – this is not the case for wild animals, especially birds. So even though a starling (*Sturnus vulgaris*) is only equivalent to one person, a flock can produce substantial amounts that get washed down the drains in towns and cities. Kay and colleagues suggest that 125,000 starlings can produce the equivalent of treated sewage from between 1.25 million and 1.25 billion humans.

The best advice is not to bath in rivers after heavy rain in summer to avoid contracting gastroenteritis, hepatitis A or typhoid. The US Natural Resources Defense Council closed over 13,000 beaches in 2001 as a precaution. Climate change will make matters worse with heavier summer storms and wetter winters.

Research by the Crohn's disease group at St George's Hospital, London, under Professor John Hermon-Taylor, suggests another hazard (Hermon-Taylor *et al*, 2000). The group found that two-thirds of water samples from the River Towy in South Wales contained MAP bacteria (*Mycobacterium avium* subspecies *paratuberculosis*), which may get washed into rivers during storms from pastures grazed by animals suffering from Johne's disease. The catchment area of the river supports over a million cattle and 1.3 million sheep. The MAP bacterium is widespread in modern dairy cattle and the group suspect that it causes Crohn's disease in humans: samples suggest that it is present in 92 per cent of sufferers.

Analyses suggest that 2 per cent of public water supplies in England and Wales may contain MAP. A sixth of samples taken from the outflow of the sewage treatment plant on Ambleside in the English Lake District tested positive (Pickup *et al*, 2006). Lake Ambleside drains into Lake Windermere where water is abstracted for public water supply. 90 per cent of sediment samples from the lakes, dating back 50 years, were contaminated. Spreading contaminated slurry on the fields recycles the bacteria. In Cardiff, Crohn's disease is most prevalent on the downwind or east side of the River Taff and the group speculates that it may be transmitted by the wind, in aerosols derived from the river.

The conclusions remain tentative and a viral source for the disease is not currently standard medical opinion, but Defra and the Food Standards Agency have already modified their recommendations to farmers and consumers.

Climate change and water-related disease

Changes in rainfall patterns, temperatures, severe storms and flooding are all likely to have a direct impact on the distribution of water-related diseases. Diseases could also be affected by changes in water demand, leading to more reservoir building and more water abstraction from rivers, which might raise water temperatures by lowering water levels and reducing the velocity of flows. Other indirect effects could result from malnourishment and weakened resistance.

Many diseases and their hosts survive better in a warmer climate. A report from the European Centre for Disease Prevention (2006) warns of a possible increased risk of chikungunya in Europe.

There are three types of possible scenario that are causing concern: disease-transmitting mosquitoes might spread as climates and habitats change; mosquitoes of the same species as those which transmit the diseases, but which currently do not do so, might become transmitters; and travellers might bring the diseases back with them and the diseases might survive in the now more amenable climate. Malaria recently reappeared in Corsica, Turkey, Georgia and Azerbaijan. *Anopheles atroparvus* and *A. plumbeus* are present in south east England and capable of hosting malaria, but they would need to have bitten a traveller returning with it. The UK Climate Impacts Programme has mapped locations in England where *Plasmodium vivax* malaria could be

Table 5.7 Amounts of faecal coliform produced by animals compared with humans, according to Kay and colleagues

Animal	Equivalent number of humans
Sheep	9.5
Pig	4.7
Cow	2.8
Starling	1.0

transmitted by *A. atroparvus* during a few months of the year under the present-day climate. Under the HadCM2 climate model's medium-high scenario, most of lowland England and Wales will be at risk by the 2080s.

Similarly, the Asian tiger mosquito, *Aedes albopictus*, which is capable of carrying malaria, dengue and West Nile fever, is present in Somerset, England. Since 1990, it has become established in Italy, Spain and the south of France, possibly imported from China, and is spreading into Belgium and Switzerland.

Malaria is already creeping up Mt Kenya as the threshold temperature of 18°C limit for breeding of the parasite reaches higher up the mountainside. It is affecting millions of people who have not previously been exposed to the disease and have no prior immunity.

Maps of predicted changes in the global distribution of falciparum malaria based on the HadCM2 model suggest that it will advance into south west USA and south east Europe by the 2050s (Department of Health, 2008). But these are based on the statistical correlation between its current distribution and the purely climatic indicators of temperature, precipitation and atmospheric humidity. In reality, the detailed picture also depends very much on the distribution and persistence of surface water and on human water management – factors that have yet to be adequately incorporated into the predictions.

Diseases caused by faecal pollution, like cholera and typhoid, are also likely to become more prevalent in warmer water. In the Netherlands, Ramaker *et al* (2005) warn that higher temperatures are likely to increase the microbial risk from older water treatment plants.

Warmer summers are also likely to increase the pathogenic risk in recreational waters from stormwater overflows (Schijven and de Roda Husman, 2005).

Conclusions

The effects of water pollution on human health and wildlife present an increasing and highly complex problem. Long-term issues like lack of sanitation and hygiene are getting worse as populations grow, people concentrate in cities and poverty spreads. The artificial fertilizers and pesticides of the Green Revolution are intensifying the impact of agriculture. And new drugs and chemicals are presenting new challenges for toxicologists.

Meanwhile, the fight against infectious diseases transmitted by animals that breed in water is meeting with mixed success. Natural disasters and climate change seem set to aggravate most situations.

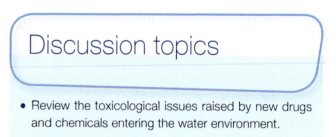

Discussion topics

- Review the toxicological issues raised by new drugs and chemicals entering the water environment.
- Explore issues of nuclear safety with special regard for mining operations.
- Examine the ways in which water management may affect the spread of disease.

Further reading

The *World Atlas of Epidemic Diseases* by Andrew Cliff, Peter Haggett and Matthew Smallman-Raynor (2004) gives an in-depth coverage of water-related diseases.

The World Health Organization's Guidelines and related publications are continually updated sources of information, and downloadable from the WHO website. Especially:

WHO 2008. *Guidelines for drinking-water quality*. 3rd edn, WHO, Geneva. Available at: www.who.int/water_sanitation_health/dwq/gdwq3rev/en/.

This book gives a good introduction to chemical pollution:

Alloway, B.J. and Ayres, D.C. 1993. *Chemical Principles of Environmental Pollution*. Glasgow, Blackie.

6 Water, land and wildlife

The Global International Waters Assessment Report found that destruction of habitat and 'community modification' is the main problem in 18 per cent of world regions, pollution in three-quarters (GIWA, 2006). Much of twentieth century development in both water resources engineering and in agriculture, forestry and urban growth ignored the close natural link between landscape and water resources. Yet the link is really so close, so enmeshed, that if water were alive it would be called symbiotic. We break the link at our peril.

The link was understood in its essentials by the nineteenth-century engineers who purchased the whole of the catchment feeding Birmingham City Corporation's Elan Valley reservoirs in Wales and put in place strict controls on farming and land use. It lies at the heart of what Mark Sagoff (2002) calls New York's 'Catskills Parable' – a tale of land acquisition that never was (see *Reed beds and compact wetlands*). However, our draining of wetlands, destruction of forests, intensive agriculture, herbicides, and pesticides, and large-scale engineering of the landscape have tended to ignore the link – and problems have ensued.

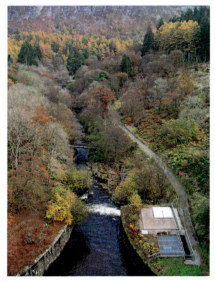

Figure 6.1 The Elan Valley Estate in mid-Wales is the largest protected landscape in the hands of a water company in the UK. It is now managed for wildlife and leisure as well as water supplies to Birmingham and, when needed, to South Wales

Fish are a good indicator of the health of freshwater systems. There are many reasons for the decline in freshwater fish stocks. Overfishing is one. But destruction of aquatic environments has played a large part. River engineering focused on flood prevention has tended to remove shallows, backwaters, and gravel shoals where fish may breed or hide from predators. Deforestation has removed shade and water abstraction has reduced the depth of water so that water temperatures rise and oxygen is degassed. Eutrophication and acidification have also played a part. In a book entitled *Silent Summer*, consciously referring back to Rachel Carson, the authors report invertebrates being wiped out for kilometres when cattle and sheep enter a river after treatment, and rivers running dry in summer through overuse by farmers and water companies in Britain (Maclean, 2010).

Blue, green, grey, black or gold?

For a colourless liquid, water has been painted many colours. Grey is the established term for wastewater, black for sewage, and Maude Barlow and Tony Clarke (2002) called it 'blue gold' to emphasize its commercial value. But the most crucial epithets are blue and green.

Traditional water resource management concentrates on rainfall and runoff. Evaporation and transpiration are considered 'losses'. The absurdity of this viewpoint is highlighted by Malin Falkenmark and Johan Rockström in their groundbreaking book *Balancing Water for Humans and Nature* (2002). 'Evapotranspiration' is literally the lifeblood of crops and all terrestrial plants. It has direct economic value in agriculture as well as sustaining the natural environment. Falkenmark and others dub precipitation, groundwater seepage and riverflows 'blue water'. The water used by the biosphere and 'lost' as vapour is termed 'green water'.

The concept is still regarded by many as 'new', but it is gradually getting embedded in policy and its value is becoming more apparent (Falkenmark and Rockström, 2006). It serves a valuable purpose in the concepts of 'virtual water' and more recently of 'water footprints', which could soon become as important as 'carbon footprints', with equally wide impacts upon daily living (see Chapter 8). The Food and Agriculture Organization (FAO) has embraced the blue-green dichotomy. So too has the International Soil Reference and Information Centre (ISRIC) at Wageningen in the Netherlands. ISRIC's Green Water Credits scheme is discussed in Chapter 15.

Blue water divides into the normal, fair weather flow in rivers, termed 'baseflow', which globally accounts for about 11 per cent of annual precipitation, and stormflow (27 per cent). From these, about 1.5 per cent of the global total is used for irrigation, of which only half is returned to the rivers. Turning to green water, in comparison crops use four per cent of global precipitation in 'rainfed agriculture', grasslands 31 per cent and forests and woodlands 17 per cent. The remaining ten per cent or so of global precipitation falls on arid lands and permanent icefields. So although some two-thirds of all the water abstracted by humans goes into irrigated agriculture, this is a paltry amount compared with nature's watering of agriculture, which could add up to more than a quarter of all the rainwater in the world. Together with commercial forestry it could account for a third of all rainfall.

The significance of green water becomes apparent when it is realized that it accounts for around 60 per cent of all water use within the global system. Nature is the largest single beneficiary.

The ecological impacts of large dams on the Volga River

Peter Khaiter

The Volga is the longest river in Europe (3700 km). The construction of eight complexes combining dams, reservoirs and hydroelectric facilities between 1937 and 1981 for flood control, improved navigation, irrigation and hydropower has converted the free-flowing river into a chain of man-made lakes (Figure 6.2 and Table 6.1).

Table 6.1 General characteristics of the Volga River reservoirs

Reservoir	Year of filling	Volume, km³	Area, km²	Maximum depth, m
Ivankovo	1937	1.12	327	19
Uglich	1939–43	1.25	249	19
Rybinsk	1940–49	25.42	4550	30
Gor'kiy	1955–57	8.82	1591	22
Cheboksary	1981	13.25	2189	18
Kuibyshev	1955–57	58.00	5900	41
Saratov	1967–68	12.37	1831	31
Volgograd	1958–60	31.45	3117	41

Industrial wastes and runoff from cities and farmland have polluted the river and many of these pollutants are retained by the reservoirs. Thus, 3000 factories dump some 42 million tonnes of contaminated waste in the basin annually, causing immense environmental and health problems. Just three per cent of surface water in the Volga river basin is considered an environmentally safe source of drinking water. Water is polluted by organic matter, heavy metals, acids and other toxic substances. Much of the watershed has been deforested, leading to increased erosion and silting. As a result, the most valuable fish species, such as sturgeon (*Acipenser sturio*), salmon (*Salmo salar*), sea trout (*Salmo trutta* m. *trutta*) and vimba (*Vimba vimba*), have either completely disappeared or at least decreased significantly. Many fish are contaminated with toxic pollutants such as heavy metals and pesticides (Khaiter *et al*, 2000).

Anthropogenic pollution of the Volga River reservoirs and the related changes in the structural organization of the three trophic levels (phytoplankton, zooplankton and zoobenthos) within their ecosystems have been reported by Khaiter *et al* (2000). Structural transformations within the communities of aquatic organisms have been found in all the Volga reservoirs. Some of them demonstrate clear signs of ecological regress.

Phytoplankton communities show a decrease of green algae in the total biomass, the dominance of blue-greens during the spring, and even trends towards the

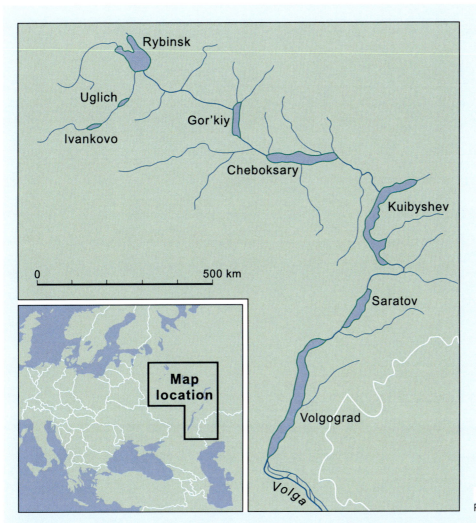

Figure 6.2 Reservoirs of the Volga River

dominance of a single species. Zooplankton communities show a decline in species variety and the development of an exceptionally high abundance of a few taxa: *Cladocera* and *Rotatoria*. Benthos communities show a constant or periodical increase in the population of *Oligochaetae*, a species typical of polluted and highly polluted waters, a decline in the total number of species and groups of species in the community, and a decrease in species abundance, including the occurrence of 'empty' samples – no life.

Irrigation and dissolved salts

Irrigation has caused salinization of soil and water since ancient times. Large areas of former desert newly reclaimed for agriculture after the completion of the Aswan High Dams have been lost through salinization over the last 50 years. Salinization has been a major problem in Iraq and the Aral Sea basin. As irrigation expands, it is now a problem in southern Europe.

Electrical conductivity is a reasonable indicator of the general level of dissolved substances in the water, although this may not always be due to agriculture. The salts may come from natural sources, from soluble rocks like gypsum, but most reports of high levels suggest human pollution. These may be due to irrigation mobilizing salts in the rocks, but high levels may also arise from industrial and municipal pollution.

The GEMS/Water data plotted in Figure 6.6 give an incomplete coverage of the world, but they do highlight some areas where wastewater management is poor, whether agricultural or urban. High levels are reported from Mexico, parts of the Middle East, especially Iran,

The drying of the Murray-Darling Basin, Australia

Simon N. Benger

The Murray-Darling Basin is Australia's largest and most developed river system. Supporting 75 per cent of the country's irrigation and providing drinking water for over three million people, the system is staggering under ever-increasing demand for limited water resources. Sustainability issues have come to the fore in recent years as decreased flows due to an unprecedented 11-year extended drought and climate change, rising salinity, declines in water quality and loss of wetlands, fish and bird stocks threaten the survival of the basin.

Situated in south-eastern Australia, and covering over one million km², or 14 per cent of the country's total area, the Murray-Darling River Basin provides over 41 per cent of Australia's gross value of agricultural production and encompasses some 30,000 wetlands. Its headwaters originate in the high country of the Great Dividing Range in the East and flow 1250 km to the mouth of the Murray River near Adelaide. A highly productive landscape in the driest continent on earth, the basin represents much of Australia's unique flora and fauna. Achieving a balance between water resource extraction and the ecological requirements of the natural environment has proved difficult.

The basin has a naturally high level of variability in precipitation and streamflows and its rivers are heavily regulated to provide water security for irrigators and urban users, for salinity control and flood mitigation. Tens of thousands of reservoirs, dams, weirs, locks and barrages have been constructed to facilitate water extraction on a vast scale. Of the almost 14,000 GL of average natural flow to the sea through the system, some 12,000 GL are diverted annually, 96 per cent of which is for irrigation. Historically, water extraction licences have been handed out to irrigators with little forethought or planning and diversion entitlements far exceed the capacity of the system. With so much water being extracted from the rivers the impacts on ecological health are immense. Large tracts of inland wetlands are dying due to reduced and less variable flows, fish stocks have plummeted as water quality declines and introduced species flourish, and water birds

are disappearing. The National Parks and Wildlife Service has been surveying water birds across the basin over the past 30 years and numbers have fallen to about a quarter of what they were in the 1980s. Low flows saw the mouth of the Murray close for the first time in 1981, and constant dredging is now required to keep it open. The adjoining freshwater Lower Murray Lakes have now been below sea level for several years and only extensive barrage structures, which once facilitated agriculture, prevent saltwater intrusion from the sea.

The large scale adoption of poorly suited European farming practices over the past 150 years has led to widespread land and water degradation throughout the Murray-Darling Basin. Excessive land clearing and poor irrigation management has resulted in excess water entering the water table causing groundwater to rise to the surface laden with dissolved salts. As well as ruining the land for agriculture, these salts are carried into the river systems, degrading river and wetland ecosystems and creating problems for water users downstream. Salinity levels in Lower Murray Lakes are now so high as to render the water unfit for irrigation, while the low water levels have exposed sediments, resulting in widespread acidification.

Figure 6.3 Acid sulphate deposits on the bed of Lake Alexandrina in the Lower Murray Lakes

Figure 6.4 Invasive polychaetes forming reefs and killing a turtle by weighing it down – the freshwater turtle population is being decimated

The Lakes and the extensive Coorong coastal lagoon, once rich in biodiversity and a Wetland of International Significance, are facing ecological collapse. Lack of flow and reduced water allocations have also decimated rural communities and industries throughout the basin. The economic costs alone of salinity currently reach into the hundreds of millions of Australian dollars annually, quite aside from the social and environmental costs. Drastic action on an unprecedented scale is required to address water sustainability issues in the basin before it is too late.

The future of the Murray-Darling River Basin may be set to improve, however. Various governments at both the State and Federal level are now working together to achieve meaningful water reform, an end to large scale land clearance and changes in farming practices. A widespread commitment to integrated catchment management, water buybacks for the environment, and reduced water usage, seeks to deliver a sustainable future for the rivers of the Murray-Darling system, their ecological communities and all those who depend upon them for their livelihoods.

Figure 6.5 The retreat of Lake Alexandrina

the west coast of India, especially in Gujarat, and the Mekong Delta. Irrigation is a major source on the Mekong. Upstream of the intense irrigation in the Delta, the water in the Mekong and its tributaries is mainly good quality. North America has a generally good record, with the exception of some rivers in the south, notably the Colorado and tributaries of the Mississippi draining the irrigated lands of the Great Plains. Remarkably high levels of electrical conductivity are also found in parts of southern Europe. Irrigation plays a large role in Spain, where it is made worse by evaporite rocks like gypsum, deposited on an ancient seabed. Over-irrigation has been common for decades and much of the excess water filters down through the bedrock before emerging in the rivers and being reused by farmers further downstream. In the Ebro basin in northern Spain, levels of dissolved solids rose by up to 10 mg/L every year from the 1950s to the 1990s, and farmers on the lower Rio Gallego near Zaragoza experienced gradual reductions in soil fertility as the soils became salinized. As a result of this experience, the latest extension of irrigation into the Monegros, south east of Zaragoza, is based on sprinklers rather than the traditional ditch and flood irrigation.

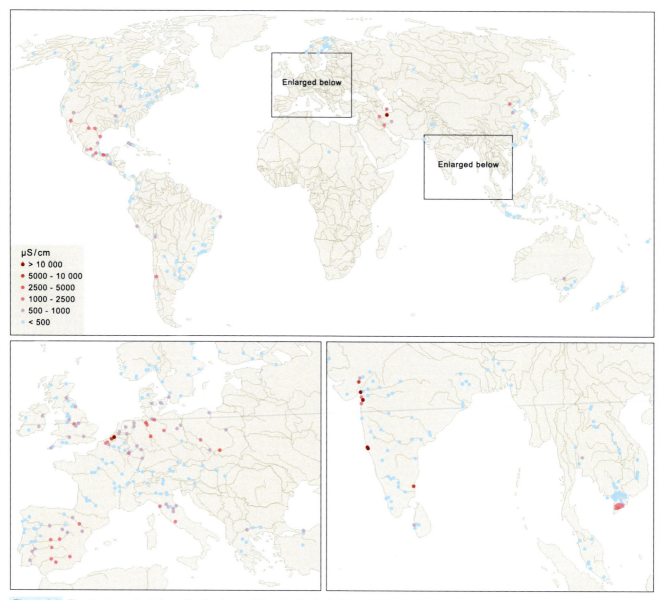

Figure 6.6 Electrical conductivity, indicating level of dissolved salts in the water

Another belt of relatively high electrical conductivity stretches across north-central Europe from the Low Countries to Poland, which probably is due more to urban-industrial sources.

Eutrophication

Eutrophication is a major problem for surface waters and a clear indication of high levels of pollution. In the USA, eutrophication is the main problem in around 60 per cent of damaged rivers and 50 per cent of damaged lake areas. It occurs when algal blooms swamp water bodies and consume the oxygen, killing other plants and wildlife. The algae are fed by excessive amounts of nitrogen and phosphorus produced by bacterial decomposition of organic waste or by artificial fertilizer leached and washed off agricultural land. Initially, high nutrient levels cause all aquatic plants to grow abundantly, but the algae tend to take over by reducing the supply of light, oxygen and nutrients to the other plants. Blue-green algae fix additional nitrogen from the air, speeding their growth. The chain of predators is also disrupted: in healthy waters, water fleas eat algae, bream eat the fleas and pike eat the bream, but when pike begin to die as excessive algal growth consumes the oxygen this natural check on algae gradually declines. At nitrogen levels of around 6 mg/L, long, waving filamentous algae float on the water, inhibiting photosynthesis in other plants, which then begin to decay. Chlorococcal algae are particularly effective at this around 20 mg/L. Bacteria feeding on the rotting plant material lower the oxygen levels further, killing the less hardy fish like trout and leaving only coarse fish like bream or roach. Eventually all higher life is killed by lack of oxygen, and the water body is engulfed in toxic, anaerobic conditions.

Droughts and summer low flow conditions can aggravate eutrophication. Shallow, slow flows cause less aeration of the water, stagnant areas and warmer water, which speeds decomposition. The situation can be exacerbated by higher abstraction rates during the summer.

Although levels of pollution in East European rivers have generally been drastically reduced as national economies come into line with Western Europe, eutrophication in the Danube Delta has increased markedly in recent dec-

ades. The Delta is the largest wetland in Europe, covering 564,000 ha, and an International Biosphere Reserve listed according to the Ramsar agreement. Galatchi (2009) largely attributes the eutrophication to the reduction of wetlands and floodplain storage areas upstream, combined with intensified agriculture. Extensive flood protection levées have reduced the filtering effect of the floodplains, notably the 800 km of embankment along the Romanian border. Dams and reservoirs have also altered the flood regime of the river, reducing the flushing of pollutants. At the same time, as countries joined the EU and benefited from agricultural subsidies, more intensive use of artificial fertilizers has increased nutrient loads. This has resulted in loss of stocks of the more valuable fish species, like carp. The Delta is still reducing nutrient yields entering the Black Sea, but it is incapable of absorbing all the nutrients that it is now receiving from upstream.

Eutrophication in Brittany's reservoirs

Alain Jigorel

In Brittany, rivers and reservoirs provide 80 per cent of drinking water. Many reservoirs were built in the 1970s and 1980s to meet increasing demand. Most of these have undergone eutrophication, because they are located in areas of intensive agriculture, which generates excessive phosphorus and nitrogen. Nitrate concentrations in Breton surface waters are often higher than 50 mg/L and sometimes even exceed 100 mg/L in wintertime. Concentrations in reservoirs are sometimes higher than the maximum authorized by the European directive (50 mg/L). Treatment plants have to be designed to remove the nitrates before it can be used for drinking. The restoration of water quality will take a very long time and to date the results are insignificant.

Surface waters are also contaminated by weedkillers and pesticides. The measured concentrations vary considerably according to the time of year and concentrations sometimes exceed the maximum concentration authorized in raw waters intended for the production of drinking water. Precautionary measures and in particular the installation of bands of vegetation along the rivers, can limit pollution effectively. In parallel, all drinking water

treatment plants were equipped to eliminate herbicides, pesticides and their derivatives.

The principal manifestation of water pollution is without question the widespread eutrophication. Slow-flowing rivers, ponds and reservoirs regularly suffer a proliferation of algae each year from spring to autumn, particularly *Cyanobacteria* (blue-green algae). The latter are receiving more and more attention because they can produce toxins – neurotoxins, hepatotoxins and dermatotoxins. Each summer, bathing and boating are prohibited in a number of water bodies because of the health hazard. Algal pollution also causes deoxygenation and changes in pH in the reservoirs. Dead algae and diatoms accumulate on the bed and cause silting up. The sedimentation can be very significant, ranging between 10 and 50 mm per year. All the eutrophic reservoirs deeper than 10 m have poor quality water, because of the accumulation in the sediments of metals, such as aluminium and iron, and nutrients (ammonia and phosphates) that accentuate the process of eutrophication.

Restoration of the aquatic environments will be a long and difficult task.

One of the main aims of the European Water Framework Directive is to eliminate eutrophication from all EU water bodies by 2015 (see Chapter 15).

Biological Oxygen Demand

The Biological Oxygen Demand or BOD is a good indicator of the level of organic pollution, measuring the amount of oxygen consumed by bacteria feeding on the organic waste. Raw sewage has a BOD of 600 mg of oxygen per litre (in a five-day lab test), compared with under 5 mg for unpolluted water. Slurry from cowsheds has a BOD ten times higher than sewage. Worst of all,

silage liquor from fermented fodder grass can have a BOD 3000 times higher than sewage and is therefore lethal to freshwater life.

In the UK, Defra recommend a BOD of under 5 mg of oxygen per litre for drinking water. By this measure, European rivers are much better than in terms of faecal coliform, with only 18 per cent of untreated riverwater sites exceeding this recommendation within the pre-2002 EU (Figure 6.7). In the Danube Delta, however, there has been a hundred-fold decrease in BOD since 1960 and quality of fish in terms of both size and range of species has declined with it (Galatchi, 2009).

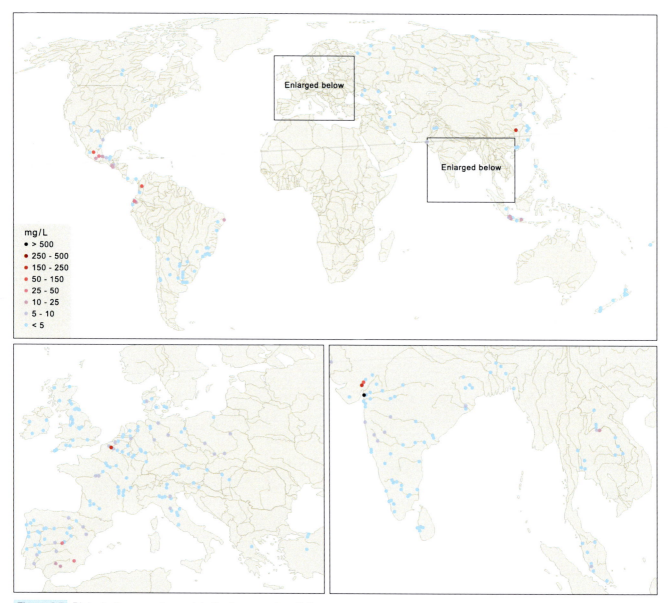

Figure 6.7 Biological oxygen demand, indicating organic pollution

Water temperatures

Aquatic life is sensitive to water temperatures and temperatures have been increasing in many rivers. An analysis of 120 years of records on the River Loire shows that temperatures have risen by 0.8°C and river discharge has decreased (Moater and Gaillard, 2006). Analyses of riverwater temperatures in Austria showed rises of 0.47°C to 1.26°C during last century (1901–1990) (Webb and Nobilis, 1995). Webb *et al* (1995) link the rise with human interference – removal of shade, heated effluents and river regulation – rather than climate change. Between 1955 and 1984, nine barrages were built on the Danube for hydropower. Channel modifications have also reduced discharges and river levels, which results in more solar heating.

It is possible that the balance may be changing this century and that climate change may become more dominant. Bonacci *et al* (2008) found rising temperatures in tributaries of the Danube in Croatia are linked to a rise in air temperatures beginning in 1988. Both Webb (1992) and Arnell *et al* (1994) conclude that by 2050 the mean temperature of most UK rivers could be over 10°C, with some rivers in the Midlands and southern England exceeding 15°C. This would mean an increase of 1.0–2.5°C in most of Britain and 4°C in the north of Scotland. Minimum temperatures are likely to rise more than maxima, so that the daily range in temperatures would be less. The probability of freezing temperatures would reduce threefold in south west England.

Higher temperatures would reduce levels of dissolved oxygen and increase bacterial activity so that BODs increase. Lower levels of oxygen combined with less dilution of pollutants in low summer discharges are likely to have a marked detrimental impact on aquatic life. Computer models of chemical and biochemical processes in rivers commonly make the assumption that temperatures remain unchanged, and the findings of Webb and colleagues suggest that these will need to be modified (Webb *et al*, 2008; Doglioni *et al*, 2008).

Acidification of surface waters

Acid rain was the research focus of the 1980s, until governments and research councils began to focus on global warming in the 1990s. But the sources of the acidification problem have not entirely gone away, the problems of rehabilitating damaged ecosystems have not been solved irrevocably, nor are the issues of climate change and acidification completely independent (see Chapter 10). Increasing acidification of the oceans due to climate change is a growing concern, but there appears to be far less attention given to similar prospects for surface waters on land, due to the same causes. Increased concentrations of carbon dioxide in the atmosphere translate into more acidic rainfall as the carbon dioxide is dissolved in the cloud droplet. One benefit from global warming, though relatively minor, is likely to be the reduction in snowfall, because snow crystals scavenge air pollutants more efficiently than liquid droplets.

The main reason why the hysteria over acid rain has subsided since the 1980s is the success of clean air legislation, which has significantly reduced emissions of acidic compounds, especially sulphates. Ironically, however, a new threat may be coming from the sea. Provided sufficient essential nutrients are available, extra carbon dioxide will fertilize phytoplankton and cause them to proliferate, releasing more of the acidic gas DMS. More immediate concern is being expressed over sulphur emissions from shipping. An EU report in 2009 notes that hundreds of thousands of giant cargo ships, often old and 'dirty', ply international waters, burning 289 million

Figure 6.8 The remains of a ditch in an old commercial forest in Wales. When the trees were originally planted, these ditches were dug to drain and aerate the soil and encourage root growth. But they have also been blamed for speeding drainage into the streams, creating more floods and directing acid rainwater more quickly to the stream, bypassing the soil that would have neutralized it more

Figure 6.9 Trees cut back from the stream bank to reduce direct dripping of acid rain into the stream. Areas of *Sphagnum* moss bog are encouraged near the stream to adsorb acids

tonnes of fuel a year with no emissions controls. The EU is proposing to control emissions in the English Channel, North Sea and Baltic by 2015.

Acid rain was first identified in Scandinavia and blamed for damage to forests and subsequently to fish stocks. Conifer forests were identified as both victim and culprit – removing base nutrients like calcium and magnesium from the soil and depositing needles on the forest floor that release weak acids as they decompose. Studies in the UK showed that fish in basins afforested by conifers were fewer and smaller than those in natural moorland basins (Edwards *et al*, 1990). This could be explained partly by lack of food (the populations of invertebrates were lower in the forest catchments), partly to lower water temperatures in the forests, and partly to frequent 'acid flushes', quick throughput of rainwater creating acidic floodflows, which in the UK situation were largely caused by forest drains that were dug at the time of planting to aerate the soil. The impact of the acid flushes could last for a long time as the fauna gradually recover from the shock. In the UK the acid flushes appear to have a greater impact than the average long-term acidity of the streamwater. The forests also act as 'brushes', filtering acid aerosols from the air in dry weather, which then get washed off the trees in rainstorms, creating a more acid event than the rain itself would have done. This is the theory behind cutting back trees away from the streams.

Sweden invested in a very expensive programme of liming water bodies. Unfortunately, this has to be an annual exercise, as either the lime is dissolved and washed away, or the coarser granules develop a coat-

ing that inhibits solution. Liming the land is more long-lasting and efficient.

The issue of rehabilitation is a vexed one. Too much liming could damage mountain ecosystems that are naturally adapted to an acid environment. Liming upland soils may also not be cost-effective for farming where soils are low in nutrient levels and the rocks are low in base cations, calcium and magnesium, and therefore have limited natural potential to 'buffer' or neutralize the acidity. The concept of 'critical loads' has been applied in Britain as an aid to planning. It measures the amount of acid deposition that the land can take before critical damage (Bull, 1991). Many mountain areas with poor soils passed this critical level years ago, and it is better to focus remedial activity on the penumbral zone, the area of advancing degradation between the 'lost' and the safely buffered zone.

From the water resources viewpoint, it is most valuable to lime the source areas of streams rather than wasting effort on the areas of the basin that do not contribute water to the streams. The pH of drinking water is corrected as standard treatment before distribution to avoid corrosion in pipes and to limit levels of heavy metals either from the environment or from corrosion in the distribution system. Nevertheless, treatment costs can be reduced if the water entering the system is less acidic and there is less polluted treatment material to dispose of.

Perhaps the biggest questions of all are: can ecosystems be effectively rehabilitated and to what degree should they be restored? Steinberg and Wright (1994) reported that attempts had been limited and very anthropocen-

Figure 6.10 An automatic limer on a stream in Wales. The system releases an appropriate size shot of lime when the float in the stream responds to an increase in water depth, indicating an acid flush event

tric – aimed mostly at restoring fish stocks. The whole community structure may have been changed by acidification. While the lower levels of life, like bacteria, may have found a bolthole in which to survive and recolonize, this is less likely for the higher animals, like fish. In reconstructing an ecosystem it is important to provide the top predators that keep the system in balance. Recent evidence shows that as species like mayfly decrease in acid water their place as algal grazers may be replaced by other species, which then make the ecosystem resistant to reinvasion once acidity is ameliorated and water chemistry is restored (Ledger and Hildrew, 2005). Experiments in the Netherlands also show limited success in rehabilitating an acidified environment (Beltman *et al*, 2001).

Disappearing forests

The FAO estimates that during the first five years of this century, there was a net global loss of forest amounting to 73,000 km². Most of this was in Africa and South America: 43,000 km² in South America and 40,000 km² in Africa. A further 3000 km² were lost in North and Central America. In contrast, reafforestation programmes planted just 10,000 km² in Asia and 7000 in Europe. Figure 6.11 highlights the areas of greatest forest loss according to the IUCN.

Although it does not feature strongly in the map, because the area of rainforest is still extensive, the Amazon basin has lost over 600,000 km² of its rainforest since the 1970s, nearly half due to agriculture and a further 14 per cent due to logging and forest fires, both natural and deliberate. In the recent past, much of the forest clearance has been for cattle ranching, but sugar cane and palm oil for biofuel production are taking over. Some natural forest regeneration has occurred here and elsewhere as farmers have abandoned the fields and left for the cities. But the net effect is minor. In China, it is urbanization and the rapid rise in the price of lumber to build houses – up 300 per cent in ten years – that is the new driver behind expanding deforestation. In Borneo, the rainforest halved between 1985 and 2005.

Forests play important roles in water resources, both positive and negative. Trees store and use water, regulating riverflows and reducing flood volumes, and they reduce soil erosion by binding the soil with their roots and reducing kinetic energy of the raindrops by intercepting the rain. The tropical rainforests not only evaporate 60 tonnes of water per hectare daily, they also store 1350 tonnes of water for every hectare of forest. Removal of forests during the western colonization of North America and Australia in the nineteenth century caused severe problems with increased floods, landslides and choked river channels. Recent increases in the fre-

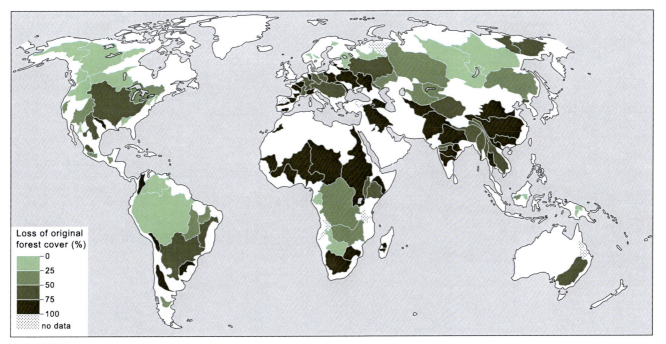

Loss of original forest cover (%)
- 0
- 25
- 50
- 75
- 100
- no data

Figure 6.11 Loss of original forests

quency of floods in Indonesia have been blamed on extensive logging. A major part of Haiti's susceptibility to hurricanes is due to deforestation for firewood and charcoal production. Hundreds, perhaps thousands, die each year from floods and landslides in tropical countries through bad management of the land and forests, including the worst for 25 years in Bali in 2005.

Throughout the twentieth century, American water engineers strove to plant forests in catchments feeding reservoirs to prevent silt reducing the storage capacity. The strategy was not totally successful, leading to a complaint from US Geological Survey hydrologist D.G. Anderson in 1970 that reservoirs should be collecting water, not sediment. However, application of the same principle in the UK had different results. The Institute of Hydrology was set up as a result of questions in parliament following experiments by a Fylde Water Board hydrologist, Frank Law, which suggested that forests waste valuable water resources and that water managers should charge forestry authorities for the loss. Subsequent research at the Institute proved that evaporation from rain intercepted on the forest canopy, together with the trees' transpiration, could reduce riverflows by up to 20 per cent. Sedimentation, on the other hand, is not a major problem in the UK climate.

Forests also have significant effects on the carbon cycle. Deforestation could be the second largest contributor to global warming after fossil fuel emissions, accounting for nearly 20 per cent of all carbon emissions, especially when fire is used. It was suggested at the Copenhagen climate conference in 2009 that countries should be paid to protect their rainforests to offset emissions. In this regard, rainforests are not alone. The UNEP/FAO/Unesco 2009 publication *Blue Carbon* suggests that mangrove forests are six times more effective in sequestering carbon than undisturbed Amazonian rainforest and salt marshes ten times more effective. Unfortunately, mangroves and salt marshes are also threatened habitats: since the 1940s, parts of Asia have lost 90 per cent of their mangroves. Experiments in Amazonia by the UK Centre for Ecology and Hydrology also suggest that mature rainforest is not a very efficient carbon sink because of the large amounts re-released in respiration. Young growing forests are more efficient.

Indirectly, forests also affect water resources through their effects on climate change. Locally, the evapotranspiration from forests contributes to rainfall. The mid-afternoon thunderstorms in tropical rainforests recycle water and maintain the forest. There is some evidence to suggest that rainfall has reduced in recent years in parts of the Amazon basin and concern that this may be due to deforestation.

The most severe drought in 50 years struck the Amazon basin in 2005 and initiated a state of emergency. Many tributaries dried up and helicopters were needed to supply food and water to many villages when the rivers became unnavigable. There was speculation that deforestation might have worsened the drought. However, the drought may have been linked to rising air over warmer water in the North Atlantic, which causes sinking air in the Amazon. Modelling at the Hadley Centre suggests that anomalies in sea surface temperatures in the Pacific and differences between temperatures in the North and South Atlantic are also likely to be responsible for drought in the Amazon (Harris *et al*, 2008). This effect could get worse with climate change, partly because of the continuing reduction in reflective aerosol pollution in the North Atlantic caused by clean air policies (Cox *et al*, 2008). The UK Hadley Centre coupled climate-carbon cycle model (HadCM3LC) predicts major forest losses later this century through lack of rainfall. Large parts of the forest might die by 2100. The UK Hadley Centre suggests that even with the lowest temperature rises, between 20 and 40 per cent of the Amazonian rainforest could disappear in 100 years: a 4°C rise could kill 85 per cent. Research by the Hadley Centre and Centre for Ecology and Hydrology suggests that droughts of the scale of 2005 could occur every other year by 2025 and in nine out of ten years by 2060 (Cox *et al*, 2008). The time delay in the

Figure 6.12 Natural recolonization by shrubs and trees in abandoned fields in the Spanish Pyrenees. Soil erosion took its toll when the fields were abandoned in the late twentieth century, but regrowth of wild vegetation is beginning to stabilize the landscape once more

forest's response to drought had masked the effect in observations on the ground, but modelling indicates that warmer temperatures will initially increase evapotranspiration. However, as the forest dries out this will reduce and eventually the trees will die, with the loss of a significant carbon sink.

Numerous plans have been advanced to contain the accelerating rate of deforestation, with hundreds of recommendations dating back to the 1985 Tropical Forest Action Plan and the Non-Binding Forest Principles approved at Rio in 1992 and endorsed by the UN General Assembly in 2007. Costa Rica pays farmers to revert to forest for its 'water service'. Carbon offset schemes allow carbon emitters to invest in rainforests to offset emissions. But it is currently a losing battle. Cambridge lawyer Dr Catherine MacKenzie says that in the face of the growing realization that internal environmental problems like deforestation can threaten international security through food and water shortages, disease and civil instability the UN Security Council might be the best placed to take on enforcing international environmental law (MacKenzie, in press).

Disappearing wetlands

Wetlands have been disappearing at an accelerating rate over the last 200 years. The USA has lost nearly half its wetlands since 1780. Germany and the Netherlands have lost more than half of their wetlands in the last 50 years. Many British wetlands were lost in the 'dig for victory' campaign during the Second World War, but more have been lost since through continued subsidies to agriculture and particularly under the European Community's Common Agricultural Policy (CAP). Overall, OECD countries have lost 13 per cent of their wetlands since the middle of the last century. Much of the destruction has been deliberate: reclaiming land for agriculture, removing supposed 'sources of pestilence' – especially mosquitoes – river engineering and flood protection, and urbanization. Reisner (1993) argues that river engineering on the Mississippi and Missouri Rivers created a twofold problem: first, channelization and wetland drainage exacerbated flooding, which led the US Corps of Engineers to build more flood-control dams and second, wetland drainage helped intensive agriculture to spread across the floodplains, which increased soil erosion and required the Corps to dredge the rivers more frequently. Interestingly, however, the IUCN's global survey of wetlands indicates that the Mississippi-Missouri-Ohio river

basin still has a relatively high percentage of wetland area compared with most of the world outside the arctic and subarctic (Figure 6.13).

Only in recent decades have the ecological and hydrological benefits of wetlands been widely accepted: as boltholes for threatened species, regulators of floods, filters for pollution and assisting groundwater recharge. The Florida Everglades are a prime example of both deliberate and inadvertent destruction. Agricultural chemicals and urban wastewater have polluted the water, and the increasing demand for water from the burgeoning resident population and tourists has lowered the water table, reduced surface flows to critical levels and allowed saltwater incursion into surface and groundwater. On the Kissimee River, flood alleviation work between 1964 and 1970 drained three-quarters of its 16,000 ha of wetland and reduced the meandering river to 40 per cent of its former length. Canals were constructed to divert water that used to flow from the Kissimee River into the Everglade wetlands. Within two years there were calls to reverse the process as fish kills occurred in Lake Okeechobee and flood problems were worse downstream. Wildfowl populations fell by 90 per cent as a result of lower water levels and agricultural pollution. The Kissimee Restoration Scheme was launched in 1984 to buy back land and naturalize it, reflooding the marshes. The programme was extended in 1992 with the aim of restoring the environment of 100 years ago by 2007. Further south, wetlands have been better valued and maintained for longer, beginning with the creation of the Everglades National Park in 1947, the Big Cypress National Preserve in 1974, and the prescription of minimum flows into the National Park by act of Congress in 1971. An elaborate flow management system maintains water levels in the South Florida Water Management District and adjacent parts of the Everglades, preventing floods and droughts and maintaining water head to recharge the important Biscayne aquifer.

Wetlands still cover three per cent of Nigeria, but they are disappearing at an alarming rate under increasing pressure on all fronts: population pressure, unprecedented land reclamation, rapid urbanization, mining, oil and industrial pollution, uncontrolled tilling for crops, overgrazing, logging, dam construction, transportation routes and other infrastructure, coastal erosion, subsidence, seawater intrusion, invasion by alien flora and fauna, sand storms, desertification and droughts (Uluocha and Okeke, 2004). Wetland destruction is threatening water resources, because the wetlands main-

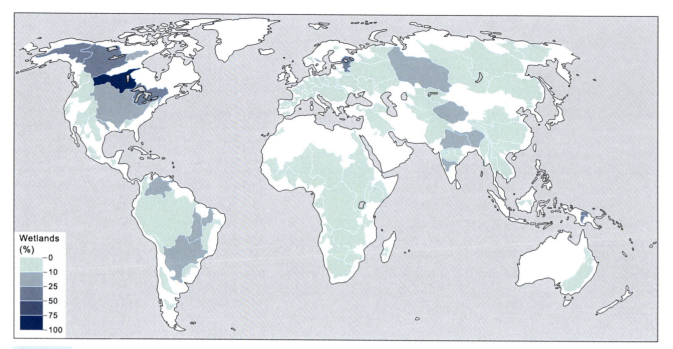

Figure 6.13 Present-day proportions of river basins still covered by wetlands

Wetlands
(%)
- 0
- 10
- 25
- 50
- 75
- 100

tain streamflow during the dry season in the semi-arid region of northern Nigeria, as well as regulate floods and improve water quality. Uluocha and Okeke call urgently for new legislation to safeguard the remaining wetlands.

Figure 6.14 Protected wetland – the Norfolk Broads, England. Centuries of drainage beginning in Roman times and culminating in Dutch engineering in the seventeenth century have drastically reduced the wetlands of East Anglia. But they are now being protected and extended

In the article 'Wetlands of the Canadian arctic', Professor Ming-ko Woo and colleagues describe the environmental and ecological role of arctic wetland systems that will be vulnerable to the effects of global warming as ground ice thaws.

Soil erosion and river sediments

Soil erosion clogs waterways and carries pollutants attached to the particles of soil into the rivers. An important factor in flooding in the Ganges Delta and on the Lower Yangtze has been the accumulation of sediments in the riverbeds caused by agriculture and deforestation upstream (Jones, 2000). This has raised riverbeds above the level of the floodplain in places and reduced the cross-sectional area of the river channels so they can hold less water before flooding overbank.

Records of average concentrations of suspended sediments in rivers (Figure 6.15) indicate high levels in many rivers in south east Asia, notably the Mekong and Ganges Delta, and in many rivers draining from the Andes into the South Pacific. In both cases, the rivers are draining from mountain areas with high rainfall and steep slopes, but erosion has been increased by removal of the natural vegetation for fuel and agriculture. India and more recently China have been engaged in reforestation programmes to check soil loss in the mountains.

Records are, however, very limited. Most monitoring is undertaken in order to assess water quality rather than for the scientific study of erosion rates and sediment transport. Records are also almost exclusively related to suspended sediments, whereas significant amounts of coarser sediment can move as 'bedload' – dragged or bounced along the riverbed. This is extremely difficult to measure. In general, it appears to be less than the amount moved as suspended particles, perhaps as low as one-fourteenth as much, because it requires large floods to transport it. However, evidence suggests that in some rivers bedload may account for 20–40 per cent of total sediment transport, in which case the records are very much underestimates (Dunne and Leopold, 1978).

One of main aims of the International Hydrological Programme's International Sediment Initiative, that was set up with Unesco collaboration in 2002, is to improve monitoring and understanding of the causes of sediment transport, and particularly the role of human activities.

Wetlands of the Canadian arctic and the potential effects of climatic warming

Ming-ko Woo, Kathy L. Young and Laura Brown

Wetlands in the polar desert environment of the High Arctic provide a special ecological niche for tundra plants, insects, birds and animals. Wherever local water supply exceeds water losses during the thawed season and the lie of the land allows water to accumulate, conditions tend to favour the formation of wetlands. With the water table at or near the ground surface for a sufficiently long time, aquatic plant communities develop. Such wetlands often occur in patches covering less than a square kilometre, supporting a lush growth of vegetation amid the barren polar desert landscape. They are frequently found at slope concavities where groundwater exfiltrates, or in depressions and valley bottoms where ponds, lakes and streams provide a reliable water source. Patchy wetlands are also common down slope of late-lying snow banks which tend to occupy topographic hollows on the slopes. The most usual hydrological rhythm for runoff on hill slopes and in streams in the High Arctic is to have high flows generated by snowmelt at the end of the prolonged winter, followed by a gradual decline in flow as summer arrives, with occasional runoff spikes generated by rainfall events, and ending in a complete cessation of runoff as winter arrives.

Despite their limited extent, patchy wetlands are a significant ecological niche for the plants and insects which support the birds, rodents and even caribou. Most wetland soils consist of an organic layer including live vegetation (mainly mosses, grasses, sedges and herbs) on top of partly decomposed peat, lying above mineral soils which range from clay to gravel. Arctic wetlands have only a brief thawed season when water exists in liquid form, but given sufficient water supply, they are self-sustaining entities in which soil saturation favours ground ice formation. The top zone or 'active layer', which freezes and thaws annually, is only 0.2–0.8 m thick. Beneath this, the ice-rich permafrost (permanently frozen ground) prevents deep percolation, so that the active layer in the wetlands is often saturated with 'suprapermafrost groundwater', which is retained by the depression.

Several processes can alter this balance, especially excessive melting of the ground ice that leads to the formation of 'thermokarst', with localized ground collapse and subsurface tunnels, which alters the drainage patterns. On a regional scale, projected climatic warming of the Arctic may extend the thawed season, enhance evaporation and eliminate the late-lying snow banks that feed some patchy wetlands. Under this scenario, the patchy wetlands of the High Arctic are considered to be highly vulnerable. The absence of large storage mechanisms, such as deep groundwater storage, makes the hydrologic responses highly sensitive to fluctuations in the climate.

The amount of ground ice strongly influences the rate of summer thaw because of the need to satisfy the latent heat requirement for ice melt (Woo and Xia, 1996). For the High Arctic wetlands, ground thaw is hindered by the abundance of ground ice in the active layer and by the insulating peat and live vegetation cover. Gradually, permafrost aggrades, with the feedback effect that a thin thawed zone above the frost table requires only a small amount of water input to saturate the entire active layer, thus sustaining a high level of wetness throughout the summer season. In this way, wetlands may survive short term fluctuations in the climate. Conversely, thawing of the permafrost underneath a wetland followed by thermokarst development, slumping and erosion can cause its demise.

Sustainability under climatic warming

A change in the regional climate accompanied by drying can disturb the equilibrium in water balance that sustains wetlands in the area. Most Global Climate Models point to significant warming of the Arctic under scenarios of increased greenhouse gas concentrations in the atmosphere. Although most of the warming is expected to occur during the winter months when air temperature would remain well below 0°C, there may be a shift in the timing of the shoulder period when air temperature crosses the freezing point, leading to an extension of the thawed season. This will negatively affect the sustainability of High Arctic wetlands. Even if summer rainfall increases, a lengthening of the evaporation period will enhance water loss. The snow will melt earlier and the many late-lying snow banks that support patchy wetlands will possibly vanish.

Under the present climate, a significantly warm summer such as 1998 caused many snow patches to shrink, but they re-established to their former size after several cool, wet years (Young and Woo, 2003). With a continued trend towards warming, the snow banks will not rebuild, leading to the demise of most of the existing wetlands.

Arctic warming will thicken the active layer and increase the extent of thermokarst. Desiccation, wind erosion and rill erosion may remove the insulating organic cover and cause the ground ice to melt and the permafrost to degrade. Permafrost plays a dual role: it concentrates water in the active layer, thus maintaining the wetland, yet the restricted storage offers little hydrologic buffer to fluctuations in weather and climate. Thus, wetland hydrology is highly responsive to short-term (seasonal and interannual) variations and long-term changes in the climate. Wetlands that receive inflow from small, local groundwater systems are quite vulnerable to climatic change since there is no reliable alternative water source to maintain a high level of wetness (Winter, 2000). Many wetlands in the High Arctic depend on precipitation for their water supply and those fed by suprapermafrost groundwater are supported by aquifers of limited extent and thickness. It is therefore surmised that these wetlands are not only sensitive to fluctuations of the present day climate, but are vulnerable under the impacts of climatic change.

Walling (2006) notes that there has been a significant increase in sediment loads in several rivers in recent years, much of which is linked to human activities, including mining as well as logging and agriculture. The trend is, however, complicated by dam building. Dams trap sediments and as the world's dam-building programmes expand, so sediment transport will be reduced, even without any intentional rehabilitation of the landscape.

To date, two attempts have been made to map the world pattern of sediment yields. Lvovich *et al* (1991) used river discharge measurements, which are more prolific than sediment measurements, and estimated sediment yields using a statistical relationship between sediment yield and river discharge (Figure 6.16). Walling and Webb (1983) based their map on actual records of sediment yield (Figure 6.17). Both have their limitations, but despite a few significant differences the maps show a general conformity. The higher rates are in rivers draining mountainous areas, or, in a few cases, draining areas with highly erodible soil, like those of the Loess Region

of China. Although the authors use slightly different classifications in the maps, Walling and Webb's map suggests lower yields in Central and Southern America than Lvovich's and higher yields in much of East Africa.

Comparing Figure 6.17 with the map of river discharges in Figure 11.25 suggests sediment yields are higher for a given river discharge in the arid and semi-arid Western USA and Mexico, North east China and East Africa, and lower than river discharges might suggest in much of the arctic and subarctic, and on the eastern seaboard of North America. The contrast between the east and west of the USA is largely a matter of vegetation cover. Lower erosion in much of the arctic and subarctic can be explained by lower topographic slopes.

Walling and Webb (1996) draw some interesting conclusions from their detailed review of research in this field. According to the Global Assessment of Soil Degradation in (GLASOD) in 1991, around a billion hectares of land worldwide suffer from severe soil degradation.

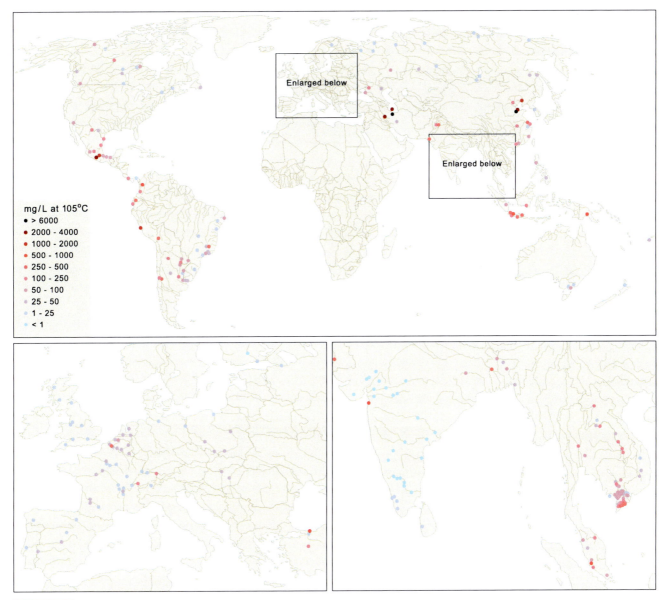

Figure 6.15 Concentrations of suspended sediments in rivers reveal a combination of natural and anthropogenic causes

Estimates of total soil loss from this area range up to 75 billion tonnes a year. Walling and Webb suggest that half this may be natural and half human-induced. Estimates of total suspended sediment in rivers suggest 13.5 billion tonnes are carried to the oceans, and another 14 billion are trapped in reservoirs. This means that just under 20 per cent of the eroded soil now reaches the ocean. There is, however, a strong possibility that the actual amounts are rather higher than the measurements suggest.

One of the most worrying consequences of soil erosion for water resources, apart from its effects on flooding, is the reduction in reservoir capacities caused by sediment trapped behind dams. A World Bank study in 1970 estimated that dams around the world were losing one per cent of capacity a year, or 50 km³, through sedimentation. In total at that time around 20 per cent of global storage had been lost through 1100 km³ of accumulated sediment. China is worse, losing over two per cent p.a. (McCully, 1996).

In many cases, lack of data on river sediments has led to developing countries taking out loans for expensive dams only to find that their lifespan is being considerably shortened by sedimentation. Many hydropower dams built in Central America in the 1970s and 1980s with loans from the World Bank and Inter-American Devel-

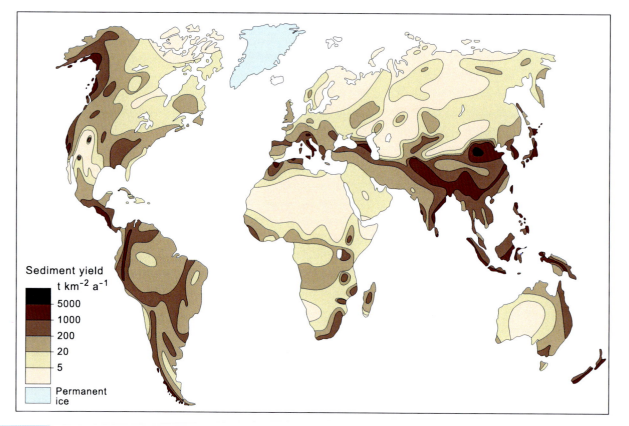

Figure 6.16 Sediment transport by rivers estimated by formula, according to Lvovich *et al* (1991). Tonnes per square kilometre of basin

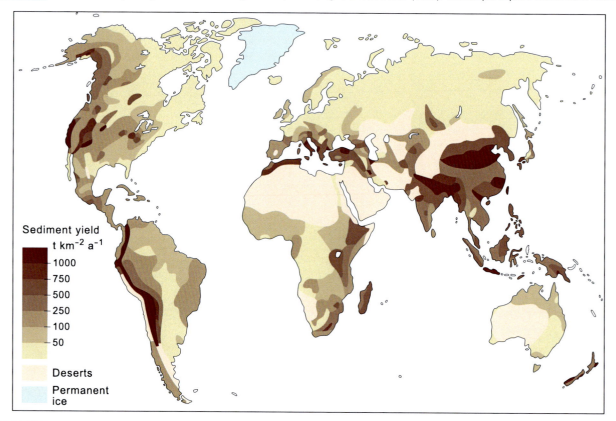

Figure 6.17 World sediment yields based on measurements, according to Walling and Webb (1983)

opment Bank are filling with sediment far quicker than expected: the US Army Corps of Engineers concluded that the life of the 135 MW Cerron Grande Dam in El Salvador had reduced to 30 years, compared with the original estimate of 350 years. All of India's large dams have been accumulating sediment faster than expected by two to three orders of magnitude. The Keban Dam in Turkey's GAP project collected two million m³ of sediment in its first ten years.

Fire, earth, water … and air

Fire was the source of the greatest technological and cultural advances of prehistory: it aided hunting by clearing forests, corralling animals and hardening spears, it cooked indigestible substances broadening the food base, and its warmth and light supported cave-dwelling, social gatherings, perhaps even language, and the great migration out of Africa. Evidence from the Transvaal suggests that even *Homo erectus* used it, perhaps some half a million years ago, though he could not make it (Roberts, 2004). Making and managing fire was one of the first great achievements of modern Man, *Homo sapiens*, who appeared a quarter of a million years ago. Eventually, it also became a weapon of war.

Natural wildfires were an obvious source of inspiration. Then, as now, they could be a threat as well as an aid, and they could be harnessed for good or evil. Australia's worst ever wildfires in New South Wales and Victoria, which killed 173 people in February 2009 and returned again later in the year, were part natural, part arson, perhaps part due to man-made climate change. More than 400 wildfires in Australia during 2009 were linked to a prolonged drought that may have been due to El Niño. Many other recent spates of wildfires in Portugal, southern France and California may be part natural, part man-made. Arson was thought to be the cause of several forest fires in northern Portugal in March 2009.

These forest fires generally follow extended droughts. Wildfires swept through wildlife reserves in Kenya in 2009 following a very prolonged multiyear drought. Fires in southern Portugal in 2005 followed an extreme summer drought and only spread more sporadically into the wetter and greener north.

Drought not only aggravates fire risk, the shortage of water also hampers fire fighting, a problem that was very evident in the fire fighting activities in Australia and California in early 2009. California's severe water shortage is partly due to historical problems with interstate transfers from the Colorado River, partly to population increases, increasing per capita demand and expanding irrigated agriculture, and partly to the increasing severity of climatic droughts. It has suffered numerous wildfires in recent years, notably in August to October 2007 and again in 2008, but not normally as early in the year as those in 2009. By the time of the 2009 wildfires, California had suffered three years of below average rainfall, estimated to have cost $3 billion in lost production and 95,000 jobs, mainly in agriculture. For two of those years, California's governor Schwarzenegger had been fighting to get increases in cross-border water allocations. The water supply system was designed for a state population of 18 million, not the current 30 million. A State of Emergency was announced following the fires in February 2009, and the governor urged a 20 per cent voluntary reduction in water use with the threat of compulsory restrictions. A second State of Emergency was declared the following May, when 18,000 people were evacuated around Santa Barbara as fires spread down parched, forested canyons.

The fourth component in the ancient philosophers' quartet of elements, air, is also an integral part of fires: oxygen is absorbed and CO_2 released. But for the spread of the fire itself, the wind is critical. Gusts stoke the fire and the fire sucks air in at gale force. During September 2009 wildfires around Los Angeles generated a rare storm cloud, a pyrocumulus. It produced heavy rain that helped douse the fires, but this was offset by lightning that set off more fires and violent gusts of wind. Again, global warming is predicted to increase the frequency of strong winds as well as droughts in most of the current fire risk regions. Strong winds exacerbated problems with the wildfires in California, Oklahoma, Texas, South Carolina and New South Wales early in 2009.

The 2009 UK Climate Projections report suggests that even Britain may become more prone to wildfires, especially on dry moorlands in summer, as global warming progresses this century.

After-effects of scorched earth

Fire may be a stimulus for renewed vegetation growth, provide fertilizer, clear dead wood and even unlock the seeds of some plant species. But it can also increase soil erosion and flood risk. Research in Spain has shown that burnt soils tend to be less water absorbent, with lower infiltration capacity and saturated hydraulic conductivity, which leads to more overland flow (Sala and Rubio, 1994). In the Mediterranean, the fires tend to occur in spring and summer and leave the soil bare and vulnerable during the autumn storms. The increased surface runoff feeds flood waters to the rivers and the lack of protection from vegetation – roots binding the soil and leaves breaking the force of the rainfall – often leads to catastrophic amounts of soil erosion. Flood risk is further enhanced by reduced evapotranspiration losses once the forest has been destroyed, so that a higher proportion of the rainfall is available to run off. This is especially so with eucalyptus forests, which have high rates of evapotranspiration. These trees also contain highly combustible oils that fuel the fires, as exemplified in the New South Wales and Portuguese fires in 2009, which bake the earth.

One positive result of burning is the enrichment of plant nutrients in the soil, especially carbon, phosphorus and nitrogen. This is one reason for deliberate burning on agricultural land. But the degree of enrichment depends on a number of factors. Phosphorus is more enhanced after moderate fires and decreases in the most intense wildfires. Moreover, the nutrients tend to be mineralized and are more easily washed away by surface runoff, especially on wildland hill slopes, increasing the nutrient status of rivers and lakes. The fire also tends to burn off the all important organic material in the soil and can lead to 'lateritization', the formation of reddish, blocky rather sterile soils similar to tropical laterite, in the more intense fires. Another common result is 'hydrophobia', in which the desiccated soil repels water. This can increase the amount of surface runoff, reduce soil moisture contents, and thus inhibit the regrowth of vegetation.

In summary, there is a wide variety of possible responses, depending on the local environment, climate and inten-sity of fires. Differences in soil texture and structure can have a major effect on response: sandy soils that develop hydrophobicity can have much higher rates of surface runoff than clay soils with good aggregate structure (Pradas *et al*, 1994). But overall the most common result of wildfires is a degradation of the soil, reduced water retention and increased flood hazard.

Conclusions

Wildlife is under threat as never before in human history. Water resources developments are part of the problem. Too many water developments have been made without consideration of the wider environment. Water resources are also the victim, when land use changes are made without regard for the impact on water. New working methods are needed encompassing a more integrated approach, especially more regard for ecological impacts – a greener approach to green water.

Discussion topics

- Follow the latest developments in solving eutrophication or acidification.
- Expand on the water impacts of forests or wetlands.
- Study the effects of soil erosion on water resources.
- Study the environmental impacts of forest fires.
- Research the issue of sedimentation in reservoirs.

Further reading

The website of the International Union for Conservation of Nature (IUCN) gives regular updates on habitat losses at: www.iucn.org.

Fuller coverage of acidification problems can be found in:

Steinberg, C.E.W. and Wright, R.F. (eds) 1994. *Acidification of freshwater ecosystems – implications for the future.* Chichester, Wiley.

A Special Issue of *Hydrological Processes* investigates the trends and processes of rising river temperatures:

Hannah, D.M., Webb, B.W. and Nobilis, F. (eds) 2008. River and stream temperature: dynamics, processes, models and implications. *Hydrological Processes*, **22**(7).

Progress on soil erosion and sediment movement can be found on the websites of:

The World Association for Sediment and Erosion Research (www.waser.cn)

The IGBP-PAGES-LUCIFS project, which takes an historical view (www.pages-igbp.org)

Unesco International Sediment Initiative (www.irtces.org/isi/)

And the International Association for Hydro-Environment Engineering (www.iahr.net), which includes the work of the International Coordinating Committee on Reservoir Sedimentation (ICCORES).

7 Dams and diversions

Dams are the most fundamental tools of water engineering. They create resources by storing water in periods of good supply, even converting flood hazard into useable resource, accumulating riverflows until needed or making a serviceable volume of water out of more meagre flows. They have been the life blood of civilization, allowing irrigated agriculture and urban development. In the modern world, they are also essential elements in large-scale river diversions and interbasin transfers. Around half the world's rivers now have one or more large dams on them and the total amount of water stored in the world's reservoir is around 7000 km^3, which amounts to nearly a sixth of the total annual riverflow in the world. Nearly two-thirds of the world's 227 largest rivers are dammed. There are now more than 45,000 large dams, with a further 1600 in progress, in over 150 countries. Two-thirds of all large dams under construction in 2003 were in China, Turkey, Japan and Iran.

About half of all dams are primarily to supply irrigated agriculture, a little over a third are principally for hydropower and the remaining one-eighth are for public water supply. The World Commission on Dams (2000) estimates that 30 to 40 per cent of the 271 million hectares of irrigated lands worldwide depends on dams, producing between 12 and 16 per cent of world food supplies.

Yet dams are controversial. They destroy as well as create. The first controversy in the canon of modern environmental protection came with a dam built in a newly designated National Park in America in the 1900s (see section headed 'Dam protests', later in this chapter). Destruction of the environment, especially endangered ecosystems, is still a powerful objection today. To this may be added displacement of people and destruction of traditional livelihoods. Estimates of the total number of people displaced vary widely and unsurprisingly there is usually a big difference between the official and the actual figures. Global estimates range from 40 million to 80 million. Monosowski (1985) estimated that more than half a million people had been displaced solely by dam projects funded by the World Bank between 1965 and 1985. The numbers said to have been displaced during the massive dam-building programmes in China and India between 1950 and 1990 range from 26 million to 58 million.

Recent evidence also blames dams and reservoirs for serious impacts on rivers and wildlife downstream: withholding sediment and nutrients from rivers and floodplains, blocking migratory fish and decimating fish life by reducing water depths, allowing temperatures to rise and causing deoxygenation of the water. The World Wildlife Fund's report *Rivers at Risk* identifies 21 rivers at particular risk. The Yangtze is most at risk, with 46 large dams either developed or planned: the Yangtze river dolphin was declared officially extinct in 2007 (see *China's National Strategic Plan: South to North Transfer*, below). Reservoirs in subarctic regions of Canada and Russia that have flooded conifer forests have suffered from the effects of rotting vegetation; hydrogen sulphide can be released by anaerobic decomposition. In northern Quebec, reservoirs suffer from a toxic build-up of mercury in the food chain and the Cree Indians no longer eat the fish. The slow rate of decay in subarctic temperatures means that problems caused by decomposition can be long-lasting: one reservoir in the Russian taiga is still suffering from decomposition problems after 80 years. Ecologists also fear that the large surface area of stagnant reservoirs will accumulate toxic pollutants from the atmosphere. Most recently, concern has centred on the release of the powerful greenhouse gas, methane, from the surfaces of large reservoirs, especially in tropical climates. The stagnant bays in the Balbina reservoir, Brazil, are one such example. Calculations such as 67.5 tonnes CO_2 being saved by the Itaipu dam in Brazil compared with electricity generated by fossil fuel ignore the methane problem.

Moreover, although reducing flood events may be seen as an advantage for homes and livestock, in many parts of the world both people and ecosystems have grown to rely on floods. For people and agriculture, the lack of new nutrient-laden silt on the fields can lead to a greater reliance on artificial fertilizers, which then leach out into the rivers and cause eutrophication, as experienced on the River Nile after the Aswan High Dam stopped the natural fertilization of the floodplain by silt deposited by the annual Nile floods. Reduced flows also mean less dilution of pollutants, which can be critical in urban areas; pollution around informal settlements on the river islands around Cairo, exposed since the building of the Aswan High Dam, is an extreme case. Some ecosystems also rely on periodic scouring and destruction to create a new environment for pioneer species. Recognition of this has led to a number of experiments with artificially created floods in Africa and North America (see *Artificial floods* in Chapter 15). In the USA, the process has gone one step further: to actually demolish dams for ecological reasons. The failure of the flood control dams to prevent the 1993 Mississippi floods also boosted the modern philosophy that integrated flood and environmental management is better than attempting outright control using hard structures. This is further adding to a decline in dam building.

One of the big mistakes of the twentieth century was to assume that engineering experience and scientific understanding in one part of the world, especially in Europe and North America, is directly transferable to developing countries. The former President of the International Association of Hydrological Sciences, Professor Kuniyoshi Takeuchi, once remarked that river engineering and flood control in Japan had suffered from applying Western approaches after the Second World War and disregarding more environmentally friendly, traditional Japanese methods: soft solutions, perhaps involving paddy fields, as opposed to 'hard engineering' with dams and flood containment. His ideas are expanded in Kundzewicz and Takeuchi (1999). In Takeuchi (2002a and b) he notes that, in 2000, the River Commission of the Japanese Ministry of Construction finally reversed years of attempts to contain floods and adopted a policy which effectively allows controlled flooding in urban areas.

The inadvertent impacts can be worse in developing countries. It is for this reason that training engineers and water scientists native to these countries, who have a better feel for the local water regimes, cultural milieux and traditional approaches, so-called 'capacity building', is now high on all agendas aimed at improving water management in the LEDCs.

The WCD (2000) reported that nearly 60 per cent of the environmental impacts of dams were unanticipated by the planners. This is partly because the need for assess-

ing environmental impacts has only been generally recognized in the last 30 years or so, and most developing countries still have no, or very inadequately applied, environmental laws. Impacts also vary widely, depending on the physical nature of the dam, its mode of operation, and the river regime. Water supply and irrigation dams, which divert water away from the river through pipes or canals, tend to have a larger environmental impact on the river reaches downstream. Water for public supplies is often not returned to the same river. Water diverted for irrigation often returns, but minus losses and plus pollution from fertilizers and pesticides. Dams designed for hydropower, flood control and river regulation tend to have lower impact, because most of the water impounded in the reservoir is subsequently released into the river.

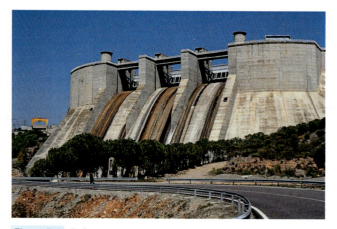

Figure 7.1 El Grado dam on the Rio Cinca, N Spain. The dam generates hydropower and diverts water for irrigation schemes, which drastically reduces flow downstream

The public water supply dams built in the late Victorian period were designed to provide clean, disease-free water, following the cholera and typhoid outbreaks in industrializing cities like London and Birmingham. They were single-use reservoirs and water companies protected the water surfaces and the catchment areas against pollution. In Birmingham City Corporation's Elan Valley dams in Wales, the city authorities bought all the farms, and controls established 100 years ago are still in place on stocking levels, limiting manure and banning artificial pesticides and fertilizers. Water is diverted to Birmingham through a 118 km gravity-fed pipeline.

Over the last 50 years a different type of public water supply dam has developed: the river regulation dam. The idea has its roots in flood control dams, like many of those built on the Mississippi since the inter-war period, whereby the dams hold back water and release it later

– after the flood wave has passed. River regulation for public water supply involves using the river itself as the pipeline, controlling the flow in order to guarantee a fixed amount available for abstraction at some point downstream. Like flood control dams they alter the time distribution of riverflows, rather than the total volume of riverflow. This means that the impact on the riverine environment downstream of the dam is more muted until the point of abstraction. There are still some impacts.

Figure 7.2 compares water and sediment discharges in two adjacent and very similar rivers in Wales, one natural, the other regulated for hydropower. The river that is dammed, the Rheidol, carries far less sediment than its natural sister, the Ystwyth, the frequency of low flows is reduced and medium flows are enhanced, as water stored in storm events is gradually released to generate power. In this instance, the power company operating the Rheidol scheme is also legally bound to prevent

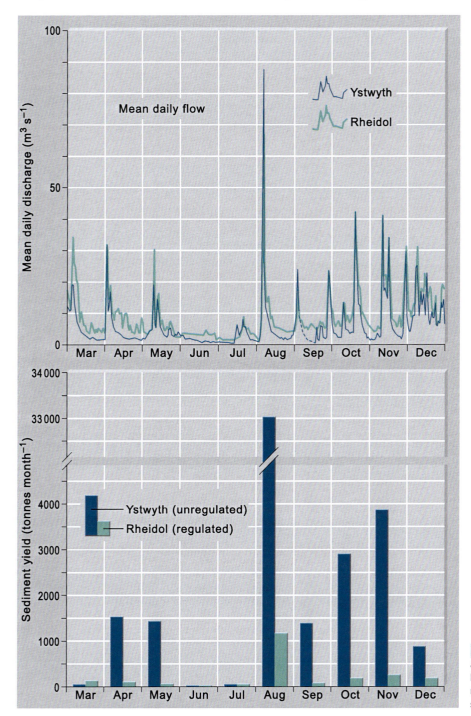

Figure 7.2 Comparison of two adjacent and similar rivers in Wales, one regulated by hydropower dams (Rheidol), the other natural. The regulated river carries less sediment and has fewer low flows

119

floods, if necessary by releasing water ahead of a storm to increase the free storage capacity in the reservoirs. For both hydropower and regulation for public supplies, the need to protect the reservoirs from pollution is much reduced. There are plenty more opportunities for pollution to enter the river between the dam and the eventual abstraction point. In Britain this contributed to two developments in the 1960s and 1970s: less environmental control above the dam and more below, opening up reservoirs for leisure use and the growth of water quality legislation to protect the rivers.

Increasingly, dams have multiple uses. A number have also undergone a change of use or a change in their mode of operation. The Elan Valley system is now a river regulation scheme as well as direct supply. The severe drought of 1976, an event estimated to happen about once in 400 years, when industry was badly affected in south east Wales, led to a plan to use water from Elan routed down the River Wye to South Wales. The scheme was implemented in time to moderate the 1984 drought, which is now the 'design drought' for most dam operations in Wales.

Some dams are also 'conjunctive', which means that they operate in conjunction with groundwater supplies. Britain's River Severn is a good example, where water released from the Clywedog and Vyrnwy reservoirs in Wales may be supplemented by groundwater pumped out of Triassic sandstone rocks in Shropshire in order to meet the legal minimum discharge at Bewdley, where water is pumped into the Birmingham water supply system. The operation of the whole scheme is controlled according to predictions from a computer model of riverflow and public demand. Groundwater pumping is a last resort, because it is costly in power and the water is of a different quality and temperature to the riverwater. Consequently, it has to be released in limited amounts and into smaller tributaries so that these differences are largely corrected by the time it reaches the main river. Temperature and water quality can also be a problem with water released from dams. Water released from deeper in the reservoir tends to be warmer in winter and colder in summer than the riverwater. It also tends to lack oxygen. Environmentally sound release therefore often requires mixing water from top and bottom by opening valves at different levels, and aerating the water, for example, by spraying it in the air.

Objections to new dams grew towards the end of the twentieth century, partly because of environmental awareness, but also because advances in dam-building technology have been making ever-larger dams feasible. As dams have become larger, the environmental and social impacts have generally increased, although location-specific factors sometimes mean that impacts are not always commensurate with the scale of the project. The trend towards more and more large dams has also been fuelled by growing populations and the desire of developing countries to underpin economic development. The World Bank has been in the forefront of funding many of these projects in fulfilment of its founding aim to eradicate poverty in the Third World. Larger dams and larger reservoirs have also displaced greater numbers of people, especially as many have been built in heavily populated regions of the developing world. Large dams also cost more, requiring more loans, and take longer to plan and build. Itaipú in Brazil, the largest before the Chinese Three Gorges Dam, took 16 years to build. The relatively small Carsington Reservoir in England took nearly a quarter century from the beginning of the feasibility study to full impounding. Kielder Dam, which was designed to form the backbone of industrial revival in northern England took nearly 15 years from feasibility to operation, and cost £167 million. It opened in 1981 in the middle of a recession. British Steel and BASF failed to come and take up the capacity they had requested. Domestic and commercial customers have been paying off the interest on the loans through their water bills ever since. This highlights another problem for dam builders – the economic environment can change radically during the long gestation periods.

The dividing line between large and small dams is somewhat arbitrary, but the International Commission on Large Dams (ICOLD) defines large dams as more than 15 m high, or between five and 15 m high but impounding a reservoir of more than three million cubic metres. The new Narmada dams in India and the Three Gorges in China are prime examples.

The World Commission on Dams was set up to investigate the advantages and disadvantages of large dams. The Commission produced a well-balanced report in 2000, which has contributed to a more environmentally aware view of new dams in much of the developed world. Large dams are often crucial sources for irrigation, public water supply and hydropower, and will remain so for the foreseeable future. But many have been built for the wrong reasons. Despotic regimes have built dams for reasons of national pride and influence, new irrigation schemes have been implemented to support agribusiness without regard for the needs of private farmers, or corruption has led to local people being deprived of their inheritance. Poor, rural, indigenous people have generally been excluded from the decision-making processes and have borne the brunt of the adverse impacts. The WCD reported that displaced people were consulted in only seven of the 34 dams in the Cross-Check Survey. And even in the 1990s, Environmental Impact Assessments (EIA) were undertaken in less than 40 per cent of newly commissioned dams.

The WCD also found that large dams generally cost far more than the original estimates, typically half again as much, and usually fail to quite meet the planned output targets. The Commission found that economic evaluation methods like risk assessment or distributional analysis were rarely used in planning: they were applied in only 20 per cent of dam projects undertaking in the 1990s. In many developing countries, the dams rarely produce enough revenue from water tariffs alone to recoup capital and running costs. This is not new. It was an axiom of the US Bureau of Reclamation during the settlement of the West in the early twentieth century that hydropower should be incorporated in dams in order to finance irrigation projects (Reisner, 1993).

The period of peak dam building is over. Figure 7.3 shows that dam building accelerated in the 1950s as the world recovered from the Second World War, and from the 1960s to the 1980s over 4000 large dams were being built annually. It peaked in the 1970s at nearly 5500 a year, but by the end of the century it had fallen to less than half that rate. Anti-dam protests and the building costs were major factors in this decline, as international finance became more selective in its funding. Nevertheless, to some extent the decline in numbers is misleading, as it conceals the increasing size of individual dams. There are also regions where building continues apace, notably China (see Changming Lui's article 'Hydropower dams and flood control in China', later in this chapter), whereas in other regions, like the USA and Europe, dam building is almost at a standstill. There are even moves to dismantle dams, sometimes because of their age, as in Spain, sometimes also for environmental reasons, as in the USA. Between 1999 and 2006, the USA dismantled 654 dams; Spain decommissioned 300 small ones. The average large dam is now 35 years old. The world recession in 2009 and its aftermath contributed to further reductions in dam building. Dams in the planning stage have been cancelled, and some already under construction have been halted or delayed (see *Debt and transfiguration? – Recession 2009* in Chapter 4).

During the last quarter of the twentieth century, the developing world took on substantial sums of international debt to build large dams, principally for irrigation and hydropower, especially in countries lacking domestic supplies of fossil fuel. The World Bank was a major source of funding, together with commercial and regional development banks. During the peak period of lending between the mid-1970s and mid-1980s over $4 billion a year were being borrowed from multilateral and bi-lateral development banks to finance dam projects – and this was probably barely 15 per cent of the total amount of investment. The WCD estimates that the total amount of borrowing from these banks since 1950 has been $125 billion. As interest rates rose, many found difficulty meeting payments. Brazil was a prime example. Much of the international debt that Brazil accrued during the 1970s and 1980s was acquired to fund large dams, designed principally to generate electricity. Brazil's payment defaults and announcement of a debt moratorium triggered the near-collapse of the international banking system in 1987. The experience gave a major fillip to the campaigns to write off the debt of developing nations. Yet ironically, in the case of Brazil it was caused by borrowing to fund the growth of big city economies, like the huge Itaipú Dam (1975–1991) to serve Rio de Janeiro, rather than to help the

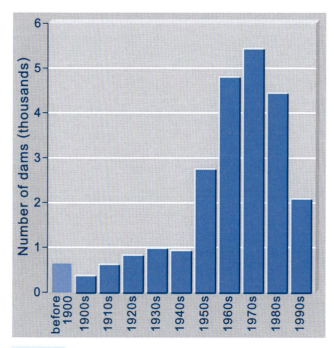

Figure 7.3 Rates of dam building in the twentieth century

rural poor, which many subsequent well-meaning public campaigns for Western countries to write off debts have focused on.

The shift in the global pattern of dam building away from its earlier focus in North America and Europe towards the developing world is largely reflected in the changing geography of hydropower generation over the last two decades (see Figure 7.5). During the 1990s, developing countries accounted for some 80 per cent of the total worldwide investment in dams, spending $20–30 billion a year on large dams.

Hydropower

Around 20 per cent of the world's electricity is currently supplied by hydropower, amounting to five per cent of all energy use. It is generally regarded as sustainable and renewable. Modern hydropower turbines are the most efficient means of generating electricity, converting 90 per cent of the available energy potential into electricity, compared with around 50 per cent from fossil fuel plants. Table 7.1 shows that hydropower has the lowest environmental impact according to a standard method of assessment.

Table 7.1 Environmental impact of hydropower compared with other energy sources

Source of electricity	Water impacts	Total weighted environmental impact* (ascending order)
Hydro – run-of-river	0	21
Hydro-dam impoundments	5.5	30.5
Wind	0	34.5
Solar – photovoltaic	0	41
Biomass	4	47
Nuclear	4.5	47.8
Natural gas	2	71–91
Coal	3.5	216.5
Oil	5	265

*Weighting emissions of solid contaminants by 10, greenhouse gas emissions by 20 and 1 for other impacts including land use, water, radioactivity and waste. Sources: CANDU Canada's Nuclear Energy Source; SENES Consultants Ltd (Canada) and Ontario Power Authority. Rating the radioactivity impacts of nuclear power as 6 and coal as 10 and weighting these by 1 is likely to underestimate nuclear impacts, considering the impacts of uranium mining (see *Radioactive pollution* in Chapter 5) and the lack of long-term solutions to waste disposal. It reflects the sources of the assessment, which all have strong nuclear interests, but it would not alter the general rankings significantly.

One in three countries derive more than half their electricity from hydropower and about two dozen countries rely on hydropower for 90 per cent or more of their supply. Twelve rely 100 per cent on hydro. Recent shifts away from fossil fuels have contributed to expansion. Between 1992 and 2001, world hydropower generation increased by 366 billion kW, or 1.7 per cent p.a. Canada tops the league table of hydropower producers with over 300 billion kW (see *James Bay Hydro – redesigning a landscape*, below), half again as much as the USA in fourth place. Brazil and China vie for second place, with China in the ascendancy, and Russia comes fifth.

Considering that more than a quarter of the world population still have no electricity and that energy demand could triple by 2050, there is going to be considerable pressure to develop more hydropower over the coming century. The UK is already planning a 900 km cable to import hydroelectricity from Norway, and Norway's hydropower industry will be one of the climate change winners as global warming increases its rainfall. A study by the Utility Data Institute in America estimates that it is feasible to develop 14,400 TWh/yr with present-day technology and that it is currently economically feasible to develop 8000 TWh/yr of this, compared with the 2600 (700 GW) currently operational and the 108 GW under construction. The hope must be that more heed is paid to the lessons of past mistakes for more equitable and sustainable development of resources.

Expanding hydropower is one possible way of reducing global warming. ICOLD (1999) report that if only half the global potential hydropower is developed this could reduce greenhouse gas emissions by 13 per cent. However, methane emissions from the reservoirs could reduce this substantially.

Unlike irrigation and public water supply, hydropower consumes very little water and returns water directly to the river. In this regard hydropower dams have considerably lower environmental impacts. However, there is an important distinction between the amount consumed, i.e. wasted or lost to the rivers, and the gross amount of water used. In the United States, hydropower is the largest *gross* user, but it accounts for only 0.5 per cent of total consumptive use, the so-called 'irrecoverable losses'. The main losses from hydropower operations are due to evaporation from reservoir surfaces. This is not to say that such losses from individual reservoirs may not be significant, hence attempts to contain them by reducing water temperatures or physically blocking evaporating water vapour (see *Controlling evaporative losses* in Chapter 17).

Even so, consumptive losses are much higher in coal-fired and nuclear plants because the cooling process is achieved mainly by evaporation. Britain's Didcot A 2000 MW coal-fired station takes 24 million m³ a year from the River Thames and returns just 17 million, as well as consuming over one million m³ of treated mains water from the river to top up the boiler circuits. Nuclear power stations use 50–100 per cent more water than coal-fired stations, hence their siting near a good supply like the sea, a lake or a large river: the UK's Dungeness power station takes in 60 m³ of seawater per second, the French station at Gravelines takes in 120 m³/s. All the nuclear power stations currently operational in the UK use seawater, but the French mostly opted for freshwater from rivers, which was probably a strategic mistake. Exceptionally low flows on French rivers, notably the Loire, during the summer of 2003 forced a quarter of EDF's 58 nuclear plants to close. Since 80 per cent of France's electricity comes from nuclear plants, the country was faced with hurriedly arranging sizeable imports of power from Britain and Germany. The average price of electricity shot up 1300 per cent and EDF lost €300 million paying for imported power. Low flows caused French stations to close down again in summer 2009 and France survived on imported electricity. Climate change is likely to increase the frequency of such events. Britain is particularly favoured in this regard by its large coastline relative to land area, but a different geographi-

Figure 7.4 Symbiosis of hydropower and nuclear power (in distance) plants at Cowan's Ford Dam (1964) on the Catawba-Wateree River system in North Carolina run by Duke Energy. Both plants utilize the same reservoir, which is backed up by ten other reservoirs upstream

cal ratio has forced the USA to develop many inland, freshwater sites. In North Carolina, Duke Energy has solved the problem by using some of the water stored in the 11 reservoirs in the Catawba-Wateree river basin, which serve its 13 hydro plants with a combined capacity of 841 MW, to provide cooling water for its fossil fuel and nuclear installations. The lakes also provide public water supply, recreation and wildlife habitats.

Figure 7.5 shows the changing geographical distribution of dam building for hydropower. In 1995, North

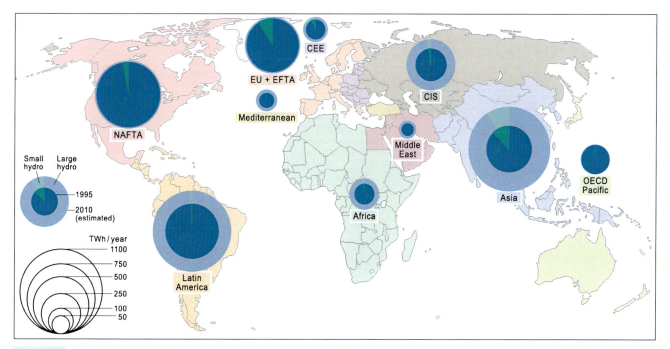

Figure 7.5 Distribution of new hydropower developments

America was the top producer of hydropower, followed by Europe and Latin America. While Europe and North America have virtually stopped large-scale developments, in many cases expanding capacity only by retrofitting generating facilities in existing dams, Latin America has continued to expand and has been joined by Asia. Both Asia and Latin America now generate about 20 per cent more hydropower than North America, and Asia is set to become the largest hydropower generator of all. The Three Gorges Dam alone was designed to provide up to 20 per cent of China's national power requirements (see the section headed *Three Gorges and China's Water and Power Needs*, later in this chapter). Africa, on the other hand, remains a relatively small participant, despite being fourth in terms of expansion since the mid-1990s and despite originally leading the way with dams like the Kariba (built in 1959) on the Zambezi and the Aswan High Dam (1970) on the Nile. To some extent Africa's lack of hydropower development is a reflection of its deep-seated problems of lack of wealth, poor governance, corruption and international fears of investment. Yet during the Cold War, the Soviet Union and the West vied for political influence through investment in dam projects. It remains to be seen whether China's growing investment in African sources of raw materials will also bring substantial Chinese investment in hydropower.

Estimating the total potential for hydropower development in the world is difficult because technology and economics are continually changing. Conventional estimates currently place 30 per cent of world capacity in Asia, largely on the great rivers that issue from the Himalayas and the Tibetan plateau. Africa holds a further 20 per cent. According to the conventional criteria, by the turn of the millennium only about 20 per cent of world potential had been developed. There is still considerable room for expansion on conventional terms. But at least as early as the 1970s technology was challenging the conventional requirements, like highwater 'heads' typical of mountain valleys with steep gradients and large river discharges, or narrow valleys with strong and impermeable bedrock foundations. The first large-scale development to challenge convention was in Canada's James Bay lowlands, where river networks covering an area the size of France have been redesigned by megadams and interbasin river diversions. These are low-head sites in a landscape where the relief is so subdued that it is often difficult to define where one river basin begins and another ends, and where the resource is mainly snowfall accumulated over the eight or nine

months of winter (see *James Bay Hydro: redesigning a landscape*). More recently, run-of-the-river and microhydro sites are making a comeback as developers eschew large dams and government agendas for expanding alternative energy sources favour small-scale generators.

Small, micro and pico hydro

A second feature of the changes in hydropower distribution is the amount of small scale hydro being generated relative to that generated in large dams (Figure 7.5). Small hydro is commonly defined as generating less than 10 MW, micro hydro less than 100 kW, and pico less than 5 kW. The amount of small hydro has grown substantially over the last decade or so. Even so, only about five per cent of the estimated world potential of 150–200 GW has been exploited to date. In 1995, Asia led in the percentage generated in small schemes, followed by Europe, with very little elsewhere. By 2010 this is still the case, but the percentage in Asia has fallen as the big powers have undertaken major expansion of large hydropower dams. In contrast, the Russian Federation is actually increasing the proportion of power generated by small schemes.

Small hydro and micro hydro have many advantages. They are cheaper and tend to have less environmental impact. By definition the dams are small or non-existent. They may be little more than a weir. Small hydro plants used to be run by some villages and minor towns in Britain, like Llandysul on the River Teifi in Wales, before the advent of the national grid in the 1950s. Britain also has many old water mill sites, possibly 20,000, and thousands of weirs that might be fitted with small, modern turbines at minimal cost. A survey in the 1980s found a total hydropower potential in Wales equivalent to 80,000 tonnes of oil in small sites capable of generating 25 kW or more, but little if any was developed despite the preferential tariffs offered by the government's 1989 Non Fossil Fuel Obligation. A quarter century later, with the UK government's new tariff incentives to cut carbon emissions, the Environment Agency (2010) has identified 26,000 sites in England and Wales capable of generating up to five MW and offers guidelines and assistance for developers to comply with the EU Water Framework Directive and the UK Salmon and Freshwater Fisheries Act. Two-thirds of all sites are in the uplands. Some 12,000 sites are rated environmentally sensitive, mainly because of migratory fish, but over 4000 sites are designated 'win-win' with no environmental problems. British

Waterways began installing small hydro in canals in 2008. The USA has also been re-evaluating small rural dam sites. The US Federal Power Commission found enough such sites to provide power for 40 million people.

Micro hydro and small hydro have been widely popular in many developing countries for some time. Micro and pico scale, private domestic power supplies may be generated from riverflow alone with no structures at all, especially in hilly areas; the water equivalent of domestic wind turbines. Vietnam has over 100,000 pico hydro turbines: a typical 300W turbine costs $20 and serves a whole family. Such domestic devices, together with small community hydro, could play a minor but significant role in the coming expansion of non-fossil fuel power sources in developed countries as well, given the right governmental support. Admittedly, rivers are not as widespread as the wind, but they are often less variable and flows tend to be highest in winter when electricity demand is also highest. The British micro hydro company Segen estimates its systems could repay the cost in ten years.

It is predicted that small-scale hydro will more than double in the UK over the next 20 years as part of the government's aim to obtain 15 per cent of electricity from renewables by 2020, while large hydro will remain more or less static. Even so, it is likely to account for only ten per cent of the UK's total hydroelectric output, excluding pump storage, by 2030, compared with four per cent in 2010. Most recent developments in the UK have involved retrofitting power turbines within existing water supply dams. Turbines were sometimes installed in dams in the 1960s to provide local backup power, but since the government's Non-Fossil Fuel Obligation legislation in 1989, retrofitting has become widespread and power is sold to the national grid at a premium. The subsequent Renewables Obligation favoured large hydro, but the 2010 Clean Energy Cashback Scheme is expected to encourage micro and pico hydro, and although the tariff for micro hydro may be only half that of solar, around ten times more energy can be generated by hydro in the British environment. The recession also encouraged micro hydro: the British Hydropower Association received over 200 applications in 2009 compared with only between five and ten a year before the recession.

Run-of-river hydro

The term 'run-of-river' is applied to non-storage or 'non-impoundment' schemes. Whereas storage schemes derive the pressure head to drive the turbines from the mass of water held back by the dam, run-of-river schemes derive the momentum from the natural, or artificially enhanced, gradient the riverwater flows down. Many small-scale hydro installations, especially pico hydro, can be called 'run-in-river' with turbines actually in the river, and they may need little or no structural interference with the flow. Run-of-river schemes can be much larger and involve barrages or small dams and channels or pipes to divert water down steeper gradients before returning it to the river. The turbines are not installed in these dams and the lack of a large dam and reservoir generally means they have less environmental impact. The impacts are largely confined to the section of river that is bypassed, where riverflow is reduced.

Some of these schemes can be very large. The high rainfall in the Coast Mountains and temperate rainforests of British Columbia, together with meltwaters from the icefields, offer a huge potential for run-of-river hydro. Developments have been underway since the BC government produced its Integrated Electricity Plan in 2006, which predicted that by 2025 the province might need to import 45 per cent of its electricity from Alberta or the Pacific North West. Run-of-river hydro schemes are either in progress or planned for rivers feeding the Knight, Toba and Bute Inlets. These will link together numerous small installations on small-to-medium rivers to create a huge potential, literally a 'Green Power Corridor' that could save some 2.8 Mt of CO_2. Plutonic Power is planning a scheme with General Electric for a series of 17 mini-dam sites on rivers flowing into Bute Inlet. It will have a combined output of 1027 MW, equivalent to two of the latest CANDU nuclear plants. The plan was filed in 2008 in response to BC Hydro's 'Clean Power Call'. Work is already progressing on a similar 196 MW scheme on Toba Inlet. Negotiations have been successfully concluded with a number of First Nation groups in the area. But one problem in Canada is that peak riverflows tend to be at melt-time in the spring and flows are low in winter and summer when power demand peaks for heating or air conditioning. There are fears that commercial pressure to extend the operating period of these installations into the breeding season could affect the salmon.

The Ghazi Barotha hydropower scheme on the River Indus immediately downstream of the mighty Tarbela Dam in Pakistan is a large, single unit run-of-river project, which involves a river diversion channel that takes the water impounded by a low barrage down to the turbines and returns it to the Indus 52 km downstream. It became operational in 2004 and generates 1450 MW. The environmental impact is reduced by returning flows to the river, but there are significant impacts on both wildlife and the livelihoods of the indigenous population along the river reach that has been bypassed. These include decimation of fish life, drastic reductions in the once numerous migratory birds that rested and fed in the riparian wetlands, less water for domestic use, especially traditional clothes washing in the river, and less dilution of pollution from domestic and industrial sources. Pollution could be worse if the economic situation allows the new industrial estate to return to full operation.

The diversion channel gains momentum by short-circuiting a bend in the river. A similar, but much smaller, 50 MW scheme has operated on the River Rheidol in mid-Wales since the 1960s. This capitalizes on what might be called a 'geomorphological accident': it short-circuits an 'elbow of capture', where a river draining directly to the sea has eroded back and captured a river that used to follow a much gentler gradient, almost at right angles. In contrast to Ghazi Barotha, however, impacts on the Rheidol are minimized by legal requirements enforced by the Environment Agency, which monitors discharges and water quality. The diversion dams must release minimum levels of compensation flows down the bypassed reach to serve farms and fisheries, and to dilute heavy metal pollution from abandoned lead mines in order to maintain minimum water quality. A fish lift, which works like a lock in a canal, and a fish ladder also allow migratory salmonid fish to get to suitable spawning grounds within the bypassed reach. The three-dam scheme must also be operated as a flood control system, releasing water ahead of critical storms, according to indications from Met Office weather radar and E.ON's own raingauges in the mountains. The Rheidol scheme is really a hybrid, with two smaller generators in standard positions in two of the dams and the other, the largest, at the outfall of the diversion tunnel.

Pump storage

Yet another variant form of hydropower plant is the pump storage scheme. This may best be described as a water power battery designed to store energy generated from other electrical sources. The aim is to utilize surplus electricity, especially off-peak at night, to pump water into an uphill reservoir and then release the water to generate hydroelectricity as the need arises. Such

schemes can be more than 80 per cent efficient. They may allow coal-fired plants to continue generating power at full potential even though the demand is not there. This way, the coal-fired plant remains operating at maximum temperature and so emits less particulate pollution into the air. The current capacity of pump storage schemes worldwide is about 100 GW, but at least 1000 GW are estimated to be potentially available for development, very probably far more. These schemes tend to use more modest sized reservoirs than conventional gravity-fed schemes and therefore the irrecoverable water losses are much lower.

Europe's largest pump storage scheme, Dinorwic, was opened in 1984 in the heart of the Snowdonia National Park in North Wales, generating 1.7 GW. At 16 seconds, the scheme has the fastest response time in the world. The scheme made use of and rehabilitated a disused slate quarry, and for aesthetic reasons within the national park the generators are housed in the largest man-made cavern 750 m beneath the Elidir Mountain. Scotland has a number of medium-sized pump storage schemes, including Ben Cruachan (400 MW), Sloy (160 MW) and Foyers (305 MW), which pumps water from Loch Ness.

The 150 MW Pont Ventoux-Susa hydropower scheme in Piedmont, Italy, is an interesting hybrid scheme that generates electricity using a mix of natural gravity feed and pump storage. The plant collects water from the Dora Riparia and Clarea rivers. At night, when electricity is cheaper and demand is lower, water is pumped from the Dora River below the turbines back up to the Val Clarea reservoir, whence it is released in daytime

Figure 7.6 The Val Clarea reservoir, the upper storage basin in the hybrid pump-storage hydroscheme at Pont Ventoux-Susa, North Italy. The reservoir has an earth dam and is lined with bitumen. The Clarea River bypasses the reservoir on the left of the photo, but the flow is used to supplement the gravity-fed part of the scheme on the main Dora River

Figure 7.7 Views above and below the hydroscheme intake on the Dora River. Sediment deposited above the intake gratings (on right of first picture) is removed and sold. Migrating fish can bypass the barrage by fish ladder

when demand returns and electricity prices can be up to 40 times higher than at night. On average, about a third of the flow is reused in this way. The Dora is a gravel-bed river and this presents problems for the turbines and increases the environmental impact. Although the gratings trap the gravel and coarser pebbles, finer material passes through the tunnels and the turbine blades have to be protected with a hard aluminium-titanium spray. Environmentally, the gratings cause a build-up of deposits above the intake, and these have to be removed three or four times a year. This material is sold for construction, depriving the river downstream of its natural sediment load. The procedure falls within environmental regulations as the gravel used to be extracted commercially before the hydroscheme was built, and a fish ladder allows migratory fish that need the gravel shoals for breeding to reach the natural river above the intake.

Three Gorges and China's water and power needs

China's Three Gorges Dam on the Yangtze River is the largest in the world. It has attracted considerable international criticism from environmental groups, including Greenpeace and Friends of the Earth. Some critics have argued that the aims could be better achieved by building numerous smaller dams, although ten smaller schemes are already underway upstream, designed to generate 12,000 MW. The demise of the unique Yangtze river dolphin, announced in 2007, has been blamed on the dam. There are undoubtedly important negative environmental effects. Poor public relations on the part

Figure 7.8 Fish ladder at the hydroscheme intake on the Dora River, North Italy

of the Chinese authorities have made matters worse by publicizing the height of the dam crest as 190 m, which has been repeated by numerous campaigning websites, when they really meant 190 m above sea level and their own environmental impact assessment actually modified this to 175 m.

Yet the dam has a number of vital roles to play in the economic development of twenty-first-century China, providing a major underpinning to the rise of China to become the largest economy in the world from the middle of the century. China's GDP is set to double by mid-century and its population is currently rising at 1.3

Hydropower dams and flood control in China

Changming Liu

China's exploitable hydroelectric resources top the world at 378 million kW, making up 16.7 per cent of the world's total. By the end of 2003, the overall generating capacity of China's hydroelectric power plants reached 92.17 million kW, about 24 per cent of the country's total power generating capacity. By 2005, the actual output reached 283 billion kWh, 15 per cent of the country's total power output.

The most promising sites for hydroelectric development are in the upper Yangtze River and the upper Mekong River (the Lancang River). The hydropower potential on the main stem of the Yangtze or Changjiang River and its tributaries is huge, totaling 268 million kW. The Three Gorges project includes one of the world's largest hydroelectric power stations with 18,200 MW of installed capacity that will produce 84.7 TWh of electricity output annually. The energy produced is projected to replace the mining and consumption of 40 million tonnes of coal every year. Among the sites already developed are Ertan (2300 MW) on the upper Yangtze River, and Manwan (1500 MW) on the Lancang River.

At present, there are more than 3100 reservoirs in the watershed of the Yellow River (Hwang Ho). Their total storage capacity is over 58 billion m³, which is 2 billion m³ more than the average discharge of the Yellow River at the HuaYuanKou station. There are ten large hydroelectric power stations.

The dams also provide flood control. About 8 per cent of the land area located in the middle and downstream parts of the seven major rivers of the country, the Yangtze, Yellow, Soughua, Liaohe, Haihe, Huaihe and Pearl River Basin, is prone to flooding.

Half the total population of the country live in these areas, and they contribute over two-thirds of the total agricultural and industrial production by value. The reserved flood control capacity of the Three Gorges Reservoir can help cut flood peaks by 27,000 to 33,000 m³/s, the biggest for any water conservancy project in the world. The dam will protect 15 million people and 1.5 million hectares of farmland in the Jianghan Plain from 100-year floods.

In the past, floods have been exacerbated by large amounts of sediment deposited in the channels of the lowland rivers, which reduces the capacity of the channel to hold the floodwater. In the Loess Plateau in the north, thousands of sediment control dams have been built and many more are planned. The Three Gorges Dam will have a similar effect. By 2003, the sediment load in the Yangtze at Datong was less than 200 Mt/yr. Upon completion of the dam, the sediment load at Datong will decrease to ~210 Mt/yr for the first 20 years, then will recover to ~230 Mt/yr during 2030–2060, and will reach ~310 Mt/yr during 2060–2110. The sediment trapping combined with the flood holding capacity of the reservoir will allow the raising of the levees in the lower reaches to be deferred.

per cent a year and is predicted to reach a peak of 1.62 billion in 2050 from the current 1.33 billion.

The contribution the dam is making is diverse. Primarily it is a hydropower scheme capable of generating 18.2 GW and providing 12 per cent of China's electrical needs. In this, it is a greener generator than the rash of coal-fired power stations that China is commissioning at great speed: a contributor to China's newly avowed commitment to controlling greenhouse gas emissions. The output will save 50 million tonnes of coal a year and be equivalent to 18 nuclear plants. Prior to the dam, 70 per cent of China's electricity was produced by thermal stations and only 19 per cent by hydropower, but officials expect that China will need to increase overall supply by 20–30 per cent to eliminate the power shortages that have been costing the economy dearly, and Three Gorges is a substantial step in this direction. The government plans to increase the overall proportion contributed by hydropower up to 40 per cent in the long term (WCD, 2000). The potential to achieve this is there – prior to Three Gorges only 15 per cent of the potential had been exploited – but to maintain the percentage while overall generating capacity increases by 25 per cent or so will require more large dams.

It is also a flood control scheme on a river that has killed millions: half a million last century alone. A total of 3.7 million people died as a result of the 1931 flood, both directly and from subsequent hunger and disease. 30,000 died in the 1954 flood. The official figure for deaths in the smaller 1998 flood is 4000, but some estimate tens of thousands. The impact of floods has got worse in recent years through a combination of deforestation, population growth and poor planning of farms and settlements. Between 1985 and 1999 forest cover in the upper Yangtze basin fell by 30 per cent.

The floods also disrupt economic life in major industrial cities like Wuhan (population 4.5 million), and destroy the important rice, wheat and lotus crops in the agricultural plains. The 1998 flood caused an economic loss of $3.6 billion in Hubei province alone and flooded 1.7 million ha of cropland. The lower basin contains about a quarter of the total cultivated land in China. Rice in particular is still a staple, although it is regarded as poor man's food and as the population becomes more affluent consumption will fall a little, perhaps down some six per cent by 2050. The reservoir is capable of holding back over two billion m³ of floodwater. One of the arguments against the dam is that the 660 km long Sanxia reservoir it creates is flooding agricultural land. In reality there is relatively little agricultural land in the Gorges themselves – they are gorges after all – and the pre-existing Gezhouba hydropower scheme (1989) had already flooded most of what floodplain there was. Figures of 41,000 up to 100,000 ha of land are quoted as being inundated, but what agricultural land there is in this region is relatively poor compared with that in the middle and lower reaches of the river. Protecting the lowland floodplains against flooding is economically and socially far more important. Agricultural production downstream of the dam in Hubei province accounts for 22 per cent of China's GDP (Morimoto and Hope, 2004). The dam will also assist this agriculture by reducing waterlogging in the rainy season and increasing riverflow during the dry season.

To what extent the Three Gorges Dam might have reduced the 1998 flood if it had been completed does not appear to have been studied (Wisner *et al*, 2004). The intention is to raise protection from the ten-year flood to the 100-year flood, but its general flood-control ability is clearly limited by the fact that many floods are generated by the large tributaries that join the river downstream of the dam.

The dam has a third role: to improve navigation and open up the hinterland to manufacturing and trade. The increased depth of water will allow 10,000 tonne vessels to carry larger volumes of coal and raw materials to cities like Chongqing and to export manufactured goods. This aims to provide alternative employment and to modernize the industries.

Yet a fourth role is water supply. China has 20 per cent of the world's population but only seven per cent of its water resources. More than 600 cities suffer water shortages and 700 million people have to drink contaminated water. By mid-century, demand for water could increase more than 40 per cent, from 2.5 billion m³ in 2006 to over 3.6 billion in 2050. At the same time, the glaciers and icefields that feed the headwaters of major rivers like the Yangtze will have melted, perhaps down to less than 40 per cent of their present size, due to climate change, creating water shortages for another two billion people.

The dam will enhance local supplies, although it will also put a premium on better wastewater treatment. Pollution was a growing problem even before the dam. By the mid-1990s, one Gt of wastewater and sewage was being released annually into the river, totally untreated. The impoundment will aggravate the problem by reducing oxygenation and the dispersal of pollutants. Unless this problem is addressed, it could cause the reservoir to become undrinkable.

The original planners also hoped to use the dam as part of the grand south-to-north transfer scheme to supply drinking water to Beijing (see the section headed *China's National Strategic Plan*, later in this chapter). This part of the scheme has received little or no publicity in the international press and is still rather tentative. The otherwise excellent probabilistic cost-benefit analysis of the scheme produced by members of Cambridge University's Judge Business School, fails to take this potential benefit into account (Morimoto and Hope, 2004). Yet if it does go ahead, it is likely to prove a very valuable resource for the north. Because the site of the dam lies at the point where the Yangtze emerges from the uplands with steep-sided gorges to contain the reservoir, the volume and reliability of water supply will be better than is possible via the eastern route, which taps the lowland reach where there is no prospect of such a large reservoir. It is also going to be cheaper than pumping water across mountain ranges in the headwaters for the western route.

Social and economic impacts

One of the greatest criticisms of the dam is the forced expulsion of up to 1.5 million (or even 1.98 million) people: the official figure is 1.2 million. Some have lost what livelihood they had, perhaps fishing or farming, and been moved out of the area completely. Others, perhaps the luckier ones, have been retrained with industrial skills and rehoused in high-rise flats higher up the hillsides. The modern flats are far superior to the old riverside houses – many were little more than slums, built with rough, unfinished brickwork, with no window glass, and sometimes with families living in rooms with a wall missing.

In the process a substantial number of towns and factories have been inundated, but the exact number varies considerably between sources and the precise meaning of 'inundated' is unclear – many will have been only partially flooded. Morimoto and Hope (2004) quote 13 cities, 160 towns, 1352 villages and 1500 factories closed, others quote 365 towns and over 3000 factories. But the old, inefficient industries have been replaced by factories producing goods for modern markets, aided by the improved facilities for river freight.

There is concern for the impact on tourism, although the gorges are still spectacular and Yangtze cruises continue to be widely advertised in the West. A number of historic monuments may have been lost, and funds have not permitted all of them to be saved or replicated, but the best have been moved out of harms way. Only 16 archaeological sites were affected, although up to 12,000 cultural antiquities were also threatened.

Environmental impacts

Despite foreign criticism of the dam, the Chinese authorities did carry out an Environmental Impact Assessment (China Yangtze Three Gorges Project Development Corporation, 1995). Admittedly, this was an official report vetted by the communist authorities, but it does not deny that there will be significant impacts and it does recommend palliative measures, many of which are being implemented. A government research laboratory is breeding migratory fish. During the construction period, fish were transported past the dam site in tanks on the back of trucks. Particular attention is paid to the valuable Yangtze river sturgeon. However, sediment movement was increased during construction and this may not have helped the river dolphins. The official EIA notes that prior to the dam the Yangtze yielded 50 per cent of China's annual aquatic produce and that it cannot guarantee that all species will survive the changing habitats.

Sediment is likely to be a big problem. The Yangtze has the fifth largest sediment load in the world, and deforestation in the upper catchment in the late twentieth century added to the problem. The upper catchment is now being reforested to reduce soil erosion and sedimentation behind the dam. This is also an anti-flood measure that will complement the dam's role, by delaying runoff and reducing sediment deposits in the lower reaches that raise the riverbed and reduce the capacity of the channel to contain floods. As with all dams, however, sediment is being trapped behind the dam, which will reduce its holding capacity. Reduced sediment transportation downstream combined with reduced flows is causing changes in scour and deposition in the estuary and allowing more saltwater intrusion there.

Wastewater treatment remains a priority. Topping (1996) predicted that unless treatment is improved waste content could increase 11-fold in some areas because of the dam, especially with some 50 different toxins being released from mines and factories in the reservoir area.

Figure 7.9 Building the foundations for the Three Gorges Dam (right) and excavating the navigational locks

Figure 7.10 Towns on the Yangtze above the Three Gorges Dam before partial flooding. New high-rise flats on the hill rehouse people displaced from lower down

Figure 7.12 Agricultural land prior to flooding – the white marker is intended to indicate the height of final flooding – a case of political spin? Peasant farmers often sowed a quick crop of rice on the small muddy areas down by the river in the hope of catching a harvest before the spring floods arrive

Figure 7.11 New roads and a bridge in one of the three gorges prior to flooding

Dam protests

There were celebrations in Tasmania in July 2008 to mark the twenty-fifth anniversary of a landmark victory for anti-dam protestors. Veteran British botanist, environmental campaigner and television personality, Dr David Bellamy, was there to recall the campaign to stop the Hydro-Electric Commission building a dam on the Franklin River and destroying a large area of ancient rainforest in 1983. Bellamy spent his fiftieth birthday in prison as a result, together with over 500 other cam-

Figure 7.13 Fishing for coke cans and plastic bottles on the Yangtze (Chang Jiang in Chinese). The river is highly polluted, especially by raw sewage from riverside cities, coal dumps and industries. The photo was taken before the Three Gorges Dam was built and shows that the river was already partly impounded as a result of an earlier hydropower dam just downstream of the new site

131

paigners, but the media exposure ensured popular support throughout Australia and worldwide. Bob Hawke's espousal of the popular cause during the 1983 national election with a pledge to save the river helped the Labour Party to return to power and his own election as Australian Prime Minister in March, as well as the success of the environmental campaign in the High Court three months later. The site is now a popular tourist destination, attracting over 150,000 people a year.

At the time of the Franklin River incident the environmental movement was still relatively young in Australia. The world's first Green Party had been founded in Tasmania in 1972 after a campaign failed to stop the damming of Lake Pedder. But the success on the Franklin River is said to have inspired a generation of environmentalists. One of the campaigners, Bob Browne, founded the Australian Wilderness Society, which has campaigned for the preservation of wild rivers, notably on the Wenlock River in the Cape York Peninsula of Queensland. The Lake Pedder experience was almost a repeat of the events in California in the early twentieth century at the very dawn of environmental objections.

The national battle to prevent a dam being built on the Tuolumne River in the Hetch Hetchy valley, California, is a classic of environmental lore. It was a foundation event for the modern environmental movement a century ago. Not only does the memory live on, but it is now the focus of attempts to restore the valley. The unlikely trigger for the battle was the San Francisco earthquake of 1906 and the great fire that ensued. Firefighters complained that they had insufficient water to dowse the flames. Up until then, the city had depended on a private water company that charged high prices and supplied water of variable quality. The fire now provided a convenient argument for the city to take control of its own water supply and institute a publicly owned water supply system, even though it transpired that a large part of the immediate problem had actually been caused by fracturing of the water mains during the earthquake. Even so, San Francisco was really getting short of both water and electricity, and a suitable site for a dam was selected in the Sierra Nevada mountains, where a gorge allows a narrow dam to hold back a large reservoir in the broader mountain meadow area of Hetch Hetchy. Unfortunately, the site was in the newly created Yosemite National Park that John Muir had spent a decade campaigning to have created. Muir is now regarded as the 'father' of American National Parks and of the modern environmental movement. He had founded the Sierra Club, the first official environmental organization, in 1892 to campaign for conservation and national parks. Now he initiated a national campaign to stop the dam using the Sierra Club, speeches and publicity pamphlets (see quotation below). The campaign delayed the project, but the prospect of large amounts of hydroelectricity from the dam tipped the balance in favour of President Theodore Roosevelt's dictum of providing the 'greatest good for the greatest number'. It was approved by federal Act in 1913, the O'Shaughnessy Dam was completed in 1923, but building of the 300 km aqueduct was slow, so that it was not until the 1930s that San Francisco got its water. Muir died broken-hearted in 1914, but it established the Sierra Club as a prominent defender of the environment.

The dam is still a vital supplier of water and hydropower for the city, but in 1987 the US Secretary of the Interior proposed restoration of the Hetch Hetchy Valley and in 1988 the Bureau of Reclamation reported to the National Parks Service that 'restoration would renew the national commitment to maintaining the integrity of the national park system'. The Sierra Club is now campaigning for this, but the aim is to try to restore the environment without destroying the water and power supplies. See the campaign website 'Restore Hetch Hetchy' at www.hetchhetchy.org and www.sierraclub.org/ca/hetch-hetchy.

> 'Dam Hetch Hetchy! As well dam for water-tanks the people's cathedrals and churches, for no holier temple has ever been consecrated by the heart of man …
>
> These temple destroyers … seem to have a perfect contempt for Nature, and, instead of lifting their eyes to the God of the mountains, lift them to the Almighty Dollar.'
>
> John Muir, Founder of the Sierra Club, in the campaign pamphlet 'Let everyone help to save the famous Hetch Hetchy valley and stop the commercial destruction of our National Parks' (1909)

Recent campaigns have been mounted in India along similar lines to Australia, especially over the plans to build more megadams on the Narmada River, which could displace at least 250,000 people – some estimates are as high as one million (see section headed *Dams and diversions in India*, later in this chapter). The campaign went international in the late 1990s when protest meetings resulted in the international prize-winning novelist Arundhati Roy being indicted in the Supreme Court.

One of the most recent protests has been even more attention-grabbing. In response to the Peruvian government's proposal to sell three-quarters of its Amazonian lands, two-thirds of the whole country, to private companies for hydropower, oil, logging and mining, the Achuar Indians threatened war. In October 2009, a number of indigenous Indian groups representing hundreds of thousands of Indians joined forces on the Pastaza River dressed in war paint and feathers. Further down the Amazon basin in Brazil, other Indian tribes in the Xinguano group finally lost their 20-year battle to stop construction of the Belo Monte hydropower scheme on the Rio Xingu in 2010, despite a high profile international campaign led by the rock star Sting. The scheme was started in the 1990s but abandoned because of protests. The 11 GW dam scheme will be the third largest in the world and flood 500 km² of rainforest, although this is now only a tenth of the area originally planned. It will not now flood land scheduled as tribal territory, but will still displace up to 12,000 people living on land not so designated – a not-uncommon loophole – the lives of up to 40,000 people could be affected by damage to fisheries and the rainforest. The government did, however, require the development company to pay $800 million towards environmental protection.

Brazil's emerging economy is increasingly desperate for electricity, despite having built some of the largest hydropower dams in the world, such as Balbina (112 MW,

flooding 2360 km², 1989) and Itaipú (12,600 MW, flooding 1350 km², finished 1991), and another 70 hydropower schemes are planned. Demand is growing faster than GDP and electricity blackouts are affecting many areas, like the one that afflicted Rio de Janiero and São Paolo in November 2009. A major power failure struck Rio soon after Itaipú came online. The decision to restart Belo Monte was made strategically ahead of a general election: politics and the economy triumphed once again. However, critics say that Belo Monte will generate less than ten per cent of its capacity during the three or four months of the dry season.

With a few exceptions like Hetch Hetchy, anti-dam protests are largely a feature of the last 50 years. A wide range of reasons lie behind this: pressure on water resources resulting in more dams, the rise of public awareness of the environmental impacts, advances in dam technology that have allowed ever larger dams to be built with correspondingly greater impact, the larger numbers of people displaced, the desire of many post-war heads of governments for highly visible illustrations of national power and progress, financial support from foreign governments keen to expand political influence and obtain contracts for their own engineering companies, and latterly from international funding sources such as the World Bank that view dams as a step towards economic revival in the Third World and may demand them as part of a development loan agreement.

The Narmada campaign marked a total reappraisal of the view expressed in the early years of Indian independence by 'Pandit' Nehru, India's first Prime Minister, that

Figure 7.14 Itaipú Dam, Brazil. The largest hydropower scheme in the world, generating 91.6 TWh in 2009 against 79.4 TWh at Three Gorges. Unlike Three Gorges the landscape is flat and the dam is 7.7 km long in order to contain the water. Right-hand photo: the spillway had to be carefully designed to prevent serious erosion in the river

133

dams are 'the temples of modern India', the basis for industrialization and improved agricultural output. The same philosophy was behind President Nasser's championing of the Aswan High Dam for Egypt in the 1950s, and behind Brazil's vast programme of hydropower dam development that contributed to a major international debt crisis in the 1980s.

Anti-dam protestors have been especially active in Spain in recent years. The trend coincides with a vast expansion of dams in and adjacent to the foothills of the Pyrenees to support irrigated agriculture on the gentle slopes of the Ebro valley. Spain now has the largest surface area of reservoirs in the world relative to the size of country. The recent National Hydrological Plan to divert water to the southern provinces is adding new pressure and eliciting objections from the European Union and international campaigns led by ecologists, including Greenpeace, the European Rivers Network, and many local organizations (see section headed *Dams in Spain*, later in this chapter).

As the cases of Jánovas in Spain and Tryweryn in Wales show, the wounds of injustice can persist for decades and there is often a feeling that the local population has suffered for the advantage of people elsewhere. Tryweryn is a particularly emotive case. It only involved the evacuation of 48 residents from the tiny hamlet of Capel Celyn and outlying farms in the valley of the Tryweryn, a tributary of the River Dee in north Wales, in order to create Llyn Celyn in the early 1960s. The reservoir is one of four designed to regulate flow in the Dee, mainly to guarantee water supply for the city of Liverpool. But

it became a major issue in Welsh politics. The nationalist party Plaid Cymru objected to the way no Welsh authorities had been consulted about the development and that a Welsh country village had been flooded for the sake of English city dwellers. It enhanced Plaid's reputation as defender of Welsh rights, even though, despite minor acts of sabotage by party supporters in 1962 and 1963, it did not officially endorse any direct action. Tryweryn also fuelled calls for a Secretary of State for Wales to represent Welsh issues at the Westminster Parliament and prevent similar Acts of Parliament being passed without consultation. And it was a fillip to the foundation of the Welsh Language Society, Cymdeithas yr Iaith Gymraeg, in 1962, since the affected population were Welsh speakers and Welsh was a dying language at the time. Hence the graffito in the photo is still regularly maintained and repainted more than half a century after the event.

As so often with popular nationalist movements, the 'stealing Welsh water' and 'destroying Welsh life' arguments were only partly correct. The Dee regulation scheme was also designed to guarantee water supplies to many Welsh customers, including residents of Llangollen, Bangor-on-Dee and Wrexham. And it is part of a successful flood control system, the only one in Wales required by law to prevent the 100-year flood, protecting the former flood-prone riparian land between Bala and Corwen, and again downstream of Wrexham. As for the cultural aspects, Tryweryn probably helped rather than hindered the regrowth of the Welsh language that sprang from the establishment of a Welsh language television channel in the early 1970s by drawing extra attention to the problem of declining speakers.

A series of recent anti-dam campaigns have arisen in Queensland following urgent plans to counter the multi-year drought, which ended dramatically in 2009 when nearly 60 per cent of the state suffered severe flooding. The government has been attacking the problem of water shortage from a number of angles, including a highly publicized water conservation campaign (see Chapter 13 Cutting demand), but also including the traditional solution of planning more dams. As the timeline in the article 'Brisbane's water crisis' (below) shows, one of the problems has been a lack of forward planning for reservoirs in the 'good' years, so that government is now faced with the dual problem of being accused of lack of forward infrastructure planning and public opposition to new dams (see box headed 'Anti-dam slogans', below).

Figure 7.15 A graffito in mid-Wales accorded iconic status. It reads: 'Do not forget Tryweryn' – a small event with national repercussions (see text)

Figure 7.16 The dam that never was. Preparatory tunnel for a reservoir in the Spanish Pyrenees which caused the population of the idyllic village of Jánovas to be evacuated. The project was one of a number that were abandoned after fierce objections to expansion of regulatory dams in the late 1990s

Figure 7.17 Former villagers and their heirs now demand the right to return to their village, more than a decade after the forced evacuation

Figure 7.18 Graffiti protesting against extension of irrigation into the conservation area in the Monegros in the Ebro basin, northern Spain. The graffiti favour the nature reserve over irrigated agriculture. The graffiti adorn the walls of an abandoned salt factory, which underlines one of the dangers of over-irrigation in the area – salinization. The main purpose of dams like Yesa on the Rio Aragon and El Grado on the Rio Cinca is to divert water south for irrigation

Anti-dam slogans: Queensland August 2006

In the midst of attempts to combat a severe multiyear drought:

Roadside graffiti:

'Be water wise, no dam disguise.'
'Educate, don't inundate.'
'Don't dam our school.'

Newspaper coverage

'"Dammed" if we don't cut back'
Headline in the Courier Mail
See also the website: www.savetheMaryRiver.com

Figure 7.19 March protesting against the enlargement of the Yesa dam on the Rio Aragon in northern Spain as part of the 1996 Hydrological Plan of the Ebro Basin, with banners that sound enigmatic to English speakers

Figure 7.20 Lake Vyrnwy, north Wales, a reservoir created to supply water to the city of Liverpool in 1888. It flooded the village of Llanwddyn, with a population of a few hundred, three public houses, a chapel and church. There was no local consultation but there was also no public protest. A new, but much reduced village was built nearby. Some 75 years later the same water company faced strong protests when it followed the same procedure as in the 1880s and flooded the much smaller settlement of Capel Celyn in the Tryweryn valley, north Wales. Nationalism was on the rise and the event had important political ramifications

Figure 7.21 Queensland graffiti

Brisbane's water crisis: a tale of politics, protests and complacency

By 2006, after suffering five years of severe drought, Queensland's water supplies were at a critical ebb and the government was actively promoting water saving by every means possible (see Chapter 13 Cutting demand).

Southern Queensland's lack of current resources can be traced to many factors, especially a history of poor understanding of a climate prone to persistent droughts, government ineptitude and lack of foresight, raids on water revenues to support state budgets, public complacency and recent public objections to dams being built in their own community. The two world wars also delayed dam proposals by diverting attention and resources even during drought periods.

In the Brisbane *Courier Mail*, Condon (2006) claimed the problems stretch right back to misjudgements made at the time of the original settlement, and he extended the list to include government fear of expenditure, advice ignored, opportunities lost, political hijacking, bureaucracy, poor data, and a community believing in limitless water. Others have pointed to the lack of independent institutions that could offer well-informed inputs into public policy debates, lack of machinery for federal government support for major resource investments, and politicizing of institutional arrangements. There have been recurrent arguments between federal and state governments and between state governments and local councils, through lack of a clear line of command and because the responsibility for building dams and desalination plants falls on local councils, yet federal and state governments provide funding and want to determine how the money is spent.

Timeline

Early nineteenth century

According to dam engineer Geoff Cousins: 'When Brisbane was settled first, they did not have a lot of experience of Australian conditions. It took a long time to realize that in the Brisbane area, our weather was a series of floods separated by droughts with not a lot in between, as opposed to the English idea of streams running continuously.' Drought-resistant gum trees were prolific, the Brisbane River short and few aborigines lived there.

Climatic cycles, political cycles and dam building

1893: Flood led to proposal for flood mitigation dam near Wivenhoe in 1899.

1915: Drought, river stopped above tidal limit, but war delayed plans for a dam.

A wetter period through much of twentieth century, culminating in the 1974 flood, coloured the experience of most people now alive, but the population was growing.

1930s: Brisbane plans its first dam, the Somerset, for operation in 1942 but delayed by war. Eventually built in 1959.

1960s: City mayor lifted water restrictions and Brisbane immediately went from the lowest per capita consumer in Australia to the highest.

1970s: Drought 1970–1. Government approved plan for a water supply dam at Wivenhoe, but the 1974 flood delayed it. Dam site identified at Wolffdene and land zoned as rural open space to limit subdivision and facilitate land acquisition.

1985: Wivenhoe Dam finally built with added flood control role: a $460 million megadam on the upper Brisbane River, 70 km NW of the city, holding five times the city's annual demand. It is to be followed by a new dam at Wolffdene in Albert valley, south of Beenleigh.

Late 1980s: Citizens Against the Wolffdene Dam lobby group 'orchestrating one of the most effective campaigns of civil defiance in recent Queensland history'.

1989: New official calculations suggest that no new source is required until at least 2030 and possibly 2050. Population projections indicate rise from 730,000 to 2.1 million by 2001 and 2.6 million by 2011 and that Mount Crosby and North Pine reservoirs will cope. The parliamentary inquiry into the proposed Wolffdene Dam favours cancellation and proposes industrial recycling using tertiary treatment instead. Albert Valley residents stage protests, Labour wins election and new government is unwilling to spend $250,000 in land acquisition costs alone as costs have risen due to re-zoning of land as residential. Wolffdene is cancelled.

Australian investment in water infrastructure falls during 1980s to 2000s as state governments take profits and fail to reinvest in water: a fraction of the investment in most OECD countries. Following the Wolffdene cancellation there is 'general amnesia on water' by all levels of government and the public (Condon, 2006). According to Cossins, Wolfdene was the most economical dam to build, its cancellation was purely political, and people relied too much on the idea that Wivenhoe was sufficient: 'As long as water flows in the taps, the public doesn't give it another thought.'

2001: Queensland government receive a plan to combat water crises suggesting business efficiency audits, subsidies for water-wise products, transparent water pricing and leakage reduction, but plan is ignored. A prolonged multi-year drought begins; water supply problems intensify in 2004–6. Proposals for new dams are vigorously opposed by residents in southern Queensland (see box headed 'Anti-dam slogans', above).

2004: Minister for Natural Resources calls for water reform and modernization of the indigenous property law to avoid crisis that could affect the Gold Coast by 2006 and the Capricorn Coast by 2016; there is a need for federal funding throughout Australia. Pipeline to supply hydropower stations with recycled water is rejected on cost.

2006: State Opposition leader says abandoning Wolffdene was 'stupid' and Brisbane is now paying the price: draconian water restrictions are the only option and

forward planning will be very expensive. July, 2006: Wivenhoe down to 30 per cent capacity, old fences breaking the water surface. Water Commission plans topping up dams with sewage recycled by reverse osmosis. Newspaper article 'Scarce drinking water wasted on excess electricity' – hydropower exported to New South Wales, where it is not needed. Pipeline planned to supply Gold Coast resorts from Brisbane. Federal Secretary for Water argues SE Queensland should look for solutions that do not depend on rainfall and cancel proposed Traverston Dam. Site of proposed Mary River Dam found unsuitable: limited bedrock, lack of river discharge but high sediment content, plus arsenic from gold mining and cattle dip. Australian Water Association says recycling could provide half the water needs and desalination could be undertaken at reasonable cost. Accusations and counter-accusations between Queensland premier and local councils over lack of strategy and politicizing of issues.

2007: May – Level 5 water restrictions reduce amount available for recycling.

2009: The drought breaks with torrential rains and floods in February. Crocodiles in the high street in northern Queensland. Flash floods April–May.

2009–2010: Summer begins with severe drought and prediction of a 'roasting summer'. Wildfires in October. New pipeline from Wivenhoe Dam to Toowoomba, the largest inland town by early 2010, where reservoir down to 8.5 per cent. But also drilling boreholes into the Great Artesian Basin – one of the largest reserves in the world. Other towns surviving on water tankers.

Dams in Spain: from a command economy to democratic objections and a National Hydrological Plan

Spain is beset by a climate in which only eight per cent of natural riverflows occur when and where they are needed. It is also beset by frequent droughts. In order to redress this, Spain has embarked on a number of major dam-building programmes since the Second World War, initially as part of the so-called 'regenerationist thrust'

under General Franco's fascist dictatorship, in which a major aim was to support the expansion of irrigated agriculture. At the time of Franco's death and the return to democracy in 1975, there were 135 major dams in the Ebro basin and the area under irrigation had expanded from 420,000 ha in 1945 to 702,000 ha. The country as a whole now has 1300 large dams, which regulate some 40 per cent of all riverflows and give Spain the biggest surface area of reservoirs relative to size of country in the world. The total reservoir capacity of 56,500 Hm³ (56.5 km³), regulating 46,000 Hm³ of riverflow a year, puts Spain fourth in the world after the USA, India and China. Overall, the reservoirs provide water to irrigate more than 2.7m ha, to generate 20 per cent of national electricity and to supply drinking water to 30 million residents and 60 million tourists. There are 30 dams for every million people, yet because of the irregular and fairly low discharges of most rivers, the reservoir capacity relative to population is still 32 per cent lower than the average for high income countries, at 1431 m³ per inhabitant.

Most dams are in the north of the country, especially on the northern slopes of the Rio Ebro basin, harnessing water from the water-rich Pyrenees. The south and centre of the country, even the southern slopes of the Ebro, are water deficient. Up to the 1990s, governments favoured dams mainly to divert water to develop irrigated agriculture in the northern half of the Ebro basin. These dams even out the seasonal fluctuations in flow, which vary widely from river to river depending of the relative importance and timing of snowmelt and rainfall sources (Figure 7.22).

Consumers in the Ebro basin use only about a third of the average natural resources available: 6 km³ a year out of a total of 18 km³. Plans were laid in the 1990s, under the 1992 Aragon Water Pact and the 1998 Hydrological Plan of the Ebro Basin, to make an additional 3 km³ available for use. The plans required more regulating dams: expansion of the Yesa and new dams in Aragon (Jánovas, Biscarrués and Santa Liestra), Navarra (Itoiz) and Lerida (Rialb).

But this was now a democratic era and public objections were raised, marches organized, graffiti like 'Yesa No' proliferated on buildings, legal action was initiated against the administration, direct action was taken against the Santa Liestra site in 1999 and Itoiz was sabotaged. According to the former chairman of the Hydrographic Confederation of the Ebro, Sancho Marco (2006), the objectors represented diverse interests, grouping together for strength but frequently acting in opposition to each other. Nevertheless, the sum total of the protest movements was to practically halt most of the left bank reservoir developments – including Jánovas.

As so often, the objections raised ranged from sound assertions to dubious statements and totally erroneous contentions, like the claim that Biscarrués would flood the Mallos de Riglos, a unique sandstone cliff that is a particularly attractive tourist feature. One of the sound objections was that the water was inefficiently used and that water-saving measures were needed before new dams. The 1996 National Irrigation Plan aimed to modernize and improve the efficiency of irrigation systems and was supported by EU funds, but much has remained unchanged. A theme that lies somewhere between the two extremes was that the administrators are not to be trusted and promote their own interests. Public scepticism of governments and administrators has continued to grow around the world, with some cause. Doubts

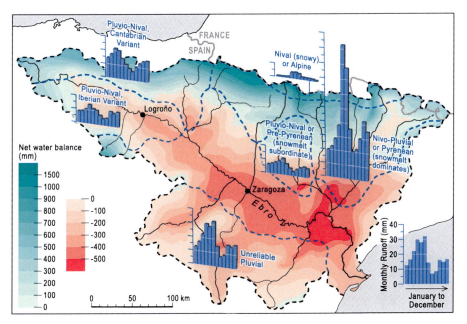

Figure 7.22 The Ebro basin showing the strong contrast between the Pyrenean rivers and the southern tributaries and the wide variety in the timing of natural flows, which the dams are designed to even out

139

were also raised about the thoroughness of surveys and safety assessments at some sites, including possible foundation issues at Jánovas, and the lack of adequate environmental impact assessments.

The National Hydrological Plan

In 2001, the Spanish Government approved a much larger plan to extend the system to the rest of the country by diverting water from the Ebro basin to supply agriculture, cities and leisure facilities in the south, where droughts and a conspicuous expansion of tourism and coastal homes during the 1980s and 1990s had put great stress on local resources. The National Hydrological Plan embodied ideas first put forward in the 1950s and 1960s to produce a 'balanced' national distribution of water. The area it aims to supply is home to 55 per cent of the Spanish population yet has only 33 per cent of the national water resources. If it is ever put into effect, the scheme would be the largest reorganization of national river systems relative to size of country anywhere in the world. It was to be completed by 2012, but its future remains in doubt.

The Plan would be expensive (£15.5 billion) and Spain sought funding from the European Union for 30–40 per cent of the cost. The EU refused on environmental grounds and even spoke of fining the Spanish Government for breaching EU directives. In response, the government commissioned an independent Environmental Impact Statement according to the requirements of EU legislation, which it approved in November 2003. The EU still remains sceptical of the proposed major river diversions on environmental grounds. Meanwhile, opposition from other sources has multiplied.

Substance of the Plan

A 912 km pipeline would carry water from the Ebro River to supply irrigation and tourist developments in the south and east of the country. This would be supported by 118 dams, 1000 km of canals, plus over 700 other works, and 41 desalination plants, all to be completed over an eight year period. The scheme would transfer over one billion m^3 a year from the Rio Ebro to the Mediterranean coast between Barcelona and Andalusia. Up to 35 other rivers would be diverted, including the Douro (Duero), Tagus (Tajo), Guardiana, Segura and Guadalquivir. In conjunction with the National Irrigation Plan, the scheme aims to irrigate more than one million hectares, representing a 35 per cent increase in the national irrigated area, nearly half of which would be supplied by water from the Ebro basin. It would also affect Portugal, as the first three rivers listed above drain across the frontier to supply Oporto, Lisbon and the water-starved Algarve.

Beneficiaries

Agriculture would take 77 per cent of all the water, including desalinized water, and 56 per cent of the diverted water. Urban areas would take 44 per cent of the diverted riverwater supplemented by 16 of the desalination plants to cover summer tourist demand.

Controversies

The northern provinces are opposed to losing their water. It is an argument that began with objections to the smaller water transfers *within* the Ebro basin: that the reservoirs and diversions are not for the benefit of the local people, the users who benefit will not be those who suffer the negative effects, and poorer regions will suffer to supply the richer south and east coastal regions.

A similar argument arose in 2008 when Barcelona and Catalonia were experiencing their worst drought since 1912. Barcelona proposed diverting the Rio Segre from Aragon to supply the city and crops. Aragon resisted. Catalonia retorted by accusing Aragon of hoarding water for 'unsustainable developments', like the Gran Scala, with 70 hotels, theme parks and golf courses, billed as the European Las Vegas. The Spanish government refused to sanction the diversion.

Critics of the National Plan say it too is more concerned with bolstering the tourist industry, including watering 66 golf courses around Murcia and Valencia. They maintain that the scheme is an antiquated response to former water supply problems, it is not needed, and that it assumes a demand that should not exist in the twenty-first century. The Spanish government says the water is needed for the agribusiness sector that covers much of southern Spain, but it is also likely to fund uncontrolled urban development and promote a further increase in tourism in a region already blighted by concrete.

Environmentalists have also mobilized international support against the scheme. Greenpeace, the World Wildlife Fund (WWF) and the European Rivers Network maintain the Plan is illegal under EU law and the Ramsar Convention on Estuaries. The WWF released a report entitled 'Seven Reasons Why WWF Opposes the Spanish National Hydrological Plan, and Suggested Actions and Alternatives'. There will be serious negative effects on fisheries, with an estimated 20 per cent decline, not only in the rivers, but also sardines and anchovies in the Mediterranean that rely on the nutrients discharged by the rivers. Over 80 conservation sites that the Spanish government proposed for inclusion within the Natura 2000 network, which was launched under the 1992 EU Habitats Directive, are threatened. The Ebro delta, one of the most important wetlands in Europe, would be severely damaged due to decreased flow and sediment supply, and increased salt intrusion from the Mediterranean Sea, affecting both wildlife and rice production. In the Pyrenees, villages would be flooded along with valleys that are the habitat of the rare Iberian lynx, and wild rivers tamed and diverted.

Locally, the Platform for the Defence of the Ebro, COAGRET (Coordinadora de Afectados por Grandes Embalses y Trasvases) and the Foundation for a New Water Culture were formed to defend people and the environment. Activists from the Platform painted whole walls of buildings with the graffito 'Transvasamentes No' (No river diversions) and a graphic water pipe with a knot in it. The Foundation was started by a group of scientists from 70 universities who favour alternative solutions. As a consequence of the intense concern for water sustainability that has been focused on the Ebro, the first ever international Expo devoted exclusively to water was held in Zaragoza in 2008. 'The Water Tribune – an Expo with no expiry date' is its legacy.

One of the scientific concerns is that climate change will make the Plan less viable. Lower rainfall, higher evaporation losses from the reservoirs and increased irrigation demand could reduce water in the Ebro by 4000 Hm^3 by 2040 and 6000 Hm^3 by 2060.

Possible alternative solutions

1 Adapt agriculture to the climate

Halt rice and sugar cane production in favour of winter crops and crops with lower water requirements, like olives and sunflowers, in autumn. Provide income support for farmers who switch: the Programme of Compensation of Income introduced in La Mancha in the early 1990s to promote such switches halved consumption in seven years, whereas the original Common Agricultural Policy subsidies they received after joining the EU in 1986 only increased production of the water-thirsty crops.

2 More water-efficient irrigation

Wasteful flood irrigation remains the main method in most water deficit regions of Spain. The WWF estimates that it wastes over 40 per cent of the water, yet this is still the method used for 80 per cent of the irrigation in Valencia, 60 per cent in Murcia and over 40 per cent in Andalusia. This is despite the insistence in the 1998 *White Book of Spanish Waters* on the necessity for Spain to distance itself progressively from the traditional agricultural model relying on flood or ditch irrigation.

Both the EU's 6th Environmental Action Programme and its Water Framework Directive emphasize the promotion of sustainable management. The Spanish government could use mechanisms under the current Common Agricultural Policy designed to promote a more sustainable use of irrigation water, for example by adding environmental conditions to olive subsidies to clamp down on new boreholes. The proliferation of illegal boreholes to irrigate olive plantations has been a major cause of aquifer depletion in southern Spain, and the National Plan has been partly justified as a way of relieving pressure on the aquifers.

3 Recycling, leakage control and improved urban water savings

Curiously, the Plan adopted in 2001 envisaged only 200 Hm^3 of recycling per year, whereas the 1995 version proposed twice as much. Recycling already slightly exceeds the Plan's provisions, so there is clearly room for improvement.

The WWF estimates that a more sizeable amount of water could be saved by modernizing urban water systems and reducing leakage losses, potentially 1500 Hm^3, of which 400 could be in the coastal regions, where until very recently housing construction has been burgeoning with very little planning control. In Catalonia, some 20 per cent of water is lost through leakage.

4 More integrated management

The WWF noted a lack of integration between agricultural policies and environmental policies and also between management of surface and subsurface waters. They showed that one of the areas in southern Spain identified in the Plan as a target for diverted water, the Júcar basin in Valencia, is not as water-deficient as originally estimated, because the Plan omitted to consider groundwater resources that the government subsequently determined are being recharged at a rate of over 1000 Hm³/yr. Similarly, some CAP agri-environment subsidies have actually been used to bolster non-profitable irrigation schemes and so increase demand rather than reduce it.

5 More environmentally friendly desalination plants

Low-energy desalination systems, which have a low environmental impact, are being promoted by the Ministry of Agriculture. One such system uses natural reverse osmosis at depths of 500 m underground and returns freshwater to the surface at ambient temperature. Solar and wind power sources are also possible. The Spanish Socialist Workers Party claims that desalination could supply 40 per cent of the water that the Plan aims to divert. Following the 2007–2008 drought, Catalonia has built a desalination plant to supply the equivalent of two months consumption a year.

6 The Rhone-Barcelona transfer plan

This scheme has been promoted since 1995 by the company that holds the concession from the French State to manage the River Rhone until 2056. It would involve a 330 km long canal transferring 15 m³/s (1,300,000 m³ a day). So far, Spain refuses to depend on another country for its water while it has plenty in its own Pyrenees. But a severe drought in Catalonia in 2008 saw Barcelona running out of water and the city agreed to import of water from the Rhone by ocean tanker. All but one of Catalonia's 15 major rivers were below emergency level and seven ships were scheduled to deliver water from Marseilles to Barcelona in May 2008. The government also considered importing water by train from elsewhere.

7 More environmentally friendly diversion schemes

The WWF and other environmental groups have proposed that the transfer plan should not be allowed to go ahead until it meets the obligations of EU laws on environmental impacts and the Birds, Habitats and Water Framework Directives. In 2003, the new Prime Minister Rodriguez Zapatero vowed that the scheme would be cancelled and any future development would only be sanctioned if it meets environmental criteria.

Large-scale river diversions

Diverting rivers is as old as building dams. The ancient Egyptians did it for irrigation and for transport. The greatest feat of the ancient world was the seventh-century Grand Canal in China connecting Beijing to the southern coastal towns for trade. Modern diversion plans are still dominated by agricultural irrigation, but they generally have a public water supply element as well, and they are on a much grander scale. However, at a time when China and India are actively engaged in planning and executing the largest river diversions ever undertaken, the concept of virtual water trading, which will be discussed in the next chapter, may be offering a cheaper and perhaps more sustainable alternative.

Diversions may be by canal, pipeline, tunnel or interconnecting rivers. Open channels are least efficient because of en route losses from evaporation or leakage. They are also most common because they are generally cheaper. Lengthy individual projects include Turkey's 180 km long Melen canal, designed to divert 1.2 billion m³/year and support Istanbul up to 2040, of which Phase 1 (270m m³/year) was completed in 2007, and the international transfers from China to Hong Kong (60 km) and Malaysia to Singapore (30 km). However, these are minor compared with the plans being realized in China, or those on the drawing board in India and Spain, and the now abandoned plans in North America and Russia.

China's National Strategic Plan: South to North Transfer

China has embarked upon the most massive water transfer scheme ever attempted. It is estimated to cost $62 billion, at least double the cost of the Three Gorges. The project aims to carry up to 54.55 billion m³ or 54.55 km³ of water a year (over twice the annual global production of desalinated water) from the water-rich south – principally the Yangtze (Chiang Jiang), China's prime river and the third-longest river in the world – to the urban industrial heartland around Beijing and the fertile but drought-ridden North China Plain.

China as a nation suffers from a major imbalance between the distribution of water and that of agricultural resources. Most water is in the south and most of the fertile, cultivable land in the north. The north has less than 14 per cent of the national water resources but is home to 45 per cent of the population and contains approximately 60 per cent of the cultivated land. The main crop in the north is winter wheat, which requires irrigation because it grows during the dry season. Yet average per capita resources in Northern China are only 650 m³, barely two-thirds of the world average, compared with 2385, more than double the world average, in the south. Nearly four million people are badly short of water in the north, crops and livestock are suffering and some cities are desperate. Per capita resources in Beijing are only a thirtieth of the world average.

The main source of water in the north is the Hwang Ho (Yellow River). Although it is China's second river, it rises on the relatively dry eastern edge of the Qinghai-Tibet massif and much of its upper reaches flow through regions with under 250 mm of annual rainfall. By comparison, the Yangtze begins life in the Tibetan Plateau among melting icefields and annual monsoons, and runs for most of its course through lands receiving over 1000 mm of rain a year. Since 2004, the Hwang Ho has failed to reach the sea for up to 200 days a year, partly as a result of over-exploitation for irrigation, but also because of a general drying trend in the northern climate. The trend in the north is aggravated by an erratic climate, with increasingly frequent crippling droughts, like that in 1995. Beijing has suffered nine years of drought since 2000. The drift to the cities and their increasing water demands have further aggravated the problem. Water tables have been falling at up to 1 m a year in the North China Plain, twice this in and around Beijing (see *Mining Groundwater* in Chapter 12). Most older wells in the city are now dry and Beijing's rapidly rising demands have led to many farm wells in the region also running dry. Over-pumping has led to local land subsidence and is widely blamed for the recent spate of sand-storms that have plagued Beijing. In contrast, the south is frequently fighting catastrophic floods. Global warming will aggravate the differences, adding to the drying trend in the north and creating stronger monsoon and typhoon development in the south. Transferring water from the south to the north is becoming more logical by the year.

China's solution, however, extends beyond diverting water from its own Yangtze. It has other countries' rivers in its sights as well: India's Brahmaputra and the Mekong, the lifeblood of Indo-China. The hydro-political implications are discussed in detail in Chapter 9 Water, war and terrorism.

Planners have identified three possible routes: the Eastern Route diverting water from the lower Yangtze near Jiangzhou; a Middle Route diverting water from the Danjiangkou Reservoir on the Hanjiang River, a major tributary that joins the Yangtze at Wuhan, and possibly with a canal extension of around 200 km to tap the Three Gorges Dam; and a Western Route tapping three headwaters of the Yangtze – the Dadu, Yalong and Tongtian – to transfer water into the Hwang. This could also involve taking water from the Brahmaputra (called the Tsangpo in Tibet) from just short of the Indian border and pumping it over the mountains to sustain the upper Yangtze and thence to the Hwang (Figure 7.23). Some of the costs of pumping the Tsangpo water some 2000 m up over the Bayankala Mountains will be recouped by hydropower generation on the other side. The Tsangpo could provide a colossal 100 billion m³. In between, there are plans to divert water from the Mekong into the Yangtze above the Sanxia Reservoir to compensate for the diversion of its headwaters into the Hwang. The Western Route will mainly serve the northern agricultural regions, while the other routes will deliver to the Beijing region. The route from the Three Gorges is slightly compromised by a reduction in the already huge height of the dam, which was ostensibly for environmental reasons, according to the chief engineer and author of the Environmental Impact Assessment (China Yangtze Three Gorges Project Development Corporation, 1995). The original aim was to have zero energy costs by using gravity to feed water from the crest at 190 m above sea level direct to Beijing at 38 m: lowering the height to 175 m now makes this more marginal. If it were to happen, though, it would also enhance the flood control role of the dam in a very cost-effective and practical way.

The scheme received the official go-ahead in 2002, exactly 50 years after Chairman Mao Tse-tung originally proposed it. Work immediately began on the 1155 km long Eastern Route, reusing the old Jiangzhou to Beijing Grand Canal for much of the way and picking up water from the Huai and Hai Rivers en route. Work was scheduled to take five years, adding a tunnel under the Yellow River and 23 new pumping stations and upgrading the seven existing ones to deliver nearly 15 billion m³.

Work on the 1267 km long Middle Route has been delayed, partly because of environmental concerns and partly because it involves a major upgrade of the old Danjiangkou Dam (1967), which is displacing 330,000 people, the largest forced eviction since the Three Gorges. Resettlement is to be completed by 2013. The Dam has

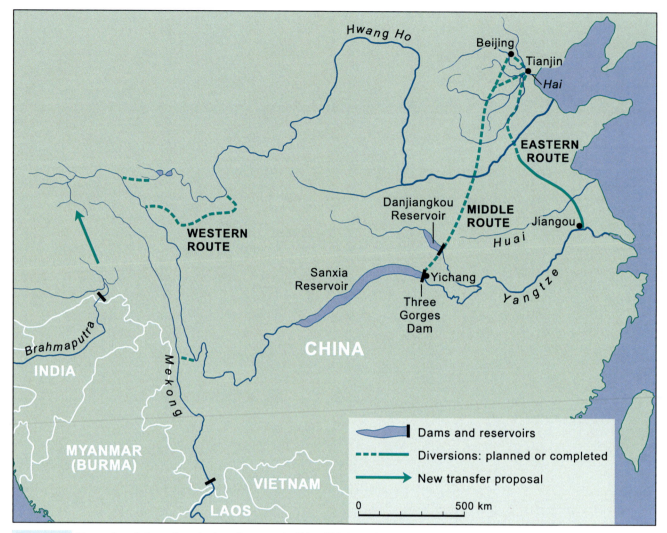

Figure 7.23 Planned south to north water transfer routes in China. This is not just an internal affair. The plan will impinge on the water resources of India, Bangladesh, Laos, Cambodia, Vietnam and, to a small extent, Burma

been raised by nearly 15 m, lifting the water level from 157 to 170 m OD. This is expanding the reservoir area by 370 km² and increasing storage to 29 billion m³ or 70 per cent of the annual flow of the river. Over three billion m³ of the storage will be available for flood control, which will benefit the oft-flooded industrial city of Wuhan. The reservoir will change from one depleted annually to a multi-year one capable of sustaining supplies through a bad year. Water is now expected to flow along the open canal route by 2014, diverting a third of the annual flow of the Hanjiang.

A four-year delay in the scheme has been partly due to the fact that the government has allowed much freer press coverage than it did with Three Gorges. Local governments, environmental and archaeological bodies, and the public have raised objections ranging from arguments over compensation for the displaced people to increased pollution in the Hanjiang River caused by lower flows and less dilution, and the lack of time and money to save important archaeological remains. The solutions include many more water treatment plants, enlarging some reservoirs downstream of the new diversion to regulate riverflows and improved compensation and relocation packages for the people – although there are complaints that these are not as good as official announcements. There are also calls for water to be diverted from the Yangtze to top up the Han River from a new dam downstream of the Three Gorges Dam.

By 2030, this central route could transfer up to 14 billion m³ a year, but this is unlikely to be achievable without additional supplies from the Three Gorges. Even then, the Sino-Italian Cooperation Program suggests that the water shortage in the region targeted by the Middle Route will rise from 16.2 billion to 31.2 billion m³ p.a. between 2000 and 2020, and the current scheme will be inadequate to support much outside the main towns.

Work was scheduled to begin on the Western Route in 2010 and be completed by 2050. It is planned to eventually carry 17 billion m³ a year. But there are currently major questions about its technical viability. The results of a survey presented at a meeting organized by the Chinese Academy of Geological Sciences in 2007 suggested that:

1 there is not enough water in the three Yangtze headwater tributaries to support such a large diversion;

2 the area is seismically active and building the six dams and seven tunnels could trigger earthquakes, rock falls and landslides;

3 that the 12 hydropower projects under way or planned upstream of the diversion will create competition and affect the amount of water available; and

4 climate change will eventually reduce glacier meltwater supplies.

The overall scheme is likely to have major environmental impacts on the donor rivers, reducing riverflows downstream of the diversions, affecting fisheries, reducing dilution of pollution, and increasing saltwater incursions in the estuaries (Liu and Zuo, 1983). Jun Xia analyses the ecological resilience of one of the possibly affected rivers, the Hai River, in the article 'Determining optimum discharges for ecological water use' in Chapter 15.

China also claims that the upstream diversions, together with a series of hydropower dams planned for the Mekong, will ameliorate floods for its neighbours: Cambodia for one might not be too happy with this seemingly innocuous argument. Similarly, the Brahmaputra diversion could also aggravate seasonal water shortages in India and Bangladesh.

US and Soviet diversion plans

Two of the largest river diversion schemes yet proposed are unlikely ever to reach fruition, each for their own very different reasons. The North America Water and Power Alliance in which the USA planned to divert Alaskan water through Canada to the water-stressed American West, and Canada's own water into the American East and Mid-west failed because no alliance was ever established – the Canadians realized there was nothing in it for them. The Soviet plan to divert Russia's arctic rivers south to the agricultural lands of the Ukraine and Central Asia failed because President Gorbachev realized that the problem facing agricultural production had more to do with the communist system of collective farms and uniform wages for everyone, irrespective of productivity, rather than lack of water. Prior to Gorbachev's decision, the river reversal scheme seemed likely to be the first grand scheme to be implemented, involving only domestic decisions within a superpower. With the ensuing breakup of the USSR and the creation of numerous independent states, many with a dislike of Russia, all chances have virtually disappeared.

American politicians return to the NAWAPA idea now and then in response to economic events. The attraction of hydropower was increased by a 1600 per cent increase in oil prices in the 1970s and the Three Mile Island nuclear disaster in 1979, and the worst drought in Californian history emphasized the parlous state of water in the west. These events led the US to reconsider the plans in the early 1980s, but by then an already expensive scheme had doubled in cost (Reisner, 1993). Since 1990, the impetus has been global warming, with fears of a drying trend west of the Mississippi. But Canada has moved on since NAWAPA was presented in the 1960s, developing its domestic hydropower schemes (see the section headed *James Bay Hydro: redesigning a landscape* later in this chapter) and husbanding its own Rocky Mountain resources to solve water shortages in the prairies, notably in the Canada DRI scheme.

The North America Water and Power Alliance

When NAWAPA was proposed in the 1960s, it was expected to take 30 years to complete and cost $100 billion (Figure 7.24). It would transfer water from Alaska, the Rockies, the Mackenzie River and other arctic basins to the Southwest via the Southwest Reservoir Canal, to the Mississippi via the Dakota Canal, and even to New York via the St Lawrence. The transfer system would provide water and hydropower, and also improve navigation.

145

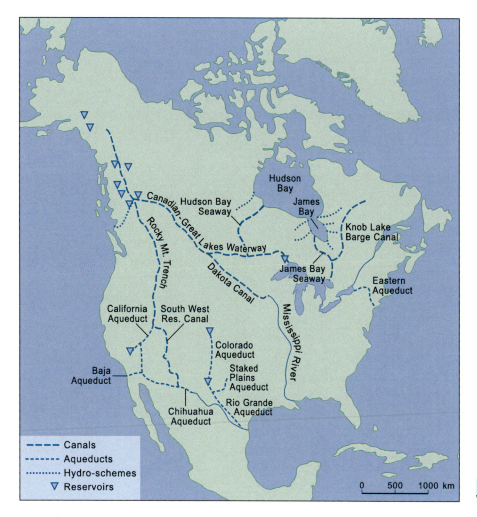

Figure 7.24 The North American Water and Power Alliance plan

The scheme would require nearly 14,600 km of canals and tunnels and numerous reservoirs. The largest reservoir would fill 800 km of the American Rocky Mountain Trench with water pumped up to an altitude of 900 m to allow water to be gravity-fed to the south-western States. An average annual runoff of 818 km³ would be diverted from a drainage area of 3.3 million km². Normally, only 20 per cent of this runoff would be used, but the plans allowed up to 40 per cent in cases of dire need.

Canada was wooed by the prospect of hydropower, more water for the drought-ridden prairies, topping up the falling water levels in the Great Lakes and improving their water quality, and by new north-south navigation routes capable of carrying heavy freight, like the iron ore from Schefferville's mines. Canada would be the main beneficiary of the hydropower, estimated at capable of providing 50 per cent of its requirements by 1980. Mexico could also benefit from the extra water in the Colorado, where intensive irrigation on the American side of the border was drastically polluting what little riverflow crossed the frontier, and in the Rio Grande, as well as from enough hydropower to service 15 per cent of its domestic needs.

Canada was not persuaded. It developed major hydroschemes in Northern Quebec and Labrador, established the Mackenzie basin as a specially protected region (see *The Mackenzie basin gained and lost?* in Chapter 5), and from 1972 improved the water quality of the Great Lakes under the joint US-Canadian pollution control programme, which limited wastewater discharges into the Lakes. Apart from some river modifications in the Mackenzie basin that resulted in a sharp increase in river freight tonnage during the late 1970s and 1980s, improving navigation has not been a major consideration: the demise of the Schefferville mines contributed to that.

Sadly, Mexico really could have used the water, as the often dry river bed of the Rio Grande demonstrates. But with nothing to give to the project, Mexico is left to its own meagre devices and a worsening social situation on its parched northern border.

Reversing Russia's arctic rivers

In 1982, the Supreme Soviet approved the most audacious plan of all, not simply to divert rivers, but to actually *reverse* the flow of all the major rivers draining into the Arctic Ocean to support agriculture in the southern republics (Figure 7.25). The decision was prompted by superpower politics as much as by the geography of water resources. The USA was controlling the amount of grain that the USSR could buy on the American market, partly to protect its own supplies, but mostly it was using the 'food weapon' to show its disapproval of the Soviet invasion of Afghanistan and also as part of a long-term policy to destabilize the Soviet Union.

The USSR was a prime example of imbalance between the distribution of water resources and population and industry. Eighty per cent of its water resources were in Asia, 70 per cent of the population in Europe, and on both continents most of the water drained into the Arctic Ocean, while most of the economic activity was in the southern basins, draining to the Caspian, Black and Aral Seas. Yet the Soviet Union was unique in having within its own national boundaries both the huge, virtually untapped discharges of the great arctic rivers and some of the world's top grain lands, as well as having a powerful command economy capable of pushing big projects through. But the grain lands in the Ukraine and Central Asian Republics were suffering from a chronic combination of natural drought and mismanagement (Holt, 1983).

The political issues began in 1972 when the USSR bought millions of tonnes of US grain, including 16 million m^3 of wheat, on the open market within a few weeks behind

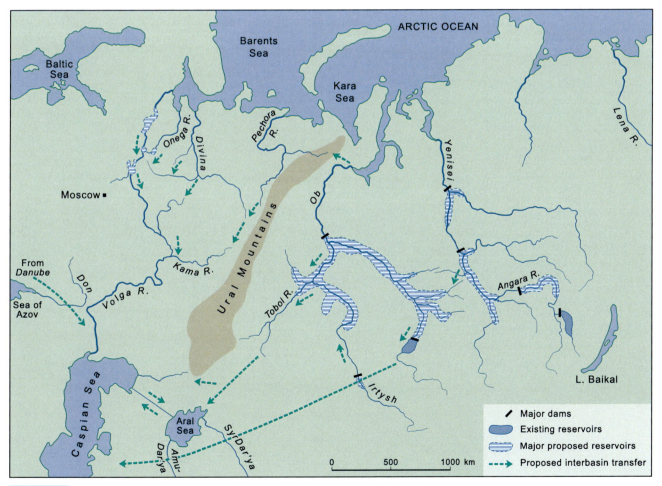

Figure 7.25 Soviet plans for reversing the arctic rivers

the back of the US government. The USSR was responding to the effects of a severe drought. What became known as the 'Great Grain Robbery' caused world wheat prices to increase nearly threefold and the price of bread in America rose dramatically. After this, the USA established a bilateral agreement with the USSR to limit the amount of food and feed grain that the Soviet Union could purchase. Another drought caused crop failures in 1975, but the amount the USSR could buy was now limited. Following the Soviet invasion of Afghanistan in 1980, the Carter administration rescinded the agreement for two years – hence the Supreme Soviet's vote to go ahead with the controversial diversion plan.

Some work had already been undertaken on the Pechora River reversal prior to 1982, damming it and blasting out a route through the mountains using controlled nuclear explosions. Following the full approval of the scheme, work was due to begin in the European sector in 1986 and in Asia in 1988. It was estimated that another 250 nuclear explosions would be needed to blast through dividing mountains. At the last moment, Gorbachev won a vote to shelve the scheme in 1986, arguing that falling yields were largely caused by the communist agricultural system and that a grand technological 'fix' would only delay making the system more efficient and introducing incentives for the workers.

The scheme would have taken over 50 years to complete and was effectively uncosted, as the Soviet monetary system was not linked to world financial systems. On completion, 60 km^3 of water a year would be diverted, beginning with diverting the Dvina and Pechora Rivers to the Volga. By 2000, 20 km^3 a year would be diverted to the Caspian, Sea of Azov and Black Sea basins, and a beginning made on diverting 27 km^3 a year from Western Siberia to the Central Asian Republics and the Aral Sea. Much of the Central Asian diversion would be via a very inefficient new open and unlined canal. This would be shadowed by 'leak collector' channels that would return about half of the 3.5 km^3 estimated to be lost through leakage by collecting groundwater seepage and local surface water. The ultimate aim was to create 2.3 million ha of new irrigated land and to replenish falling water levels in the Caspian and Aral Seas (see *Aral Sea – lost and regained?* in Chapter 15).

Environmental impacts

Southern agriculture would undoubtedly have profited from the increased water supply, while the north would have fewer snowmelt floods, which could be either an advantage or disadvantage depending on the degree to which the spring freshets were needed for local water supply. Fisheries based on migratory fish like salmon would also suffer.

Greatest international concern, however, focused on the potentially global consequences of depriving the Arctic Ocean of huge volumes of freshwater. According to the most established western view, this would reduce the seasonal growth of sea ice, which would feedback to the atmosphere, causing the average winter latitude of the Ferrel jetstream and its attendant depressions to shift northwards over the North Atlantic and so aggravate rainfall shortages in southern Europe. The argument centres on the fact that the freshwater freezes more readily than saltwater and being of lower density it floats on top of the seawater and so encourages ice formation.

However, an alternative argument held that reduced freshwater discharges from the colder continent, leading to warmer sea surfaces and more evaporation, resulting in more snowstorms that would chill the water and provide freezing nuclei that stimulate ice growth. But the ice might be much thinner and melt earlier, because more snowcover and cloud cover would reduce radiative cooling. Computer models in the 1980s were hampered by lack of knowledge and by the complexity of the interactions that might be involved, but the scheme was halted for political rather than environmental reasons before the scientific arguments were fully resolved. Ultimately, the route nature might have taken depends mainly upon the critical balance between sea surface temperatures and the amount of reduction in freshwater discharges, with windiness playing a supporting role in mixing the surface layers.

In the end, the Aral Sea was the main victim of the axing of the scheme.

James Bay Hydro – redesigning a landscape

Soon after Canada rejected the NAWAPA proposals, it began developing the most ambitious hydroscheme to date in the north of Quebec Province (Figure 7.26). Hydro-Québec's $50 billion 'project of the century' was launched in 1971 and aimed to develop 26,000 MW over the next half century from the rivers of the James, Hudson and Ungava Bays, of which 16,000 MW is currently installed. It was the first mega-scale hydroscheme in the subarctic. It is a low-head scheme, using the large quantities of snow that accumulate in northern Quebec and capitalizing on the ease of river diversion in a glaciated lowland with subdued catchment divides. The main technical detraction is that the rivers are highly seasonal, with flows dominated by the spring snowmelt, so that the reservoirs need to be exceptionally large. This is a major source of environmental impact.

The scheme was devised in an era when social and environmental impacts were of secondary concern. But as the scheme evolved so did public opinion. The Cree Indians bore the brunt of the development and formed the nucleus of the opposition (see the section headed *Social impacts and anti-dam politics* later in this chapter).

The electricity was intended to support Quebec's economic growth, but the amount generated would exceed Canada's needs and the surplus ten per cent or more is intended for export to the USA. The scheme was spurred on by Newfoundland's successful exportation of power from Churchill Falls through Quebec to New York, and by Quebec's own Separatiste politics.

The first phase began on the La Grande River in 1972, diverting water from the Eastmain, Opinaca and Caniapiscau Rivers into five reservoirs on the La Grande River. By completion in 1986, this had a capacity of 10,300 MW and involved building 215 dikes and dams, including three enormous ones, flooding 11,300 km² and diverting 19 billion m³ of annual discharge from seven rivers.

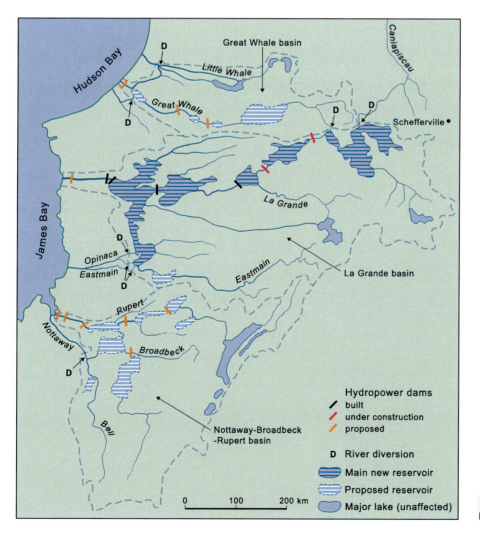

Figure 7.26 The James Bay Hydroscheme

The second part of the scheme began in 1989 with three new reservoirs, including two on the Laforge River, and five new power stations designed to raise the total capacity to over 15,000 MW by 1996. During this stage, Hydro-Quebec proposed an additional scheme on the Great Whale River in the north. In all, another ten rivers in the Great and Little Whale basins in the north and the Nottaway, Broadbeck and Rupert basins in the south were to be diverted. However, the scheme was halted in 1994 due to environmental concerns, protests by First Nations, and fears that the extra electricity supply would outstrip demand. Perhaps the most critical factor was New York State's withdrawal from the electricity purchase agreement due to a decrease in energy requirements and the high profile public protests. In this, the First Nation campaign was key, including the Grand Chief's canoe trip from Hudson Bay to the Hudson River in New York State. The scheme was only partially resumed after a new agreement was signed by the Cree in 2002 (see 'Social impacts and anti-dam politics', below).

On completion, the whole project would have reversed or drastically altered 19 major rivers, transformed the landscape over an area of around 350,000 km², and submerged 176,000 km² of forest and tundra. The hydrology and the general environment of an area approaching the size of France would be redesigned and a fifth of Quebec's surface water, or more than 1.5 per cent of the world's liquid freshwater, harnessed.

Social impacts and anti-dam politics

The social response has been so extreme and so powerful that it has fundamentally affected the development with delays and cancellations, financial and legal compensation, and established a legal requirement for Environmental Impact Assessment. New road and air links to the south and increased contact with construction workers from southern Canada have been a major culture shock for the previously isolated local people and increased problems of social deprivation. Together, exposure to a different culture and disruption of traditional hunting and trapping pursuits have destroyed time-honoured lifestyles.

The anti-dam movement has involved an alliance between First Nations and environmentalists, using a combination of legal challenges and wide publicity. The long-running campaign was always more than anti-dam; it was the beginning of the fight back by aboriginal peoples against the loss of their heritage lands. Feelings were inflamed at the outset when road-building began before the indigenous peoples were informed about the project. The 5000-strong Cree Indian population continue to be the most vociferous in their objections, despite some significant victories. The 3500-strong Inuit population in the Great and Little Whale catchments in the north has also been involved, but has not always agreed with the Cree. Some Inuit have tended to favour development in their area under the final phase more as a source of increased wealth.

The Cree-Inuit alliance first challenged the development in court in 1972. They initially won the case to halt the scheme on what they regarded as their traditional hunting grounds, but then lost on appeal. The 1975 James Bay and Northern Quebec Agreement awarded the Cree and Inuit nations about $250 million in compensation for loss of their aboriginal land claims. This was the first such agreement in Canada. It gave the Cree unprecedented powers of self-government as a 'First Nation', but it was signed under duress and there have been lingering feelings that the developers and the Government of Quebec have paid too little regard to their concerns. The Cree have been particularly outspoken in their opposition to Phase II and beyond, especially the damming of the Great Whale River. In 1990, they obtained a federal court order requiring public hearings and a full EIA, which was entirely lacking in the original plan, before Phase II could go ahead. This played a major part in successfully blocking the Great Whale development in 1994. The Grand Council of the Cree subsequently signed a deal, 'La Paix des Braves', in 2002 allowing development of the southern part of Phase II on the Rupert and Eastmain rivers. This led to disagreements among many Cree chiefs. Despite this, the Grand Council went ahead and signed a formal end to litigation in 2004, much to the chagrin of the Grand Chief newly elected in 2005, who publicly championed wind power as an alternative. Cynics have also suggested that agreeing to development of the southern part of the final phase (in Cree territory) and blocking the Great Whale (in Inuit territory) is a partisan victory for the Cree.

Environmental impacts

The project has already had an unprecedented impact on the hydrology of Northern Quebec. The Caniapiscau-Koksoak River system is the first arctic river to be partially reversed. The river diversion at the outfall of Lake Caniapiscau redirects flow west-southwest into the La Grande River, cutting the discharge from the Koksoak into Ungava Bay in the north by a third. The discharge of the Eastmain River into James Bay is reduced by 90 per cent. In contrast, the discharge of the La Grande into James Bay is doubled, from 1700 m³/s to 3400, as a result of river diversions concentrating flow into the La Grande megadams. The difference is even greater in winter with a tenfold increase from 500 to 5000 m³/s. The extra flow is caused by release of water stored from the summer in the series of gigantic reservoirs. In the natural state, winter riverflows are restricted by the lack of liquid precipitation and by ice growth on the rivers and lakes. A modest but significant portion of these limited natural winter flows may be caused simply by the build-up of snow weighing down the ice cover and displacing water from the lakes, as Jones (1969) discovered in a small headwater of the Caniapiscau. Diversion of the Rupert River into the La Grande will halve its outflow into the Rupert Bay arm of James Bay, but add another 420 m³/s to the La Grande's discharge.

Increased water levels on the La Grande create additional fears. Two thousand five hundred Cree were advised to leave their traditional township of Chisasibi on Fort George Island in the mouth of the La Grande, because hydraulic engineers feared that the La Grande scheme would cause the sandy island to erode. The higher winter discharges also destabilize the river ice, so that it no longer forms a sound bridge across the river for the Indian hunters. Worse still, decomposition of the vegetation flooded by the reservoirs has caused a toxic build-up of mercury in the food chain and the Cree no longer eat fish from the reservoirs. There has been concern that the reservoirs might accumulate atmospheric pollutants as well, either through direct precipitation or by concentrating snowmelt runoff: snow is good at scavenging pollutants because free ionic bonds at the ends of the crystal lattices attract aerosols in the air and wet snow on the ground also traps aerosols. Improvements in air quality as a result of Clean Air Acts have reduced this risk in recent decades.

The reservoirs and diversions also disrupt migratory mammals. An unexpected flood on the Caniapiscau River in the spring of 1984 during construction of the river diversion caused the death of 10,000 migrating caribou, despite an emergency helicopter airlift. The extension of Lake Bienville on the Great Whale in Phase II would risk disrupting a caribou calving area. However, the per-

manent reduction in flows on the Caniapiscau since the diversion has reduced the risk for migrating caribou.

Offshore, increased sediment yields during construction caused concern for the valuable lobster population of James Bay, and long-term changes in salinity in Hudson Bay may threaten the Beluga whales. Concern has also been expressed over possible geological shifts due to compression from the weight of the impounded water, especially for the La Grande-3 reservoir, which covers the glacial deposits to a depth of 90 m.

Dams and diversions in India: Narmada and India's National Perspective Plan

Annual rainfall in India as a whole is well above the world average. But, like its Asian twin, China, it has severe problems with the geographical distribution of its water resources, which are exacerbated by a burgeoning population. The north and east are water-rich, but the south and west are water-poor. Rainfall in the Himalayan Mountains in north east India is some of the highest in the world at around 11,000 mm a year, while in the arid west it can be barely 100 mm, a massive 100-fold difference. This means that the water resources available per person in the Brahmaputra basin are as high as 13,000 m³, while in the Sabarmati basin in drought-ridden Western Gujarat and the Pennar basin in the south they are just 300 m³.

And as in China, the rain is also highly seasonal. The annual monsoon typically supplies 75–80 per cent of resources in just ten to 16 weeks. In the north, this is supplemented by meltwaters from the glaciers and snow-cover in the Himalayas in late spring. Climatic variability adds to the range, exaggerating floods and droughts. In 2004, a third of India suffered floods, while another half suffered from drought. The severe flood on the Kosi River in Bihar in 2008 displaced four million people. The late arrival of the monsoon in 2009 caused the driest June since 1926, with rainfall 46 per cent below normal. June to August rains averaged 30 per cent below normal, but in the Punjab, India's 'breadbasket', rains were down 50 per cent and some areas in the west, like Maharashtra State, received only 60 per cent of normal. Agriculture suffered badly and so did many of the 700 million people that depend on it. Drought emergencies were declared in parts of Manipur and Jharkhand and the national government banned exports of wheat. Public water supplies failed, there were street fights over water and water tankers had to have police protection in Madhy Pradesh.

The drought did for India what the world recession had not done: reduced GDP some 1.5 per cent (see *Debt and transfiguration? – Recession 2009* in Chapter 4). Agriculture accounts for 18 per cent of GDP. The Mumbai stock market plummeted over fears that consumer demand would fall and millions of poor subsistence farmers would lose their livelihood.

Meanwhile, India's population is growing rapidly. It rose 21 per cent in the decade to 2001 and is predicted to rise from the present 1.17 billion to at least 1.6 billion by 2050. Per capita water resources are dwindling in tandem: 6008 m^3 at the time of independence to around 1700 now. The estimated effects of population rise for coming decades range from a conservative per capita resource of 1140 m^3 by 2050 to an alarming drop to below 1000 as early as 2025. The latter would plunge the second most populous country in the world into the realm of water scarcity (Bandyopadhyay and Perveen, 2004).

Hunger is already a problem. In 2008, the International Food Policy Research Institute ranked India 66th out of 88 countries on the Global Hunger Index. It rated hunger as 'serious' in four of the larger states and 'alarming' or 'extremely alarming' in the other 13. Irrigation provides critical support for Indian agriculture and the World Bank says India is on the edge of an 'era of severe water scarcity'.

India's solution is very similar to China's: megadams and diversions. But it is also more complex in both aims and administration. What China is doing with three routes, India is proposing to do with 30. While two-thirds of China's routes focus on the capital city, India's canals will serve a scattered and predominantly rural population. Where China has central control, India is a democracy and development is being devolved. China is just beginning to experience the effects of allowing local objections. In modern India, local voices have their say.

The Narmada Megadams

As India gained independence in 1947, the new Prime Minister Pandit Nehru declared that dams are the 'temples of modern India' and championed a scheme to construct a series of hydropower dams on the Narmada River. But it was not until 1961 that he laid the foundation stone for the first megadam, the Sardar Sarovar.

Even then, the scheme did not go ahead and it was not until this century that the dam was completed, following repeated delays caused by anti-dam campaigners. At the time of Nehru's death in 1964, there were serious fears that the population would outgrow the food supply, the Green Revolution was only just beginning, and the need for more irrigated land seemed paramount.

The Narmada is one of the largest rivers in north west India, running 1245 km from the centre of the subcontinent to the Arabian Sea at Bharuch, some 350 km south of the Thar Desert. Back in 1901, the British Irrigation Commission considered building a barrage at Bharuch, but decided that the soils were not suitable.

Sixty years later the motivating forces had changed. However, disputes over the relative share of the costs and benefits broke out among the riparian states – Gujarat, Madhya Pradesh and Maharashtra – as soon as the scheme was resurrected in the early 1960s. Prime Minister Indira Gandhi established the Narmada Water Disputes Tribunal in 1969. Gujarat enlisted the non-riparian state of Rajastan, which is dominated by the desert, to boost its claims for a larger share of water resources to help the drought-prone three-quarters of Gujarat State. Agreement was not reached till 1979, by which time the only sizeable new dam, the Tawa Dam (1973), had been built. The subsequent Narmada Valley Development Plan envisaged a chain of 30 large dams, including megadams at Sardar Sarovar and Narmada Sagar, 135 medium and 3000 small dams on the main stream and 41 of its tributaries. The Sardar Sarovar in Gujarat forms the bottom step of the staircase of dams that will become the largest river modification scheme in the world. The megadam has a planned hydropower capacity of 1450 MW and has become the second largest in the world after successive Supreme Court judgements between 1999 and 2006 gave permission for it to be raised from the initial 80 m to 122 m. If and when the whole scheme is completed, it will turn the river into a series of lakes and displace 250,000 inhabitants (about one per cent of the current population in the basin).

Anti-dam campaigns

Local opposition began to form in the mid-1980s and was consolidated into the highly effective Narmada Bachao Andolan (Save the Narmada Movement) in 1989. In the 1990s the opposition became international. The World Wildlife Fund, the International Union for Conservation of Nature (IUCN) and the International Rivers Network joined the opposition, and the Environmental Defence Fund gave financial support to the NBA. The World Bank withdrew its support in 1990. Indian author and environmental activist Vandana Shiva, and Booker Prize-winning novelists Salman Rushdie and Arundhati Roy joined the international publicity campaign. Roy's book *The cost of living: the greater common good and the end of imagination* (1999) includes an extended essay on Narmada. The films *A Narmada Day* (1995) and *Drowned Out* (2002) spread to the world the message of the suffering of the indigenous tribes being forcibly evacuated. Opponents claim that India has a poor record on resettlement. To Hindus the publicity also emphasized the sacrilegious impact on one of their most sacred rivers: bathing in the Ganges purifies the soul, but merely looking at the Narmada does the same.

Environmental impacts

The scheme received approval without a proper EIA. This is despite the fact that the Indian Central Water Commission produced such guidelines in 1975 and the Department of Environment produced its 'Guidelines for Environmental Assessment of River Valley Projects' in 1985. Since 1987, more environmental reports have been produced for the Narmada than for any other scheme in India, but these come after the scheme was designed and are often based on incomplete evidence. Kothari and Ram (1994) claim that the wide-ranging assessment by the internationally respected team at HR Wallingford (1993) was based on an over-estimate of downstream riverflows, either accidentally or deliberately supplied to them.

Environmental concerns include:

1 destruction of forests. Nearly 14,000 ha of forest are being flooded by the Sardar Sarovar alone. Compensatory afforestation is planned, but in Kutch, which is a very different environment. Biodiversity is threatened, so too are the traditional tribal uses of the woodlands.

2 premature silting-up of the Sardar Sarovar. While construction of the upstream dams, especially the Narmada Sagar, is delayed, soil eroded from the wider catchment will be concentrated in the lower reservoir. Only 27 per cent of the critically erodible land is earmarked for remedial measures. The World Bank would like complete coverage, but there is currently no plan for this.

3 destruction of the marsh crocodiles' breeding grounds by inundation (*Crocodylus palustris* is on the IUCN's Red List of endangered species).

4 waterlogging and salinization of soils, particularly in the reach between the two megadams.

5 saltwater incursion in the estuary owing to the lower river discharge.

6 destruction of traditional fisheries, including the giant freshwater prawn.

7 downstream reduction in the quantity and quality of water available for drinking.

8 increase in malaria vectors with an attendant increase in water pollution, as the main method of control is spraying with pesticides.

Many of these problems could be alleviated given the right management. The World Bank's current view is that it is now a 'reasonably well structured programme', but 'still exhibits weaknesses'. Some excuse has been made that this was an entirely new scale of project for India – for both resettlement and environment – and that lessons have been learnt along the way. This is nothing unique. And it is not just scientific understanding that is advancing. Public views are also changing around the world. But one wonders how much remedial work might have been commissioned without the public outcry.

In positive terms, the scheme has been dubbed a lifeline for drought-ridden Gujarat. Rajastan may also benefit if diversions extend that far. About 18,000 km² of new irrigated land is planned, which should be a major step in India's fight for self-sufficiency. And the electricity is much-needed as India vies with China to be a leading world economic power. But with estimates suggesting the population in the Narmada catchment alone is likely to increase by more than 50 per cent to over 40 million by 2021, here, as elsewhere in India, it is a catch-up race that will be hard to win.

The National Perspective Plan – a 'Garland of Hope' or Hype?

Following severe drought and floods in a number of states in 2001–2002, the Government of India created the 'Task Force on Interlinking of Rivers in India' comprised of specialists and members from both water-surplus and water-deficit states. The Task Force was mandated to evolve a road map for implementation of a scheme to link the rivers from the Himalayas to the southern tip of India within 12 years. In essence, the scheme is intended to harness the tremendous volume of water available during the annual monsoon in the Himalayan region, which causes regular flood problems, and redirect it to the drier regions.

The idea of interlinking rivers by canals was first put forward during the British Raj in the nineteenth century, but the purpose then was navigation. The Minister for Irrigation proposed a single Ganga-Cauvery link to carry water to the south in 1972, and five years later a private proposal suggested a double link consisting of a Himalayan Canal and a Central-Southern 'Garland Canal'. Both were rejected on grounds of cost and technical feasibility. Yet in 1980, the Ministry of Water Resources produced a much broader plan, the National Perspective Plan, and government set up the National Water Development Agency (NWDA) to prepare feasibility studies for the scheme in 1982.

The Plan has two components: the Himalayan Rivers Development and the Peninsular Rivers Development. The NWDA identified 31 river links on 37 rivers (Figure 7.27). The Himalayan component would transfer water from the Ganges (Ganga) and Brahmaputra towards the water-deficit regions in the west. The scheme proposes transfer of 1500 m^3/s from the Brahmaputra and 1000 m^3/s from the Ganges, out of average flood discharges of 60,000 and 50,000 m^3/s respectively. However, the government has been holding detailed information as

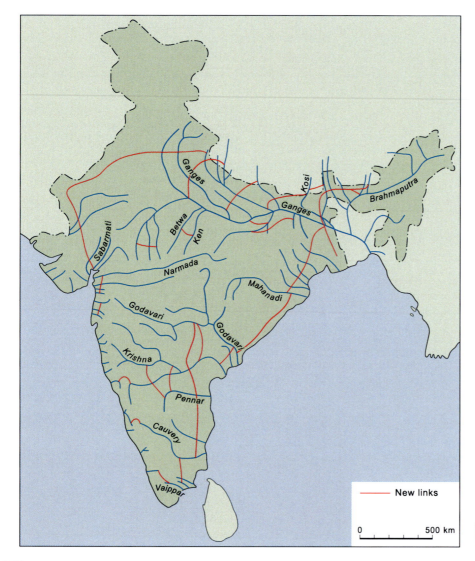

Figure 7.27 Interlinking Rivers – India's National Perspective Plan

New links

0   500 km

confidential because of the international implications, particularly the injurious effect on Bangladesh. It would include agreements with Nepal and Bhutan to build retention reservoirs on some of their headwater tributaries.

The Peninsular component comprises 16 links in four groups, the largest of which runs down the east coast and links up with the Himalayan section, completing a transfer very much along the lines of the discarded Ganga-Cauvery plan.

The scheme is now more than just about transferring water to water-deficit regions for irrigation and public water supply. It includes hydropower production, improved navigation, flood moderation and artificial groundwater recharge. The Central Ground Water Board plans to use some 71 billion m³ a year of water diverted from the Sarda River on the border of Nepal to recharge aquifers beneath the Thar Desert (Figure 7.28).

The regional governments of Uttar Pradesh and Madhya Pradesh agreed to proceed with the first link, between the Ken and Betwa rivers, which is seen as a trial run for the Plan, in 2005. It will divert 1020 million cubic metres a year from the Ken. A Duke University study published in 2009 tested the Ken-Betwa scheme using a Super Reservoir Model – treating all the storages as one super-reservoir – and modelled the scheme under a variety of scenarios for water balance and irrigation demands (Mysore, 2009). It showed that the original feasibility studies had greatly underestimated the risk of 'dry-downs' in the planned system: the risk could reach 35 per cent in the worst scenario. The study illustrates the need for more sophisticated computer modelling in complex systems like this.

Obstacles and opposition

The chairman of the Task Force on Interlinking of Rivers in India, Suresh Prabhu, called the scheme a 'Garland of Hope', the solution to India's current and future water crisis. Dunu Roy, director of the Hazard Centre in New Delhi, has branded it the 'Garland of Hype', pointing out that the various aims are conflicting, such as flood control versus hydropower, or the need for irrigation water to be released in summer when the power companies are wanting to store water to generate power during the dry season. The scheme will also require considerable power to pump water across numerous high watersheds, like lifting the planned 1200 m³/s up 116 m over from the Mahanadi River to the Krishna. To shift 37.8 km³ of

water a year from the Ganges to the Deccan Plateau will require 3.9 GW.

Because the administration of water in India is devolved to state governments, there are frequently fierce arguments between states as they support their own interests on the relative costs and benefits. Individual states have also expressed divergent views about the studies and feasibility reports prepared by the NWDA. Uttar Pradesh has voiced fears about the loss of water and hydropower potential to Madya Pradesh in the latest, Ken-Betwa scheme, and these fears are writ large in interstate relations throughout India.

Much of the public opposition comes from the poor record on resettlement of displaced people. There is a history of neglect, some of it clearly affected by tribal and caste differences. One official estimate predicts that just under half a million people will be displaced, but other estimates are as high as 3.5 million. This is on an unprecedented scale. Yet the pilot Ken-Betwa transfer scheme was approved in 2005 without any agreed plan for relocating even the relatively small number of 8550 people likely to be displaced.

One of the main environmental concerns is the flooding of large tracts of forest and farmland. Reservoirs alone are estimated to inundate over one million hectares and at least a further 700,000 ha are likely to be affected. Other strong objections focus on the impacts on the ecology and biodiversity of both the donor and receiver rivers, including the introduction of waters with different chemical signatures, depriving donor rivers of the scouring and depositional effects of floods that bring nutrients to the floodplains and new habitats for pioneer biological communities. In many respects, the science needed to adequately design such water transfers is very much in its infancy (see *Restoring and protecting the aquatic environment* in Chapter 15).

However, the greatest obstacles have yet to manifest themselves in the form of international disputes with China, Bangladesh and the Himalayan nations, should the Plan be fully implemented.

Some conclusions

Dams are the foundation tool of water management. As populations increase, so dams proliferate to convert floodwaters into usable resource and to store supplies for drier periods. They are an essential and cannot be avoided. Yet no dam and no river engineering can be undertaken without some environmental impact. In the modern world, this often means human impacts as well. The art is to minimize these impacts.

Ever larger dams and ever more ambitious river diversions – interbasin transfers on almost continental scales – are now making these impacts immense. The technology to do this is widely available. The political and economic forcing factors are growing. Yet the scientific and regulatory frameworks for minimizing the impacts are generally deficient.

In developed countries, far more regard is now given to protecting the environment. While the emerging nations of India and China are still actively building large dams, dam-building is stalling in North America and Europe, where managers and planners are beginning to seek alternatives, like controlling demand, and even to demolish dams or rewrite the operational rules, like recreating floods artificially. The current fate of Spain's national water transfer scheme is a marker for the new era of legal environmental controls in Europe.

As the new millennium is unrolling, the plans of the twentieth-century superpowers, Russia and America, are consigned to the waste bin. In contrast, the plans of the potential superpowers of the twenty-first century, China and India, seem more set to achieve fruition. If they do, both schemes are going to cause some international ructions with the neighbours.

In the domestic arena, public protest movements continue to play an important role in controlling and questioning developments. As a democracy, India is home to both active popular protest and the common view of officialdom that concern for environmental impacts is 'a kind of luxury' that a developing country cannot afford. As a federation of states with devolved powers, India is

hostage to interstate arguments that may yet prove the crucial factor that stops the national plan being implemented. As a command economy, China is going to be the first country to implement its nationwide plans, which are likely to be achieved before creeping liberalization finally allows public and regional objections to have a substantial effect. Whether the diversion plans are sufficient to stem the looming problems of increasing demand and climate change, and to see the country through to superpower status is highly questionable.

The case of Australia is yet another manifestation of the good and the bad elements of both democratic protest and democratic governance. Australia is beset by a hostile climate that seems to be getting worse. Administrations have been slow to recognize the fact and equally reluctant to confront anti-dam protest in recent years. There seems little doubt that Australia will need to build more dams and maybe implement some old-style interbasin transfers. But one positive outcome is that Australia is way ahead with alternatives and recent campaigns to control demand are examples for the world (see Chapter 13 Cutting demand).

Discussion topics

- Consider the advantages and disadvantages of large dams.
- Assess the environmental and social impacts of large-scale river diversions.
- How effective are mass protests against large dams and river diversions?

Further reading

General coverage of the pros and cons of dams can be found in:

Brink, E., McLain, S. and Rothert, S., 2004: 'Beyond Dams – options and alternatives.' Report by American Rivers and International Rivers Network. www.internationalrivers.org/en/the-way-forward/water-energy-solutions/beyond-dams-options-alternatives.

International Commission on Large Dams (ICOLD), 1999: *Benefits and concerns about dams.* www.icold-cigb.org/PDF/BandC.PDF

World Commission on Dams, 2000: *Dams and Development: a new framework for decision-making.* WCD.

World Wildlife Fund website: http://www.wwf.org.uk/what_we_do/safeguarding _the_natural_world/rivers_and_lakes/dams_and_infastructure.

8 Trading water – real and virtual

Even before water was designated as an economic good, water was traded. Before refrigerators, ice was a valuable trading commodity. It still figures in international trading. The Food and Agriculture Organization's (FAO) trading statistics show ice being exported from India to the EU. But two developments in the late twentieth century caused a major acceleration in trading; one was real, the other virtual.

Real water exports expanded quite suddenly with the development of the bottled water industry. This was accompanied by the growth in bulk shipping of water to islands, especially islands that benefited from the growth in tourism, or to strategic outliers like Gibraltar. These have often used decommissioned oil tankers, although shipping in large plastic containers and tug-hauling water bags are increasing in popularity (see *Imports by sea – water bags, drogues and cigars* in Chapter 14). In recent years, Turkey has signed an agreement with Israel to provide water in large containers. Turkey has also considered exporting water to North Cyprus in water bags rather than by pipeline or tanker.

The consumption of water in bottles or large containers varies markedly from region to region (Figure 8.1). Western Europe is the largest overall consumer and has the largest per capita consumption. Eastern Europe has the lowest total consumption, but by no means the lowest per capita consumption. The lowest per capita consumptions are in Asia, Oceania, Africa and the Middle East, largely reflecting general levels of affluence.

However, international trading in real water is literally a drop in the ocean compared with 'virtual' water. All foodstuffs that are traded have required water to produce them, the unseen so-called 'virtual water'. The water is needed to grow crops and fodder, for animals to drink, and a substantial amount may also be used in food processing. If a country imports food, it is effectively using other countries' water and saving its own. So there may be a strategic advantage in saving domestic

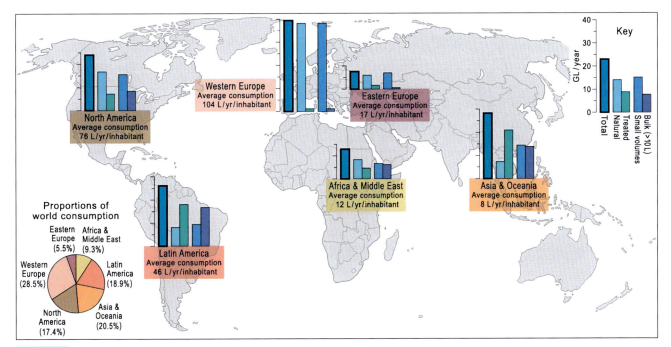

Figure 8.1 World consumption of water traded in bottles or containers

water resources by importing food, especially for a water-scarce nation.

It is not only food that 'contains' virtual water – almost every product that is traded has used water in the production process. So the burgeoning of international trade over the last half century also represents a burgeoning in virtual water trading. This has important geopolitical ramifications. It means that internal security may be increased by importing food and goods, but it also means that external security might be reduced by becoming too beholden to foreign suppliers that could cease trading.

Taking virtual water imports into consideration, 62 per cent of the Britain's effective total water consumption comes from abroad. This gives a whole new insight into the relationship between economic growth and water consumption. It means that little more than a third of Britain's economic consumption has a direct impact on domestic water resources. Put another way, the quality of life in Britain would require water consumption to more than double if it depended solely upon domestic water resources. The impact on the environment would be colossal and the plight of water resources in SE England would require major long-distance water transfers. The UK's total, import-based 'external water footprint' is around 46 billion m^3 (46 km^3) a year (see the section *Trading virtual water*, later in this chapter).

International water trading

Trade in real water has grown nearly sevenfold since 1960. Figures 8.2–8.4 show the geographical pattern of this trade. The 12-nation European Union (EU12) is by far the largest importer and exporter of water, but it exports more than twice as much as it imports. This trade is to a considerable extent driven by the well-established taste for bottled water and the resultant large bottled water industry with a number of world players like Perrier and Evian (Figure 8.5). Despite its abundant water resources, Switzerland is a net importer. Other cases reflect the balance between domestic resources and demand more closely. Canada exports almost eight times more than it imports, with the USA being the main trading partner. The USA is very much a net importer, importing ten times as much as it exports. Australia is a net importer with a very similar import/export ratio to the USA, only overall volumes are ten times smaller.

Russia is another net exporter, but China imports and exports very similar amounts. The levels of trade in the Middle East are very much smaller, but most countries are net importers, with the curious exception of Saudi Arabia. Elsewhere in the world, trading is on a much smaller scale.

Several factors are now appearing to conspire to check and maybe even reduce trade in real water for the first time since the 1960s. One is the success of campaigns to reduce bottled water sales and the better provision of safe public water supplies. The second is the credit crunch and the reduction in disposable incomes, with pay cuts and unemployment. The collapse in world trade in 2009, which certainly affected virtual water trade, may also have some impact as international shipping suffers cutbacks and price increases. The longer-term impacts remain to be seen.

Bottling it

'It borders on morally being unacceptable to spend hundreds of millions of pounds on bottled water when we have pure drinking water, when at the same time one of the crises that is facing the world is the supply of water.'

Phil Woolas, UK Environment Minister
on BBC TV *Panorama*, February, 2008

'It's a completely unnecessary product and will fall away without the hard marketing behind it.'

Giles Coren, *The Times* restaurant critic,
London, 2009

For many reasons, the bottled water industry has expanded rapidly worldwide since the 1980s. Between 1997 and 2004 alone global consumption nearly doubled and export trade grew by a third (Figure 8.6). For equally numerous reasons, there are now signs of it falling out of favour, especially in North America and Western Europe. Both the expansion and the contraction are in part due to arguments that are unfounded. Fashion swings both ways. But there are also some very sound arguments both for and against bottled water in different circumstances.

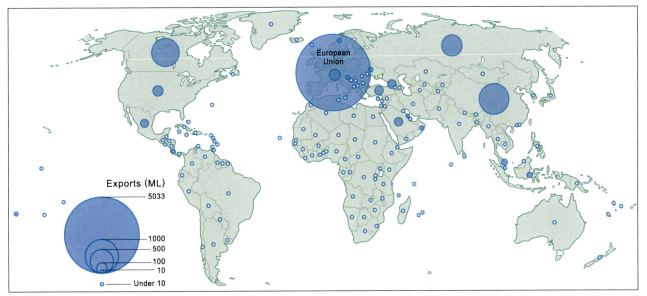

Figure 8.2 Worldwide exports of water, 2002. See Figure 8.4 for EU details

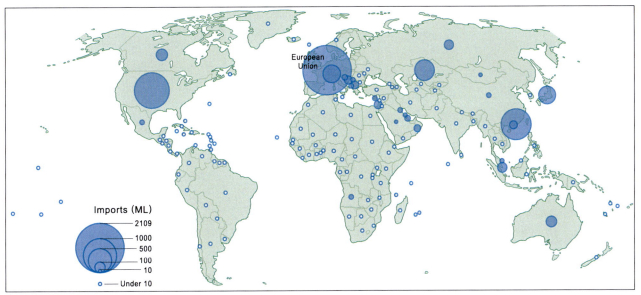

Figure 8.3 Worldwide imports of water, 2002. See Figure 8.4 for breakdown within the EU

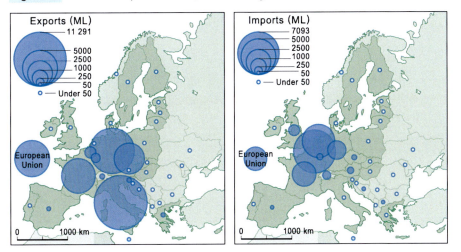

Figure 8.4 Export and import of mineral water in the EU, 2002

159

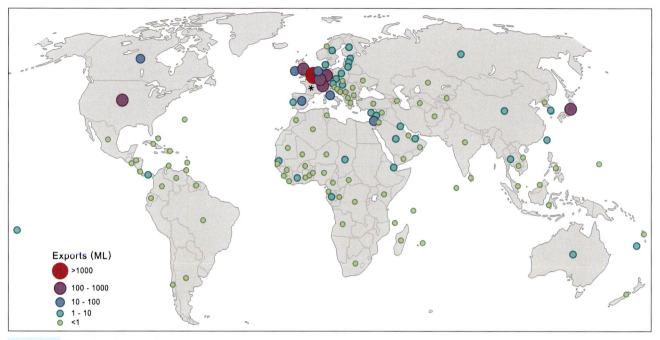

Figure 8.5 French water exports

The original impetus that drove bottled water was safety when most public supplies were potential sources of infection. This was a major factor in Europe at least until the final quarter of the twentieth century, and the habit has persisted. The average Frenchman or Italian still consumes five times as much bottled water a year as an Englishman. The notion that the water is not only safe but health-giving further bolstered trade. Bottled spa waters, even slightly radioactive, or water with a high mineral content are still considered health-giving in some parts of the world, yet there is little evidence to support this. Perhaps a higher salt content can be helpful in a warm climate where the body loses salts by sweating, but analysts found Vichy Saint Yorre water seven times over the recommended limit, which could be dangerous for someone on a low-salt diet. More recent advertising has focused on supposed advantages for women's health, for dental hygiene, to counter obesity (by reducing intake of sweetened drinks), and on the sensitivity of the brain to dehydration – therefore a boon for students in exams.

The recent decline in sales is largely the result of arguments from environmental pressure groups that have been taken up by mains water companies and governments. UK sales peaked at 1.3 billion litres in 2007, having risen from just 30 million litres in 1980. The UK Food Standards Agency and the Consumer Council for Water are among official bodies that have led the campaign against it. The campaign has been supported by restaurants that have decided to bow to customer demands and revert to tap water, waiving the large profit that drove the hard sell in the 1990s. It also owes something to the 2008–9 recession. UK sales fell by over 15 per cent between 2007 and 2009. In the first half of 2009, sales of Nestlé brands – Perrier, Vittel and Buxton – fell by nearly four per cent, also due partly to the strength of the Swiss franc. As a result of the downturn, tens of millions fewer bottles were sold.

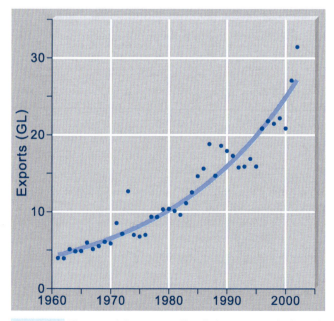

Figure 8.6 The growth in exports of bottled water over the last 50 years

Not all parts of the world are following the same pattern, however. The rise of the middle classes in Asia is set to reinvigorate world consumption, which is predicted to reach 280 billion litres by 2012. Between 2010 and 2015, India's consumption is expected to double. China's middle class is expected to approach 350 million by 2011. The rapid increase in affluence could cause consumption in China to rise faster than it did in Europe and the USA during the 1990s. The poor quality of public supplies is adding to this. A UN survey in China found that over half of tap water samples had unacceptable levels of bacteria. The 2009 Global Bottled Water Report by Zenith International predicts that global sales will rise 18 per cent between 2009 and 2012, others estimate 25 per cent. Small wonder that the big multinationals Nestlé, Danone, Coca-Cola and PepsiCo are all establishing bottling businesses in developing countries. Along with this goes the production: Thailand's Indorama Polymer could soon become the second largest producer of plastic bottles in the world.

One of the biggest objections to bottled water is the plastic bottles. Of the 13 billion plastic bottles sold in the UK in 2008, only a third were recycled. Millions end up in landfill. Millions more end up in rivers and in the sea, like the gigantic garbage gyres in the Pacific Ocean. The plastic slowly breaks down into small gravel-sized pellets that birds ingest, blocking their digestive tracts. The Belu Natural Mineral Water company has donated funds towards a 'rubbish-muncher' to extract bottles from the River Thames. In China, fishing for bottles is a money-making business.

Bottled water is frequently criticized on grounds of cost, perhaps 800 to over 5000 times more expensive than tap water, or the fact that the quality of water is only checked officially at infrequent intervals, that it does not contain any long-acting disinfectant and that over-extraction of groundwater can lower water tables over wide areas. On cost, Dr Barry Groves calls it 'one of the most profligate schemes ever devised to deprive the gullible of their money' (www.second-opinions. co.uk). On quality, there is no legal limit for bacterial levels in Britain, except that none must be introduced during treatment or bottling, and Chester Public Health Lab reported high levels in many bottles. The legal limit in tap water is 100 bacteria per litre and it is mostly less than two. Similarly, no tests are required for pesticides or organic chemicals in bottled water. However, the carbon dioxide added to sparkling water is a bactericide.

Further objections are being raised in the guise of carbon footprints. In Britain, transporting the water alone is estimated to have cost over 33,000 tonnes in CO_2 emissions in 2009, to which must be added the energy used in processing and in producing plastic bottles. Bottling water produces nearly 400 times as much CO_2 as tap water. However, it fares much better when compared with other bottled drinks: one litre of still bottled water creates around 165 g of greenhouse gas, whereas fizzy drinks produce three times more and juices eight times.

To counter the anti-bottle campaigns, the UK bottling industry formed the Natural Hydration Council in 2008. The Council points out that British water companies lose more water in a day through leakage than the bottled industry produces in a year. And it claims that over 1100 tonnes of chlorine are added as a disinfectant to tap water annually in the UK. Chlorination is the most widely used form of disinfection. It is very effective against bacteria, though not against protozoa cysts like *Giardia lamblia* and *Cryptosporidium* (see *No-plumbing disease* in Chapter 5). However, it can become toxic to humans in large quantities and especially when the water being treated contains high levels of organic substances. Trihalomethanes and haloacetic acid can then form, which are carcinogenic. The risk is particularly acute where reservoirs are fed from upland peaty soils and it was a concern when chlorination levels were increased in Britain during the 1980s to meet EU drinking water standards. The taste of chlorine is a major factor turning drinkers away from tap water. Better filtration, improved flocculants or alternative methods of disinfection are the answer. Restaurants offering tap water in developed countries now commonly use their own purification systems.

A common objection to tap water is that it offers a route for mass medication. This generally means fluoridation at present, but there are greater concerns. For years, arguments have been ranged both for and against fluoridation of public water supplies as a prophylactic against dental caries. Most professional medical and dental authorities support it, but many individual practitioners do not and have questioned the scientific evidence supporting it. The head of the British Chartered Institution for Water and Environmental Management (CIWEM) recorded his objections to fluoridation in 2009 on the grounds that it is a form of mass medication. It is undemocratic in that individuals may not be aware they are taking a medicine and they have no choice other than not drinking the water. It is also a case of 'one size fits all' and there are fears of toxic overload. The fluoride in the water can add to fluoride in toothpaste, in the air and in food, potentially causing a variety of symptoms from fluorosis, darkening teeth, contributing to brittle bones, and increasing the uptake of lead and behavioural disorders in children (see box 'The health effects of iodine and fluorine in drinking water in China' in Chapter 5). The 'fluoride' is actually fluorosilicic acid, which is classified as a toxic waste by the US EPA: it is not organic fluoride, which would be medically more effective. In addition to disputing the benefits, critics point out that fluoride is a waste product from the fertilizer industry that would cost the industry a considerable amount to dispose of safely otherwise.

Perhaps a greater fear is that this could be a first step on the road towards more mass medication. Some doctors in America have suggested that statins be added to public water supplies to reduce cholesterol and heart disease. So far, this has been rejected.

And then there is the fear of accidents, sabotage or terrorism. People are still suffering from aluminium poisoning after a technician mistakenly poured the flocculant aluminium sulphate into the treated water instead of into the water to be treated in Camelford, Cornwall, UK in 1988 (see *Terrorism* in Chapter 9). Alternative polymer flocculants have been widely introduced since the incident. Failure by both operators and equipment was blamed for 295 cases of fluoride poisoning in Hooper Bay, Alaska in the 1990s. The potential for deliberate poisoning of the water supply is discussed in Chapter 9 Water, war and terrorism. Bottling companies are not necessarily immune from sabotage. There were fears that Perrier might have succumbed in 1990 when a North Carolina lab found excessive levels of benzene in the water, but Perrier decided it was due to operator error during filtration. Perhaps of more concern is that pollutants derived from the environment may not be tested for in bottled water. High levels of uranium have been found in a number of European bottled waters due to the groundwater filtering through rocks containing uranium (Krachler and Shotyk, 2008) (see box 'Uranium pollution in South Africa' in Chapter 5).

Bottling tap

There has been strong criticism of bottling companies that market repackaged tap water. Public revulsion forced Coca-Cola to abandon plans to introduce the practice into the UK at its Sidcup plant in 2004 and withdraw its Dasani brand. The debacle was largely a question of disastrous marketing, although there was a breach of the law. In its American homeland, Dasani is tap water that is purified by 'Nasa-style' processes, including three filtrations, ozone disinfection and added minerals. Extra filtering, dechlorination, UV disinfection, and the addition or removal of minerals for taste or nutrition is common in purification processes. Much of the bottled water in America is purified tap. PepsiCo's Aquafina is the most popular bottled water, followed by Dasani. Most US supermarket home brands are the same: not so in Britain. Forty per cent of all bottled water in the world is purified tap. It has its uses, especially where the public supply is unsafe. But this is not the normal European tradition, and oddly Coca Cola bows to this on the continent: in France and Belgium Dasani is spring water. But there are a few purified waters. Purefect 95 is produced in Manchester and claims to be 'from upland water sources utilising the latest treatment techniques', i.e. purified mains water. Crystal Spring is London tap water with the chlorine taken out, but analysis has found more copper, zinc and 10,000 times more bacteria.

The trouble with Dasani (or 'Sidcup tap' as it came to be known) was mainly in the presentation. Trading standards officials investigated Coca-Cola for misrepresentation by using the word 'pure'. The water supplier, Thames Water, objected to the implication that their water was not pure. Customers objected to being sold tap water when they were expecting spring water and to paying 3000 times more for it than out of the tap. Timing was also against it – memories of an edition of the popular television sitcom *Only Fools and Horses* in which the trickster lead tried to sell bottled tap water were still vivid. The fact that Dasani was actually *purified* got missed. But the really critical issue was that tests revealed a level of bromide, a carcinogen, at more than double the permitted limit (22mg/L compared with a 10 mg/L limit), which resulted in the recall of half a million bottles. The bromide was introduced during the processing when a batch of calcium chloride containing it was added to raise the calcium level to regulation concentration. That problem was rectified, but the damage to brand image caused by less than explicit labelling was long term and expensive – at least £10 million.

Only water that is labelled natural 'mineral' or 'spring' water is strictly chemically untreated. In the UK, mineral waters are tested for 13 chemicals but there is no requirement for daily or weekly testing, compared with

90 chemicals for public supplies constantly monitored. There are no such regulations for spring water. However, Nestlé is trying other routes to combine satisfying taste and safety in one: its Pure Life comes from a natural spring in Derbyshire but is then purified, while its Aquarel comes from seven springs that are mixed to taste – this also offers security of supplies.

Ethical bottling

In recent years the bottled water industry has sought to improve its image. The rise of ethical bottled water companies reflects this trend in a very practical and well-meaning way. It fits in with former UN Secretary-General Kofi Annan's call for a Global Contract for Business at the same time as being a boost to marketing, like fair-trade and recycling. Large bottled water companies are now adopting ethical principles, like biodegradable bottles and gifting to charitable water projects, that were pioneered by a small number of ethically based private companies. AquAid, a fast growing British water cooler company established in 1988, was one of the larger pioneers. It has operated in partnership with Christian Aid since 1997 and donates 40p to Christian Aid water projects for every 19 litres of bottled water sold. In southern California, the founder of the Ethos Water company, Peter Thum, was so disturbed by the poverty in South Africa that he decided to contribute to non-profit water aid organizations and to use the water bottles to remind people of the global water crisis.

The Belu Natural Mineral Water is the first to use biodegradable water bottles made from corn as well as recyclable glass. It was launched in 2004 and was the first non-profit UK charitable company. All its net profits go to clean water projects, like building pumps in Tamil Nadu or Mali. It focuses marketing on expensive restaurants in the UK.

The 'One' company was founded during Live 8 in 2005 and sells through most major British supermarkets. It does not have charitable status, but 100 per cent of its profits on an annual turnover of around £1.5 million go to water aid projects in Zambia, Mozambique, Malawi and South Africa. It is introducing biodegradable bottles and has the idea of fitting a 'qode' to bottles that will link mobile phones to a website where donations can be given for PlayPumps in Africa (see *Elephant pumps, play pumps, treadle pumps and rope & washer pumps* in Chapter 14). It has been building a new play pump every fortnight, helping tens of thousands of people.

The Frank company was also founded in 2005 and has sought charity status. It uses bottles made of recyclable PET and glass. It had a £100,000 turnover in 2007 and ploughs all profits into sustainable clean water projects in places like Andhra Pradesh. The smoothies company Innocent also uses only recyclable or biodegradable bottles.

The success of these small private ventures has led to interest from the large multinationals. Nestlé claims to have reduced the amount of plastic in its bottles by a third in ten years and reduced the amount of water used in its processing plants by six per cent. Multinationals are also seeking to buy up the small companies. Starbucks bought the Ethos water company in 2005. Proctor & Gamble tried unsuccessfully to buy Belu. Cynics have dubbed it 'greenwash', the attempt to buy respectability by purchasing ethical water companies. Nevertheless, ethical investment is a rapidly growing business, with annual growth rates of around 15 per cent p.a. In the UK it broke through the £10 billion barrier in 2004.

Emergency supplies

Whatever the continuing arguments about quality, cost, carbon footprint and bottle disposal, there are two situations in which bottled water is valuable, even essential. It is a valuable product for those on the move, whether in sport or travel. Portable hydration can even be live-saving. But it is absolutely essential in emergency relief, in natural disasters or war.

Public water supplies are among the first services to be disrupted in earthquakes. Bottled water is one of the prime essentials in emergency aid. After the Indian Ocean tsunami in 2004, bottled water and portable water treatment plants were among the first international aid to arrive. The main problem then was the influx of seawater into wells and surface waters. In the relief of Port au Prince after the Haiti earthquake in January 2010, the problem was destruction of homes and pipelines. Bottled water supplies are often a first step in restoring public order and confidence after a war, as seen in Iraq and Afghanistan in recent years. Bottled water would have alleviated some of the woes of the Palestinians of Gaza after the Israeli offensive in 2009 if more aid had been allowed through. Sadly, bottled water can also help corrupt, despotic governments to flourish while their people suffer from lack of safe water. This was the case with President Mugabe's Zimbabwean government when death and disease spread after the National Water Authority turned off the public supplies through lack of funds to buy chemicals in November 2008.

Bottled water is also a convenient means of supply during droughts, water rationing or breakdowns in public supplies. Part of the projected increase in demand in Asia is ascribed to households stocking up for droughts, which climate change will aggravate.

Lastly, bottled water is a fundamental of modern warfare. It is to modern armies in battle what tinned food was for the Duke of Wellington's campaigns against Napoleon. There is no simple answer in the dispute between 'tap and cap'.

Trading water-rich drinks

Trade in juices, wine and liquor also expanded significantly over the last half century. These contain significant amounts of real water as well as consuming water in their manufacture and in the growth of their constituents, fruit and grain – their virtual water content. The average bottle of wine is 87.5 per cent water. Spirits (liquor) have around 57–62 per cent water. Since the late 1990s, a new category of beverage has appeared, 'enhanced waters', which lie between mineral water and soft drinks. Coca-Cola's Vitaminwater is one such, an over-diluted squash with reduced calories. These drinks range from flavoured water to 'power drinks' designed as energy-giving fashion items. The general aim is to capture new consumers by marketing a lifestyle and open up a new market for water. New drinks are continually appearing and provide a substitute for flagging mineral water sales in the West.

There are very marked imbalances in the world trade in beverages (Figures 8.7–8.9). Mexico is a top exporter of beverages despite its severe regional water shortages. China is also a major net exporter, despite its own water shortages. The USA is the biggest importer in the world by far, five times bigger than the next, its neighbour Canada. Yet Canada is a rare case with practically a trade balance – very different from the marked imbalance in its trade in mineral water. Japan and Russia are major net importers.

Beverage exports are clearly driven by economic gain rather than superior water resources. In many cases, exporters ignore local water shortages. Similarly, imports are driven by wealth rather than necessity. Increasingly, the trade is controlled by large multinational beverage companies, many of whom are also involved in the bottled mineral water trade. Coca-Cola has been buying up small Indian producers since the early 1990s. Its production plant in Rajastan, one of the driest regions of India, was blamed by local farmers for a drastic fall in the local water table. The company has since become more mindful of environmental impacts and the power of the green consumer. In 2006, more than half the water used by the company was used in production processes, not bottled. In 2007, Coca-Cola made a £10 million commitment to the World Wildlife Fund to reduce its water consumption with a target that it replaces every drop by 2008 and it expanded support for reforestation, rainfall harvesting and more water-efficient farming. Drinks manufacturers operating in some developing countries may find that it is in their own interests to invest in

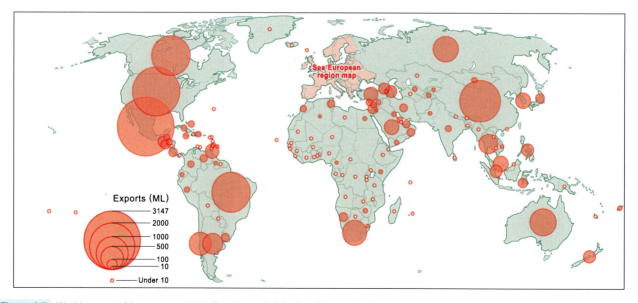

Figure 8.7 World export of beverages, 2002. See Figure 8.9 for breakdown within the EU

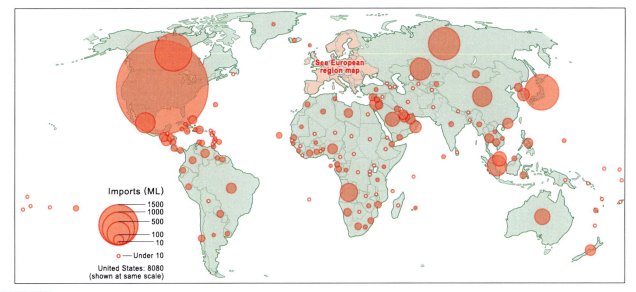

Figure 8.8 World import of beverages, 2002. See Figure 8.9 for breakdown with the EU

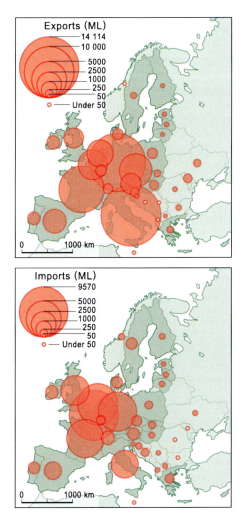

Figure 8.9 Exports and imports of beverages in Europe

local clean water infrastructure, because without clean water it is impossible to make their products, and without healthy local populations there are no customers. So there can be some advantage for local communities.

Trading virtual water

'Between 15 and 20 per cent of the water problems in the world can be traced back to production for export to consumers elsewhere in the world.'

Arjen Hoekstra, Director,
UNESCO-IHE Institute for Water Education,
The Netherlands

Virtual water is the water that is used to produce a product. The concept was originally proposed by Professor Tony Allan of London University's School of African and Oriental Studies (Allan, 1998). He suggested that the water shortage problems of the Middle East and North Africa (MENA) region might be solved by dropping aims for national self-sufficiency in food production and relying instead on imported food. The MENA was the first substantial region to face a water deficit and this has been hampering its economic development.

Professor Arjen Hoekstra and his team at the University of Twente and UNESCO-IHE Delft have quantified the international flows of virtual water and have extended the concept to cover all types of product. Table 8.1 lists

165

some average values. In general, water consumption is greater in drier climates. The latest crop calculations take into consideration the differences in the water requirements for similar products produced in different environments using the FAO's CROPWAT program to calculate crop water requirements and FAO statistics on yields and world trade (Hoekstra and Hung, 2005).

The team have also developed the idea of a 'water footprint', which is the total amount of water used in the production of goods and services (Chapagain and Hoekstra, 2004). This can be applied nationally or for individual producers. Warner (2003) argues that it could usefully be applied to individual regions or river basins as well. The national water footprint can be divided between internal or domestic and external or international trade footprints. The concept is analogous to the carbon footprint calculations being introduced as part of the fight against greenhouse gas emissions. The team provide short instructional courses in the method and have recently produced a Water Footprint Manual for use by individuals and companies. The article 'The globalization of water' by Arjen Hoekstra, later in this chapter, summarizes the concept and its implications.

The water footprint concept reveals the wide differences in consumption around the world and also the extent to which consumption in one nation has an impact on the water resources of another. While the average world citizen consumes around 1240 m³ a year through drinking, washing, food and other goods, Americans consume twice as much and Chinese presently consume a little over half as much, although this is rising rapidly (Figure 8.10). In more everyday terms, people in America consume over 6000 litres of virtual water daily, compared with less than 2000 in China.

World trade means that all countries are to some degree dependent on others. Consumers of imported goods in Europe, Japan and the USA have impacts, sometimes serious ones, on water resources in the developing world and emerging economies, especially in Latin America, where Brazil and Argentina are major exporters. Africa is a net exporter, particularly to Europe, yet much of the continent suffers water shortages and more people are starving in Africa than on any other continent (see Róisín Murray-William's article 'Valentine flowers from Kenya – a perverse trade?', later in this chapter). The USA, Canada and Australia are also major exporters, but their domestic resources can generally sustain the trade with backing from reciprocal imports.

Table 8.1 Virtual water needed to produce some common products

Product	Litres of water
Manufactured goods	
Car (steel only in average car)	32,000
Microchips 1 kg (for cleaning during production)	16,000
Bio-diesel 1 litre (from soybean)	14,000
Leather shoes – pair	8,000
Cotton T-shirt	4,000
Bio-ethanol (sugar beet [lowest] to sugar cane and maize [greatest])	1,400–2,600
Synthetic rubber 1 kg	460
Copper 1 kg	440
Steel 1 kg	260
Plastic 1 kg	41
Cereals, fruit and vegetables	
Bananas 1 kg	850
Rice 1 kg	3,400
Sugar (cane) 1 kg	1,500
Wheat 1 kg	1,350
Maize 1 kg	900
Apples/pears 1 kg (single 100g = 70 litres)	700
Cabbage 1 kg	200
Potatoes 1kg	160–250
Tomatoes 1 kg (single 70 g = 13 litres)	180
Lettuce 1 kg	130
Animal products	
Beef 1 kg	15,500
Pork 1 kg	4,800
Eggs 1 kg (single 60 g = 200 litres)	3,330–4,700
Chicken 1 kg	3,900
Processed food	
Chocolate 1 kg	24,000
Hamburgers 1 kg (single 150 g = 2400 litres)	16,000
Cheese 1 kg	5,000
Bread (from wheat) 1 kg	1,300
Drinks	
Coffee, large 250 mL cup (to grow, package and ship the coffee)	280
Milk 250 mL glass	250
Orange juice 250 mL glass	210–250
Wine 250 mL glass	240
Apple juice 250 mL glass	240
Beer (from barley) 250 mL glass	75
Tea 250 mL cup	30
Bottled water 250 mL (processing and plastic)	0.75–1

Sources: Hoekstra and Chapagain (2008); Rijssenbeek (2007); Gleick (1993); www.h20conserve.org

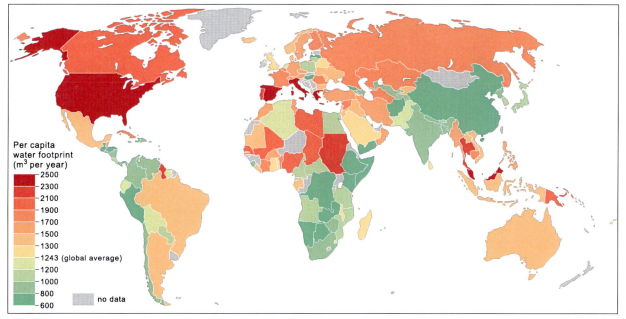

Figure 8.10 Average national water footprint per person, coloured according to above (redder) or below (greener) the global average of 1243 m³/capita/year for washing, drinking and the production of food and other consumer goods

Looking at the total water footprint of nations, not accounting for population size, India and China loom largest, followed by the USA (Figure 8.11). But whereas American consumption relies heavily on imports to sustain it, consumption in India remains, for the time being, largely reliant on domestic sources (Figure 8.12). China, however, has begun to import substantial amounts of goods from abroad and its external water footprint is now comparable to that of Japan, Germany, Italy and the UK.

The wider implications of virtual water trading

The implications of these concepts are extremely important, not simply for water resources, but also potentially for national security (Hoekstra, 2009). International trade has the potential to alleviate national water scarcity, but it can also increase national water dependency – that is, dependency on other nations that may or may

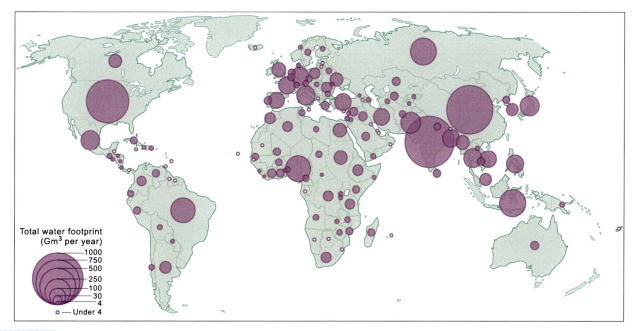

Figure 8.11 The total water footprint of nations

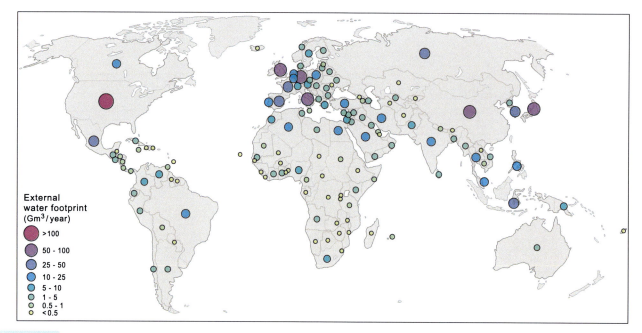

Figure 8.12 The external water footprint of nations: the proportion of national water footprints due to imports

not remain friendly or even capable of continuing to supply the products. Figure 8.13 shows the extent to which Western Europe, Japan, Korea and the Middle East are dependent on virtual water imports.

What has been called the 'virtual water hypothesis' maintains that trade, especially in foodstuffs, can allow a country to overcome the restrictions imposed by the

water resources available domestically. This may also reduce international conflict over shared resources. Green Cross International's 'Water for Peace' initiative follows this philosophy (www.gci.ch/). The Middle East effectively ran out of water in the 1970s, but according to Tony Allan (2001) more (virtual) water now 'flows' into the region than actual water flows into Egypt down the Nile. Despite a continuing policy to strive for self-

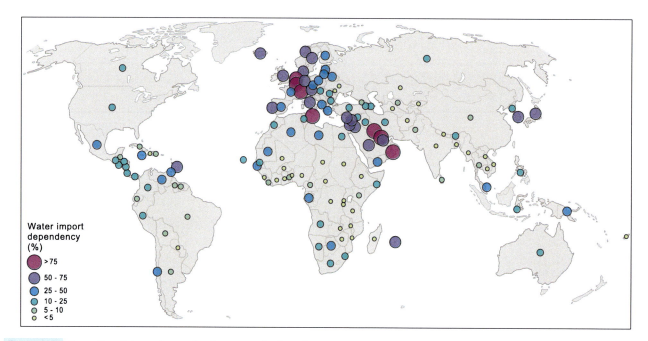

Figure 8.13 Proportional dependency of nations on water imports

sufficiency as far as possible, Egypt still imports the equivalent of 14 billion m³ a year on top of the 65 billion it withdraws domestically. Jordan, on the other hand, is totally dependent on imports, which, at the equivalent of five billion m³ a year, are five times more than the water it withdraws within its own territory.

It can, however, work the other way. Over-exploitation of shared resources to produce export goods could conceivably trigger international disputes: perhaps real water wars caused by virtual water.

Some writers have called the virtual water trade 'perverse' in a number of different ways. Tony Allan suggests it is perverse because the availability of virtual water imports has slowed the pace of water policy reforms intended to improve water efficiency (Allan, 2003). He maintains it has also distorted international relations in places where there are conflicts over shared water resources. The Jordan basin is an example, where competing countries have been able to defer or even deny the onset of severe water shortages by benefiting from low world grain prices, which are the result of the perversion of normal economic forces by subsidized agriculture in Europe and North America (see *Israel and Palestine* in Chapter 9). Warner (2003) suggests that it may also be perverse because it redistributes not only water use but also water stress and insecurity in ways that might heighten social conflict. Many African countries could fall into this category.

'A flower is 90 per cent water. We are one of the driest countries in the world and we are exporting water to one of the wettest.'

Severino Maitima,
Ewaso Ngiro Water Authority, Kenya,
talking to John Vidal, The Guardian, 2006

International trade can aggravate problems by encouraging countries to produce products for monetary gain at the expense of their own water resources and those of other water users who share the resource. Flowers from Kenya could be a case in point (again, see the article 'Valentine flowers from Kenya – a perverse trade?'). Here is a multimillion pound industry using vast amounts of water purely for export of a non-edible product in a country officially suffering long-term water scarcity (Figure 8.14), plagued by multiyear drought, where millions are starving. Until the rains finally returned in the autumn of 2009, causing floods, landslides, cholera, malaria and hypothermia, Kenya had suffered four years of devastating drought. Four million Kenyans were on the brink of starvation, together with nearly 400,000 starving Somalis who had crossed the border for safety from drought and civil war. Nearly 20 million more were starving in East Africa as a whole, many migrating to city slums, added to which food aid to Kenya was corruptly appearing in Sudan. East Africa and the Sahel have long been prone to drought. It used to recur around once every decade, but it seems to be becoming more persistent (see *Drought in the Sahel* in Chapter 3).

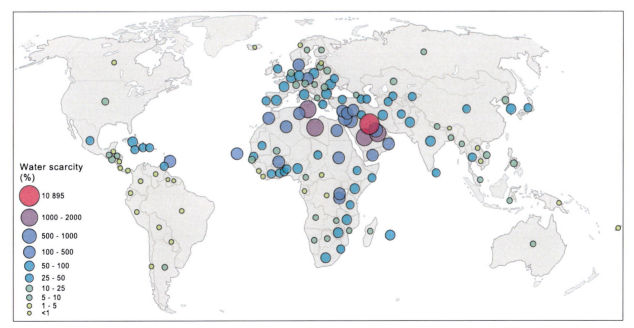

Figure 8.14 Water scarcity

The globalization of water

Arjen Y. Hoekstra

Although the river basin is generally seen as the appropriate unit for analysing freshwater availability and use, it becomes increasingly important to put freshwater issues in a global context. International trade of commodities implies flows of 'virtual water' over large distances, where virtual water should be understood as the total volume of water that is required to produce a commodity. 'Total volume' refers to the sum of rainwater (relevant in the case of crops) and ground and surface water. Virtual water flows between nations can be estimated from statistics on international product trade and the virtual water content per product in the exporting country. The latter is the sum of the crop and/or process water requirements per unit of product. As an example, the virtual water content (m³/ton) of a livestock product is calculated based on the virtual water content of the feed and the volumes of drinking and service water consumed during the lifetime of the animal and the process water requirements to process the animal products.

International virtual water flows

In the period 1997–2001 the global volume of virtual water flows related to the international trade in commodities was 1625 billion m³/yr. Figure 8.15 shows the average virtual water balances at the level of the 13 world regions and the biggest net virtual water flows between those regions. About 80 per cent of these virtual water flows relate to the trade in agricultural products, while the remainder is related to industrial product trade. An estimated 16 per cent of the global water use is not for producing domestically consumed products but products for export. With increasing globalization of trade, global water interdependencies and overseas externalities are likely to increase. At the same time liberalization of trade creates opportunities to increase global water use efficiency and physical water savings.

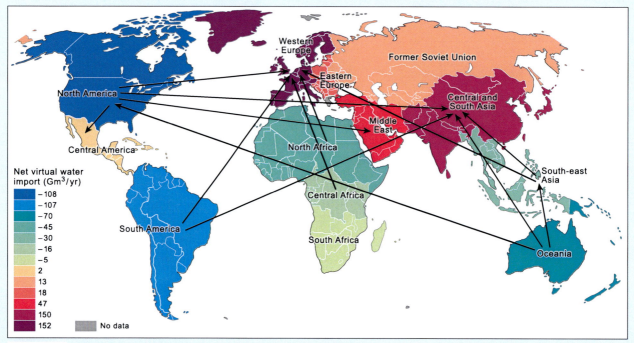

Figure 8.15 Regional virtual water balances and net interregional virtual water flows related to the trade in agricultural products (1997–2001). Only the largest net flows, greater than ten Gm³/yr, are shown

Global water saving

Many nations save domestic water resources by importing water-intensive products and exporting commodities that are less water intensive. National water saving through the import of a product can imply saving water at a global level if the flow is from sites with high to sites with low water productivity. It has been estimated that the countries in the world together use a total volume of 1253 billion m³ of water per year to produce agricultural products for export. If the importing countries would not have been able to import those products, they would have required a total volume of 1605 billion m³ per year to produce those products domestically. As a result of international trade in crop and livestock products there is thus a global water saving of 352 billion m³ per year (Figure 8.16). This saving is equivalent to 28 per cent of the international virtual water flows related to the trade of agricultural products and equal to six per cent of the global water use in agriculture. National policy makers are however not interested in global water savings but in the status of national water resources. In the period 1997–2001, Mexico imported nearly 500,000 tonnes of husked rice per year from the USA and in doing so saved 1.06 billion m³ of water per year. However, due to the higher water productivity in rice production in the USA, if compared to Mexico, there was also a global water saving.

The rice exported to Mexico was produced in the USA using 0.62 billion m³ of water per year, which means that the global water saving amounted to 0.44 billion m³ per year. International trade can also be less efficient from a global point of view. For instance, in the period 1997–2001 Indonesia imported more than 400,000 tonnes broken rice per year from Thailand and thus saved 1.3 billion m³/yr of its national water resources. However, Thailand used 2.3 billion m³/yr of its water resources to produce the rice exported to Indonesia, which implies a global water loss of 1 billion m³/yr. Water use for producing export commodities can be beneficial, as for instance in Côte d'Ivoire, Ghana and Brazil, where the use of green water resources (mainly through rain-fed agriculture) for the production of stimulant crops for export has a positive economic impact on the national economy. The opportunity cost of the rainwater applied does not outweigh the benefits gained. More doubtful is the use of 28 billion m³/yr in Thailand for the production of export rice, because a significant fraction of this water is water abstracted from ground and surface water, which is scarce and has high opportunity costs.

Water footprints

The impact of consumption by people on the global water resources can be mapped with the concept of

Global water saving = 352 × 10⁹ m³/yr

Figure 8.16 Global water savings associated with international trade of agricultural products (1997–2001). Only individual savings of over five billion m³ a year are included

the 'water footprint'. The water footprint of a nation is defined as the total volume of freshwater that is used to produce the goods and services consumed by the inhabitants of the nation. The internal water footprint is the volume of water used from domestic water resources; the external water footprint is the volume of water used in other countries to produce goods and services imported and consumed by the inhabitants of the country. The water footprint deviates from traditional indicators of water use in the fact that the water footprint shows water demand related to *consumption* within a nation, while the traditional indicators (e.g. total water withdrawal for the various sectors of economy) show water demand in relation to *production* within a nation. Furthermore, the water footprint visualizes how consumers in one particular part of the world indirectly employ the water resources in various other parts of the world. The citizens of the USA have a very large water foot-

print, 2480 m³/cap/yr in average. The Chinese people have an average footprint of 700 m³/cap/yr. The global average water footprint is 1243 m³/cap/yr. The average per capita water footprints of all nations are shown in Figure 8.10. The four major factors determining the water footprint of a country are: volume of consumption (related to the gross national income); consumption pattern (e.g. high versus low meat consumption); climate (growth conditions); and agricultural practice (water use efficiency). The total water footprint of a nation or region indicates the total water use for all consumption categories, including crop products, livestock products industrial products and domestic water consumption. The above figures on water footprints exclude water requirements related to water pollution. With an example for cotton-related water requirements, however, Figure 8.17 shows how the water footprint concept can be extended through quantifying the impacts of pollution as well.

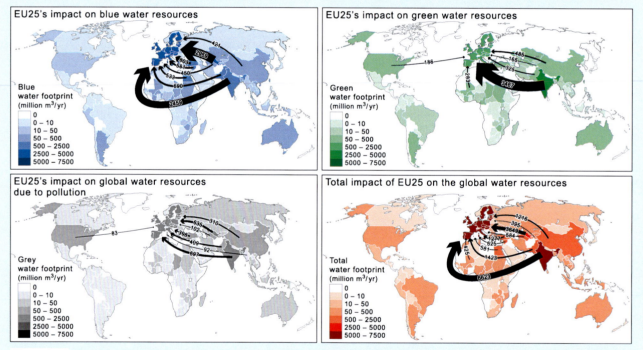

Figure 8.17 The water footprint of people in the enlarged European Union (EU25) related to their consumption of cotton products (1997–2001)

Unsustainable production for export can also harm the environment, as seen in the Aral Sea (see *The Aral Sea: lost and gained?* in Chapter 15). Chapagain *et al* (2006) take this argument further by suggesting that European consumers of cotton goods have been indirectly responsible for 20 per cent of the drying up of the Aral Sea. Hoekstra points out that the Japanese consume large quantities of American cereals and soybean, and so

contribute to the mining of aquifers, like the Ogallala, emptying rivers, like the Colorado, and increasing evaporation losses from irrigation (Hoekstra and Chapagain, 2008). Similarly, the 6.5 million tonnes of goods China exports to the UK alone each year are contributing to the falling water tables in the North China Plain and the drying up of the Hwangho.

Valentine flowers from Kenya – a perverse trade?

Róisín Murray-Williams

Flower exports are, for Kenya, the largest earner of foreign currency after tea, amounting to around $250 million a year. Around 70 per cent of Kenya's exports go to The Netherlands, where most of the flowers pass through the Dutch flower auctions and may be re-exported around Europe, and nearly 20 per cent goes direct to the UK. In the first six years of this century, exports increased 250 per cent. A fifth of these annual earnings come from St Valentine's Day flowers: little wonder that armed guards were deployed around the flower farms during the tribal clashes in February 2007. Kenya's flower industry supports over 1.5 million people directly and indirectly and produces 90 per cent of all flower exports from Africa.

Yet it is a voracious water user in a country where water is scarce: in 2006 the floriculture industry consumed 7.5 million m^3 of blue and green water. It takes 32.5 litres to grow just one bunch of 12 roses. It requires a further 16.5 litres of grey water to dilute fertilizers and pesticides to an environmentally safe level. In terms of the annual transfer of virtual water, trade with The Netherlands is equivalent to over 5 million m^3 of water and a further 1.5 million m^3 goes direct to the UK. Although the industry consumes well under one per cent of Kenya's national water resources, it has a substantial impact on society and environment within and downstream of the flower-growing areas.

To support this production, fertilizer use increased by a fifth between the mid-1980s and mid-1990s, and high levels of pesticides have been blamed for the death of raptors in the River Ngiro. Chapagain (2006) calculates that 44 m^3 of clean water are needed for every tonne of flowers grown to dilute wastewaters to safe levels. Based on this estimate, the volume of water needed to dilute the agrochemicals in the wastewater from Kenya's flower industry to permissible levels rose from about 326,568 m^3 a year in 1980 to 3.8 million in 2006.

The water level in Lake Naivasha, which lies downstream from Kenya's largest floriculture region, began falling in the 1980s just as the floriculture industry was established. The lake is particularly sensitive to changes in river inflows because the underlying groundwater mainly drains away from the lake. The River Ngiro that runs through the Naivasha flower growing area now stops nearly 100 km short and the head of the Ngiro Water Authority claims the big flower firms are 'stealing' water in unsustainable quantities (Vidal, 2006). Conflicts over water have broken out between the large international horticultural companies that control most of the production and the small, independent pastoral farmers.

The economic imperative is strong. The economic return per cubic metre of water increased from about £2.75 in 1980 to £25.5 in 2002. In the early 1980s, Kenya was anxious for foreign exchange earnings and then in the 1990s the prices for its other major exports, tea and coffee, dropped. The area under floriculture around Naivasha and Thika increased tenfold over this period, and the water footprint increased proportionally.

Using the water footprint methodology outlined by Orr and Chapagain (2006), blue water use alone in the Kenyan flower industry increased nearly tenfold up to 2.5 million m^3 in 2006. Blue and green water combined increased from 643,834 m^3 in 1980 and 7.5 million in 2006. To this must be added nearly 4 million m^3 a year if agrochemicals are to be diluted to safe levels.

The Kenya Flower Council attempts to reduce the environmental impact by encouraging use of electronically-controlled drip irrigation and restricting the use of agrochemicals. Members of the Council account for almost 70 per cent of all exports. However, the increased efficiency in water use has been more or less balanced by the rapid growth in the industry.

There is concern that Kenya's loss of its Least Developed Country status in 2007, under which it could avoid import duties, might eventually lead to the international companies moving production out of the country, perhaps to Ethiopia, Uganda or Tanzania. International companies are inherently more footloose and will try to benefit from every economic incentive and lower regulatory controls whenever possible. Production continues to increase for the time being, although the highest year-on-year increase (26 per cent) was back in 2002.

Real water transfer in agricultural crops– a footnote

While virtual water trading is by far the largest element in international water trade, every agricultural product contains a certain amount of real water, less so for dried products. This water is eventually released into a new environment. The thought is interesting academically, even though the quantities involved are miniscule in global terms. However, for individuals, eating water-rich food is a good substitute for drinking water. Indeed, it may be medically preferable, as it contains more nutrients and so causes less dilution of the saline level in body fluids than pure water; it may even increase the level. It could be as life-saving as drinking water, perhaps more so.

A tomato is 95 per cent water. All uncooked soft fruit is high in water. Cooking alters the amount of this water that is ingested. However, from an import/export viewpoint, all organic products are transferring water: a potato is 80 per cent water, a live cow 74 per cent. This water content is, of course, part of the so-called 'virtual water' that is being traded, because it is part of the water that was used to grow the crop or rear the animal. So not all 'virtual water' is actually virtual.

Conclusions

Growing world trade is increasing the interdependency of nations in terms of water resources as well as economies. This has the potential to be a force for the good: to sustain economies that suffer from water scarcity and to reduce international tensions by providing an alternative means of acquiring goods or reducing water demand in shared river basins or transboundary aquifers.

In theory, trade in virtual water could create a more equitable use of world water resources by allowing water-intensive commodities to be produced where it is most efficient – where sufficient resources are available without damaging the environment or depriving other users, and where water use per tonne of product is at a minimum. World trade fully optimized in terms of water is, of course, a utopian world that will never be. For many reasons, such as national security or profit, countries will act in their own interests rather than the common good. But although economic and political considerations warp the ideal, the virtual water trade *is* reducing the hegemony of physical resources and it is helping to alleviate extreme water scarcity: the Middle East is a prime example.

Even so, it also has the power to pervert: to encourage overexploitation for monetary gain. It can transmit stress just as it can alleviate it. The same may be said of bottled water.

Discussion points

- Try calculating the virtual water trade between two nations of interest to you.
- Consider how you might calculate the amount of real, as opposed to virtual, water being transferred by international trade in agricultural products. Try a test case.
- Evaluate the Water Footprint Manual and try applying it to your own household or business.
- Calculate your own water footprint at www.waterfootprint.org.

Further reading

The concept of virtual water and its consequences for national security are available in numerous publications from UNESCO-IHE in The Netherlands. It is easy to sign up to the free newsletter at:

www.waterfootprint.org

An excellent introduction and overview is provided by:

Hoekstra, A.Y. and Chapagain, A.K. 2008. *Globalization of Water: sharing the planet's freshwater resources.* Blackwell, Oxford.

The earlier book is also clear and interesting:

Chapagain, A.K. 2006. *Globalisation of Water: opportunities and threats of virtual water trade.* Taylor and Francis, London.

A detailed introduction to the calculation of water footprints is available in:

Chapagain, A.K. and Hoekstra, A.Y. 2004. *Water Footprints of Nations.* Value of Water Research Report Series No. 16, UNESCO-IHE, Delft, the Netherlands.

The Water Footprint Manual is available from the Publications section of the Water Footprint Network at:

www.waterfootprint.org

A new book from the Pacific Institute on bottled water:

Gleick, P.H. 2010. *Bottled and Sold.* Shearwater.

9 Water, war and terrorism

'Water scarcity threatens economic and social gains
and is a potent fuel for wars and conflict.'

Ban Ki-moon, UN Secretary-General,
at First Asia-Pacific Water Summit
Beppu, Japan, December 2007

War

The universal need for water and the often critical time limits to its healthy storage have made freshwater supply a strategic resource in armed conflict throughout history. A secure water supply has always been one of the first prerequisites for protecting a city in wartime. One of the more notable achievements of the ancient world was the kilometre long tunnel engineered by Eupalinos of Megara, which diverted spring water to the city of Samos and sustained the Greek city through a critical Persian attack around 550 BC (Hart-Davis, 2004). Although some of the earliest urban settlements, like the Neolithic town of Çatalhöyük (7000–5500 BC) in modern Turkey, show little evidence of water engineering and were established away from major rivers, the 'hydraulic civilizations' of Mesopotamia, the Indus and Egypt seem to have developed greater sophistication. On the other hand, there is ample evidence of war and violence in these civilizations, whereas no such evidence has been found at the apparently peace-loving town of Çatalhöyük (Jones, 2004).

Sustainable water supplies are still open to threat from war and terrorism, either directly or indirectly. Both forms of violence have used disruption or poisoning of water supplies as a weapon, and both may cause collateral damage to water supplies. The distinction between war and terrorism can be arguable. Special Operations forces may use some tactics similar to those employed by terrorists, with similar repercussions for water systems, but the main practical distinction is between formal military actions and more informal, smaller scale guerrilla-style activities that may be directed more at civilian personnel and designed to engender fear and panic as much as any specific damage.

Some acts of 'water violence' have been executed by retreating forces, either in an act of peek or to protect their rear. In the First World War, German troops poisoned wells as they retreated from Windhoek after it was captured by South African troops in 1915. Retreating from Holland in 1945, the defeated German troops blew up the dyke protecting the newly reclaimed Wieringermeer polder, flooding the land and four villages for no military purpose: 'bloody-mindedness' said the Dutch. As Iraqi forces retreated from Kuwait in 1991, they opened the valves on scores of oil wells, releasing tonnes of oil into the Gulf. It is believed the goal was to close down Saudi Arabia's giant desalination plant at al-Jubail, the prime source of water for Riyadh.

'It has long been recognized that among public utilities, water supply facilities offer a particularly vulnerable point of attack to the foreign agent.'

J. Edgar Hoover,
Director of the FBI,
to the American Water Works Association, 1941

More commonly, water has been used directly as a weapon of war, especially withholding or destroying water supplies as a means of wearing down resistance. The Romans used dead carcasses to pollute enemy water supplies (Goldsworthy, 2002). During the 1991 Gulf War, a plan was proposed for Turkey to cut off the flow of the Euphrates to Iraq, but never implemented (see section headed *The Tigris-Euphrates basin*, later in this chapter). After the war, however, Saddam Hussein ordered the supreme modern example of hydraulic

175

warfare – the draining and poisoning of the Tigris-Euphrates marshes in a barely disguised and highly successful attack on the Marsh Arab population (see box 'War against the Marsh Arabs' later in this chapter). Thankfully, two decades on, following Hussein's deposition after the second Gulf War, land and people are being restored (see box 'Restoring Iraq's Mesopotamian marshlands' in Chapter 15).

In 2001, Palestinians destroyed the water pipeline of Yitzhar and Kisufim Kabbutz as retribution after Israelis destroyed a Palestinian water cistern, blocked tanker deliveries and attacked material to be used in a wastewater treatment project. In August 2006, as part of their rebellion against the Sri Lankan government, the Tamil Tigers shut off the water supply from Mavilaru reservoir to 65,000 mainly Sinhalese inhabitants of Trincomalee, which was seen as an act of 'ethnic cleansing'.

Large dams can be prime military targets as well as the focus for international influence. The Kajaki Dam in southern Afghanistan represents both and is probably the most fought-over dam in the world. It was built by the Americans in the early 1950s as a Cold War showcase, subsequently taken over by Soviet forces and never completed. It was intended to irrigate 1800 km^2, but salinization set in early. The dam was later taken over by the Taliban. The USAF bombed the dam's powerhouse in 2001 and repeatedly attacked the dam in 2007 to try to dislodge Taliban fighters. Having regained the area, a force of 4000 ISAF troops escorted the final turbine to the dam in September 2008. It remains vulnerable to Taliban attack.

Water bombs

Occasionally, excess of water has been used as a weapon. During the Second World War, the British 'dam buster' aircraft attacked the dams of the Ruhr and Eder valleys with bouncing bombs, bursting the Mohne and Eder Dams in 1943. Other dams were destroyed by the flood wave for 50 km downstream from the Mohne. The

War against the Marsh Arabs

Encouraged by public announcements from President George Bush, Sr., that they might expect US military support, Shi'ite Muslims in southern Iraq, including the Marsh Arabs, rose up against the remnants of Saddam Hussein's military in 1991. No help came and the regime proceeded to commit near-genocide under the guise of draining the marshes and reclaiming the land for agriculture. The marshes provided an impenetrable refuge for the military, but operations began with poisoning the water. Many Marsh Arabs fled to Iran. Subsequent draining and causeways enabled artillery to be moved in. During 1992 and 1993, 40 rivers that used to flow into the al-Amarah Marsh were diverted and the Mother of All Battles Project completed the canal dubbed the Fourth River. Barely 85,000 of an original population of five million or more are thought to have remained, and they were forcibly resettled. UNEP estimates that 90 per cent of the 20,000 km^2 marshlands was drained.

Ostensibly, the move was intended to complete an established plan to reclaim more land for agriculture, begun in 1984. Some 80 per cent of land in the Tigris-Euphrates basin has suffered salinization and 33 per cent has been totally abandoned. Ironically, the food shortage was aggravated by UN sanctions after the war. The scheme aimed to cleanse 1.5 million ha of salinized land, but there are serious doubts about its efficacy. Some international consultants believed the rehabilitation could be successful, but the World Conservation Union predicted that most of the drained area would become desert within 10 to 20 years.

The UN called it the worst ecological disaster of modern times. By 2003 most of the former marshland was barren. A unique ecology was destroyed, topsoil used for dykes and salinization soon began to affect the newly exposed land. The Indian porcupine, smooth-coated otter and grey wolf became extinct and many species that evolved in the local habitat, like the Basrah reed warbler, were threatened. The marshes used to be a winter haven for two-thirds of all the wildfowl in the Middle East as well as for many others from as far away as western Siberia. A 5000-year-old culture, described in the explorer Wilfred Thesiger's 1964 classic The Marsh Arabs, had been all but destroyed.

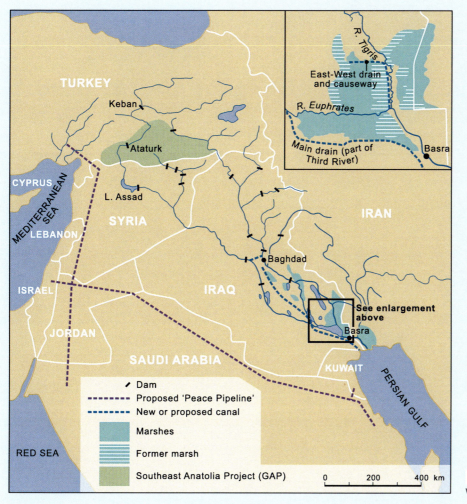

Figure 9.1 Mesopotamia and the war on the Marsh Arabs

13.4 MW hydropower plant in the Eder Dam, then the largest dam in Europe, was destroyed. The industrial city of Dortmund, together with Mulheim, coal mines, ironworks and factories were inundated and some 1500 people killed (see Notes).

South Korea has long feared North Korea might use its control of cross-border rivers to release unexpected floodwaters as 'water bombs' to attack the South. Panic first struck in 1986 when the then military government in South Korea warned, in a pamphlet entitled 'Water bomb over Seoul', that the North could use the new dam it was building on the Han River and perhaps other dams in this way, killing millions and possibly to disrupt the 1988 Olympics in Seoul. The government suggested that a flood release could wash away a number of dams and flood Seoul, and it raised money to build a protec-

tive dam on its side of the border to guard against such a catastrophe. In the event, nothing happened and it came to be dismissed as propaganda to increase hatred of the communist regime in the North.

However, the spectre surfaced again in September 2009 when an unexpected flood wave 2.3 m high struck on the Imjin River, killing a number of people, with no reported rainfall. Blame focused on the Hwanggang Dam or one of the smaller dams on the North's section of the river basin, releasing an estimated 40 million m³, one tenth of the Hwanggang reservoir's total capacity. Was it a deliberate 'hydraulic attack' or just human error? It could certainly have fitted in with a series of military provocations being made at a time when North Korea was under intense international pressure to abandon its nuclear weapons programme.

177

Water in collateral damage

Water resources were both a weapon and an accidental victim in the Balkan war during the 1990s, from land-mines used to blow up dams to the accidental pollution of soil and water caused by allied bombing of electricity facilities (see Mladen Picer's article 'Effects of war in the Balkans on water resources and water quality', below). Heavy bombing and shelling can disrupt water distribution and sanitation systems and are likely to leave a long-lasting 'war footprint', specially in the environment, for example in pollution from explosives or materials released from the factories or buildings under attack. The mining or bombing of dams, distribution pipelines and treatment plants, the disablement of management systems involving human beings or computer control systems, and general environmental pollution are the main risks. Examples might be depleted uranium from shells or PCBs from electricity stations, for example in Bosnia. Depleted uranium has been blamed for severe medical effects following the second Gulf War. In 2010, the Iraq government advised women around Falluja, the site of some of the heaviest resistance to US troops, not to have any more babies because of a spate of congenital abnormalities (including hydrocephalus, heart defects, neuro-spinal difficulties and tumours).

A rather different type of collateral damage arose from the Cold War between the USSR and the West. As a result of the intense competition during the space race of the 1960s to the 1980s, water resources have been severely polluted over a large area on the flight path between the Baikonour Cosmodrome and the Altai Mountains. Farm animals are dying and large numbers of children suffer from nervous disorders, jaundice and anaemia.

At least one modern war has been waged in order to secure water supplies rather than destroy them. Israel's Six-Day War in 1967 aimed to gain control of the headwaters of the River Jordan and aquifers of the West Bank, in response to the 1964 Syrian-Jordanian plan to dam and divert water from the Yarmouk River. Former Israeli Prime Minister Ariel Sharon stated in 2001 that

Effects of war in the Balkans on water resources and water quality

Mladen Picer

The Balkan countries have suffered for many years from neglect of the environment and the lack of maintenance and improvement in water supply and sewage systems; some of these systems have even been completely destroyed by the effects of war. These issues are of immediate and critical concern for the region, affecting economic development and stability. Research on the environmental impact of warfare has shown that many types of highly stable organic contaminants have entered the environment. The karstic region of former Yugoslavia warrants particular attention because of its exceptional ecological vulnerability. Landmines have severe environmental consequences, introducing poisonous substances into the environment as their casings degrade. This substantially decreases the productivity of agricultural land and increases an area's vulnerability to water and wind erosion, which in turn can add sediment to rivers, adversely affecting water habitats. The Balkan wars affected the now independent states of Croatia (mainly from 1991 to 1995) and Bosnia and Herzegovina (1992 to 1995), together with 78 days of NATO bombing in Serbia and Kosovo (1999).

Croatia

During the war between Serbia and Croatia, the bombing of industrial sites, refineries and other installations caused the release of polluting materials into streams, groundwater and rivers. Destroying hydroelectric power stations can cause water contamination and the effects of such contamination last over a long period of time. Over half of Croatia's national territory was directly affected by the war. The area affected was home to 36 per cent of the population and covered more than 30 per cent of the nation's economic activities. It is estimated that by 1994 total war damage amounted to US$27 billion, which amounts to US$5650 per capita.

The precious water areas directly affected by the war include the Plitvice Lakes and Krka National Parks, Kopacki Rit Nature Park and Lonjsko Polje. Waste materials in the war zones and damage due to the bombing of chemical industrial complexes resulted in pollution of groundwater, soil and rivers. Many freshwaters were very heavily contaminated with polluting substances, e.g. oil from the Sisak oil refinery observed in the Danube delta near the Black Sea in 1994. In the Lika region, in particular, the contamination of groundwaters will be extremely costly to rectify, especially bearing in mind the very considerable migration via fast flowing underground rivers, even as far as the coast. During the armed conflict no remedial measures were feasible and the remaining water contamination, e.g. contamination of wells with oil, pesticides, paint and animal remains, will continue to affect drinking water resources for many years to come.

There are also a number of modern examples of the use of water as a weapon. The Peruca Dam was damaged in the Croatian War in 1993, endangering the townspeople of Omis. A positive side effect however was that the destruction of industry meant little or no polluting discharges from these industries entered the rivers. As a result, at Pakrac, the Pakra river, which had been heavily polluted for decades, was again supporting fish in 1993.

Bosnia and Herzegovina

The events of the 1992–1995 war in Bosnia-Herzegovina resulted in many water resource management facilities suffering damage. Since the outbreak of the war investment in new flood protection facilities has stopped, and the resources for maintaining existing facilities have been negligible. This is particularly true for towns along the Sava River. The consequences of floods resulting from exceptionally high waters in this area would be immeasurable. The situation is not much better in other parts of the country, as evident from the floods in the Tuzla Canton in June 2001, in which major damage was inflicted on crops, housing and infrastructure, estimated at more than $45 million. Some of the installations were also used for military purposes and damaged so badly during the war that they no longer function properly. The structure of some dikes has been damaged as a result of bunkers being built inside the dikes. Mines have been laid around some flood protection installations. During the war, five of the seven wastewater treatment plants were closed due to war damage, stripping of equipment and installations, lack of maintenance or shortage of electricity. The plants in operation today are all very small, and more than 95 per cent of the municipal wastewater is discharged directly into water bodies without any kind of treatment.

Although irrigation systems are not very common – only two per cent of the total arable land of about 1,123,000 ha is irrigated – those irrigation systems that do exist have been seriously damaged due to a combination of poor maintenance and the war.

The quality of drinking water is on the whole mediocre, and for the nearly 50 per cent of the population who do not have access to public water supply systems the water quality is probably even more questionable.

The high level of PCBs now affecting the Klokot 2 spring near Bihać, which feeds a commercial fish farm may adversely affect the use of the fish for food. However, Klokot 1 is free of PCBs, which is fortunate because its water serves as the main water-supply source for almost 50,000 inhabitants in Bihać.

Serbia and Kosovo

Many of the industrial zones targeted by NATO planes were located along the Danube and tributaries like the Sava, Morava and Lepenica rivers. As a result, tonnes of petroleum products, heavy metals and other toxic substances poured into the rivers from burning industries hit in the bombing raids. Moreover, Europe's largest waterway was blocked for a long time by the wreckage of bridges that NATO planes destroyed.

Scientists point to the following war-related destruction:

- The bombing of the Prva Iskra chemical plant in Baric released 165 tonnes of hydrofluoric acid into the Sava River.

- The destruction of electrical transformers in Bor resulted in uncontrollable combustion of PCBs and pyralene.

- The bombing of a storage tank in Ripanj caused 150 tonnes of used transformer oil to spill into the Topciderska River, a tributary of the Sava.

- Damage was also sustained to the water supply system in the Zemun and Bistrica hydroelectric power station in Polinje.

179

'While the border disputes between Syria and ourselves were of great significance, the matter of water diversion was a stark issue of life and death'. Israeli airstrikes caused the scheme to be dropped (see section headed *Israel and Palestine* later in this chapter).

Emergency services

The recent Nato Advanced Workshop on Threats to Global Water Security concluded that the present arrangements for securing water supplies during and immediately after an armed conflict are 'rather inadequate and amateur', especially as a large part of the work is left to NGOs like the Red Cross and Red Crescent (Arlosoroff and Jones, 2009).

Emergency work has two aims: to provide adequate volumes of water for drinking and, most importantly, sanitation, and to prevent poisoning and the outbreak of epidemics. The Workshop concluded that the order of priority should be:

- **First priority:** *Provision of* any *water* – quantity is more important than quality, especially for sanitation, because the most immediate danger is from disease caused by inadequate sanitation.

- **Second:** *Continuity of supplies* – restoring and securing electricity supplies is essential for pumps and control systems.

- **Third:** *Attend to water quality* if the emergency supply was unsafe. This may involve importing bottles or containers and the provision of portable water treatment systems.

There should also be a long-term 'community plan', which aims to prepare people for emergencies – and this applies equally well to natural disasters. This should aim to educate people on what to do prior to or as early as possible in a conflict, including safe storage of water, methods of purifying and disinfecting, and the dangers of specific diseases. The plan should also include provision of cheap, basic equipment to secure supplies.

One of the more controversial recommendations of the Workshop is that the armed forces should be formally involved in restoring safe water and sanitation at the very early stages when it is too dangerous for aid workers to enter the field of combat. The idea was taken up by the future UK Prime Minister David Cameron in March 2010, a short time before the General Election. His remarks elicited immediate condemnation from the charities. Save the Children maintained that the military should not do humanitarian work because it blurs the distinction between the combatants and civilians and so puts civilian aid workers at risk. However, the Department for International Development supported the idea, saying that military and civil organizations need to be combined as security is a huge problem for aid organizations.

The Workshop proposed the creation of special sections of the military or under military control that are more specialized in water and sanitation provision and restoration than the current military engineers, and noted that NGOs have a multitude of other aims and interests and are funded on a non-professional and potentially less reliable basis, mainly by donations. The hope was also expressed that in the long-term the UN might develop a formal and specific set of international rules concerning responsibilities for the protection and restoration of water supplies and sanitation during armed conflicts.

Military control of the weather – a future scenario?

'While some segments of society will always be reluctant to examine controversial issues such as weather-modification, the tremendous military capabilities that could result from this field are ignored at our own peril. From enhancing friendly operations or disrupting those of the enemy via small-scale tailoring of natural weather patterns to complete dominance of global communications and counterspace control, weather-modification offers the war fighter a wide-range of possible options to defeat or coerce an adversary.'

Weather as a Force Multiplier:
Owning the Weather in 2025.
Research paper presented to the US Air Force
by Colonel TJ House *et al*, 1996

General Dwight D. Eisenhower once said that all that the Germans would have needed to keep control of Normandy during the Second World War was permanent, really bad weather (Fuller, 1990). The only weather modification method available at that time was fog dispersal (see Chapter 17 Controlling the weather), but the events of the war clearly sowed the seeds in the military mind to aim for greater capability. Artificial rain-making methods came just too late for deployment in that war, but they were used extensively during the Vietnam War (1967–1972) to bog down the Vietcong. The aim of Operation Popeye was to seed clouds with silver iodide to extend the monsoon season over North Vietnam, Laos and Cambodia, causing floods and landslides, mainly to degrade and block the Ho Chi Minh Trail. The USAF flew sorties twice a day for up to 45 days a year over a five-year period. Unfortunately, the Vietcong were skilled in operating during the monsoon.

This is old technology. The American military is thinking far more radically. A research paper entitled 'Weather as a Force Multiplier: Owning the Weather in 2025' produced by the US armed forces outlines some of the possibilities and purports to offer a 'road map to weather-modification in 2025' (House, 1996).

The water resources aspects focus on precipitation, fog and storms. The aim of interfering with precipitation could be longer term than currently possible and include the prospect of inducing starvation through drought (Table 9.1). The report cites evidence that fog may be induced, at least on a small scale (100 metres), to conceal troop movements. And it notes that while hurricanes, like Hurricane Andrew that destroyed an airbase in Florida in 1992, are far too powerful to be controlled, smaller storms might be controllable. It cites research using black carbon dust as an aerosol in order to increase evaporation and convection (Gray *et al*, 1976). This involves creating an aerosol of fine dust over a large water body. The dust absorbs solar radia-tion, warming the air and increasing evaporation and convection, eventually leading to a rainstorm. The process is similar to the natural mechanism that produces the lakeside snowfall around the Great Lakes.

The report sees all technologies improving by 2025, but three key predictions stand out, centring on improved computing power and mathematical techniques, smart materials based on nanotechnology, and a Global Weather Network. It notes that weather forecasting is currently at best only 85 per cent accurate over 24 hours, but it expects dramatic improvement by 2015. However, the authors predict that between 2015 and 2025 increasing population pressure will magnify the effects of natural disasters, so that governments will turn more to weather modification. They expect rainmaking techniques to improve so much that by 2025 rainfall can be control-led over a period of several days. Part of the advances will incorporate nano-scale smart-seeding materials that adjust their size and dispersal to individual situations, which will deliver artificial weather 'made to order'. The third development, the Global Weather Network, is seen as a hyper-fast data transmission system that will deliver real-time measurements from a denser and more accurate observing network of highly improved sensors to forecast centres from anywhere in the world, fed into improved computer models based on the latest nonlinear mathematical techniques – a faster version of the present World Weather Watch (see Chapters 18 and 19).

There is no doubt that these developments are taking place despite the concern at the UN General Assembly in 1977 that resulted in the 'Convention on the Prohibition of Military or Any Other Hostile Use of Environmental Modification Technique (ENMOD)'. The Convention committed the signatories to desist from military or hostile use of weather modification. The US military claim the Convention has slowed the pace of development and caused it to focus on suppression rather than intensification of weather (House, 1996).

Table 9.1 Uses of weather modification in war, according to the US military research paper 'Weather as a Force Multiplier: Owning the weather in 2025'.

Modification	Aggressive	Defensive
Precipitation	Deny freshwater, induce drought	Maintain troops' comfort, control visibility, Improve/maintain level of operational capability
Fog and cloud	Removal for airfield operations, increase concealment of operations	Increase concealment
Storms	Choose battlefield environment	Hamper enemy operations
Virtual weather	Mislead enemy with false information	Hamper enemy operations

Cyber wars

'I don't think we can win in cyberspace – it's like the weather – but we need to have a raincoat and an umbrella to deal with the effects.'

Rear Admiral Chris Parry,
Head, Development, Concepts and Doctrine Centre,
UK Ministry of Defence, 2006

Modern water control systems are already heavily reliant on automated, computer control, and most of the improvements underway in water security rely on computers. Hackers can be very sophisticated and even the best antivirus software can be circumvented and firewalls can be penetrated, even the Pentagon's, as recent events have proven. Cyber attacks are relatively easy to initiate from anywhere. Systems that control water distribution have traditionally operated in a very open environment, but they now need more protection. The UK government has said that companies are not doing enough to protect themselves. The head of the British Army, General Sir David Richards, said in 2010 that threats against UK infrastructure require less new weaponry and more computer security. Attacks are most likely to come via the internet, so a lot can be achieved by holding critical systems offline.

Cyber attacks may be state-sponsored, acts of terrorism, or commercial espionage and they are increasing. The USA recorded over 37,000 attempted breaches of government and private systems in 2007 and Congress cited Chinese espionage as 'the single greatest risk to the security of American technologies'. The actual number of attacks is probably much higher as the statistics do not include attacks on control systems that do not have firewalls or intrusion detection systems.

The USAF has a new unit of 40,000 staff preparing for cyber war. MI5 raised the alert on China's cyberspace spy threat in Britain in 2007: 300 businesses were warned. The UK government openly accused China of espionage on government and financial institutions and posted a letter on the Centre for the Protection of the National Infrastructure website alerting critical businesses, like banks and water and power companies, of the threat from Chinese state organizations. At the 2008 meeting of World Economic Forum in Davos, the Director of the FBI urged governments, companies and individuals to assess their vulnerability to 'cyber attack' by electronic spies. During 2008, the UK suffered thousands of cyber attacks every day, aimed mainly at electricity, telecoms and water companies. Thames Water was named as a possible target. A NATO team led by former US Secretary of State Madeleine Albright reported in 2010 that cyber attacks on one country's power networks or other critical infrastructure that result in casualties and destruction similar to a military attack would be sufficient to mobilize the whole of NATO under current rules.

Most attacks appear to come from China, Russia, North Korea and India, although attacks have been traced to 120 other countries. To date they appear to have been primarily concerned with commercial espionage, collecting data via 'custom Trojans' hidden in emails. But they could just as well be aimed at disabling a computer-controlled system. The Pentagon and British sources have identified the Chinese telecoms company Huawei as a particular threat. Clandestine software introduced over telephone lines can sit on PCs unnoticed until activated by a remote controller and can then take control of security webcams and other systems.

Too much or too little water could be released; detection, purification and warning systems could be degraded or shut down; intruder alarms could be disabled and raw sewage could be released into the environment. State-controlled espionage is unlikely to take that route, but terrorists might, and with less risk of being caught in the act.

Even commercial espionage can affect the water industry, especially private companies. In the modern world of international takeovers, prior information on a rival's bid for a water company can be a crucial tool in negotiations (See *Privatizing water* in Chapter 4).

Terrorism

'The water supplied to US communities is potentially vulnerable to terrorist attacks by insertion of biological agents, chemical agents, or toxins … The possibility of attack is of considerable concern … These agents could be a threat if they were inserted at critical points in the system; theoretically they could cause a large number of casualties.'

The US President's Critical Infrastructure Assurance Office, 1998

Terrorism is as old as warfare itself, but the events of 9/11 in New York and Washington in 2001 awakened the USA and many western countries to the developing threat of Islamist extremism and the increasing variety of methods and enemies. There were 12 thwarted terror plots in the USA in 2009: more than any year since 9/11. Terrorists may be foreign or home-grown, political and/or religious, part of an organization or solitary.

They may have a philosophy or simply a grudge. Devising effective counter-measures is complicated by the fact that the impacts may not be immediate. Immediate impacts might be spectacular, but long-term fallout, for example, continuing to affect public health could be a greater threat: anthrax could cause widespread deaths and persist for years.

Following 9/11, many US water companies immediately began to remove potentially sensitive information from their websites and Congress set up a Directorate of Homeland Security to oversee protective measures. Following the discovery of a computer in Afghanistan containing structural analysis programs for dams, the FBI's National Infrastructure Protection Center issued a bulletin in 2002 warning of 'Terrorist Interest in Water Supply and SCADA [Supervisory Control And Data Acquisition] Systems' (Shea, 2003). The National Infrastructure Protection Center now sends warning messages to the Association of Metropolitan Water Agencies' Information Sharing and Analysis Center (ISAC). ISAC then delivers early-warnings to water officials throughout the country.

One ineluctable geographical problem is the extreme concentration of possible targets: just 15 per cent of all large drinking water and sewage plants in the USA supply more than 75 per cent of the population. There are big hydropower plants as well. The Grand Coulee Dam on the Columbia River alone produces 25 per cent of US hydropower. The Hoover Dam on the Colorado River provides seven per cent of all public electricity in the west USA. A bypass was completed in 2007 to divert traffic away from the Hoover Dam and public tours inside the dam have been stopped – as they have in many dams in the USA, Britain and elsewhere.

Although these represent the largest targets and would cause maximum disruption, they will be best protected. On the other hand, there are hundreds of smaller water targets that could be less protected. They would not have the same physical effect, but they could have a disproportionately high scare effect.

The task of providing effective protection is complicated by the very large range of possible aims and methods of attack: whether toxins are chemical, biological or radiological; where and how the toxins might be introduced into water systems; whether the terrorists' aim is to kill, infect or simply create fear and panic; whether the aim is to poison or to physically disrupt the water supply system by breaching pipelines to cut off supplies or by breaching dams to cause a flood; whether the attack is physical or cyber; or whether the attack is indeed aimed at the water system itself or just using the distribution network to attack other buildings, as in the possible case of using sewers to plant bombs beneath key buildings. Sewage plants and sewer networks are also theoretical targets. Such attacks may not be so spectacular, but they could lead to widespread infection.

The list of vulnerable activities also extends well beyond municipal drinking water supplies. It includes bottled water production and food and beverage manufacture. Companies using their own private water sources could be especially vulnerable to commercial terrorism: the recall of thousands of bottles of Perrier Water after some were found to contain traces of benzene in 1990 illustrates the potential commercial damage that can be done, whether this is intentional or accidental. In that particular case, Perrier attributed the problem to filters designed to remove the natural benzene that had not been renewed.

Prevention of waterborne disease has been hindered by ageing infrastructure and outmoded analysers in many water plants. The USA Environmental Protection Agency (EPA) established a water protection task force in 2001, charged with helping federal, state and local partners to improve protection of drinking water supplies from terrorist attack. And the EPA's National Homeland Security Research Center initiated the Environmental Technology Verification (ETV) programme to examine rapid screening tests that could detect water terrorist acts as early as possible. The ETV study shows that the Aqua Survey IQ-Tox Test™ performed extremely well for the detection of toxic agents or 'threat contaminants' in drinking water. The tests focus on both chemical and biological toxins, such as aldicarb, colchicine, cyanide, dicrotophos, thallium sulphate, *Botulinum* toxin, ricin, soman and VX. Britain has also spent tens of thousands of pounds on improving the security of water systems, and the Severn Trent Eclox system performed well in EPA tests. The international Bentley Institute is marketing a wide range of software to help design more secure water systems, including WaterCAD, WaterGEMS and WaterSAFE™, offering a full detection, prediction and response system developed by Haestad. Starodub (2009) gives a detailed up-to-date review of progress with instrumental systems to detect threat contaminants in water. He notes that despite the 1972 Biological Weapons Convention, many countries continue to produce agents of biological warfare, ranging from biochemical

183

toxins to viruses and bacteria, which may find their way into terrorists' hands.

Chemical agents are generally the easiest for terrorists to get and use. Biological agents like mycotoxins are harder to get and more expensive, and the effects may be delayed for weeks, which may or may not serve the terrorists' purpose. However, *Botulinum* bacteria can be found naturally and botulism is a powerful, short-term infection that could kill 100,000. In a report to the USAF Counterproliferation Center, Hickman (1999) noted that an individual with an internet connection and $10,000 could build a biological fermentation capability, producing trillions of deadly bacteria that do not require any sophisticated delivery system. Fortunately, the more deadly agents like anthrax, smallpox or radioactive materials like polonium-210 and caesium-137 are not so easily obtained, although water soluble caesium chloride is obtainable in powder from hospital supplies. On the other hand, mixtures of a number of different threat contaminants could have a deadly effect and they will be more difficult to eliminate than single agents. Detection kits cannot be expected to cover every possible contaminant, so there is still room for the time-honoured practice of running water through tanks of fish, especially *Gnathonemus petersii* (elephantnose fish) or *Daphnia*: disturbed reactions from the animals give an immediate warning about water quality.

In June 2002, the federal government passed the Public Health Security and Bioterrorism and Response Act (H.R. 3448), which requires all public water systems serving more than 3300 people to submit a security assessment and an appropriate emergency response

Figure 9.2 Fish used to monitor the quality of water taken from the River Po at the Turin water treatment works. Disturbed activity will indicate the presence of toxic pollutants

plan. Water companies were instructed to prepare for the worst case scenario, as this would automatically provide cover for lesser attacks. No defence system can ever give total protection from every possible type of attack, and it is impossible to monitor every part of the system continually, but firms were required to install threat detection kits at strategic locations and to take all reasonable measures, such as: improving security checks on employees and visitors; limiting public access to infrastructure; fencing, illuminating and installing cameras at all facilities; enhancing security on computer systems; establishing backups and alternative sources; securing access points to systems; sensitizing the public to the dangers, to increase vigilance and reporting of suspicious events; and improving communications, warning systems and coordination with the emergency services.

Toxic substances could be introduced into reservoirs, at water intakes, at treatment plants or within the distribution system. Some are more effective than others. Reservoirs are generally regarded as the least effective because of the large amount of dilution of any chemical toxin, but, as natural outbreaks of *Cryptosporidium* show, biological contaminants can multiply and invalidate the argument. Radioactive contaminants may also be disproportionately effective, despite dilution. Fifty employees at the Kaiga nuclear power station in India fell ill after a colleague poisoned a water cooler with tritium in November 2009. The radiation from tritium is relatively weak (it occurs naturally in the environment) and cannot penetrate skin. However, as with all radioactive isotopes, ingestion raises it to a different level, risking cancer and genetic abnormalities. Inside jobs like this are worrying and require more careful screening of employees, including psychological profiling – but grudges can always develop.

The complexity of assessing vulnerability and deploying protective measures is illustrated by the case of the Washington Suburban Sanitary Commission (WSSC). The WSSC is one of the ten largest water utilities in the USA, with four reservoirs in the Patuxent and Potomac watersheds, 60 storage tanks, 14 pumping stations and some 8300 km of water mains. It serves 1.6 million people in Montgomery and Prince George's counties and sells some water to the Howard County Bureau of Utilities. Ironically, the WSSC found itself having to withhold information from colleagues in the industry, for fear that crucial information might fall into the wrong hands.

The costs entailed in assessing and improving systems have been substantial, but most improvements will also provide protection again natural pollution events and against more minor cases of vandalism. Nevertheless, federal aid was made available immediately following the Bioterrorism Act, and the EPA spent $53 million to help 400 large public utilities across the country examine and improve their systems.

The American College of Preventive Medicine produced an online guide for the medical profession, jointly sponsored by the EPA and the American Water Works Association entitled 'Physician preparedness for acts of water terrorism'. The College points out that, although detection methods in catchments and water plants are improving, the first indication that a contamination event has occurred may be a change in the patterns of disease and illness, potentially involving a cluster of water-related cases of toxicity or a community-wide outbreak. Doctors and healthcare workers are likely to be the 'front-line responders' and need to learn to recognize unusual trends in patients' diseases and the early warning signs that may result from biological, chemical or radiological terrorism. The medical community could play a vital role in minimizing the impacts by maintaining a heightened level of suspicion. The guide recommends close cooperation between healthcare providers, public health and water utility practitioners, law enforcement professionals and community leaders, including constantly updated information.

To date, there have been very few successful attacks on water systems. A religious cult managed to contaminate the public water supply in The Dalles, Oregon, with *Salmonella* and infected over 750 people (Clark and Deininger, 2000). An attempt in 2002 to poison the water system in Rome was foiled when police apprehended terrorists in possession of 4 kg of a cyanide compound and maps marking the city's water pipelines, a reservoir and the entrances to the US embassy. Both companies and governments have generally presented the argument to the public that attacks are unlikely to be very successful physically, mainly because of the sizeable amounts of poison needed, especially if it is introduced into a large reservoir, and because most toxins and pathogens are either neutralized by chlorination or ozone or else filtered out in water treatment plants. But the introduction of the substances in the transfer pipes post-treatment is likely to be more effective. Water quality is not generally monitored in distribution pipes.

Two acts of cyber terrorism stand out. In 2000, a discontented former employee at a treatment plant in Maroochy Shire, Australia, was able to access the controls of the sewage plant remotely and discharge around 1000 m³ of untreated sewage into the environment. And in 1994, a hacker successfully broke into the computer system of the Salt River Project in Arizona, one of Arizona's largest water suppliers and the third-largest public power utility.

The most potent weapon of the water terrorist, however, is fear, generating public panic, and the best defence is denial of publicity. This was the strategy adopted by the British government in 1999 when it blocked media coverage of the Irish Republican Revenge Group's threat to introduce poison via public water hydrants. Even so, if the poisoning had taken place, it could have been serious. This type of 'backwater' introduction of toxic agents is a serious threat, with the potential of making ill or killing large numbers of people. A small pump fitted to a domestic water tap can counteract the mains water pressure and allow agents to flow backwards into the system (Arlosoroff and Jones, 2009). Scottish terrorists also threatened to use backwater introduction via street fire hydrants in 2007. It would be impossible to monitor over a million hydrants in England alone, but just 10 kg of poison introduced in this way is reported to have the potential to kill 200,000 people.

In 2002, the Astronomer Royal, Lord Martin Rees, laid a bet of $1000 in an internet competition for the magazine *Wired* that by 2020 an instance of bio-terror or bio-error will have killed a million people. And in his book *Our final century?* (2003/4), subtitled *How terror, error, and environmental disaster threaten humankind's future* in the American version, he suggests that 'the odds are no more than fifty-fifty that our present civilisation on Earth will survive to the end of the present century without a serious setback'. The prospect of terror due to *error* affecting water supplies was clearly illustrated by an incident in Camelford, Cornwall, in 1988, when a driver deposited 20 tonnes of the water-cleansing coagulant aluminium sulphate into the wrong tank. It caused levels of the compound in tap water serving 20,000 homes to reach up to 620 mg/L – the EU guideline is 0.2 mg/L. It caused intense psychological trauma at the time and there were official attempts to alleviate concern, but ten years later, analyses revealed damaged cerebral functions that could not be attributed simply to the trauma (Altmann *et al*, 1999). Some people suffered bone disease, and the fear of developing Alzheimer's disease.

185

The Nato Advanced Research Workshop on Global Water Security (Arlosoroff and Jones, 2009) made the following important recommendations:

1 Priority should be given to advance information from national intelligence agencies in order to be forewarned.

2 Priority should also be given to continuous, automated online monitoring, increasing the effectiveness of the instrumental systems and the range of analyses, including DNA analysis, luminescence and live animals, and reducing the currently very high costs of around $100,000 per instrument. This raises the question of the possible willingness or ability of private water companies to afford the equipment – who should control security measures?

3 Improving the education and preparedness of the medical profession and the public.

4 Developing improved and cheaper domestic-scale in-house safeguarding systems, like UV irradiation and carbon filters with pre-programmed automatic responses when overloaded or used up.

5 Improving the security of computer systems.

The final solution

'Dialogue is the way forward. Terrorism is a symptom, not a root cause.'

Terry Waite, former kidnap victim and Archbishop of Canterbury's Special Envoy to the Middle East, 2007

'With the terrible weaponry of the modern world; with Muslims and Christians intertwined everywhere as never before; no side can unilaterally win a conflict between more than half of the world's inhabitants. Our common future is at stake.'

Open letter from 138 Muslim scholars to Christian churches, October 2007

Terrorist threats are unlikely to subside until the root causes of discontent are removed or negated. This seems to be a very long way from being achieved with the Islamist issue. But there is no question that this must be the ultimate aim, as both the open letter of Muslim scholars to Christian churches and the Archbishop of Canterbury's special envoy have said. This means fighting poverty, inequality and injustice and also religious fanaticism.

From the water security point of view, the quantitative impact of such acts, at least in terms of current technology, is likely to be very limited and local. But as Rear Admiral Parry observed, any technological advantage developed to deal with the threats is unlikely to last (Almond, 2006). It is a game of cat and mouse and nothing remains static.

Can virtual water prevent water wars?

One interesting suggestion from the originator of the concept of 'virtual water', Tony Allan, is that international tensions over water in the Middle East have become less tense, not more, since population growth caused demand in the region to outstrip available resources in 1970 (Allan, 2003). The reason for the irony, according to Allan in his article 'Virtual water eliminates water wars', is that water has been available on the international market in the form of 'virtual water' (see Chapter 8). This now means that countries that can afford to import grain and other products do not need to utilize their own scarce water resources to produce them. Grain is the largest user of water in the region and by 2000 the MENA region was importing 50 million tonnes. He predicts that the remaining ten per cent of water needed for drinking, domestic and industrial use may soon be met by low-cost desalination. His only regret is that the market has delayed the introduction of much needed improvements in water efficiency.

The Green Cross Water for Peace Project has adopted the virtual water hypothesis. But while virtual water may have reduced some international tension, there is little sign of its effect in the Israeli-Palestinian conflict, and it seems to have little relevance to international terrorism, which is more driven by poverty, inequality and religious idealism.

Transboundary waters

The World Water Council warned of possible 'water wars' in 1992. The main focus then was on river basins that are divided between a number of countries, and where the upstream state has the first call on the water, often to the detriment of the downstream state. Less obvious but equally contentious resources lie in aquifers that cross international frontiers. Wolf (1998) noted 261 major international rivers, whose basins cover nearly half of the total land area of the globe, and 'untold numbers of shared aquifers'. The Intergovernmental Council of Unesco's International Hydrological Programme endorsed the setting up of the Internationally Shared Aquifer Resources Management (ISARM) programme to try to find ways to reduce tension and achieve sustainable management, in response to the 'Ministerial Declaration of The Hague on Water Security in the 21st Century', at the World Water Forum in 2000. International law has been almost totally inadequate for solving disputes. This and recent improvements are discussed in Chapter 20.

Around 60 per cent of African rivers are transboundary and 65 per cent of Asian rivers. In the case of rivers, the problem arises from the historical habit of choosing rivers as natural boundaries, aggravated, especially last century, by colonial powers and treaties that drew straight lines on maps. The USA is a prime example, sandwiched in large measure between the Rio Grande and the 49th parallel. Treaties regulate use of the main cross-border rivers and it is inconceivable that a water war could break out between the NAFTA countries, but there have been disagreements and tense moments. Mexico remains dissatisfied with the quality and volume of water it receives from the Colorado, and with the depletion of the Rio Grande. The USA has had issues with Canada over its use of the Columbia, over the proposed North American Water and Power Alliance, and recently over the terms of the North American Free Trade Agreement.

Transboundary resources in Europe are perhaps even better regulated. The International Commission for Navigation on the Rhine was established after the 1815 Congress of Vienna, the most recent update being in 1963. The International Commission for the Protection of the Rhine was set up in 1950 and legally incorporated in 1963. And the International Committee for the Hydrology of the Rhine was established in 1970 under encouragement from UNESCO to promote sustainable development. Political history has delayed similar agreements on the Danube, especially the Cold War, but the International Commission for the Protection of the Danube River was set up in 1998 and its Danube River Basin Management Plan for sustainable management

under the EU Water Framework Directive was endorsed by a Ministerial Declaration in Vienna in February 2010.

Outside of Europe and North America, however, agreements are limited. The Mekong and the Southern Africa Development Community's Regional Strategic Water Infrastructure programme are two of the more notable.

The Middle East has already witnessed a number of water wars and could see more as resources dwindle. The region suffers political, ethnic and religious conflicts in addition to national boundaries that place the headwaters of many important rivers – like the Nile, Tigris, Euphrates and Jordan – in the hands of countries that are already at variance. As demand already exceeds supply by up to fourfold in much of the region, some of the most rapid population growth in the world – predicted to double in the next 25 years and even possibly quadruple before stabilizing – coupled with the prospects of diminished resources due to climate change, will add to conflict.

The Tigris-Euphrates basin

As a result of the settlements following the First World War, which broke up the Ottoman Empire, Turkey holds the headwaters and Iraq is the downstream state (Figure 9.1). In between, the Euphrates passes through Syria and some major tributaries of the Tigris are in Iran. The break up also left the Kurds without a state of their own, split mainly between Turkey and Iraq, and their continuing fight for independence has also affected the course of water disputes.

As many Middle Eastern countries began to reach the limits of their water resources in the late 1960s, Turkey embarked upon ambitious plans to make the headwater region the 'bread basket' of the Middle East by the twenty-first century. The south east Anatolia or GAP Project aims to build 22 large dams to irrigate 1.6 million ha to grow the equivalent of half Turkey's rice and vegetables, to introduce cotton as a cash crop, and to generate 26 billion kW pa, or 70 per cent of Turkey's needs.

The project has been delayed by lack of funding, as the World Bank and Western countries perceived the potential destabilizing effect on regional politics. Turkey decided to proceed on its own. The Project began displacing tens of thousands of Kurdish peasant farmers. Many fled to Iraq, the Kurdish PKK terrorist organization was formed and a series of bombing campaigns ensued.

In the 1980s, Turkey advertised plans for a 3000 km 'Peace Pipeline' to supply water from two of its Mediterranean rivers, the Ceyhan and Seyhan, through Syria and Lebanon to Israel, Jordan, Saudi Arabia and Kuwait. It would carry between six and eight billion m³ a year, priced at a third of the contemporary cost of desalinated water. The aim was to raise funds and curry political favour. The proposed pipeline pointedly avoided Iraq, which was the source of much of the armed Kurdish resistance. The project fell into abeyance through numerous political obstacles, not least the inclusion of Israel, and its vulnerability to the rising tide of terrorism. Syria supported the PKK, the Kurdish Workers Party, by allowing it to run terrorist training camps in its territory. Turkey appears to have opted to apply physical pressure on Syria by cutting off the Euphrates: when the first megadam, the Atatürk, was completed in 1990, there was zero discharge in the Euphrates for a month as the reservoir filled up. Syria and Iraq laid plans for a joint military retaliation. Turkey officially denied applying deliberate pressure and had warned Syria and Iraq of its plans beforehand. However, it gave Turkey leverage in secret negotiations, under which Syria agreed to close the PKK training camps and extradited rebels in return for an assured water supply with a guaranteed minimum of 500 m³/sec. However, that supply was only around 60 per cent of the original average discharge. Iraq fared worse. When Syria's own damming and irrigation programme began in the 1970s with the al-Thawrah (Revolution) Dam, which created the Assad reservoir designed to irrigate 640,000 ha, Iraq had threatened war in support of the more than three million farmers whose livelihoods were affected. War was only narrowly averted in 1975. Now, as a result of GAP on top of Syria's diversions, Iraq receives only a third of its original supplies from the Euphrates. Turkey's claim that the dams remove the seasonal drought and flooding problems for Iraq hardly compensate.

The Atatürk Dam incident gave the Allies in the first Gulf War of 1991 a precedent, and a proposal was advanced to cut off water to Iraq as part of the war effort. The plan was never implemented as it would also harm a key ally, Syria, and would probably have had more effect on civilians than on the Iraqi military machine. However, the UK government was more sympathetic to Turkey by the late 1990s and offered export credit guarantees of £200 million for the construction company Balfour Beatty to work on the proposed Ilisu Dam on the Tigris. The support was short-lived. A two-year protest campaign led by environmental groups including Friends of the Earth caused Balfour Beatty and Sweden's Skanska to pull out. The UK government stated that all future guarantees would require environmental impact assessments. Construction began in 2006 with support from German, Austria and Switzerland, only to be withdrawn again in 2008 after Turkey failed to reach the required environmental and social standards, especially on the resettlement of displaced Kurds and the destruction of the 10,000-year-old Kurdish town of Hasankeyf. Turkey restarted it yet again in July 2009.

Israel and Palestine

The main point of contention between Israel and its Arab neighbours has been over the use of the River Jordan and its tributaries, but groundwater resources have been emerging as equally controversial in recent years, extending outside the Jordan Valley and into Gaza (Figure 9.3).

Israel is the most efficient user of water in Asia – it has to be! A quarter of Israel's water is reused and it aims to recycle 430 million m³ in 2010. But there is a strong contrast between the 300 litres per day per head consumption in Israel and 80 litres in Jordan. Moreover, Israel and Jordan are using 120 per cent of the 'safe yield' from the Jordan Valley. The River Jordan is reduced to a heavily polluted trickle for parts of its length. Water tables are falling in Jordan and the West Bank, springs are drying up, groundwater is turning salty in Amman and over-pumping on the Mediterranean coast, including Gaza, is causing saltwater incursions.

Israel fought the Six-Day War in June 1967 in large part to secure water resources, as well as to gain more defensible borders. The main aim was to pre-empt plans by Syria and Jordan to dam the River Yarmouk, which they achieved first with tank fire and then airstrikes. Syria and its allies had begun three years previously by diverting some Jordan headwaters, first the Banias River to the Yarmouk, followed by Lebanon connecting the Hasbani River to the Banias, aiming to divert 20–30 million m³ pa to the Yarmouk. Capture of the Golan Heights not only gave Israel command over Syria's Bekaa Valley, but also extensive coverage of the Jordan headwaters, including the Yarmouk.

Syria's actions were partly in response to river diversions by Israel. The 'Jordan Valley Unified Water Plan'

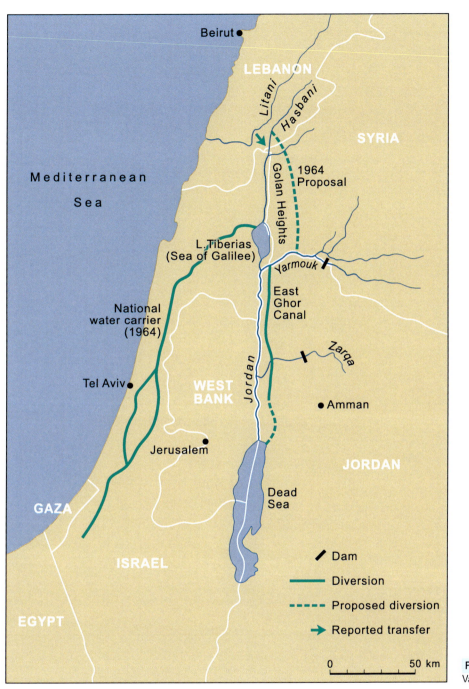

Dam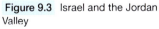

Diversion

Proposed diversion

Reported transfer

0   50 km

Figure 9.3 Israel and the Jordan Valley

advanced by the US ambassador Eric Johnston in the mid-1950s had given Israel the right to channel water from the Jordan River to the Negev Desert in the south, in return for Israel dropping its claim to water from the Litani, a river draining to the Mediterranean wholly within Lebanon. Under the Plan, Israel was to receive 400 million m³ p.a., nearly 40 per cent of the Jordan River's average flow, Jordan would get 480, Syria 132 and Lebanon 35m m³. But the Plan was rejected by the Arab League.

Israel went ahead nevertheless, diverting headwaters of the River Jordan around Lake Tiberias (the Sea of Galilee) in the 1950s and channelling water from the Lake to the coast and southern Israel via a National Water Carrier completed in 1964. The scheme included diverting unwanted saline springs into the Jordan River, beginning its slow degradation. The National Carrier aimed to redress the balance in water resources between the north with 93 per cent of the nation's water, and the south with 65 per cent of the arable land (Hillel, 1992).

189

When Israel established a buffer zone in southern Lebanon in 1978 to counter attacks from Hesbollah, it gained control of the Hasbani and the Wazzani springs that feed it. A plan to formally annex south east Lebanon was rejected by Israel's ally, the Army of Free Lebanon. However, Israel began pumping water from the Wazzani and appears to have annexed the Litani as well. According to the UN Economic and Social Commission for Western Asia in 1994, Israel diverted over 200,000 m^3 pa from the Litani and Wazzani Rivers, leaving large areas of southern Lebanon short of water for irrigation and drinking.

Water is also an important element in West Bank politics. Prior to the Six-Day War, Israel took 60 per cent of the water abstracted from aquifers that straddle the border between the West Bank and Israel proper. It has taken some 80 per cent since. Now, 40 per cent of Israel's water comes from West Bank aquifers. Palestinians are forbidden licences to sink new wells, but not so the newcomers in the illegal Israeli settlements in the West Bank. Since the building of the boundary wall between Israel and the West Bank, intended to keep out terrorists, and also ruled illegal by the UN, the plight of ordinary Palestinians in villages near the border has become worse, with many villages and farms cut off from their wells by the wall and with no rights to drill new ones (Pearce, 2007). The plight of Gaza's water supplies is even worse since the Israeli invasion early in 2009, which destroyed much of the infrastructure and Israel's subsequent blockage that bans the importation of cement and steel products. With no rivers and no alternative to abstracting water from the increasingly salinized aquifer, the prospects are bleak.

Israel for its part can hardly feel secure enough to comply with international demands to relinquish the West Bank to an independent Palestinian state without alternative sources of water. It signed an agreement with Turkey, its only friend in the region, to import up to 400 million m^3 a year by ship. It has treated this as an option of last resort, but since the disastrous murder by Israeli commandos of eight Turkish aid workers on the relief flotilla heading for Gaza in June 2010 and severe condemnation from Turkey's Prime Minister, it seems unlikely this option will be available any more. Israel's only other serious option is desalination, which it has begun, despite elements in the National Water Corporation saying it is too expensive.

An extremely elaborate and highly controversial scheme has also been commissioned from the British architectural firm of Norman Foster, dubbed 'Red to Dead'. The scheme envisages transferring water from the Gulf of Aqaba on the Red Sea over the mountains to the Dead Sea. It would require a considerable amount of electricity to pump the water across, but hydropower would be generated in the descent into the Dead Sea, and since the Dead Sea is 400 m below sea level the hope is to generate more electricity than it uses. Some seawater could also be desalinated for water supply. The level of the Dead Sea has been falling as a result of diversions from the River Jordan and the sea has been getting more salty. Opponents say the salinity of the Red Sea will not match and have an adverse effect on the special ecology. However, as the cost would be so great and the gains small, except perhaps for tourism, the scheme is unlikely to go ahead.

Meanwhile, Syria is now going it alone with plans to build 20 dams on Yarmouk, which will severely impact on flow reaching the Jordan.

Egypt and the Nile

'In the next few years the demographic explosion in Egypt, in Kenya and in Uganda will lead to all those countries using more water; unless we can agree on the management of water resources we may have international or intra-African disputes.'

Former UN Secretary-General
Boutros Boutros Ghali

The Nile basin covers ten countries with widely differing ethnic, religious and economic backgrounds (Figure 9.4). Four of those countries are among the ten poorest in the world, with per capita incomes of under $200 a year. Many have been ravaged by civil wars over the last 30 years which have contributed to famine, disease, refugees and some broken economies, from Ethiopia and Sudan to Uganda and Rwanda. While these have undoubtedly reduced demand for water, prevented more investment in water infrastructure, and deferred international conflicts over water, the effects are being countered by a 'demographic explosion'.

The basin is currently home to 160 million people and this is expected to double in the next 25 years. Egypt's rate of population increase is approaching three per cent p.a. and demand is rising. The World Bank estimates

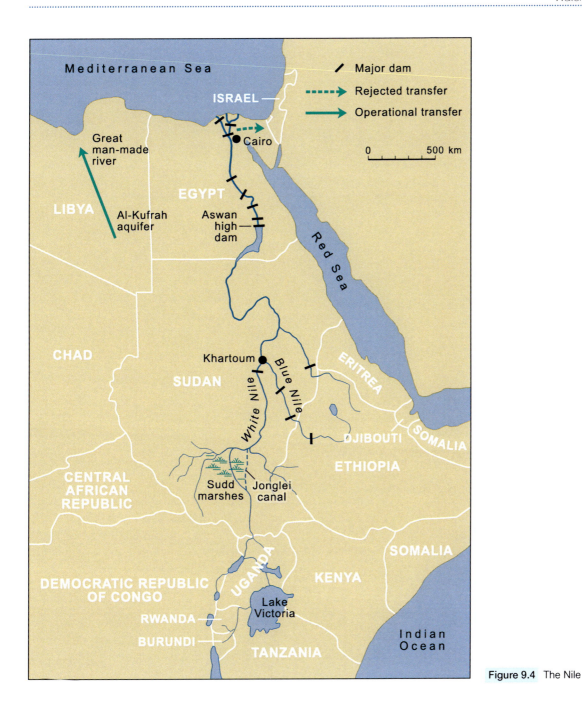

Figure 9.4 The Nile

that Egypt's population of 80 million could reach 97 million by 2025. Many upstream countries are experiencing similar population rises. Egypt relies on the Nile for 90 per cent of its water and it is extremely exposed to water resource developments in the countries upstream, especially to the plans of Ethiopia and the Sudan. Both Egypt and Sudan have objected to any dams being built by upstream countries.

The Nile is also such an attractive proposition in the midst of a water scarce region that both Israel and Libya have harboured desires to get Nile water. A plan was advanced by Zionists in 1903, well before the modern state of Israel existed. For a while around 1980 it seemed this might happen under international pressure aimed at solving the Palestinian problem. With no such pressure to support it, Libya's proposal was quickly rejected.

Sudan has longstanding plans to drain the Sudd Marshes with the partially built Jonglei Canal, which would halt the massive evapotranspiration losses from the marshland and release a vast amount of water for new

191

irrigation projects. Only frequent and prolonged civil war has delayed work. Ethiopia, on the other hand, has been free of civil war for longer and is now beginning to implement its plans for a series of large hydropower and irrigation projects on the Blue Nile. The Tana Beles Dam was finally inaugurated in May 2010 with aid from the European Investment Bank and particularly from Italy, its one-time colonial power, in support of the Italian construction company Salini. Egypt's response is awaited. In 1970, President Sadat had threatened war if dam building began which could divert 85 per cent of flow in the Blue Nile (Postel and Wolf, 2001). Ethiopia has plans for up to 40 dams on the Blue Nile system. More recently, Egypt objected to Tanzania's proposal to pipe water from Lake Victoria for drinking water and irrigation and threatened to bomb the construction site if it went ahead.

For many years, the only formal agreement on the Nile was the 1959 Nile Waters Agreement between Egypt and Sudan, which guaranteed Egypt 55.5 billion m^3 a year. This was an agreement obtained under duress. Egypt sent a military expedition into territory disputed with Sudan between January and April 1958, ahead of the negotiations and prior to Egypt's construction of the Aswan High Dam, which created Lake Nasser whose waters would back up into Sudan. Sudan agreed to take just 18.5 billion m^3. But Egypt has continued to expand irrigation and has regularly 'borrowed' extra water from Sudan in support of massive irrigation programmes, especially in the Nile Delta. Uniquely, Kenya actually volunteered to limit their development of the Nile to help, even to its own disadvantage.

As a result of the Camp David Accord brokered by President Carter and the subsequent Peace Treaty with Israel in 1979, the Egyptian government went further, acceding to Israel's request for one per cent of Egypt's allocation of Nile discharge to be diverted into the Negev and connecting to the National Water Carrier in exchange for establishing full Palestinian sovereignty in East Jerusalem and the return of Sinai, which Israel had occupied since the Six-Day War. Egyptian President Sadat proposed extending the 'Peace Canal' to Lebanon and Jordan as a way of defusing tensions and acting as the basis for regional cooperation. In addition, he proposed to use the diversion to reclaim land for agriculture in the Sinai, especially to irrigate 2500 km^2 in the North Sinai Agricultural Development Project. Even so, this was a step too far for many in the Egyptian public. It was also too much for fellow

Arab countries, who had regarded Egypt as the only country that could put real pressure on Israel over the Palestinian issue. Egypt was forced out of the Arab League for years. The deal to share water with Israel was dropped, ostensibly when Israel's promise over East Jerusalem did not materialize. But it also followed threats to President Sadat's life, which sadly came to fruition soon afterwards with his assassination by a member of the Egyptian military.

The plans also enraged other states in the Nile Basin, who regard unilateral disposal of Nile waters as an illegal act. Nevertheless, Egyptian internal plans to divert Nile water to the Sinai continue. Since 1987, water has been diverted to the west bank of the Suez Canal, and in 1996 plans were unveiled to tunnel under the Canal to carry water into Sinai up to 40 km short of Gaza. Israel is reportedly still trying to get Nile water and may be negotiating with Sudan (Bleier, 2009).

Although no full-scale water wars have yet broken out, Egypt is reported to have trained élite army units in both desert and jungle warfare, the one in case of disagreement with Libya and Sahelian countries, the other for countries further south. There is still some concern over a possible conflict with Libya. Some Egyptian hydrogeologists fear that Colonel Gaddafi's 'Great Man-Made River' could reduce the Nile's discharge because the Nubian Sandstone aquifer system it exploits extends into the Nile basin and the scheme might draw off water from the river (see box 'Libya's Great Man-Made River' in Chapter 12), Egypt is also exploiting the system and recent assessments do suggest that particularly intense water extraction could lower water tables up to 70 km beyond the Egyptian-Sudanese border. Further extraction in the Siwa oasis near the Libyan border could also draw saline water into the aquifer, especially if coupled with more extraction from Jaghbub on the Libyan side of the frontier (Salem and Pallas, 2001).

There are, however, some encouraging developments in collaboration. The Nile Basin Initiative (NBI) was officially launched in 1999. Its projects include: the Nile Trans-boundary Environment Action Project; Water Resources Planning and Management Project to improve water efficiency in agriculture; and the Equatorial Lakes Subsidiary Action Project (www.nilebasin.org). In May, 2010, members of the NBI, with the notable exception of Egypt and Sudan, signed the Agreement on Nile River Basin Cooperative Framework. The Framework aims for fully equitable water rights for all the riparian countries in order to achieve sustainable socio-economic development. It is due for full ratification by May 2011. Egypt and Sudan are objecting to any revision of the 1959 agreement (Zahran, 2010).

India, Pakistan and Bangladesh

The partition of British India between India and Pakistan in 1948 left the Indus basin divided in a particularly convoluted fashion, and Pakistan divided between East (modern Bangladesh) and West. Disputed territory in Kashmir has continued to cause military conflict, with disputes over irrigation water exacerbating tensions and bringing the two countries to the brink of war. The Indus Waters Agreement was reached in 1960 after 12 years of negotiations led by the World Bank, but disputes still arise.

There has been less agreement in the East, where Bangladesh broke away from Pakistan in 1971. Bangladesh is a classic downstream user. Eighty per cent of the country is formed by the delta floodplains of the Ganges, Meghna and Brahmaputra Rivers, but less than ten per cent of the total catchment area of these rivers lies within Bangladeshi control (Figure 9.5). The rest is in the hands of India, China, Nepal and Bhutan. India has taken more and more from the Ganges and built scores of dams to regulate dry season flows to supply its rising population. India's diversion of the Ganges at the Farraka Barrage, just 18 km short of the Bangladesh frontier, has been a major source of contention since 1975. There was no long-term agreement to guarantee water to Bangladesh. The UN managed to broker the Ganges Waters Agreement in 1977, which guaranteed 60 per cent of Ganges discharge to Bangladesh at the height of the dry season, but India allowed it to lapse in 1988. Bangladeshi estimates suggest that crops worth £2 billion have been lost as a result. Only 260 m³/s entered Bangladesh in the Ganges in 1993 compared with nearly 2000 m³/s being diverted at Farakka through the Hugli to Kolkata (Calcutta), in part to flush silt out of the harbour as well as for water supply. By 1995, a third of the country was threatened with economic and environmental catastrophe because the pre-monsoon low flows had been reduced by 80 per cent. As a result of lower flows since Farakka, saltwater has spread up the distributaries of the delta, fishermen have lost their livelihoods and river salinity at Khulna has increased 58-fold. Because of less sediment from the Ganges, the seaward edges of the delta are eroding and the parts of the largest mangrove forests in the world are dying in the Sunderbans National Park. Perhaps one of the worst results is the lowering of water tables in the delta which has aggravated problems of arsenic poisoning (see the section headed *Arsenic in groundwater* in Chapter 5).

In contrast, India has maintained accords with Nepal, whereby Nepalese dams regulate flows to improve public supply its citizens in Northern India and generate hydropower for India. The World Bank did refuse to fund an extra Indian-sponsored hydropower scheme in Nepal because of the Ganges dispute, and Bangladesh successfully blocked an Indian plan to divert the Brahmaputra wholly through Indian territory to join the Ganges outside Bangladesh. These pressures did get India back to the negotiating table and in December 1996 it signed a 30-year agreement, which guarantees Bangladesh 50 per cent of the Ganges discharge during the critical March-

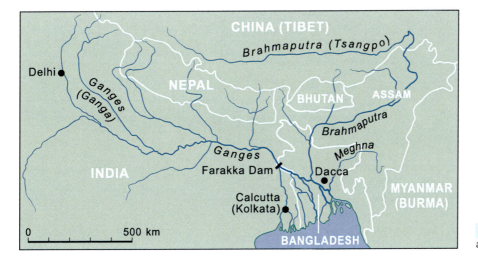

Figure 9.5 The Ganges, Brahmaputra and Bangladesh

May period, rising to 80 per cent in the event of a severe drought.

Bangladesh's main worry now could be China, which has plans to divert the Brahmaputra or Tsangpo from Tibet to feed the Yellow River (see section headed *China's National Strategic Plan* in Chapter 7).

The Mekong

The UN rightly regards the Mekong Accord signed between Laos, Cambodia, Thailand and Vietnam in 1994 as a major success. The UN set up the Committee for Coordination of Investigations of the Lower Mekong Basin in 1957, but for decades it failed to get

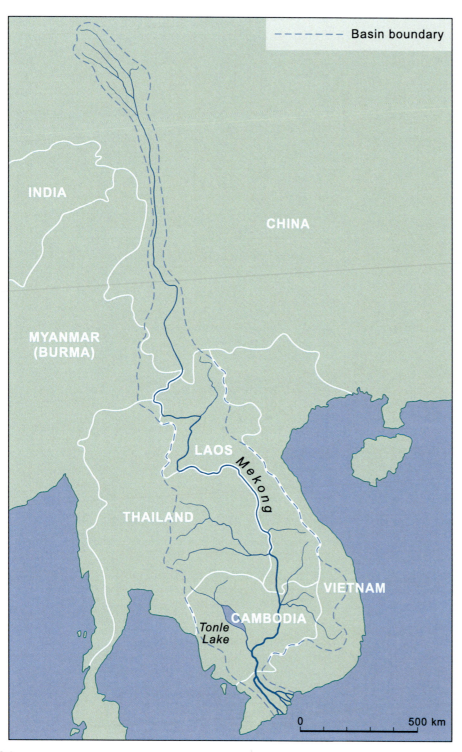

Figure 9.6 The Mekong

a multinational agreement. The Committee produced a plan in the 1960s for 100 dams to regulate the 30-fold range in seasonal discharges along the river and to serve irrigation, hydropower and navigational needs for all the riparian states. But conflict in Indo-China, especially the Vietnam War, delayed progress. Only the six main projects were completed in the 1960s and these mainly benefit Thailand, since this was the only peaceful state in the region. In the 1980s, although relative peace had returned except for the Khmer Rouge in Cambodia, the Committee began to bend to environmental pressures and favour small-scale and non-structural solutions over large dams. Particular concern was expressed then and now for the impact of engineering developments on the Tonlé Sap or the Great Lake of Cambodia. This is a unique water feature on a tributary of the Mekong that has acquired UNESCO World Heritage status. When the Mekong is in full flood during the monsoon season, the floodwaters cause the Tonlé Sap River to back up and create a 13,000 km² lake. It is vital to local villagers who fish the lake and, when the waters recede and the lake returns to cover just 2850 km², they grow rice in the new silt on the lake bed. It is also a wildlife haven.

It seems almost inevitable that dams and diversions will eventually kill this spectacle. Plans were revived in the 1990s for a series of major hydropower dams. Laos is actively developing hydropower for export. The dams will transform the wild river and reduce the annual floods and the silt that so much of the agriculture depends upon. Vietnam's economy is especially vulnerable to these developments, because 60 per cent of its agricultural produce comes from the Mekong Delta. But a third of the population of the four signatory nations lives within the basin and largely subsists on fishing and rice production.

However, the situation is going to deteriorate further and the seeds of the downfall were there at the signing of the original Accord. The headwater states of China and Burma (Myanmar) never signed. Now China plans to divert water to the Yangtze and Three Gorges Dam, as well as building perhaps hundreds of dams on the Mekong itself. It is working with the Burmese regime on building hydropower dams on their territory.

These plans will destroy the traditional resources of all the signatories and no one is going to oppose China. When flow in the Mekong reached a 50-year low in the first half of 2010, the immediate response was to blame China. China, of course, denies it is to blame, but it is only time until it is.

The Senegal River – international cooperation and ethnic violence

Cooperation between the three riparian states of the Senegal River basin – Senegal, Mali and Mauritania – was initiated in response to the Sahelian droughts of the early 1970s. The Organisation for the Development of the Senegal River (OMVS) was established in 1972 to coordinate water management, reduce economic vulnerability and accelerate economic development. It began with building two dams, the Manantali and Diama, to irrigate new farmland, generate 800 GWh a year and provide a navigational connection to the sea for Mali.

Unfortunately, it overlooked some crucial social aspects that resulted in ethnic violence. The boost given to farming along the river resulted in rising land values. A powerful elite in Mauritania acquired property rights and many villagers were forced to move and resettle among people who resented them. Moors and black Africans quarrelled. Two Senegalese peasant farmers were killed by Mauritanians in a quarrel over grazing rights in 1989 and ethnic violence broke out in both countries, leaving hundreds dead. Order was eventually restored by the Mauritanian and Senegalese armies, but sporadic outbreaks recurred till full diplomatic relations are resumed in 1991. Even so, the displaced population received no help or compensation.

The OMVS aims to ensuring equitable sharing of the costs and benefits of the water management programme, to strengthen community-based management units and to maintain an Environmental Impact Mitigation and Monitoring Programme. However, critics point to unresolved inequities in the sharing of costs and benefits at sub-national and community levels and a lack of coherence between the different water management units, partly due to historical evolution.

A water union in southern Africa

The Regional Strategic Water Infrastructure Development Programme of the Southern African Development Community (SADC) is perhaps the greatest success story of all. It is totally homegrown and did not involve any violence or war. The SADC had developed as a loose grouping of nine nations attempting to lessen their dependence on apartheid South Africa in 1980 and became a formal development community in 1992. With the end of the apartheid regime, the newly enfranchised South African government of Nelson Mandela brokered the regional water cooperative agreement between 12 nations of southern Africa, together with a hydropower agreement with the Democratic Republic of the Congo (called Zaire at the time) in 1995. The overriding aim was to mitigate the effects of the frequent droughts that had been gripping the region. The agreement established a new joint water administration for the region based on the principle of Equitable Utilization (see Chapter 20). Expertise and information are shared and plans developed for a water grid to divert water from the Zambezi to drier parts of the region. The Dar-es-Salaam Declaration on Agriculture and Food Security in the SADC Region accelerated the development programme for transboundary resources and management, including interbasin transfers, in 2004. Plans published in 2009 include doubling the area under irrigation by 2015 and halving the number of people without access to safe drinking water and sanitation. New dams are planned at Movene (Mozambique-Swaziland), the Kafue Gorge (Zambia-Zimbabwe) and Metolong (Lesotho). The Community now comprises 15 nations – South Africa, Mozambique, Swaziland, Zambia, Zimbabwe, Namibia, Botswana, Lesotho, Angola, Malawi, Tanzania, the Democratic Republic of the Congo, and the islands of Mauritius, the Seychelles and Madagascar.

The SADC plans to become a 'European Union' of southern Africa, progressing from Free Trade Area (2008) to Common Market by 2015, and a single currency by 2018. But there is one significant difference: water. The integration of water resources management goes well beyond the EU's Water Framework. The plans are offering a peaceful solution to a region with marked differences in national water resources.

'From Potential Conflict to Cooperation Potential' (PC-CP) facilitates multi-level and interdisciplinary dialogues in order to foster peace, co-operation and development related to the management of transboundary water resources.'

Mission Statement of the PC-CP programme, Unesco-Green Cross International (2001+)

Conclusions

War and terrorism can have a huge impact on water security. Water insecurity can also trigger conflicts. The issues will never be solved while inequality and lack of mutual understanding and cooperation exist. It is a tall order. In this, the work of Unesco and Green Cross International in their programme 'From Potential Conflict to Cooperation Potential' (PC-CP), which is helping countries manage their shared water resources in a peaceful and equitable manner, is a vital first step. And the efforts of the Southern African Development Community are developing into a model of peaceful cooperation.

Discussion topics

- Research the roots of terrorism and the threats it holds for water security.
- Follow the progress of any one of the regions of water conflict outlined in this chapter.
- Follow the development of SADC's water programme.

Further reading

The Unesco/Green Cross International PC-CP programme can be followed at: www.webworld.unesco.org.

The US National Research Council (2004) reviewed the EPA's technical support action plan and a number of reports to the US Congressional Research Service assess the terrorist threat to critical infrastructure (Shea, 2003; Copeland and Cody, 2005).

Hickman, D.C. 1999. *A Chemical and Biological Warfare Threat: USAF Water Systems at risk*. The Counterprolifera-

tion Papers Future Warfare Series No. 3, USAF Counterproliferation Center, Air War College, Air University, Maxwell Air Force Base, Alabama.

A general introduction to the various agencies overseeing international cooperation on the Rhine is provided by:

Frijters, I.D. and Leentvaar, J. 2003. 'Rhine case study.' UNESCO-IHP, PCCP Series Publication, 33pp. http://www.unesco.org/water/wwap/pccp/case_studies.shtml.

10 The threat of global warming

'Tackling climate change globally will demand political will and leadership, and strong stakeholder engagement. Adaptation to the changes expected is now a global priority. Improved monitoring is needed, and it is urgent to enhance our scientific understanding of the potential tipping points beyond which reversibility is not assured.'

'Global Environment Outlook' report (GEO-4),
UN Environment Programme, 2007

Global water resources are highly sensitive to temperatures, not only through the obvious effects on evaporation, but also because of the effects on wind patterns, convection and rainfall distribution. The evidence for global warming over the last two centuries is now overwhelming. There is also broad agreement among all the major Global Climate Models that the warming is likely to continue over the coming century, so long as the concentration of greenhouse gases continues to rise in the atmosphere. The Intergovernmental Panel on Climate Change stated in its Fourth Assessment (2007): 'Warming of the climate system is unequivocal, as is now evident from observations of increases in global average air and ocean temperatures, widespread melting of snow and ice, and rising global average sea level.'

All three major data sets of global temperatures – Nasa's Goddard Institute for Space Studies (GISS), the American National Oceanic and Atmospheric Administration (NOAA) through its National Climatic Data Center (NCDC) and the Climatic Research Unit (CRU) at the University of East Anglia, which processes data from the British Met Office – agree that average temperatures have risen since the early nineteenth century, despite numerous fluctuations and a levelling off in recent years. The events surrounding 'Climategate', when hackers published private emails from the director of CRU after the Copenhagen climate summit in December 2009, which appeared to indicate a conspiracy to exaggerate the warming trend, led to a number of valuable in-depth reviews of the data by scientific institutions and expert assessors. One of the principal sources of bias in the records that all analyses and re-analyses have been at

pains to eliminate is changes in land-use and urbanization. This is a difficult issue. Most met stations are near cities that have grown considerably over the last two centuries, many are also at airports where jet engines might affect temperatures. However, data centres like that at the UK Met Office collect around 1.5 million observations a month not only from some 2000 land stations around the world, but also at sea from over 1000 drifting and moored meteorological buoys and the 4000-strong Voluntary Observing Ship programme. Satellite data are also used.

As part of the drive to improve transparency, all three international databases are now available online. The WMO also made the data available and the UK Met Office released the computer code used in the analysis into the public domain and published a detailed re-analysis (Stott *et al*, 2010). The Met Office analysis uses a procedure called 'optimal detection', whereby their Global Circulation Model (GCM) is run with different natural and human causes of climate change – including solar radiation, El Niño, volcanic eruptions, and greenhouse gas, dust and soot emissions – and the predicted warming trend is then compared with the station records. The evidence included Arctic sea ice records and global precipitation patterns as well as sea and air temperatures. Arctic sea ice cover is now 40 per cent less than in 1972 and winter snowcover in the northern hemisphere is also slightly less. The leader of the Met Office review, Peter Stott, concludes: 'The wealth of evidence shows that there is an increasingly remote possibility that climate change is being dominated by natural factors rather than human factors'. Independent analysis of the data

197

did reveal a few errors in the original Met Office data, but not enough to alter the overall trend. One key finding in the Met Office reassessment is that rainfall patterns appear to be changing more quickly than expected, implying that future changes may be greater than current GCMs are suggesting (Table 10.1). The reasons are as yet unknown.

Table 10.1 Key results from the 2010 UK Met Office reassessment of evidence for climate change

Feature	Trends
Temperature	Global average temperatures increased by about 0.75°C over the twentieth century: 2000–2009 was the warmest decade on record. Human influence detected on every continent.
Rainfall	Wetter regions generally getting wetter (mid- and high latitudes in N Hemisphere, tropical regions). Drier regions getting drier. Some evidence that rainfall patterns are changing faster than expected.
Humidity	Atmospheric moisture increased over last 20–30 years, according to ground and satellite observations. This increases risk of extreme rainfall events and floods.
Ocean temperature	Warming over last 50 years in Atlantic, Pacific and Indian Oceans. Cannot be attributed to variations in volcanic, solar or ocean current activity.
Ocean salinity	The Atlantic is becoming saltier in the sub-tropics due to increased evaporation. High latitude ocean regions expected to become less salty in long term due to increased ice melt and rainfall.
Sea ice	Arctic Ocean summer minimum declining by 600,000 km²/decade. Can only be explained by human influences. Antarctic – small overall increase since satellite record began in 1978, consistent with increased greenhouse gas and reduced ozone layer. This can explain increases, e.g. in the Ross Sea, and also some decrease, e.g. in Amundsen-Bellingshausen Sea.

Source: Stott *et al* (2010) and www.metoffice.gov.uk/

Although not differentiating between natural or human causes, satellite data analysed by Nasa also confirm that the first decade of this century was the warmest on record, and that the rate of warming has increased during the last 30 years, with a rise of 0.2°C/decade in mean global temperature compared with an overall rise of 0.8°C between 1880 and 2010. Satellite measurements are capable of providing a more complete and unbiased sample of surface temperatures than individual met stations. All the top ten warmest years in the last century and a half occurred during the last two decades: such a concentration of warm years is statistically most unlikely to be random. The European Centre for Medium Range Forecasting also issued a new set of records covering the last 30 years incorporating satellite data, which confirm the warming trend.

'With more than 450 lead authors, 800 contributing authors, and 2,500 reviewers from more than 130 countries, the IPCC 4th Assessment Report represents the most comprehensive international scientific assessment ever conducted … in proportion to the sheer volume of research reviewed and analysed, these lapses of accuracy are minor and they in no way undermine the main conclusions … we should continue to be critical but constructively so and in ways that openly recognize the strengths and limitations of the scientific process itself.'

Statement by ICSU on the controversy around the 4th IPCC Assessment
International Council for Science (ICSU), 23 February 2010

Causes of climate change

Part of the temperature rise *is* due to natural causes. In the early 1800s, the world saw the final throws of the Little Ice Age (LIA), which had caused varying degrees of cold since the fifteenth century. The River Thames frequently froze over in London and 'frost fairs' were held on the ice, the last in 1814. However, although these events are often taken as climatic indicators, they were only partly due to the climate. River engineering played a part. The narrow arches of the old London Bridge obstructed flow and the river never froze up after the new bridge was built in 1831 and the new city embankments speeded up the flow.

The last big cold snap in the 1810s coincided with Napoleon's strategic underestimate of the severity of the Russian winter in 1812 and the wet weather that bogged his cavalry down at Waterloo in 1815. Since then, the world has been climbing out of the Little Ice Age. The exact origins of the LIA are unknown, but it was probably partly due to reduced solar activity and partly to processes on Earth, especially volcanic eruptions.

Solar activity

Sunspot activity was exceptionally low during the Maunder Minimum (1645–1715) and dipped again in the Dalton Minimum (c. 1810s–1840s). Incoming solar radiation, i.e. visible sunlight, UV and infrared (part of the wide spectrum of 'electromagnetic radiation'), could have been a few tenths of a Watt per square metre less during the Maunder Minimum than now (Houghton, 2009). Solar activity has gradually increased since then. Changes in the amount of ultraviolet light reaching the Earth's atmosphere create more or less ozone in the stratosphere. More stratospheric ozone traps heat being lost from the surface of the Earth and reradiates heat back to the surface. It also affects circulation in the stratosphere, whence there is some evidence of links to the troposphere, our weather zone. Scientists have suggested that the ozone hole over Antarctica during the 1980s and 1990s may have cooled the continent and masked the effects of global warming there until the recent recovery in ozone levels following the phasing out of the destructive manmade CFC gases under the provisions of the 1987 Montreal Protocol (Scientific Committee on Antarctic Research, 2009). The Committee's report predicts that gradual repair of the hole over the next 50 years could raise temperatures on the continent by 3°C, although it points out that this is unlikely to have much effect on the centre of the ice sheet as a rise from −55°C to −50°C will not have much effect on melting.

It is not simply a matter of minute changes in sunlight. Sunspots are accompanied by solar flares, which are the main source of the solar wind – a stream of charged particles that cause the Northern Lights when they strike and heat up the ionosphere. The strength of the solar wind could affect the weather via a range of mechanisms, as yet not all fully understood. Gusts in the solar wind affect the electrical properties of the upper atmosphere and the effects may somehow be transmitted to the lower atmosphere. But perhaps most important is that when the wind is weakest, the Sun's magnetic field, which shields the Earth from cosmic radiation from the galaxy, is also weakest and this allows more galactic cosmic radiation to penetrate the atmosphere and may cause more low-altitude clouds to form, which reflect more solar radiation and so cool the Earth's surface (see

Clouds and rain-forming processes). When the solar wind weakens, the ionosphere is cooler and it shrinks, as in the past decade. Incoming cosmic rays alter the electrical conductivity of the atmosphere, with potential implications for lightning strikes. The rays may also cause more nitrogen oxides to form in the atmosphere, which affects the radiation balance of the Earth.

Some of the falling off in the warming trend during the 2000s (which caused concern and bolstered climate change deniers prior to the Copenhagen Climate Conference) might be explained by the period of quiet Sun since 2002 that reached a 100-year low in 2007: the solar wind reached a 50-year low and radio emissions a 55-year low. There were months without any sunspots and 'recovery' was unusually delayed till beyond the end of the decade.

The verdict of the Intergovernmental Panel on Climate Change (IPCC) 2007 report is that there is now more evidence for solar forcing of climate change, but that the role of galactic cosmic radiation (GCR) remains in dispute. However, science is advancing rapidly and the CLOUD Project at the European Centre for Nuclear Research (CERN) is reporting fascinating results from cloud chamber experiments and from analyses of new data on global cloud cover from Nasa's CERES satellite (Svensmark *et al*, 2007, 2009; Kirkby, 2008, 2009). Svensmark found a correlation between annual cloud cover and levels of galactic cosmic radiation. The CERN team find that higher levels of GCR are associated with a cooler climate, fluctuations in precipitation at high latitudes, a southward shift in the Intertropical Convergence Zone (ITCZ) – the junction between the southern and northern hemisphere circulations (Figure 11.5), which is responsible for some monsoons, most notably in West Africa – and weaker monsoons in general in SE Asia and Africa. They speculate that the mechanism may involve:

1 ion-mediated nucleation encouraging cloud-formation – brighter clouds and longer-lasting clouds reflecting more sunlight back to space (see *Clouds and rain-forming processes* in Chapter 11)

2 producing highly charged aerosols and cloud droplets, cloud condensation nuclei or ice particles

3 converting gases like sulphur dioxide and nitrogen oxides into aerosol particles that scatter and absorb electromagnetic radiation, or

4 through indirect effects on the global electrical circuit.

Long-term, polar ice has increased during periods of high GCR, like the LIA, although the current retreat of arctic ice does not seem to fit this pattern (see Chapter 12 Shrinking freshwater stores).

Kirkby (2009) even sees an effect on sea level rise. The rate of change is lowest during low sunspot periods, varying roughly between +7 and −3 mm/yr over the cycles, which can be related to changes in the rates of thermal expansion of the ocean and melting of land ice. The problem for research is that the range or 'modulation' in sea level rise is much greater than could be due to changes in incoming solar radiation, which points to one or more of the less visible and currently speculative processes being in control.

Volcanic activity

Volcanic eruptions emit carbon dioxide that adds to the greenhouse effect and sulphur dioxide (SO_2) and dust particles that cool the atmosphere. The carbon dioxide has a longer-term effect. The sulphur dioxide and dust raise the planet's albedo, but for a shorter period; the dust may reflect more sunlight back to space, it may also act as nuclei for cloud formation, as does the sulphur dioxide, which converts to sulphate particles under sunlight (see *Clouds and rain-forming processes* in Chapter 11). Thus, colder periods can be associated with a concentration of volcanic eruptions over a few years, but a long-term warming trend may result from an extended period of eruptions, maybe over millions of years, as postulated for the end of 'snowball Earth' 650 million years ago (see Chapter 11 The restless water cycle). In contrast, several powerful volcanic eruptions between 1638 and 1643 helped cooling at the beginning of the Maunder Minimum, the period of quiet Sun, and contributed to poor harvests and political turmoil across Europe during the 1640s. The fact that European harvests recovered after 1660, when there were fewer eruptions, and remained good for 30 years despite the low solar activity, gives a measure of the importance of the eruptions.

It is not just a question of the frequency of eruptions, but also of size and type of eruption, the height at which material is injected into the atmosphere and the relative composition of gases and ash. Many individual volcanoes in the 'Ring of Fire' around the northern edge of the Pacific have produced eruptions large enough on their own to have worldwide effects on temperatures and precipitation. The final cold snap of the LIA, during the Dalton Minimum, was augmented by the single, huge eruption of Tambora in Indonesia in 1815, that was followed by two exceptionally cold years around the world: in Canada and New England 1816 became known as 'the year without a summer'. Krakatoa (Indonesia, 1883), El Chichón (Mexico, 1982) and Pinatubo (Philippines, 1991) all affected temperature and rainfall as far away as the UK. Pinatubo put more SO_2 into the stratosphere than any eruption since Krakatoa and UK temperatures dipped 0.5°C below the 1982–2006 average in 1993. El Chichón probably played a part in the 1°C dip in UK temperatures in 1985–6. One reason why the 1980 Mount St Helens eruption in Washington State had little effect outside North America was that it largely erupted sideways, devastating the local area, but not injecting much into the stratosphere. Once material is injected into the stratosphere it can circumnavigate the globe in a week or two and linger there for far longer before it finally returns to earth. Dust from El Chichón was still visible in the stratosphere over the UK some five years after the eruption (Thomas *et al*, 1987).

Even without an exceptional eruption or concentration of eruptions, volcanoes are currently responsible for around 25 per cent of all the SO_2 in the atmosphere. This means that it is not only an important player in the radiation and heat balance of the Earth, it is also a significant element in acid rain.

The eruptions from the Icelandic volcano beneath Eyjafjallajökull in 2010 raised the possibility of yet another cooling mechanism. An international expedition organized by the National Oceanography Centre, Southampton, was launched to investigate whether iron in the fallout might feed phytoplankton growth, which would absorb more CO_2 (see *The biosphere* below). The region of the North Atlantic south of Iceland and east of Greenland is important for deep water formation, which could take some of this extra phytoplankton down with it (see *The ocean* below). If true, this might provide one of the longer term influences of volcanoes on climate.

Human activities

The burning of fossil fuels since the industrial revolution is widely blamed for the current global warming trend. But there are other human causes. Perhaps the largest is deforestation. One calculation suggests that this was the main human factor up to the mid-twentieth century, linked particularly with western colonization of the New World. Its effect is twofold: where the ancient forests are used for firewood or burnt to clear land for farming, carbon that has been sequestered over decades or centuries is suddenly released into the air, and once the trees are dead they no longer act as a sink for atmospheric carbon.

The other human causes are largely related to population growth – our sheer numbers, our need for food and our concentration in cities. The effects of urbanization and the 'urban heat island' are well known: the concentrated release of heat from homes, offices, traffic and factories, high rise buildings that intercept sunlight and reflect it into the 'urban canyons' of streets, and air pollution (Barry and Chorley, 2010). A much smaller amount of heat comes from human beings themselves – each individual must dissipate metabolic heat of the order of 100W, so 6.6 billion people are releasing 660 billion W into the environment. But this is still microscopic compared with the 341×10^{15} W reaching the Earth's surface from the Sun. While the vast expansion of cities in recent decades has undoubtedly affected *actual* mean air temperatures, and this is exaggerated by the fact that meteorological stations are overwhelmingly concentrated in and around cities, these effects are thought to have been largely eliminated from the main international climatic data analyses (IPCC, 2007).

Far more significant is the effect of agriculture through irrigation and animals, both of which can release methane as well as carbon dioxide and water vapour. Further expansion of irrigated agriculture and increasing meat consumption in the large emerging economies could have a major effect on the warming trend. Irrigation must increase, at least in the shorter term, in order to feed the world. To what extent this follows traditional water-profligate practices or not, and whether it feeds water-hungry varieties of crop or not, remain unknown. Similar questions hang over the expansion of reservoirs. The World Commission on Dams (2000) calculated that reservoirs account for up to 28 per cent of all greenhouse gas emissions. This severely dents the conventional view that hydropower is necessarily a totally 'green' source of power. However, the Commission did admit to wide variations in emissions from reservoirs. In practice, there is a dearth of data on reservoir emissions and an urgent need for coverage of different climatic environments.

Methane and nitrous oxide

Methane (CH_4) is a powerful greenhouse gas: although much rarer than CO_2, it is up to 20 times more effective molecule for molecule. It is the main component of natural gas and arises from the decomposition of vegetable matter in wet conditions. It is currently responsible for a little under a fifth of the warming effect of all the non-water gases, or around a third that of CO_2. Its lifespan in the atmosphere is only a tenth that of CO_2, but emissions are increasing. Over the last 2000 years, its concentration has more than doubled. Human activity, especially irrigation, must be a major factor. Rice paddy fields are estimated to produce between 30 and 90 tonnes a year. Methane emissions were increasing at about 10 ppb a year in the 1980s, but fell back to 5 ppb in the 1990s and showed little increase during the first five years of this century (Houghton, 2009). However, it began rising again in the late 2000s, so that by 2008 it reached 1800 ppb and measurements from the arctic in 2010 indicate 1850 ppb. This suggests that the rate has returned to near that of the 1980s. The reasons for these fluctuations are unclear. The range of future scenarios adopted in the IPCC Special Report on Emissions Scenarios (SRES) (2000) showed a great range of possible trends from a reduction of 25 per cent to another doubling over the twenty-first century.

This great uncertainty is even worse when the future of methane hydrates is considered. These are peculiar compounds that have a shell of water around a methane core, known chemically as clathrates. There are huge quantities frozen in the permafrost at high latitudes: they are really an unusual form of ice and melt at 0°C. This means they are unstable immediately the frozen ground melts, and there are concerns that this could lead to a strong positive feedback to global warming (Lipkowski, 2006). Destabilization of methane hydrates is believed to have been a principal mechanism of global warming at the end of the last ice age. Similar hydrates are known to exist on the ocean bed. These are probably too deep to be readily susceptible to ocean warming, but the potential remains, especially for hydrates buried in shallower sediments in the arctic.

Nitrous oxide is another greenhouse gas on the increase. It accounts for about six per cent of the effects of the non-water gases and has increased by about 15 per cent since the eighteenth century. Unlike methane, it is long-lived (about 120 years) and over its lifetime its effects are 300 times greater than CO_2. Most appears to come from the ocean. The main human source is nitrate fertilizers and this appears to have been underestimated until recently. It has another role: it is the main destroyer of the greenhouse gas ozone. So once again, there is a lack of data combined with ambivalent effects, which complicates modelling. But it appears to fluctuate less than methane.

Aerosols and global dimming

Soot, smoke and sulphur dioxide gas from fossil fuels, smoke from cooking and wildfires, deforestation and stubble burning, and dust from agriculture – principally ploughing in semi-arid regions – have all affected the radiation balance of the atmosphere over the last century. The effects are complicated. Broadly speaking, until very recently they have been considered to have a predominantly 'anti-greenhouse' effect, contributing to what has been termed 'global dimming'. Any light-coloured material, like clay particles or smoke from forest fires, tends to reflect sunlight back to space. When sunlight acts on sulphur dioxide gas it creates solid particles that act in a similar way. During the latter part of the twentieth century, over half of all SO_2 emissions were man-made. Dimming dominated between 1950 and 1980, and contributed to the post-war dip in the warming trend. Maps of the effect of sulphate particles show a net cooling over large areas downwind of the main urban-industrial regions, especially Europe, Japan and the eastern USA of the order of 2–6 W/m². The pattern is now changing. Less SO_2 is now emitted from the old industrial centres as a result of clean air legislation, but more is being emitted from the developing world. China and the Himalayas are now among the worst affected. In contrast, Europe has experienced the most 'brightening'. Overall, global sulphur emissions have reduced from over 18 megatonnes (Mt) a year in 1980 to just 4 Mt in 2002.

This dramatic reduction in SO_2 emissions has not only added to global warming, but it will also have had an effect on clouds and rainfall, because sulphate particles are good condensation nuclei and dissolve in cloud droplets to form acid rain (see *Acidification of surface waters* in Chapter 6 and *Clouds and rain-forming processes* in Chapter 11). By stimulating cloud-formation and creating denser clouds, SO_2 therefore has a double cooling effect, assuming most of the clouds affected are low cloud. By contrast, dark aerosols like soot tend to absorb radiation, warming the atmosphere.

Recent research has complicated matters further. So-called 'brown clouds' have been identified that are capable of travelling thousands of kilometres, carrying aerosols from America to Europe, Europe to Asia and China to the USA (Haag, 2007). Ramanathan *et al* (2007) used unmanned drones to measure the effects of brown clouds over India. They calculate that soot aerosols in these clouds have added to the effects of greenhouse gases over the Himalayas in roughly equal amounts. Between them they have raised local air temperatures by 0.25°/decade since 1950, double the global average, and been sufficient to account for the recent melting of the Himalayan glaciers. Smoke from cooking fires in Central Asia appears to be a major source of the aerosols in the brown clouds that cross the Himalayas in winter and spring. Factories and dung-heap fires add to brown clouds over southern Asia from Tibet and Pakistan to Vietnam and Beijing. Unlike greenhouse gases that last hundreds of years and readily mix more or less evenly around the globe, the effects of black carbon aerosols are relatively localized because they remain airborne for just weeks or months. The effects may also depend on the type of carbon. Hansen *et al* (2005) report that soot aerosols from fossil fuels have a warming effect, whereas organic carbon from biomass burning has a cooling effect.

Further complications are emerging from research on air pollution and global dimming that are potentially important for water resources. Mercado *et al* (2009) conclude that increased air pollution enhanced plant productivity by up to a quarter between 1960 and the millennium, and led to an extra net ten per cent of carbon being stored in the soil. It has been assumed that plants grow best in sun, but this research shows this is not always the case – forests and crops can thrive in hazy or 'diffuse' sunshine because scattered light bathes more leaves. Plants absorb less carbon in clean air. So aerosols and low clouds work together to reduce global warming through the biosphere as well as the atmosphere. Does the enhanced photosynthesis, however, mean greater water use?

Modelling and predicting future climate change

Global Climate Models or General Circulation Models (GCMs) are extensions of daily weather forecasting models, but they are more complicated. Where short-term weather forecasting is basically an exercise in transferring and transforming a body of air over a few hours or days based solely on its internal properties, predicting changes for longer periods involves complex exchanges and feedbacks between the air and the surface. They have been developing for two decades and are now highly sophisticated. But there are still significant gaps in scientific knowledge and limitations as to what can be modelled. Computer capacity has expanded dramatically, but it is still impractical to model most convective storm systems within a global scale model. The British HadCM3 GCM has a spatial resolution of 2.5° latitude by 3.75° longitude, equal to around 295 by 278 km at a latitude of 45°, which is far too coarse. There is voluminous literature on the question of 'downscaling', that is, methods for interpolating climatic values for river basins that are smaller than the grid square resolution of GCMs. These include statistical methods in which local meteorological measurements are correlated with synoptic-scale weather patterns during present observations and this statistical link is then used to downscale from the GCM synoptic patterns (Pilling *et al*, 1998; Wilby *et al*, 2002; Jones *et al*, 2007). The problem may also be addressed by 'nesting' a Regional Climate Model (RCM) or a Limited Area Model (LAM) within part of the global model. These may offer grid cells of 50 km square or less. These models allow better simulation of convective storms and also the effects of mountains: HadCM3, for example, cannot model the effects of the British mountains on rainfall.

HadCM3 does, however, include important elements that were lacking in earlier models: a sophisticated model of ocean circulation and an initial attempt to model biospheric interactions. The fourth generation Hadley Centre model, HadCM4, with further improved modelling of the ocean and biosphere, is due to be released shortly.

Further uncertainty is introduced by our lack of knowledge as to how population will grow, what greenhouse emissions will be like and how effective the adoption of carbon-reducing technologies might be. For this reason, the IPCC devised a set of 'scenarios' of possible change, ranging from business-as-usual to full adoption of effective carbon reductions. These are the Special Report on Emissions Scenarios (SRES) scenarios (Nakicenovic and Swart, 2000). The results of GCM simulations are not forecasts like daily weather forecasts. They are simulations of what might be if social, economic and technical developments were to follow the storyline of one of these SRES scenarios. Although the latest GCMs model the gradual progression of increasing greenhouse gases and produce daily values, again these are not real forecasts of daily changes. Rather, they provide samples that allow a statistical summary to be made for selected decades.

This is extremely valuable for estimating the risks of extreme events. Finally, no results are published without strict quality control that involves taking the average of a large number of simulation runs. Summaries, like those produced by the IPCC, often use averages of a number of different GCMs, such as HadCM3, ECHAM5 (Germany), NCAR-DoE (USA) and CSIRO (Australia). See *Further Reading* for introductions to GCMs.

We will concentrate on just two aspects: the role of water in climate change and the impacts of climate change on water.

The role of the hydrological cycle in global warming

Water vapour is the most important greenhouse gas (GHG) of all. On average it accounts for about half of greenhouse effect, maintaining air temperature at a liveable level: without it we would freeze to death. However, the actual effect of water in the atmosphere is far from easy to model. So long as the water remains as vapour, it acts as a greenhouse gas. Once it condenses or forms ice crystals, its radiative effects become more complicated, even reversed.

Even the so-called 'radiative forcing' due to the vapour can be difficult to assess because many of the wavelengths of infrared radiation that water vapour intercepts are shared with other greenhouse gases, especially CO_2. So the radiative effect of water vapour alters according to the amount of other gases present in the atmosphere that share the same wavelengths. In combination with other GHGs, the contribution might fall to around a third; more alone it could rise to nearly three-quarters.

The second issue is the amount of water in the atmosphere. The overall amount of water vapour is likely to increase as the climate warms up. This is partly due to an increase in evaporation and partly to the increased water-holding capacity of the warmer air. A three per cent increase in the water vapour content of the air could add 0.5°C to greenhouse warming. Ultimately, however, the water-holding capacity provides a limit, the saturation point when the air is holding the maximum amount of vapour possible at a given temperature (determined by the Clausius-Clapeyron relationship between capacity and temperature). Beyond this point, extra vapour is forced to condense, or at sub-zero temperatures to sublimate as ice crystals.

There is some controversy over exactly how water vapour concentrations will change as the atmosphere warms up. The normal assumption in GCMs is that relative humidity (RH) will remain constant. (A relative humidity of 100 per cent means the air is saturated.) However, Minschwaner and Dessler (2004) produced evidence to suggest that RH will not increase as much as previously assumed. This is because warmer surfaces will not only increase evaporation but also increase convection, which will carry some of the extra moisture with it higher into the atmosphere, where the air is colder. There the vapour will 'freeze-dry', forming ice clouds and reducing vapour. They modelled their hypothesis and found that data from NASA's Upper Atmosphere Research Satellite confirmed this pattern. They argue that GCMs are overestimating the warming effects. However, as we shall now see, ice crystals themselves also trap radiation from the Earth.

Clouds are one of the most complicated parts of the system. Their role in the radiation budget depends on their altitude and composition as well as their amount. In general, high clouds (above 7 km) like *cirrus* are thin and predominantly composed of ice crystals. These tend to let sunlight through but are very efficient at blocking outgoing heat loss from the Earth. Their net effect is to add to global warming. They are a significant factor as they currently cover about a fifth of the globe daily. Air pollution, especially from high-flying aircraft, can stimulate their growth. This is an important part of air travel's effects on global warming – it is not simply a question of carbon emissions. Low clouds (below 2 km) tend to be thicker and liquid and reflect more sunlight back to space. The resulting increase in daytime albedo is quantitatively more important than the amount of heat loss they block, even though they raise night-time temperatures. Thus, an increase of just three per cent in average global low cloud cover could negate a 2°C forcing by greenhouse gases. Until the introduction of satellites in the Clouds and the Earth's Radiative Energy System (CERES) from 1997 onwards, this small difference was close to the error margin for surveillance (see *Satellites* in Chapter 18).

Researchers from CalTech and the New Jersey Institute of Technology have used these satellite cloud cover data in conjunction with measurements of 'earthshine' and proved the close relationship between the Earth's albedo and cloudiness. The research shows that the Earth became less cloudy between 1984 and 2000, rather more cloudy 2001–2003, and then reverted to pre-1995 levels. These data are helping refine climate models.

Early computer experiments by the UK Met Office also demonstrated the importance of internal cloud physics. Making different assumptions about ice content caused a 2°C difference in mean global temperatures, ranging between 2.7 and 4.5°C (Mitchell, 1991). Although ice crystals are more efficient infrared absorbers, they tend to fall out of clouds that have weak updraft currents – like *cirrus* – faster than water droplets. This interrupts the Bergeron precipitation-forming process (see *Clouds and rain-forming processes* in Chapter 11) and can leave a more liquid cloud (Rowntree, 1990). It is still difficult to get the necessary observational data on ice crystals in clouds on a global scale. Radiative properties are affected by the shape and size as well as the number or density of crystals. A recent technological breakthrough that addresses these needs – an optical scattering instrument designed to fit on aircraft – is described by Kaye *et al* (2008), but global coverage by aircraft is clearly limited.

What is apparent from satellite coverage, however, is that *cirrus* clouds themselves have been increasing in the last two decades at rates of around two to three per cent on land and four to six per cent over the ocean, linked primarily with air traffic (e.g. Stubenrauch and Schumann, 2005). A field experiment by the UK Met Office over the North Sea in 2009 clearly demonstrated how contrails (condensation trails) from aircraft can expand into broad cloud cover. This is part of the argument blaming air travel for contributing to global warming, although as Stubenrauch and Schumann point out, only five to ten per cent of situations favour *cirrus* formation.

The concentration of water vapour, that is the humidity of the air, varies widely over space and time. Whereas carbon dioxide molecules can stay in the atmosphere for more than a century and so become well-mixed around the globe, the average water molecule is resident in the air for less than ten days. So on average the point of deposition, predominantly as rain, is not so distant from the point of evaporation. This is demonstrated in the afternoon thunderstorms of the Amazon rainforest. It is why deforestation in the Amazon basin is reducing rainfall – perhaps by 20 per cent in places over the last decade, although this awaits scientific confirmation. The answer to the oft-maligned hypothesis that forests make rain is 'yes, large forests do' locally, but all forests do in the long run. A similar argument has been advanced for the grand, but quite impractical, proposed scheme to resuscitate the fast diminishing Lake Chad by importing water from the Congo to help irrigate agriculture: that the water lost by evaporation would be returned to the basin and maintain the lake. However, the authors of Unesco's atlas of the world water balance (1978) calculated that water evaporated from the planned increase in irrigation in the Ukraine would eventually be returned in the Central Asian Republics, at that time all within the Soviet Union, but that Australia is too small to benefit from returns from its irrigation. If such calculations are correct, then we must point to differences in weather processes in different parts of the world. For example, water vapour is likely to move shorter distances in a climate dominated by rainfall generated by vertical convection than where rain comes mainly from large, frontal storms such as mid-latitude depressions.

The exact balance between these often opposing factors is extremely difficult to compute. It is clear that a warmer world will mean more evaporation and more water in the atmosphere. The warmer surfaces will also increase convective activity, which will cause the water vapour to rise higher, condense and create clouds. Water vapour reaching higher altitudes will also freeze and sublimate, so a warmer atmosphere could have more ice, not less. These factors seem likely to increase the rainfall intensities. To what extent this cloud cover cools the Earth remains a moot point.

The ocean

Oceanic processes are extremely important in regulating climate. The great advance in the current generation of GCMs has been to couple atmospheric models with oceanic models and to extend modelling of ocean circulation to the deep ocean. HadCM3 models 20 levels in the ocean to a depth of 5 km.

The ocean is not only the main source of atmospheric moisture (Figure 11.3), it is also a major storer and transporter of heat. In the traditional view, the ocean delays climate change by storing large amounts of heat and releasing it only gradually. We now know that it is more complicated and that certain elements of the ocean react quite quickly to atmospheric change.

Until recently ocean transport was thought to account for up to one-third of all poleward heat transfer, but oceanographic evidence now suggests the proportion may be slightly more than half (Bryden *et al*, 1991; Chahine, 1992). This is because water has a higher heat capacity than air and the ocean covers most of the tropical Earth, and also because warm ocean currents are free

to transport heat out of the tropics everywhere except the northern Indian Ocean. These ocean currents affect regional patterns of evaporation and rainfall (see Chapter 11 The restless water cycle).

Current interest is focusing on two specific aspects of ocean circulation that have only come to light in recent decades: the thermohaline circulation and El Niño. These operate at opposite ends of the time spectrum. Whereas the thermohaline circulation is continuous and takes centuries to complete, El Niño lasts less than a year and recurs roughly every five years (see *Oscillations and teleconnections* in Chapter 19). Both are being incorporated in GCMs.

As its name implies, the thermohaline circulation is driven by a combination of differences or 'gradients' in heat and salt (Broecker, 1989). The Gulf Stream and its extension the North Atlantic Drift are part of it and are responsible for the milder and wetter climates in Western Europe. The Stream is driven by the thermal gradient in the ocean between the Caribbean and Greenland. This has been known for some time. But it now appears that it is also 'sucked' along by water sinking south of Greenland. By that time the excessive amount of evaporation has left it very salty (Figure 11.16). It has also become cold, and the cold, dense water sinks. Similar sinking points occur off Antarctica. One off the end of the Mertz Glacier in East Antarctica produces 20 per cent of all the world's 'bottom water'. There were fears in 2010 that a newly calved iceberg the size of Luxembourg (2550 km²) might settle over this area and block bottom water formation for decades. This would reduce oxygen supply to life in the ocean deep as well as disrupt the thermohaline circulation.

There are longer-term fears as well. Increased meltwaters from the polar icecaps and higher river discharges entering the Arctic Ocean caused by increased rainfall in Canada and Russia will dilute the salty water. The freshwater will also float over the surface of the salty water because it is less dense. This will insulate the saltier water from evaporation. The net effect will be to reduce the formation of extra salty bottom water and slow down the North Atlantic Drift and perhaps divert it slightly southwards. In response to fears that this might mean that global warming will cause Europe to get colder, the Hadley Centre simulated this in their GCM (Gregory *et al*, 2005). The conclusion was that it will not be strong enough to halt warming in Europe, but it will reduce the rate of warming a little.

Direct observations in the North Atlantic reveal the complexity of the processes. In general, the surface water in the North Atlantic has grown saltier over the last 50 years, which would be expected from increased evaporation and is matched by warmer sea surface temperatures. This should promote the thermohaline circulation, but there have been marked fluctuations, notably in the mid-1960s when increased precipitation, melting land ice and icebergs escaping through the Fram Strait sent a freshwater pulse into the Atlantic. Studies by Bryden *et al* (2005) have suggested that the 'meridional overturning circulation' slowed by a third in recent decades compared with the 1950s, although the Gulf Stream remains fairly constant. The deep water also seems less salty now. However, the most recent evidence suggests that rapid fluctuations are normal and that the limited data available so far is insufficient to infer any trend (Cunningham *et al*, 2007). Observations since 2004 have shown fluxes in the deep water varying from 35 million tonnes per second to just four million (see Chapter 18 Improved monitoring and data management).

The importance of El Niño is also becoming more apparent. This is a leading example of strong coupling between the ocean and the atmosphere. Incorporating it into recent GCMs and medium-term weather models has improved predictions. There is also some evidence that El Niño events have been stronger in recent decades. They appear to be both a giver and a taker in climate change.

El Niño peaks around the end of the year, towards Christmas, hence the Spanish name meaning 'little boy' (see *Oscillations and teleconnections* in Chapter 19). Its name is Spanish because the first effects were noticed on the west coast of Central and South America, where every few years the waters warm up and cause catastrophic decimation of the valuable anchovy fisheries. It also creates heavy rains in the Americas and drought in Indonesia and India. This is created by a reversal of the surface current in the equatorial Pacific bringing warm water to the Americas. It is linked to a similar reversal in the atmosphere, the Southern Oscillation, which contributes to it through frictional contact with the ocean surface. The atmospheric process begins with a weakening of the easterly Trade Winds that normally push warm water westwards across the equatorial Pacific and encourage the cold Humboldt Current to spread further up the west coast of South America from the Antarctic and its nutrient-laden waters to upwell off Ecuador.

This is accompanied by an eastward movement of a low pressure area from its 'normal' position in the eastern Indian Ocean towards the eastern Pacific, which carries the warm water with it. This warm water spreads out from the largest 'warm pool' in the world in the western Pacific.

Satellite observations have revealed a second mechanism at work: wave motions, not ocean waves in the common sense, but much longer waves – some are thousands of kilometres wide – that bounce around the ocean. These waves are Rossby waves generated by the Earth's rotation, like the Rossby waves that form in the mid-latitude jetstream and control the routes of depressions. These travel slowly westwards across the Pacific. When they hit the western rim of the ocean, the momentum is reflected and translated into waves of warm water called Kelvin waves, which carry water eastwards from the warm pool, raising sea surface temperatures by up to 7°C and sea levels by 5–10 cm.

There is a certain degree of regularity in the whole El Niño/Southern Oscillation (ENSO) sequence, but only so far. The cycle ranges from three to seven years. El Niños last between 12 and 18 months. In between, 'normal' events peak as 'La Niña' (little girl) events, when cold water reaches its maximum extent across the Pacific. Both the weakening of the Trade Winds and the migration of the low away from the eastern Indian Ocean are fitful. Only the more persistent changes seem to translate into full-scale reversals.

ENSO events have almost global scale effects and improving monitoring and modelling have become key endeavours (see *Satellites* in Chapter 18 and *Oscillations and teleconnections* in Chapter 19).

The ocean is also a major carbon sink, both chemically and biologically. Carbon dioxide dissolves in the water. This is behind fears of acidification of the oceans, which may kill many ocean species, especially coral polyps. As CO_2 increases in the atmosphere, rainfall over the ocean will become more acidic. But the carbon also forms shells and precipitates as carbonate minerals, both of which end up on the ocean bed and provide an extremely long-term depository for atmospheric carbon. However, gasses dissolve less in warmer water and once water is chemically saturated it can absorb no more. This should mean that the ocean becomes a less effective carbon sink as temperatures rise. However, warming should also increase windspeeds, which will generate turbulence in the water and mix the surface water deeper in, diluting the carbon concentration. And biology should take over where chemistry leaves off.

The biosphere

The response of plant life to global warming is one of the most complex problems for climate modellers. The effects are threefold: plants cycle water and carbon, and they can affect surface albedo.

Evaporation and transpiration from plants are inherently more difficult to predict than precipitation because they depend on plant species, soil properties, and weather both current and in recent days. Substantial amounts of precipitation may be intercepted by leaves and plant architecture, whence it may evaporate or drip. Transpiration is more complex (and usually only calculated in the combined term 'evapotranspiration'). Plants generally transpire more when they are growing, temperatures are higher, air humidity lower and winds stronger, but they may also limit transpiration losses by closing stomata to protect themselves against desiccation. Plants may also exude fluid when the air is saturated simply to maintain sap movement. The degree to which the air contains less than the maximum moisture it can hold at a given temperature, that is its relative humidity, is fundamental for calculating both the potential and the actual evapotranspiration. Unfortunately, this is still not well predicted in GCMs, although the other critical components, windspeed and radiation balance, are rather better predicted.

A considerable amount of research has been carried out by botanists over the last two decades into understanding how plants react to climate change. But there are still significant aspects that require more research and much still remains to be done to incorporate the knowledge into GCMs and to simulate the reaction of an assemblage of plants in a given area. There are still issues of scale – how to generalize a landscape and upscale processes from individual plants or plant species to mixed vegetation or a mixture of patches of vegetation. The problem of extrapolating from the vegetation 'patch' to the river basin or GCM grid square, is matched by the complementary problem of downscaling from GCM grids to the scale of river basins (Bass *et al*, 1996). The Biospheric Aspects of the Hydrological Cycle (BAHC) project within the International Geosphere-Biosphere Programme made substantial progress on upscaling models for different types of landscape and ecosystem, building up to regional models that couple land cover with regional climate systems (Kabat *et al*, 2004). SVAT (Soil-Vegetation-Atmosphere) models are a basic tool for this, representing the transfer of water, energy and carbon through the combined system.

Carbon dioxide is a fertilizer. But different plant species respond differently. Growth rates tend to be greater in response to higher CO_2 levels in plants with a so-called C3 metabolism, like wheat, than in plants of the C4 group, like maize and many tropical species. Different plant species also respond differently to drought or 'moisture stress', and this response may vary according to CO_2 concentrations. New research by Mercado *et al* (2009) also underlines the further complication caused by aerosol pollution in the atmosphere (see *Aerosols and global dimming*, above). The net effect on the exchange of water between plants and the atmosphere is extremely varied and not wholly understood.

The ocean biosphere could be the most important of all, yet it is also among the least understood. Phytoplankton absorb CO_2 in photosynthesis like land plants and they are expected to flourish in a CO_2-rich world. But phytoplankton response could be reduced by genetic damage from more solar UVB radiation, especially if the shielding effect of the ozone layer is further reduced (Hardy and Gucinski, 1989).

Predicted changes in water resources and riverflows in the twenty-first century

Researchers are still working hard to improve the output from GCMs in terms of hydrology. This is partly a matter of scale. It is also a question of complexity. GCMs are much better predictors of temperature and pressure than of rainfall and evaporation. And predicting riverflows is still more complicated, because riverflow is also a product of the landscape – the rocks, soils, relief and vegetation. Instead of the smoothly changing patterns of rainfall, riverflow is a checkerboard. Rivers right next to each other can respond very differently to the same rainfall. This makes it much more difficult to generalize. It also requires a lot more information in order to model any changes.

The difference is illustrated in Figure 10.1. Here the relatively simple patterns of rainfall and evaporation become more complex when riverflow is modelled.

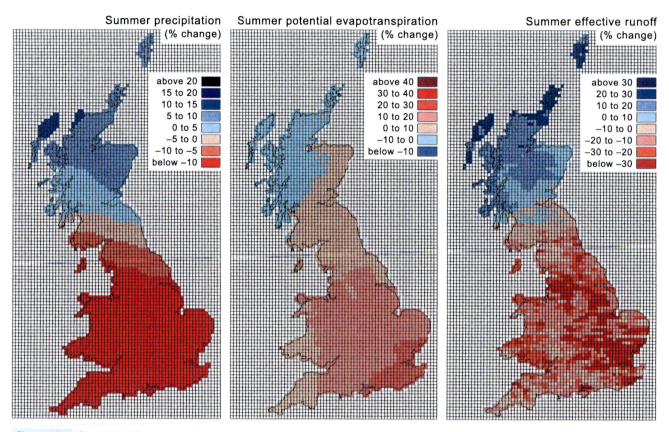

Figure 10.1 Changes in the patterns of summer rainfall, potential evaporation and riverflow in Britain for 2065. Changes are shown as a percentage of present-day values based on a hydrological model run for each 10×10 km grid squares with its specific landscape properties. (After Pilling and Jones, 1999)

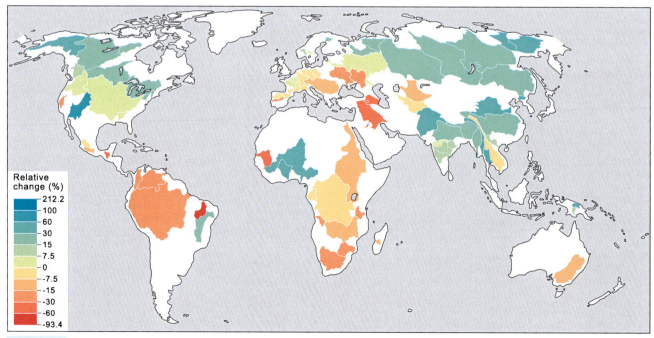

Figure 10.2 Changes in global riverflows by the 2050s relative to the 1960–1990 average. Based on the WaterGAP model and IPCC A2 climate change scenario

Global changes in average annual water resources

Palmer *et al* (2008) use the WaterGAP model to predict changes in annual riverflows for the 2050s. Figure 10.2 plots the results using the Hadley Centre HadCM3 climate change model and the IPCC A2 scenario in basins around the world. A2 is one of the more extreme scenarios, with rapid growth in population and world economies resulting in one of the higher estimates of greenhouse gas concentrations. It is generally taken to represent a reasonable 'worst case scenario', though not as extreme as A1 (IPCC, 2007).

The WaterGAP model includes consideration of whether the rivers are dammed or not. This is the most sophisticated modelling procedure yet applied to the problem. Table 10.2 summarizes the changes based on 291 basins.

The broad global pattern is a familiar one: drier regions getting drier and wetter regions getting wetter. Broadly speaking, global warming is likely to reduce water resources in the lower mid-latitudes and sub-tropics, roughly equatorwards of 50°N and 35–40°S, and increase resources polewards of 55°N and 40°S. The 'winners' are likely to be North America, northern Europe and Asia. The 'losers' are Central and Southern America, most of Africa and southern Europe. Much of Australasia may benefit, but not Australia itself, where the Murray-Darling basin fairs particularly badly: not welcome news following the six drought years that have afflicted much of eastern Australia in the late 2000s (see *The drying of the Murray-Darling basin* in Chapter 6).

Note that the overall 8–9 per cent reduction in riverflows shown in Table 10.2 appears to be at odds with the theoretical increase in global water resources that should result from warmer seas and more precipitation. This is, of course, a summary of only a limited sample of rivers, although the selection is intended to be fairly representative. However, there is also a real suggestion that the enhanced river resources will be disproportionately concentrated in just a few regions, where large increases will take place – here estimated as 24 per cent in North America and 35.5 per cent in northern Asia.

Some earlier predictions suggested that equatorial regions will experience an increase in resources (e.g. Arnell and King, 1997), and this would be consistent with increased convective activity in the Hadley Cell producing more rainfall (see Chapter 11 The restless water cycle). However, the latest modelling, which takes more account of landscape properties, suggests that much of equatorial Central and South America is set to become significantly drier. Central America as a whole tops the list of losers, followed by South America (see Table 10.2). In Africa, only the West African regions that receive rains from the summer migration of the ITCZ, notably the Niger and Volta Rivers, appear to get more resources.

The fact that Europe as a whole shows little change masks the strong contrast between a wetter Scandinavia and Baltic, and a drier Mediterranean and Eastern Europe. This will boost hydropower resources in Scandinavia, but will be stressful for the larger populations in the south, damage Mediterranean agriculture and compound the problems for seasonal tourism. The pattern of reduced resources can be seen in the Tagus and Danube, extending to the Don and Dniepr in the Ukraine and southern CIS. This dry belt extends further into the Middle East and the Central Asian Republics – the Tigris-Euphrates and on to the Caspian and Aral Sea basins.

Similarly, even though Western Europe as a region may show little change in river resources, numerous studies show strong variations within the region. An early example is Bultot's sensitivity study of three rivers in Belgium with different underlying geology (Bultot *et al*, 1988). This showed that with a uniform increase in rainfall, the greatest increases in riverflow are in a clayey, lowland basin, and how outflow from an aquifer in a sandy, lowland basin maintains flow in summer better than in the rocky, upland basin in the Ardennes.

In Britain, studies suggest that the present gradient in water resources between the wetter northwest and the drier southeast will intensify, increasing pressure for water transfer schemes (e.g. Arnell, 1992; Arnell and Reynard, 1993). Arnell found that changes in runoff amplify rainfall changes by up to threefold in SE England. Modelling by Pilling and Jones (1999) suggests that riverflows in SE England could fall some 25 per cent or more by the second half of this century. The only salvation is that basins with greater groundwater storage, which are more common in the lowland southeast, may benefit during summer from higher winter rainfall. In Wales, Holt and Jones (1996) concluded that short-falls are likely to start as an autumn problem and move progressively earlier into the summer in later years as the oceanic system warms up. Jones *et al* (2007) compare results from a variety of scenarios and modelling approaches and find them remarkably consistent. A

Table 10.2. Changes in riverflows by 2050 under the IPCC A2 global warming scenario. Based on data from Palmer et al (2008)

Region	Dam impacted				No dams				All rivers		
	Total change† km³	Overall per cent change‡	Sample size*	Average river per cent change**	Total change† km³	Overall per cent change‡	Sample size*	Average river per cent change**	Total change† km³	Overall per cent change‡	Average river per cent change**
NORTH AMERICA	+321.0	+13.1	39	+14.5	+172.5	+25.3	36	+33.7	+493.5	+15.8	+23.7
CENTRAL AMERICA	−114.3	−46.1	9	−41.7	−43.0	−60.1	4	−59.4	−157.3	−49.3	−47.2
SOUTH AMERICA	−1337.2	−16.5	18	−16.1	−170.9	−29.9	20	−20.9	−1508.1	−17.4	−18.6
EUROPE	−0.8	0.0	36	−2.4	+42.5	+21.8	5	+15.8	+37.1	+2.5	−0.1
Northern Europe	+47.3	+13.2	15	+10.1	+42.8	+21.0	4	+20.8	+90.1	+16.5	+12.1
Southern Europe	−48.1	−5.1	21	−7.6	−0.3	−4.9	1	−4.9	−53.0	−5.6	−7.5
ASIA	+1225.3	+18.8	49	+13.2	+205.5	+24.6	29	+22.9	+1430.3	+19.5	+16.0
North Asia	+444.5	+23.6	4	+23.5	+158.5	39.5	15	+38.7	+484.0	+21.2	+35.5
South Asia	+780.8	+16.8	45	+7.6	+46.5	+10.8	13	+5.9	+827.3	+16.3	+10.8
AFRICA	−71.4	−3.4	15	−4.1	−10.5	−4.0	9	−5.5	−81.9	−3.5	−4.6
AUSTRALASIA	+9.7	+9.3	6	+5.8	+109.5	+15.8	17	+7.4	+120.2	+15.1	+6.9
WORLD	−1571	−7.9	172	−7.7	−224.7	−6.6	119	10.1	−1800.3	−8.4	−8.7

Key:
Red indicates a marked reduction; blue a marked increase.
† Addition or reduction in the total volume of flow for the selected rivers.
‡ Change as a percentage of total flows in the selected rivers.
* Number of rivers included in the selection. The number used in the 'All rivers' columns is the sum of the dammed and undammed samples.
** Average change in individual rivers. This is different from the overall average because a large change in a small river has little impact on the overall amount of change.

report published by the English Environment Agency in 2008 suggests that households need to reduce consumption by at least one-third over the next 40 years to compensate for a 10–15 per cent reduction in available resources in southern Britain. For the south of England it is not just a question of seasonality, more in winter, less in summer; annual resources are also likely to be less, whereas Scotland's annual resources will increase.

There are similar regional contrasts within North America. Although North America as a whole is second only to northern Asia in the winners' league, this hides an increasing contrast between the contiguous States of the USA and Canada. This is bound to have an impact in coming years on the already rather strained relations between those two countries over the tradable status of water – essentially a one-way trade out of Canada (see *The North America Water and Power Alliance* in Chapter 7). The WaterGAP simulation shows an overall increase of 11 per cent in US river resources against a 20 per cent increase in Canada. But if the Alaskan rivers are removed, the increase in US rivers falls to just four per cent, while the Alaskan rivers show a massive 34 per cent increase.

At least this run of WaterGAP suggests that most of the USA will gain in resources, whereas some earlier models suggested that the Midwest and particularly the grainlands of the Great Plains will lose out, which would have a major impact on the amount of grain available for global food aid. The exception could be California as here represented by the Sacramento River. Transfers from an enhanced Colorado River might partly offset this, but the volume of water available from the Colorado is very limited: the 80 per cent increase in discharge suggested by the model is on a current annual discharge of just 1.3 km^3, one of the lowest among the large rivers of the world.

In contrast, the Indian subcontinent and most of southern Asia fair better in this model than in some other predictions. There have been fears that the Indian monsoons might weaken (e.g. Arnell and King, 1997), but this modelling suggests otherwise. Even so, India's central and southern states show less improvement in resources than the north – a situation that could help justify India's grand north-to-south river diversion plan (see *The National Perspective Plan – a 'Garland of Hope' or Hype?* in Chapter 7). The sole exception in southern Asia identified by WaterGAP is the Mekong, but this is only a one per cent reduction on a current annual discharge of 422 km^3 and well within the limits of reasonable error. Overall, the outlook for India and China is better than suggested by other modelling experiments or by extrapolating recent trends. The most notable differences are the increased riverflows in the Himalayas and a 39 per cent increase in the Yellow River basin. These are in contrast to current fears that glacier melting will eventually reduce riverflows in the Himalayas and a well-recognized drying trend in northern China. It demonstrates some of the uncertainty that still surrounds these predictions. Unfortunately, this uncertainty applies to an extremely critical part of the world, potentially affecting the lives and livelihoods of billions of people.

Despite the uncertainties in southern Asia, the WaterGAP modelling agrees with other models and with extrapolations of current trends in all northern latitudes. All rivers in or near the Arctic Circle have enhanced flows, even in Siberia, in line with predictions of increased rainfall in northern latitudes.

Seasonal changes

One of the key changes found in numerous studies in the UK is an increase in 'seasonality', wetter winters and drier summers. In Wales, Jones *et al* (2007) show that even though most climate change scenarios lead to little if any change in annual water resources, there will be an increase in river discharges and a greater risk of flooding in winter and a greater risk of drought in summer. This presents a problem for water managers, which may require new or enlarged reservoirs later this century.

Changes in seasonal resources have so far received less attention than average annual resources. The season of maximum change and the degree of seasonal concentration will vary around the world and will present its own particular problems for local water management. An attempt to assess seasonal changes in Europe based on early results for a double-CO_2 scenario from the GISS GCM suggested that much of the Mediterranean would experience lower rainfall throughout the year except in autumn, perhaps due in part to a build-up of convective activity over warmer seas (Jones, 1996). The problems for the Mediterranean could be severe. Desertification is already creeping into southern Spain, Sicily and Italy, and many areas are under increasing pressure from tourist demand during the summer. Devising a blueprint for sustainable water resources in the region is a major concern of the Euro-Mediterranean cooperative project Plan Bleu (www.planbleu.org).

Perhaps the greatest concern of all is what will happen to the monsoons. Both the Indian and West African monsoons have been very variable of late, with a number of weak annual events. The worst cases in which the Indian monsoon either failed or was badly delayed occurred in 1965–66, 1972, 1987 and again in 2009, with severe economic and social effects. The monsoon supplies 80 per cent of India's annual rainfall over a four-month period. In June 2009, the rainfall was 46 per cent below normal – the worst since 1926 – and during June to August some areas received 60 per cent less. In Maharashtra State, over 80 per cent of wells were dry in August. Agriculture and the lives of the 700,000 mainly poor people (60 per cent of the national population) who depend on it were severely disrupted. With agriculture accounting for 18 per cent of India's GDP, the rate of growth in the economy was set back, probably by a quarter from the predicted 6.8 per cent rate. The monsoon had a greater effect on GDP than the global credit crunch.

Most current GCM predictions suggest that Southeast Asian monsoons will become more unreliable. The events of 2009 in India may be a foretaste. The West African monsoon is also showing signs of changing as the Intertropical Convergence Zone (ITCZ) has not been penetrating as far inland in recent decades (see *Drought in the Sahel* in Chapter 3).

The shift from snow to rain

There will also be significant changes in the timing of riverflows as a result of a shift from snow to rain. The effects could vary widely. Table 10.3 lists some possible changes in snowfall in the Alps and Bavaria. In Finland, the combination of more annual rainfall and less winter snow could reduce the period of snowcover by two or three months, snowmelt freshets and springtime floods will be less pronounced, snowmelt events in mid-winter will become more frequent, but river levels will be higher for the rest of the year (Vehvilainen and Lohvansuu, 1991). Activities that rely on winter storage and delayed release of meltwaters are likely to be adversely affected, e.g. hydropower (cp. *James Bay Hydro – redesigning a landscape*). In contrast, higher temperatures could maintain or even enhance spring flood levels on the Rhine and Danube, but they would occur earlier in the year (Kwadijk, 1993; Gauzer, 1993).

The presence of glaciers will have a strong influence on meltwater response. Less snowcover and warmer temperatures will increase the exposure of the glacier ice,

which will then melt faster because it has a lower albedo (Chen and Ohmura, 1990). Glaciers will retreat not just because there is less snow to replenish them, but also because they will have less protective cover. Studies on the upper Rhone suggest that riverflows in basins with a large amount of ice are more sensitive to temperature changes, but those with less ice are more sensitive to changes in snowfall (Collins, 1989).

Changes in extreme events

It is now commonplace to read news reports of predicted increases in the severity of extreme weather events like floods, droughts and hurricanes linked to global warming. It is true that GCM results tend to suggest that extreme events will become more frequent in a warmer world, but it is impossible to say that individual present-day weather events are indicative of man-made climate change or even that a series of such events is.

Many of the individual events that trigger these media reports are either part of the natural spectrum of events

Table 10.3 Changes in the reliability of snowfall with a 2°C warming. Based on the 2006 OECD report

Location	Per cent chance of snow		Per cent reduction[*]
	Now	+2°C	
Germany			
Bavaria (Oberbayern)	18	3	83
France			
Haute-Savoie (Chamonix, Val d'Isère)	35	18	49
Austria			
Tyrol (St Anton, Obergurgl, Lech, Ischgl)	75	45	40
Salzburg (Kitzbühel)	35	24	31
Italy			
Alto Adige (Livigno)	31	20	35
Piedmonte & Aoste (Courmayeur, Cervinia)	18	15	17
Switzerland			
Graubünden (Davos, Engelberg)	36	35	3
Valais (Verbier, Zermatt)	49	49	0

[*] Calculated from the OECD (2006) data

that occur within any climate or else they are part of what is termed 'climatic variability', periodic departures from the average that may involve solitary extreme events or a cluster of events in a cycle that eventually returns to the average and may even swing the opposite way. In other words, they may be part of a natural random variation or part of cyclical variations that oscillate around the average. The fact is that climate is, in the words of statisticians, 'non-stationary': forever changing. Climate is defined by statistics. The World Meteorological Organization (WMO) defines climate as the average over a 30-year period of records, but every 30-year period will give different statistics. Rodda (1970) presented a classic example from the rainfall record at Oxford. Records going back to 1881 show that the heaviest daily rainfall to be expected once in 50 years was nearly twice as high during the wettest period as in the driest. What might be done to improve the assessment of magnitude and frequency of extreme events will be taken up in Chapter 19.

Nevertheless, meteorological theory supports the results of GCM simulations of future trends. The extra energy available in a warmer world will feed convective storms, rainfall intensities are likely to increase, and changing wind and pressure patterns will expand the areas prone to drought. The daily weather sequences for different scenarios that are produced by the latest GCMs, while not real 'forecasts' for actual days, provide a large ensemble of possible days that allows statistical analysis of the risks of extreme events. These do show that greater extremes are likely (Jones *et al*, 2007).

However, there is intense controversy over any evidence that this trend has already begun. The IPCC (2007) report says the world has 'suffered rapidly rising costs due to extreme weather-related events since the 1970s'. The 2006 Stern Report, which was commissioned by the UK government following a bout of severe flooding, took a similar view, citing a two per cent per annum increase in economic losses since 1970 over and above increases in wealth, inflation and population growth. Projecting the trend suggested that losses could reach between half and one per cent of global GDP by mid-century.

There are, however, a number of pitfalls in these arguments. First, flooding problems in Britain have been aggravated by new housing and commercial developments on floodplains, resurgence of sewage from old combined drainage systems and in some cases blocked or inadequately maintained drains; not just heavy rain. Secondly, both the IPCC and Stern misused the work of Robert Muir-Wood of Risk Management Solutions in London, which was commissioned by IPCC member Roger Pielke. Actually, Muir-Wood found no net increase in losses from 1950 to 2005 if economic growth is taken into account, and the two per cent annual increase between 1970 and 2005 was all due to strong hurricanes in 2004 and 2005 (Miller *et al*, 2008). The vice-chairman of the IPCC report, Professor Jean-Pascal van Ypersele, has promised a revised assessment.

Tropical storms, hurricanes and cyclones

There is still considerable debate over whether the frequency and intensity of hurricanes is increasing due to global warming. A spate of extreme cyclonic storms in the first decade of this century, like Hurricane Katrina that hit New Orleans in 2005 and Typhoon Nargis that hit Burma in 2008, seem to suggest that they are increasing. During the 2008 season Atlantic hurricanes caused damage estimated at $54 billion. Among these, Hurricane Kyle was the worst to reach Nova Scotia for nearly two decades in September 2008. However, 2009 was unusually quiet, but this may have been related to the developing El Niño, which generally seems to dampen activity in the Atlantic.

The IPCC (2007) report suggests that South Asia is likely to warm up more than the world average. This could fuel tropical storms. Already, tens of thousands of people are driven from their homes and thousands die every year in India and Bangladesh as a result of tropical storms, particularly by coastal floods. This could rise to millions. Mumbai, Kolkata and large areas of Bangladesh are very exposed to these coastal floods. Typhoons and hurricanes release torrential rains. A stationary hurricane can deliver 500 mm a day, a moving one 200 mm, creating flash floods. Even a lesser 'tropical storm' can deposit substantial rainfall: in 2007 Erin yielded between 75 and 200 mm in southern Texas, reaching as far as San Antonio some 200 km inland. These storms begin at sea and once they cross the coastline and are starved of water supply hurricanes normally die back to the strength of a tropical storm within 36 hours.

Whether there is a trend currently developing or not, physics gives a clear indicator for the future. GCMs predict generally increased activity. Warmer sea surface temperatures will fuel more intense tropical storms, cyclones and typhoons (hurricanes). Hurricanes need two factors to develop: sea surface temperatures over 27°C, that provides sufficient evaporated moisture, and a Coriolis force (caused by the Earth's rotation) strong enough to give the spin to the winds, which means locating at least 5° from the equator. An additional factor helps: being on the western side of the ocean basin, because, for reasons related to the way the Trade Winds descend out of the subtropical high pressure zones on the east side, the tropical inversion layer is lower on the east side, limiting the height convective storms can grow to on the east. Height is all important to allow sufficient condensation and release of latent heat to energize the storm (Jones, 1997).

Saunders and Lea (2008) find that hurricane activity in the North Atlantic is extremely sensitive to sea surface temperatures. They calculate that 40 per cent of the increased activity during 1996–2005 relative to 1950–2000 was due to sea temperatures. They predict that a further 0.5°C rise could cause a 40 per cent increase in activity. Research at the National Center for Atmospheric Research (NCAR) in Boulder supports this view. However, an interesting complication has been discovered by Vecchi and Soden (2007). They find that while warmer seas in the vicinity of the storm do fuel activity, warmer seas elsewhere can dampen it. This is because warmer seas in other areas of the tropical ocean warm the atmosphere, so that when the winds carry this air around the globe there is a lower vertical gradient in air temperature, which means weaker convection. Hence the authors conclude that regions where warming is above the tropical average have the greatest potential for hurricane development.

There is evidence of increase in the North Atlantic. The long-term average from 1950 to 1990 was ten tropical storms a year, of which five became hurricanes, but in the period 1998–2007 the average was 15 tropical storms with eight becoming hurricanes. The upsurge in the North Atlantic began in the early 1990s. In the view of the IPCC, the increased activity since 1995 is 'more likely than not' to be linked to man-made climate change. However, Nyberg *et al* (2007) conclude that stronger vertical wind shear (i.e. stronger winds aloft) before 1995 tended to destroy developing hurricanes. By reconstructing a 270-year history of hurricanes from geological evidence – corals and sediments – they also suggest that the recent upsurge in activity is not unusual, so questioning the view that it is predominantly anthropogenic.

The Pew Center for Global Climate Change suggests there is little or no evidence of a similar increase in other regions (www.pewclimate.org). However, there have been occasional storms in recent years in places that do not normally experience them. The South Atlantic is not normally considered a hurricane region, yet in March 2004 Hurricane Catarina hit Brazil. The Hadley Centre GCM only predicts hurricanes around there by around 2075. Another southern hemisphere cyclone, a category 3 hurricane named Fanele, hit Madagascar in January 2009. In 2007, the Mosquito Coast in Central America experienced two maximum strength, category 5 hurricanes for the first time since reliable records began in 1928: Dean in August and Felix in September. At the same time as Felix, Henriette began unusually in the eastern Pacific and lashed the Baja coast: east Pacific storms are so rare that five have been called Henriette since 1983 without fear of confusion. Norbet, an extremely dangerous category 4 hurricane, hit the Pacific coast of Mexico in October 2008. In July 2008, Hurricane Bertha developed unusually far east off the Cape Verde islands, and also very early. The North Atlantic hurricane season normally peaks August to mid-October when seas are warmest. But hurricanes have been occurring outside the normal hurricane season, as in 2003.

Tropical storms are not all bad. They can break droughts. The floods can also wash substantial quantities of carbon and other nutrients into the sea. This can cause a blooming of phytoplankton, which enhances carbon sequestration (Wetz and Paerl, 2008), as too can deposition of the carbon directly onto the seabed. A single typhoon in Taiwan washed as much sediment into the sea as a whole year's rainfall. Similarly steep catchments in the Caribbean could have a comparable effect, but more research is needed.

Impacts on groundwater

Predicting the impact on groundwater is even more complex than for surface water. This is partly because of lack of information on the aquifers – limited monitoring of groundwater, limited information on rock properties, especially cracks and fissures – that forms the basis for numerical modelling, and partly because of the great variety of physical situations – mechanism of recharge and drainage, and feedbacks with local climates – which make it difficult to extrapolate from one region to another.

Recharge can be sensitive to the intensity of rainfall as well as the amount. High intensity tends to cause more rapid runoff: a higher proportion runs away on the surface and less infiltrates into the ground. The rate of recharge does not increase in proportion to the intensity. This was the situation during the multiyear drought in SE England in the early 1990s, when intense rainstorms failed to recharge the depleted aquifers. However, if the higher riverflows flood the floodplain, then aquifers beneath the floodplain will receive more recharge: this is commonly a major source of recharge for floodplain aquifers. In semi-arid and arid regions, increased rainfall can lead to more plant growth, which intercepts, evaporates and transpires more of the rain, limiting recharge (Scanlon et al, 2006).

The decline of snowfall in favour of rainfall may also affect recharge. Snowmelt is a major contributor in the Sierra Nevada and modelling predicts that snowfall will decline 33–79 per cent by the end of the century (Dettinger et al, 2004). Rainfall will not recharge the groundwater so well as it tends to run off quicker, whereas snowmelt provides a more steady, lingering infiltration. In contrast, aquifers in Alaska and Siberia may get more recharge as permafrost and ice covers melt and allow more infiltration through the soil (Hughes et al, in press).

The IPCC (2007) reports the results of modelling the impact on global groundwater recharge patterns by Döll and Flörke. They used the A2 and B2 scenarios and climatic outputs from the HadCM3 and ECHAM4 GCMs and compared results for the 2050s with the standard 1961–1990 baseline. These suggest that SW Africa, NE Brazil and parts of the Mediterranean will suffer marked reductions in recharge. However, the authors noted that the differences between the GCMs are greater than the differences between the two emissions scenarios – which should not be the case. It underlines the need for further refinement in climate modelling.

Coastal aquifers are vulnerable to saltwater intrusion. Over-exploitation allows seawater in, as in Gaza. Rising sea level also increases saltwater incursion. Ranjan et al (2009) calculate that under the A2 scenario significant reductions in fresh groundwater resources will occur throughout the coastal areas of the Caribbean, the Gulf of Mexico and Central America from the southern USA to northern Brazil, the Mediterranean, Australia, the Yellow Sea and much of Africa. Islands built of coral or porous volcanic rocks are particularly vulnerable

and can be slow to recover when a drought reduces the freshwater lens. Some saltwater intrusions may extend a number of kilometres inland, others only a few metres. Models of the response have to make a number of assumptions, especially whether the advance of saltwater is driven by the head of seawater or the flow of the groundwater and local models are needed to inform sound management practices (Hughes et al, in press).

Sea level rise

Sea level rise threatens many low-lying islands and coastal regions with inundation. It may also cause pollution of coastal aquifers by saltwater incursion or the build-up of pollutants in river estuaries through slacker drainage.

The IPCC (2007) report estimates that average global sea level will rise by between 18 and 59 cm by 2100, depending on the economic and emissions scenario used. A rise of just 40 cm in the Bay of Bengal would flood more than a tenth of the coastal lands of Bangladesh and force up to ten million people to become refugees. Already, a thousand inhabitants were evacuated from Carteret atoll in Papua New Guinea in 2005 because of sea level rise. Some of the more extreme projections for sea level rise would have greater worldwide effects. A one metre rise in sea level is calculated to displace 100 million people in Asia, mostly in China, Vietnam and Bangladesh. Fourteen million people in Europe would also be at risk and there would be major economic costs: $156 billion to protect the coastal regions of America according to one estimate. A 1.4 m rise, though most unlikely, could displace a tenth of the world's population.

Evidence from tide gauges suggests that sea level has been rising at a rate of 1.8 mm a year over the past century, but this seems to be accelerating. Between 1993 and 2003, tide gauges and satellite altimetry indicate a rise of 2.8–3.1 mm a year. The IPCC estimate is based on a continued rise of about 2 mm p.a. or an acceleration to just over 6 mm p.a. The large range in the estimate reflects not only uncertainties about the amount of temperature rise, but also the large uncertainty about the processes causing sea level rise.

The main process causing sea level rise last century is thought to be the melting of small glaciers. But this has now been overtaken by thermal expansion of the ocean; water expands as it warms up. Whereas small glaciers probably contributed about half of the overall rise between 1961 and 2003, and thermal expansion 40 per cent, the latter contributed to over 60 per cent of sea level rise between 1993 and 2003, and glaciers only around 25 per cent (Houghton, 2009).

If the world's small glaciers melt away, but there is little change in Greenland or Antarctica, then the rise will level off. The reasons for the recent acceleration are unknown, but may be related to melting in the polar ice sheets. There is broad agreement that Greenland is set to become a major contributor. There is less agreement about Antarctica, although evidence is accruing to suggest that it too will melt faster than it is fed by new snowfall (see *Shrinking land ice* in Chapter 12). The IPCC estimates the rise could be 10–20 cm more if polar ice sheets continue to melt, but the report points to the current uncertainties about trends and lack of knowledge on processes.

A number of other factors may also affect sea level to a lesser degree, for example, if there is a change in the relative balance between evaporation and precipitation – warmer sea surface temperatures will increase evaporation, but there are many factors that conspire to cause proportionately more precipitation to fall on land (see *Clouds and rain-forming processes* in Chapter 11). The net result is expected to be more riverflow. Most of this riverflow will drain back into the ocean. Global warming will increase the amount of river discharge entering the Arctic Ocean. That is, unless rivers are reversed or dammed more (see below, and *Reversing Russia's arctic rivers* in Chapter 7).

Some scientists regard the IPCC estimates as too conservative. The Dutch Delta Commission (2008) recommended planning for rises of 0.55–1.3 m in regional sea level for the Rhine Delta region, but this included a factor for continued land subsidence. There is currently little evidence to support more extreme views like those of Vermeer and Rahmstorf (2009) who extrapolated the recent acceleration to get a rise of 1.88 m by 2100 given a 6.4°C temperature rise – the most extreme of the IPCC (2007) scenarios compared with 75 cm for the least. Their estimate received widespread criticism for being over-simplistic when it was released to coincide with the Copenhagen climate summit. It is based purely on a mathematical relationship between sea level and temperature developed from analysis of data for 1881–2001, and the view that the rate of rise increases at higher temperatures. Much of the twentieth-century

sea level rise was due to the melting of small glaciers, which will be contributing less by 2100. There is even less support for earlier suggestions that the West Antarctic Ice Sheet (WAIS) might collapse and raise global sea levels by 5–6 m over the next century (e.g. Tooley, 1991). Large parts of the WAIS rest on rock below sea level and it is the most unstable ice body on Earth (see *Shrinking land ice* in Chapter 12). However, computer models developed by Bamber *et al* (2009) based on the latest glaciological information suggest that it is more stable than previously supposed and that even if it did collapse it would only raise global sea level by 3.3 m. Nevertheless, contributions from polar icemelt seem likely to increase (see *Shrinking land ice* in Chapter 12).

Lastly, measuring and predicting sea level rise is complicated by the fact that the sea is not 'flat'. It is affected not only by the gravitational pull of the Moon and Sun, but also by changes in the Earth's own gravity around the globe and by the spin of the Earth. As Antarctica melts and the mass at the South Pole reduces so the Earth's gravity in the southern hemisphere will be reduced and seawater will pile up more in the northern hemisphere. Bamber *et al* (2009) predict that it will also affect the rotation of the Earth, which will add to sea level rise in the northern hemisphere with potentially critical effects in the Indian Ocean and around North America. They calculate that sea level around North America will rise by 25 per cent more than the global average, increasing flood risk for major cities like San Francisco and New York. The Baltic Sea is also a good example of how wind alone can affect sea level, piling water up in some of the bays.

To this must be added tectonic movement of the land. One of the few tide gauge records that does not confirm sea level rise comes from Stockholm. This shows a steady fall in sea level relative to the land because Sweden is still rising 'isostatically', due to the gradual rebound as the Earth's crust continues to adjust to the removal of the overburden of ice at the end of the last ice age. The land is in fact rising faster than the sea. The flipside of this is that south of a fulcrum around southern Denmark, the Netherlands and the southeast corner of England, including London, are sinking and the apparent sea level rise is greater there – a factor affecting the Thames Barrier and the Delta Project.

There is, however, one important negative effect that human activity is having on sea level rise: reservoirs. Reservoirs stop and delay riverwater reaching the sea. Chao *et al* (2008) calculate that reservoirs have reduced sea level rise by 30 mm over the past 50 years by impounding and withholding a cumulative total of 10,800 km³. In perspective, this is equivalent to a quarter of the annual riverflow in the world. By the onset of the Second World War only around 400 km³ had been impounded, but the rapid rise to present levels began in the early 1950s when cumulative impoundments stood at around 700 km³. Small though the 30 mm reduction is, without the reservoirs mean sea level would be 3 m further up on the average beach, with some implications for coastal erosion and storm surges. It also means that 'natural' sea level rise is greater than estimated.

A bigger issue though is what will happen to reservoir volumes this century. In effect, the trend is still upwards and rising demand will ensure this continues (see Chapter 7 Dams and diversions). Building more dams could be part of a geoengineering solution to sea level rise. But should building stall, the sea will rise more.

There is also an important scientific issue. It underlines the gaps in our knowledge of the processes driving sea level rise. The IPCC (2007) report admitted that adding up all the known natural causes does not account for all of the measured rise. Adding reservoirs into the equation only makes this worse. Moreover, Chao *et al* found that when they added the amount of water impounded progressively back to the sea over the last 80 years, global sea level rise was almost constant at +2.46 mm a year, whereas the conventional view based on the observed records is that the rate of rise has been variable. There is a strong implication here that a significant part of the variability might be explained by rates of reservoir building.

Yet another case of human interference that does not appear to have been quantified at all is 'groundwater mining', that is, exploiting deep groundwater that is not being renewed in the current climate (see *Mining groundwater* in Chapter 12). This is bringing ancient water out of its isolation back into the active hydrological cycle: Libya's scheme alone is capable of raising 2 km³ p.a. (see *Libya's great man-made river* in Chapter 12). More such schemes are planned as water scarcity strikes the semi-arid regions.

Storm surge floods

One of the worst aspects of sea level rise is storm surges. While sea level rise will be slow and more predictable, these are occasional extreme events that require more sophisticated forecasting and warning systems. They occur when a severe storm breaches sea defences and floods the land. They can cause destruction and temporary flooding well above sea level. The English Environment Agency sets a 'blue line' at 5 m above sea level to delimit land that could potentially be affected. Storm surges are usually due to a combination of high sea levels and intense storms. The worst events occur when a storm strikes the coast during high tide, especially if it is a spring tide. Strong winds can pile water up against the shore and also create large waves, and the low atmospheric pressure can temporarily cause an additional slight increase in sea level.

Global sea level rise will aggravate the situation in two important respects: it increases tidal height and the deeper water reduces the friction on the seabed or beach, which slows down the waves. So the waves strike the coast with greater force. Coastal erosion is therefore likely to increase. Two aspects of climate change are thus likely to combine to make storm surges more frequent and more damaging. One is sea level rise. The other is increased storminess (see *Changes in extreme events*, above).

One of the worst storm surges last century occurred in the North Sea in 1953. Over 1850 people died in the Netherlands and 280 in eastern England. It led to the Dutch Delta Project, a chain of dykes, dams and storm surge barriers down the Dutch coast, which was finally completed in 2000. More than a fifth of the Netherlands is already below sea level, but it is well-protected and a model for how the risks of sea level rise might be contained. The Dutch design their coastal protection works for the one in 4000 year extreme surge event. Although the Delta Project is a success as regards flood protection, however, there are unfortunate side-effects on pollution and wildlife (see *Cleanup in The Netherlands* in Chapter 15).

Sea ice and sea level rise

Sea ice is not commonly used as a water resource and it has been generally assumed that its potential demise will not affect sea level rise – after all, surely most sea ice floats and displaces only as much water as it contains, including any snowcover. The main significance of sea ice reduction is in lowering the regional surface albedo: shrinking sea ice is a major reason for greater climate warming in the arctic, perhaps four times that of the tropics.

The area covered by arctic sea ice has been shrinking for the last three decades and reached a record low in summer 2007 at 4.27 million km^2, 39 per cent below the 1979–2001 average. In September 2007 the coverage was half that of a typical September in the 1960s (see Table 10.1). Interestingly, 2007 also saw a record-breaking amount of river discharge entering the Eurasian section of the Arctic Ocean. Shiklomanov and Lammers (2009) report a total annual discharge of 2254 km^3 from the six largest Russian rivers, compared with the next highest of 2080 in 2002 and a long-term average of just 1796 km^3. They also note that there is a trend of increasing discharges even disregarding 2007, which would fit GCM predictions for changing patterns of precipitation. The extra freshwater discharge should make ice formation easier – but it didn't.

Even so, by summer 2009, ice coverage had recovered to over five million km^2 and by winter 2009–10 had returned to 2001 levels, perhaps due to the Arctic Oscillation (see *Oscillations and teleconnections* in Chapter 19). Nevertheless, the long-term trend is clear. On average, arctic ice coverage in summer has been decreasing by around 600,000 km^2 per decade.

Model results reported in the IPCC (2007) assessment predict that the Arctic Ocean will be ice free in summer by 2060 to 2080, though some estimates say it will be quicker because recent observations show the ice is also thinning more than estimated. Satellite observations show a sudden downturn in the winter of 2007–8 when ice thickness over Arctic Ocean as whole was ten per cent below the average of the five previous winters and the famed Northwest Passage opened up in 2008 (Giles *et al*, 2008). Measurements of ice thickness taken by Peter Wadhams (Professor of Ocean Physics at Cambridge University) from a nuclear submarine using an upward echo sounder, showed that in March 2007 the ice was 50 per cent thinner than in 1976. And explorer Pen Hadlow's sea ice trek for the 2009 Catlin Arctic Survey showed that ice in the Beaufort Sea, that is normally multiyear ice, was mainly only one year old, which means it is more vulnerable to melting. Wadhams believes the arctic could be ice free in summer in 20 years (www.catlin.com/cgl/media/press_release/pr_2009/209-10-15). Wang and Overland (2009) predict ice-free conditions by 2037 based on projections from six IPCC models. Nasa-funded modelling by Bruno Tremblay and colleagues at McGill also suggests the arctic coastline could become ice free in summer by 2040 and he believes that melting could reach a dramatic tipping point soon. A new and interesting feature

of this work is that half of all the different runs of models showed periods of sudden, large reductions in ice cover, up to two-thirds in a decade, during the twenty-first century. These are presaged by growth in areas of open water between the thinning ice, where the lower albedo captures heat, which is then conducted into the ocean (Holland *et al*, 2006). Retreat is particularly abrupt when ocean currents are also transporting heat into the region.

However, more recent research suggests that melting sea ice and ice shelves may also affect sea level, contrary to received wisdom. This is because the meltwaters both cool and dilute the seawater, altering its density. Modelling has shown that the volume of meltwater from an ice shelf can be nearly three per cent greater than the original displacement because of the difference in salinity (Noerdlinger and Brower, 2007; Jenkins and Holland, 2007).

Falling lake levels

Lake levels can be sensitive indicators of changes in regional water balances. There are widespread reports of falling lake levels. Many cases are blamed on climate change, but there is usually a combination of factors, including overuse, especially for irrigation. One of the rarer cases in which water levels are rising is in proglacial lakes, which are being fed by glaciers melting at accelerated rates.

Nearly 8000 km^2 of lakes and wetlands have dried up in North China in recent decades as a result of overexploitation of surface and groundwater combined with a drying trend in the climate (Xia, 2007). The recent history of falling levels in Lake Sevan in Armenia is almost entirely a result of local human activities rather than climate change (Vardanian and Robinson, 2007). A key factor in a 17 m fall there was over-deepening of the river draining the lake to feed a hydropower scheme.

The reasons behind the recent fall in water levels in the Great Lakes remain an enigma. The lakes contain 20 per cent of the world's freshwater, but 99 per cent of it is derived from the last ice age and is non-renewable. Levels have been falling since the late 1990s, the longest downturn in their recorded history. Lake Superior and Georgian Bay have fallen 60 cm in the last decade, Lake Ontario 20 cm and Erie 10 cm.

The US-Canada International Joint Commission on the Great Lakes is due to report in 2010 on whether the fall can be attributed to climate change. Environment Canada's models predict continued falls of up to 1.2 m by 2050 despite increased precipitation. Evaporation losses are a major factor. Water temperatures in Lake Superior rose 2.5°C 1979–2006, which has meant less lake ice cover and more evaporation in winter as well as summer. But there is also an engineering element. The St Clair River was dredged in the 1960s. This initially drained an extra 3.2 billion litres a day into Lake Erie, but rampant channel erosion has now raised this to an estimated ten billion, hence Erie has fallen much less. Containing the erosion would be a major corrective move.

Conclusions

There is no doubt that global warming is under way and presents critical problems for water resources that need to be addressed sooner rather than later. As ever, it is those parts of the world that are already stressed and least able to cope that will suffer worst. International collaboration is essential, including redoubling of technology transfer and capacity building to assist developing countries. GCMs are forming the first line in the battle for awareness.

Discussion points

- Explore GCM predictions for your area.
- Compare the different effects that the various SRES scenarios will have on water resources, globally, regionally or locally.

Further reading

The UK Met Office online offers regular climate change updates and background at: www.metoffice.gov.uk/climate change/science/explained/.

The WMO Guide lists the standard methods of extreme event analysis and a detailed discussion of the pros and cons can be found in Jones (1997):

WMO, 1994. *Guide to hydrological practices*. 5th edition, WMO Pub. No. 168, Geneva, WMO, 735pp.

WMO, in press. *Guide to Hydrological Practices*. 6th edition, Geneva, WMO.

Jones, J.A.A. 1997. *Global Hydrology: Processes, Resources and Environmental Management*. Harlow, Longman.

The impacts of extreme climatic events and analyses of trends and projections are well covered by a number of papers in:

Diaz, H.F. and Murnane, R.J. (eds) 2008. *Climate Extremes and Society*. Cambridge University Press, Cambridge.

Part 2 Nature's resources

11 The restless water cycle

Humankind entered the world when the atmospheric water cycle was in one of its most changeable phases. It is possible that this played an important role in our evolution as a species. It has certainly led to invention and resourcefulness in managing or adapting to change, day to day, season to season, year to year, decade to decade and beyond. Today, we face the prospect of global warming with concern. The trends seem real and the climate models that predict marked change by the end of the century are ever more soundly based. However, unlike our ancestors, we have forewarning and we have an expanding range of solutions for adapting and managing.

For the last two million years, the Earth has been locked in an ice age sequence. We have no reason to suppose that this has ended. It has consisted of major periods of expanded glaciers and ice sheets, typically lasting around 100,000 years, interspersed by shorter, warm 'interglacials' lasting 10,000 years or so. We are in one of these interglacials now and maybe towards the end of it.

Many of the factors thought to be responsible for this climatic aberration are still with us. The oscillations in the Earth's orbit around the Sun remain essentially the same. A 100,000-year stretch shifts the orbit between more elliptical, placing Earth further from the Sun at certain times of the year or more circular, while a changing tilt and a wobble in the polar axis points different hemispheres more towards the Sun, or away. This fundamental pattern has probably changed little since the Earth was created, but the dramatic shift in water storage between ice and liquid over the last 1.8 million years appears to be due to changes in global geography.

Continental Drift has moved the continents into critical positions and created mountains where landmasses collide. Polar regions remain unusually cold because the Antarctic continent blocks the penetration of warmer oceanic waters in the south, and the Eurasian and North American continents block them in the north. One calculation indicates that, if continental drift continues at present rates, in 50 million years Europe and America will be far enough apart to allow sufficient warm water into the Arctic Ocean to halt this cycle. Mountain-building may also have had its effect. The raising of the Antarctic mountains will have made the continent even colder. The raising of the Himalayas may have accelerated the sequestration of carbon from the atmosphere: carbon dioxide dissolves more in snow and cold water and the rejuvenated rivers may carry more away to be deposited in the ocean as limestone. This process would reduce the greenhouse effect and cool the atmosphere (Raymo and Ruddiman, 1992).

At the same time, the expansion of snow and ice surfaces also cools the atmosphere by reflecting more sunlight back into the air: while the average grass surface may absorb 85 per cent of the sunlight, new snow may absorb only ten per cent. Because of this, deglaciation lags behind the forcing factors in the Earth's orbit. Conversely, the beginning of glaciation is delayed by the vast amount of heat stored in the oceans. There is an in-built resistance to change, but in the current geography, change wins through.

There is strong evidence for numerous ice ages in the more distant geological record. There may have been up to five ice ages in the Neoproterozoic (late PreCambrian) more than 500 million years ago. Russian climatologist Mikhail Budyko calculated from a simple energy balance model that there could be a tipping point if ice advanced to within 30° of the equator, whereby the enhanced albedo of the Earth would cause run away cooling. Suggestions that the whole Earth may have become ice-covered – a 'snowball Earth' – and the hydrological cycle halted completely remain controversial. If the cycle did stall, the lack of heat-conserving low-level cloud formation would add to the cooling. Some modellers suggest the tropics may have retained significant amounts of open water – a 'slushball Earth' with a drastically weakened hydrological cycle. Budyko's climatic model did not contain a mechanism for melting the ice, but subsequent theories have proposed that a build-up of greenhouse gases could occur as volcanic eruptions add CO_2 to the atmosphere over millions of years, which is not removed by precipitation and rock weathering. Professor Ian Fairchild likened the situation to a baked Alaska pudding, cold on the inside and hot outside, and presented new evidence supporting the idea of an atmosphere high in CO_2 ending these Snowball Earth events (Bao *et al*, 2009).

These shifts in temperature and water have caused vast changes in the distribution of plants and animals. For the last two million years humankind has survived by following these shifts. We are still hugely dependent upon these climatic patterns, but is this perhaps about to change? Are we capable of defying nature in this as our ancestors have in countless smaller ways, from building houses to irrigated agriculture?

The water cycle now

The world's freshwater resources are currently stored mainly as snow and ice or groundwater (Figure 11.1).

Although rivers and lakes are often the first to be exploited, they hold only a small fraction of the total. They are especially sensitive to changes in supplies from the atmosphere.

The water cycle is dominated by the oceans. The oceans are the largest store of water on Earth and the principal source of evaporation and precipitation (Figure 11.2). 97 per cent of all water on Earth is stored there and 86 per cent of all evaporation occurs over the oceans. Together with the Sun's radiation and the Earth's orbital characteristics – its spin, tilt and orbital track – the size and location of the oceans control the pattern of world climate and the distribution of water resources. Estimates now suggest that oceanic circulation transfers more of the surplus heat from the tropics towards

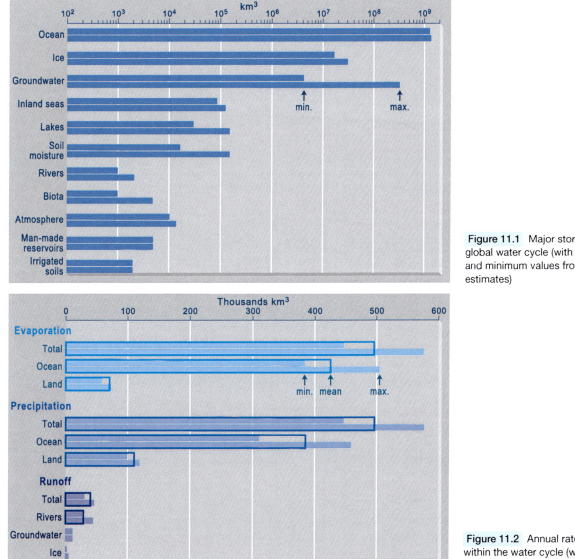

Figure 11.1 Major stores in the global water cycle (with maximum and minimum values from a variety of estimates)

Figure 11.2 Annual rates of exchange within the water cycle (with maximum, minimum and mean estimates)

higher latitudes than does the atmosphere. The oceans also have one of the slowest rates of turnover in the whole system, on average second only to the ice caps. The longest residence time is found in the ocean deeps. A raindrop falling in the ocean might expect to stay there for more than 10,000 years if it enters the deep, but if it remains near the surface its component molecules could be wending their way through the air again in a very short time. The depth of the oceans and the long residence times combine with the high heat specific heat and relatively low thermal conductivity of the water to make ocean temperatures slow to change. This has an ameliorating effect on coastal and maritime climates on scales from seasons to centuries.

Evaporation, unlike the oceans, is an essentially invisible component in the water cycle, yet it is the key to our existence in more ways than one. It too has a crucial role in the world's climate. Water vapour is the main greenhouse gas. By capturing and returning much of the Earth's infrared heat loss, the natural greenhouse effect maintains the average surface air temperature at 33°C warmer than it would otherwise be. About 30°C of this is due to water vapour. Since the global average mean annual temperature is only 14°C, it is clear that life based on the internal circulation of liquid water could not exist as now – over much of the Earth – without the heat-conserving effect of atmospheric water, or an effective substitute gas. In addition, water vapour plays a crucial role in transferring heat from the tropics towards the poles. Heat is absorbed in the tropics to evaporate surface water. This latent heat of vaporization is subsequently released when the air cools and the water condenses to form clouds and precipitation, predominantly at higher latitudes. Recent estimates suggest that nearly a third of the energy that drives atmospheric circulation is transported and released as latent heat. Sadly, this process is also the driving force for some of the most destructive storms, hurricanes or typhoons, because it is the concentrated release of the latent heat that causes the intense convection around the eye of the storm.

If a cycle has a starting point, then evaporation is the logical beginning of the water cycle. It provides the essential link between the land and the sea, allowing water to be transferred to the land and making the water fresh in the process. The net transfer of water evaporated from the oceans and deposited on land is the principal source of river flow and of water resources for humans and freshwater ecosystems. It is also the principal sculptor of the land surface on which we live. In round figures, this currently amounts to about 40,000 km^3 a year (Figure 11.3). This is the amount by which precipitation on land exceeds evaporation from the land surface and terrestrial water bodies, respiration from animals and transpiration from terrestrial plants in an average year.

Humans currently use a mere 15 per cent of this resource globally, and most of this is returned to the environment within days, albeit variously polluted. Terrestrial ecosystems, including freshwater aquatic systems, require a further very substantial supply for healthy life, although the exact amount is currently difficult to determine (see Chapter 6 Water, land and wildlife). An additional one per cent of this resource gets locked up in permanent or semi-permanent terrestrial ice bodies every year, most of it in remote polar regions, where it may remain outside the active water cycle for centuries or millennia. This is partially replaced by glacial meltwaters that add a small but locally significant supply to rivers (Figure 11.4). Meltwaters provide a third of all the water used in irrigation worldwide and are especially important for irrigation prior to the annual monsoon along the great rivers draining the Himalayas. Overall, this would seem to suggest that there is no shortage of freshwater in the world.

However, averages can be misleading. Its distribution in time and space depends on the winds that carry the rainclouds. The global pattern of winds is set by the distribution of incoming solar radiation and the Earth's deflecting spin (Figure 11.5). This creates the greatest precipitation in equatorial regions, where the trade winds carry moisture evaporated from the tropical oceans into the convective low-pressure belt at the thermal equator, and local evaporation is rapidly recycled as rainfall. This is a haven for nature: it is the most species-rich region on land. But it is inimical to humankind. The Amazon basin has the largest liquid freshwater resource on Earth, but one of the lowest population densities outside the drylands. The secondary precipitation peak in mid-latitudes occurs in far more amenable temperatures, and this has been the locus of most human development. Here, the convergence of warm, moist air from the tropics and cold, dry air from polar regions causes the warm air to rise, generating low pressure 'depressions' and frontal precipitation at the weather front where the two air masses meet. In warm periods, this is supplemented by precipitation developed by local convection.

Between these two precipitation peaks lies a region of low precipitation, largely generated by the high-pressure belt, which forms the poleward arm of the tropical

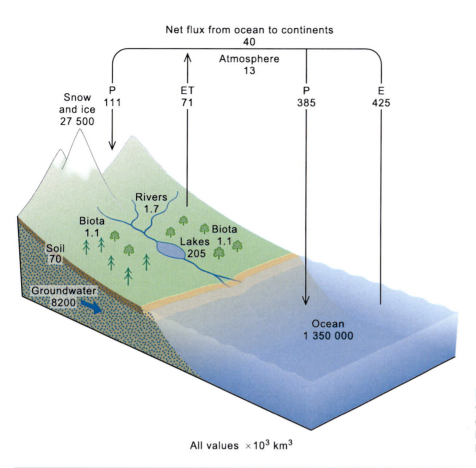

Net flux from ocean to continents
40

Atmosphere
13

Snow and ice 27 500

P 111   ET 71   P 385   E 425

Rivers 1.7

Biota 1.1

Biota 1.1

Lakes 205

Soil 70

Groundwater 8200

Ocean 1 350 000

All values ×10³ km³

Figure 11.3 Storage and exchange within the average global water cycle (in thousands of cubic kilometres) P = precipitation, E = evaporation, ET = evapotranspiration

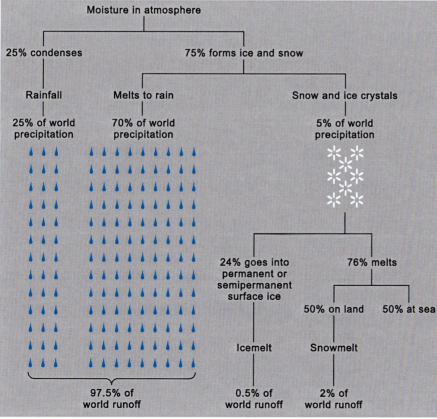

Moisture in atmosphere

25% condenses

75% forms ice and snow

Rainfall

Melts to rain

Snow and ice crystals

25% of world precipitation

70% of world precipitation

5% of world precipitation

24% goes into permanent or semipermanent surface ice

76% melts

50% on land   50% at sea

Icemelt

Snowmelt

97.5% of world runoff

0.5% of world runoff

2% of world runoff

Figure 11.4 Pathways of rain and snow feeding the world's rivers

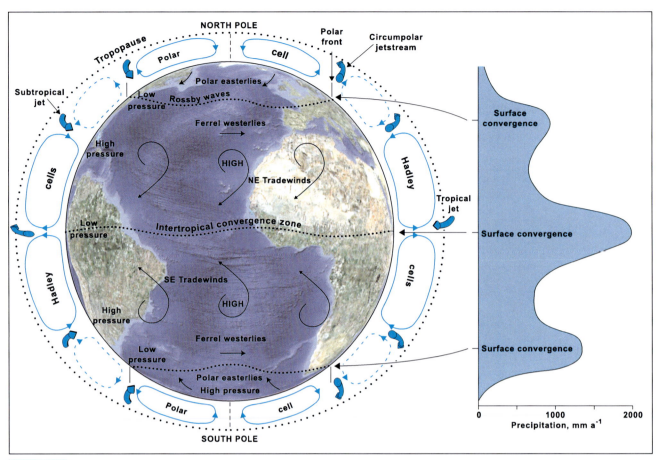

Figure 11.5 Global wind systems and the pattern of precipitation

'Hadley cells' (Figure 11.5). Here, descending air stifles convection and inhibits the build-up of storm clouds, even though in some areas like the Rajputana desert in Rajastan the air might be as moist as a tropical rainforest. In the northern hemisphere, this region extends well into Central Asia as 'continentality' takes its toll: the moisture content of the westerly winds blowing to the north of the Himalayas reduces as they progress further from the sea.

Polewards of the mid-latitude precipitation peaks, the precipitation potential is reduced by colder seas, colder air and the sinking air over the poles.

Superimposed upon this average pattern, temporal fluctuations beset most regions. Atmospheric water transport is a fragile system, partly because the volume of water stored in the air is the smallest of all, providing little capacity for smoothing over fluctuations in the amount of input from evaporation in warmer or colder periods. Partly, it is because the storm tracks shift. Location and intensity change with the seasons, with longer-term fluctuations in atmospheric pressure gradients, with warm-

ing and cooling trends in the Earth-atmosphere system, and even due to shifts in warm or cool ocean currents.

The ever-changing components

Precipitation

Precipitation is the driving force for water resources. Figure 11.6 shows the average annual pattern of precipitation. Note the belt of low precipitation across the subtropics in both hemispheres, which is caused by the subtropical high-pressure zones where descending air stifles convection (see Figure 11.5). This is most marked in the northern hemisphere, which has more land mass in the subtropics, where it extends into Central Asia and is now spreading into North China.

Figures 11.7 to 11.10 show the seasonal march of precipitation around the world, based on a new analysis of recent climatic data. In January, the dry area in the northern hemisphere is far larger than the annual average, covering most of North Africa and Asia and central

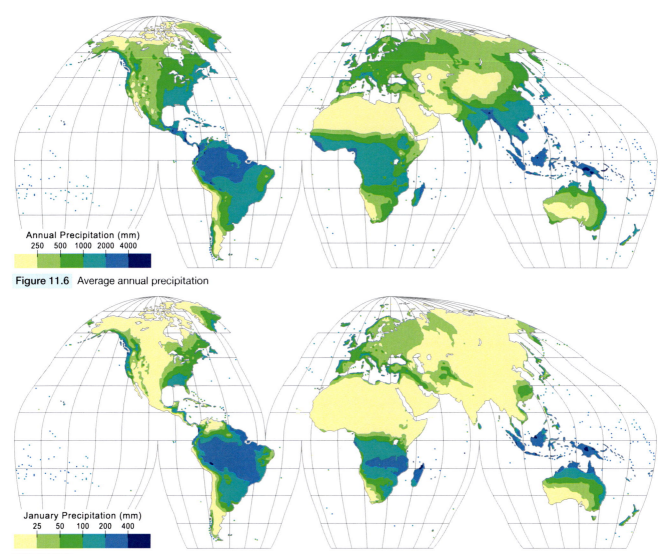

Annual Precipitation (mm)
250 500 1000 2000 4000

Figure 11.6 Average annual precipitation

January Precipitation (mm)
25 50 100 200 400

Figure 11.7 January precipitation. Winter crops in the Middle East require irrigation. Extensive dry areas cover most of Asia, North Africa and central North America

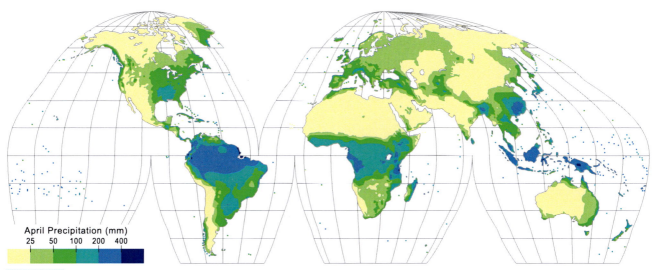

April Precipitation (mm)
25 50 100 200 400

Figure 11.8 April precipitation. Rains begin to reduce the dry areas in the northern hemisphere, but drought extends in the southern hemisphere, especially in Australia and southern Africa

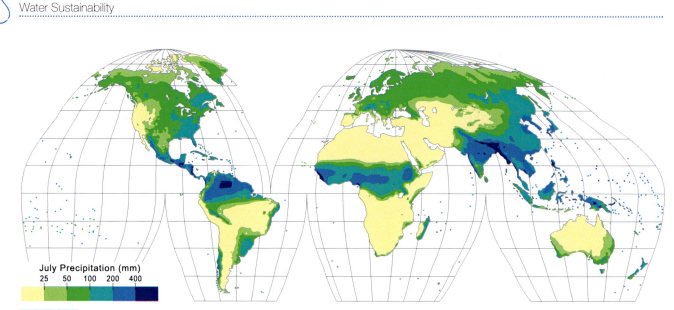

Figure 11.9 July precipitation. Expansion of low precipitation across the southern hemisphere (winter) and through the Mediterranean into the Middle East (summer). Monsoons in SE Asia

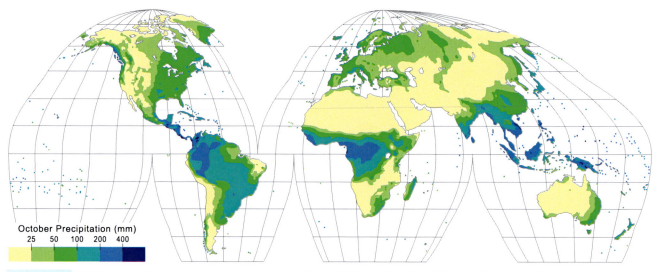

Figure 11.10 October precipitation. Asian monsoons die away, but tropical rains in central Africa and South America expand

North America. Winter crops in the Middle East require irrigation from groundwater. This is the wet season in the southern hemisphere mid-latitudes. By April, the rains in the southern hemisphere are generally on the retreat and rain is returning to Southeast Asia. In July, the southern hemisphere is in the middle of winter, oceans are cooler and precipitation is near its lowest ebb. The Mediterranean also receives less rain as the depression tracks migrate polewards. Rain expands across North America east of the Rockies as the westerlies suck warm, moist air from the Caribbean up the Mississippi-Missouri basin. In South Asia, the monsoon rains appear. By October, the rains increase in Western Europe as tropical air laden with moisture from the warm ocean current – the Gulf Stream/North Atlantic Drift – meets air from a cooling

arctic. But the rains die away in North China and Central Asia along with the retreat of the monsoons in South Asia. Rain is returning to the southern hemisphere as the Sun warms the ocean surfaces.

The traditional solution to these seasonal surpluses and deficits has been to build dams to hold back the water from one season to the next. Figure 11.11 shows that seasonality is highest in the subtropics and especially around the world's deserts. It is also high in the prairies and Great Plains of North America and in North China. Both of these areas are key agricultural regions now suffering long-term water management problems.

Variations in water resources from one year to another generally present more problems. Inter-annual vari-

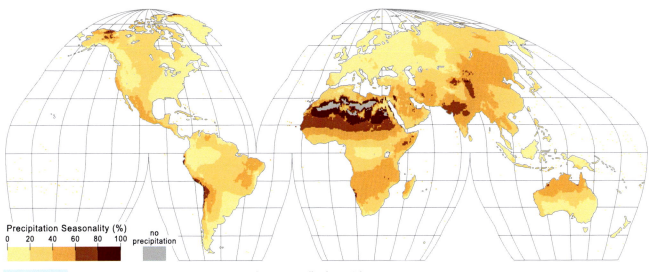

Figure 11.11 Seasonality of precipitation (see Notes for seasonality formula)
Seasonality problems are worst in the subtropics, especially around deserts

ability is measured in Figure 11.12 by the coefficient of variation. Once again, the desert margins in both hemispheres stand out as highly variable. The drought that threatened in Niger in 2005, though aggravated by human mismanagement, was part of the same high level of variability that has caused a series of Sahel droughts and mass starvation, attracting large-scale international aid from the time of the great drought of 1968–72, through Bob Geldof's Band Aid Trust in 1984 and the subsequent Live Aid concerts to Live 8 in 2004 (see 'Drought in the Sahel' in Chapter 3). A key problem for West Africa is the degree to which the rain-bearing Inter-tropical Convergence Zone (ITCZ) – the meeting

point between the atmospheric circulation in the two hemispheres – penetrates inland in the summer. There is some evidence that penetration has been weaker in recent decades and that there are cycles in its movement.

Figures 11.13 and 11.14 show extremes of wet and dry years. They plot the wettest or driest year to be expected once in 20 years, so that only five per cent of years are likely to be as wet or as dry. Note that the probability relates to each site on the globe. The extremes never occur everywhere across the globe at the same time, so these maps do not show real years viewed globally, but

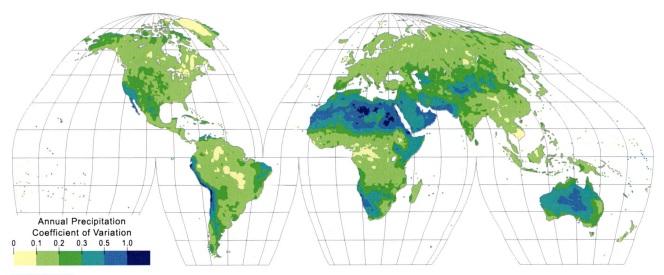

Figure 11.12 Year-to-year variability of annual precipitation, as measured by the coefficient of variation. Inter-annual variability can be more difficult to manage than the seasonal cycle. Desert margins are most vulnerable

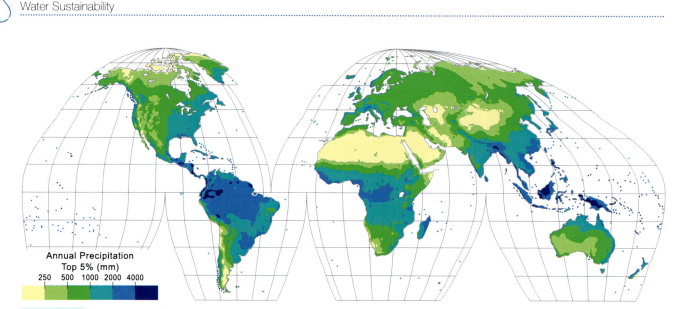

**Annual Precipitation
Top 5% (mm)**
250 500 1000 2000 4000

Figure 11.13 Very wet years. The map shows the largest annual precipitation to be expected once in 20 years at each location. Note that this is not a world map of a particular year, but of the once-in-20-years event at each point on the map. Wet years can be problematic in equatorial and monsoon climates, whereas the western USA, Central Asia and Australia fare better, with lower extremes

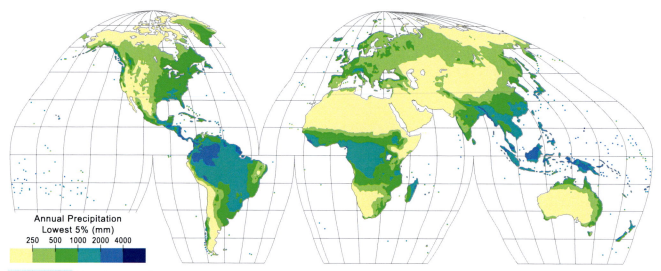

**Annual Precipitation
Lowest 5% (mm)**
250 500 1000 2000 4000

Figure 11.14 Very dry years. The map shows the lowest annual precipitation to be expected once in 20 years at each location, as in Figure 11.13. Note the marked expansion of dry areas in the Sahel, Central Asia and Australia compared with the map of average years

they represent a measure of 'worst case' for each site. These show that the western USA, Central Asia and Australia fare significantly better in the 'very wet years', but in the very dry years, the dry areas expand markedly in the Sahel, especially in the Horn of Africa, and in Central Asia.

Three factors – low mean annual precipitation, high seasonality and high year-on-year variability – have been plotted together in Figure 11.15, in order to distinguish the main sources of water management problems. For clarity, only the worst ten per cent of land area afflicted by these problems has been mapped. The most difficult

water management cases are where two or three of these indicators combine. This again highlights the desert margins. The most problematic areas are in North Africa, the Middle East and North China. The Australian interior suffers particularly from highly unreliable rainfall year to year, but does not generally rank as extreme on the other indicators as those other areas. Southern California, however, experiences the dual combination of ranking high in both inter-annual variability and low precipitation.

The worst problems occur where two or three of these combine. Again, desert margins stand out.

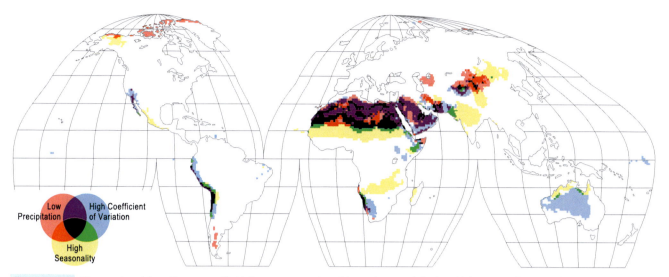

Figure 11.15 The most problematic areas with triple management problems: low precipitation, high seasonality and high inter-annual variability

Evaporation

Although precipitation is generally more important than losses due to 'evapotranspiration' for determining the local water balance in most of the world, evaporation, especially from the oceans, drives terrestrial precipitation. The transpirational component is also the key to plant life and food production. Two-thirds of continental precipitation is returned to the atmosphere as water vapour, of which at least half is transpired by plants. Water taken up by the plant roots is broken down in the process of photosynthesis and combined with carbon to grow 'biomass'. Some of this water is reconstituted and evaporated away in the process of respiration. The remainder acts as a transporter of nutrients through the plant and is eventually returned to the atmosphere as transpired water vapour. Falkenmark and Rockström (2004) have called the water that plants return to the atmosphere as vapour 'green' water to distinguish it from 'blue' water, which takes the liquid route back to the sea via rivers or seepage on the coast (so-called 'runout'). They estimate that 60–70 per cent of human food production is rainfed, using 'green' water, as opposed to artificially irrigated with 'blue' water, and that 1200 m³ of this green water are required to produce enough food to adequately feed one person for a year.

Figure 11.16 shows the pattern of actual evaporation and evapotranspiration in the average year. The lowest rates are found over the deserts, both warm and cold (ice covered). The highest rates occur over subtropical seas, asso-

ciated with clear skies in the subtropical high pressure belts. Warm ocean currents heading polewards on the western margins of the oceans extend high rates of evaporation into mid-latitudes. These provide major sources for precipitation in mid-latitudes. The Gulf Stream and North Atlantic Drift raise ocean temperatures and evaporation rates across the North Atlantic, feeding rainfall over Western Europe. Evaporation off the east coasts of the USA and Japan is also enhanced by the 'oasis effect', whereby the westerly winds blowing over the warm Gulf Stream and Kuro Shio currents take up moisture more readily because they have been dried during their long passage over the continents. In contrast, cold currents tend to reduce evaporation on the eastern margin of many oceans, notably on the west coast of the Americas and southern Africa, contributing in some regions to coastal deserts. Low rates of actual evapotranspiration in mid-latitude continental interiors, despite high summer temperatures, are principally caused by the lack of rainfall.

Evaporation rates are slightly reduced over equatorial waters by the cloud cover in the ITCZ. In contrast, evaporation on land peaks in the equatorial rainforests. Actual evaporation rates at sea are typically two to three times greater than on the adjacent land, although greater differences are found around subtropical deserts of the subtropical high-pressure belt and the least difference occurs on the margins of equatorial rain forests.

As with precipitation, there are wide differences between seasons and between years (as shown in Fig-

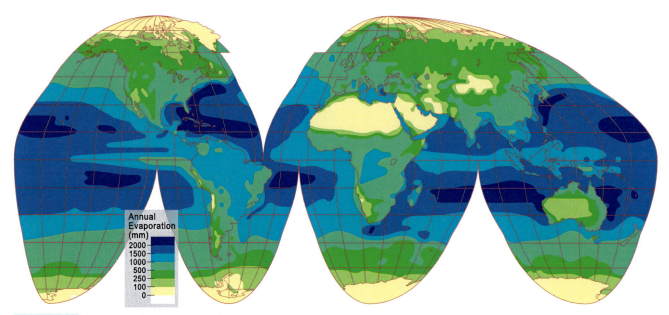

Figure 11.16 Mean annual actual evaporation

ures 11.17 to 11.20). January sees generally below average evaporation over the oceans, except around the warm currents in the Caribbean/Gulf Stream and the Japanese Kuro Shio (Figure 11.17). Evaporation is reduced over much of the northern continents, except western Europe. Interestingly, some 'negative evaporation' appears over snow-covered surfaces in arctic and subarctic regions as a result of sublimation of atmospheric moisture on the snow. High evaporation rates spread into mid-latitudes in parts of the southern conti-

nents during the summer. By April, oceanic evaporation tends to reach a minimum worldwide. On land, rates are beginning to reduce in the southern continents and to increase in the northern continents. July sees some return of higher rates in the subtropical seas of the southern hemisphere and a marked increase on land in the mid-latitudes of the northern hemisphere. Rates have decreased in both these regions by October, but rates remain high over Western Europe and the Mediterranean, and notably in the Sahel.

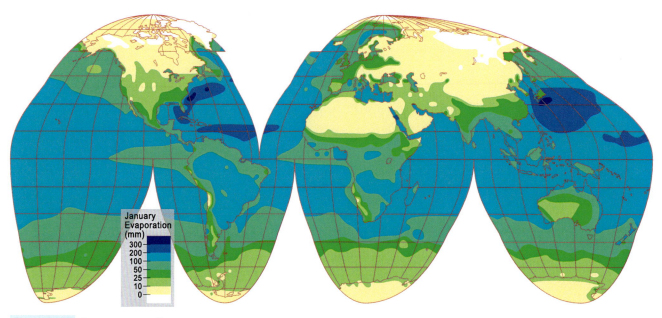

Figure 11.17 January evaporation

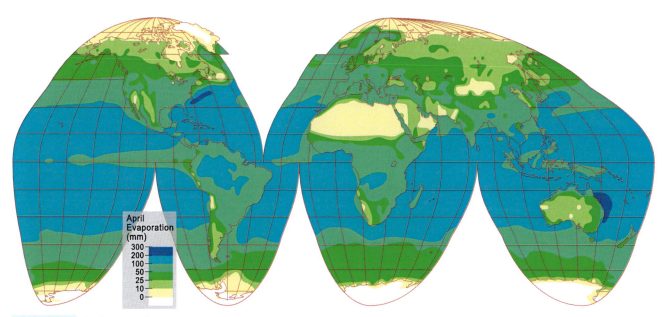

Figure 11.18 April evaporation

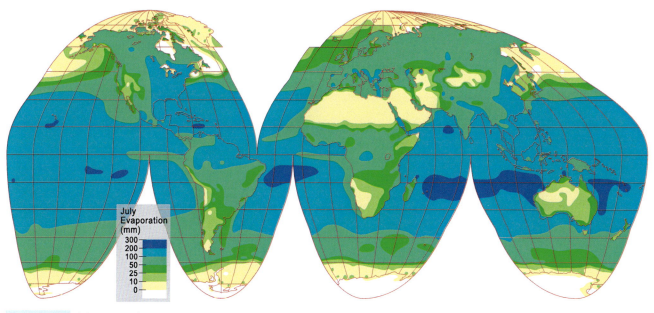

Figure 11.19 July evaporation

The coefficient of variation mapped in Figure 11.21 indicates highest variability is to be found in and around the deserts, again both hot and cold deserts. The subtropical oceans show generally low variability. One very notable feature occurs in the equatorial Pacific, where higher variability is caused by the El Niño Southern Oscillation.

The greatest year-to-year differences in the *total volume* of evaporation are found in the oceans (compare point locations in Figures 11.22 and 11.23). However, in percentage terms, some of the greatest inter-annual variation occurs on land (Figure 11.21). Once again, it is the desert margins that stand out. Here, the high level of variability in evapotranspiration losses compounds the problems created by high variability in rainfall. Some of the lowest variability is found in the high mid-latitudes of the northern hemisphere: much of north eastern North America and the

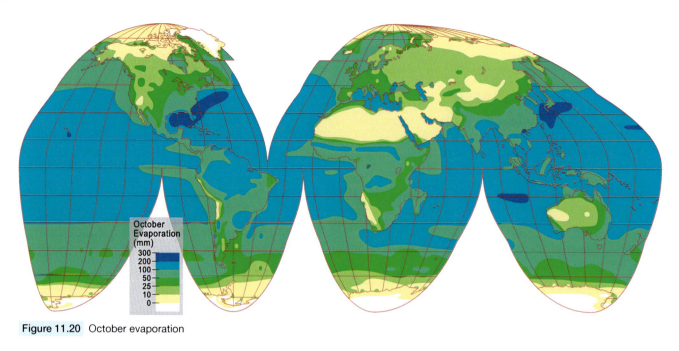

Figure 11.20 October evaporation

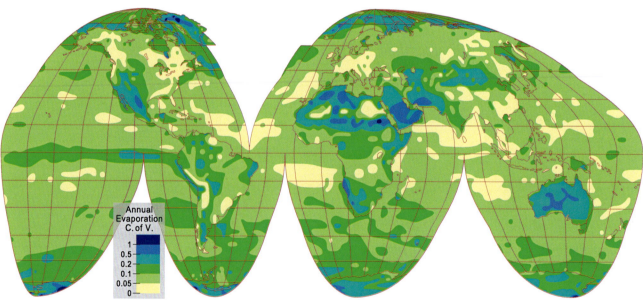

Figure 11.21 Inter-annual variation in evaporation. The coefficient of variation is particularly high in desert margins

central belt of the EU from France to Poland have the lowest degree of variation.

The high inter-annual variation in evaporation in the eastern equatorial Pacific, is due to the quasi-regular, cyclical reversal of the normal east-west flow of the ocean surface current. Every five years or so the so-called El Niño current flows from west to east, bringing warmer waters (up to 7°C above normal) to the coast of Ecuador, accompanied by high evaporation rates and torrential rains and floods on the mainland.

Net water balance

The total amount of water resources available is obviously determined by the balance between precipitation receipts and evaporation losses. Figure 11.24 shows the mean annual balance around the globe. The bluer the area on the map, the more positive the balance and the greater the local water resources. Conversely, the redder the area, the lower the level of resources. The map highlights the water scarce regions that extend across North Africa and the Middle East and on into Central

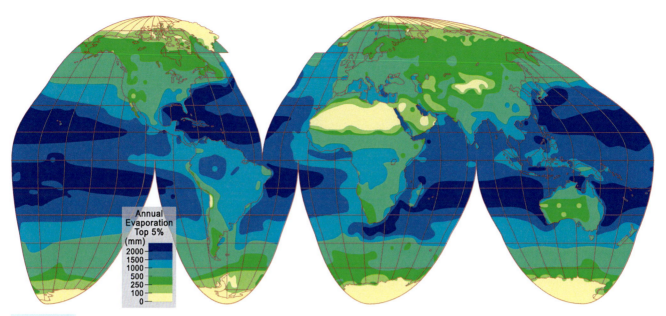

Figure 11.22 Highest annual evaporation in 20 years at each location. Evaporation is considerably higher in desert margins and equatorial rainforests than in the average year

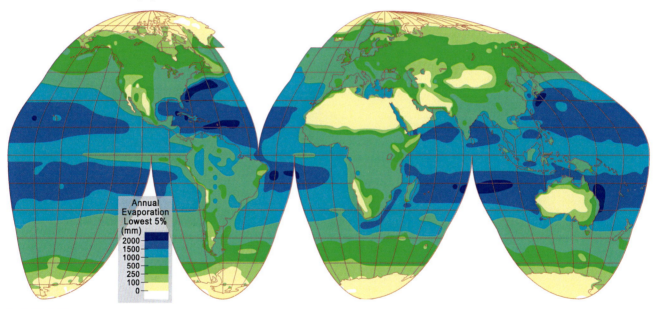

Figure 11.23 Lowest evaporation years. Again, the greatest deviations from normal are in desert margins and equatorial rainforests

Asia, and also the areas that have been called the 'west coast deserts' in South America, southern Africa and Australia, where cold currents from Antarctica reduce oceanic evaporation and provide less moisture for the onshore winds.

The Humboldt Current off the west coast of South America is primarily responsible for the Atacama Desert, but its importance is far more extensive. Fluctuations in the strength of the current are an integral part of the El Niño phenomenon, which has now been shown to have

an influence on weather patterns over a large part of the world's mid-latitudes, including America and Western Europe (see *El Niño/Southern Oscillation (ENSO)* in Chapter 19).

These cold currents chill the air and reduce its vapour-holding capacity. But although these effects reduce rain-storms, they can produce extensive coastal fogs, which local wildlife benefits from. For example, a species of beetle in Namibia collects moisture by raising its fore-wings to catch fog droplets.

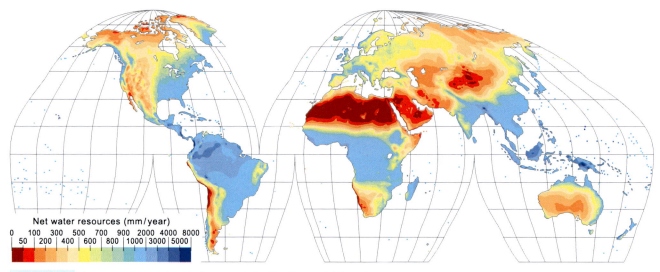

Figure 11.24 Average net annual water balance, indicating the available water resources

In fact, fog and dew are a much under-researched water resource. They do not appear in conventional assessments of water receipts, which are based on rain gauges. Yet measurements made decades ago suggest that Table Mountain in South Africa receives half again as much moisture from dew as it does from rainfall. We are still a long way from being able to assess the amount of water that certain ecosystems receive from these sources of so-called 'occult precipitation', but recent experiments in South Africa are leading the way towards possible exploitation of fog for human water resources (see *Harvesting fog and dew* in Chapter 14).

Runoff

About one-third of continental precipitation drains to the sea as liquid water. Surface runoff provides the bulk of the water used by humans worldwide and is the lifeblood of freshwater aquatic ecosystems and much of terrestrial wildlife. Despite 'water crisis' concerns, average riverflows around the world are estimated to be some seven times greater than the total amount of water currently extracted for human use. Unfortunately, only about one-third of this is 'stable' flow. The rest occurs in various levels of floodflows, which generally require artificial regulation by dams or weirs to convert them into usable water resources.

All too obviously, however, the extremes of high and low flows create some of the most devastating environmental hazards. Floods are responsible for about 40 per cent of the damage caused by all natural disasters. They are estimated to have claimed the lives of over 75,000 people a year during the twentieth century – a rate of 34 people per million p.a. (Goklany, 2007). Droughts killed even more: over 130,000 p.a. – 58 per million.

However, floods are not all bad. Floods can clear and clean rivers and deposit valuable nutrients and silt on the land, renewing channels and floodplains for new life to develop. The annual flood on the Nile was the mainstay of life in Egypt from the earliest civilizations to the twentieth century. Nature conservationists have recently championed the use of artificial floods as a means of restoring habitats destroyed by dam-building. Notable experiments with artificial flood releases from dams have been undertaken on the Colorado River and in Africa (see *Artificial floods* in Chapter 15). Even lesser fluctuations in flow can have valuable ecological functions.

The average pattern of riverflow shows a very skewed distribution, with most of the continental land area generating less than 100 mm of runoff per square millimetre per year, and a few hot spots generating in excess of 4000 mm (Figure 11.25). The lowest flows are found in the subtropics, western North America and central Asia, associated with low rainfall and high evaporation.

Marked seasonal shifts are visible in Figures 11.26 to 11.29, with particularly noticeable poleward movements in tropical regions associated with summer in both hemispheres. January is mostly drier than the annual average, except in Western Europe, eastern North America and equatorial regions. During April, riverflow generally increases in Europe and North America, fuelled by snowmelt and increasing sea surface temperatures, but

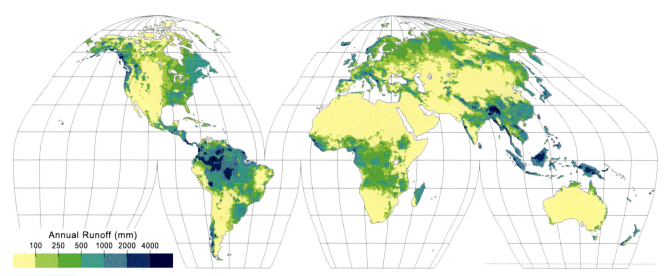

Figure 11.25 Average annual runoff. The subtropics and continental interiors have the lowest riverflows

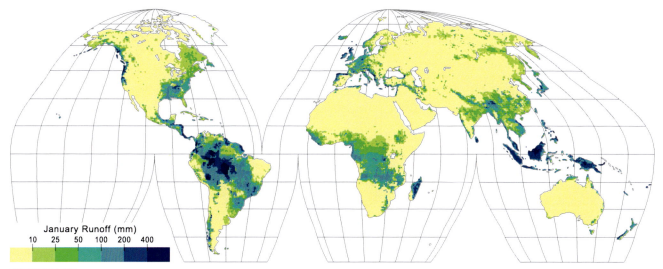

Figure 11.26 January runoff. Most regions are drier than average except western Europe, eastern North America and equatorial regions

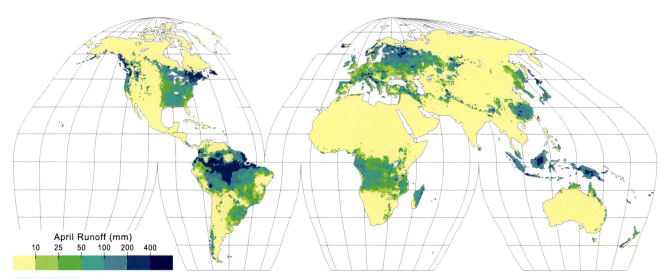

Figure 11.27 April runoff. Expansion in the area of higher runoff in Europe and North America; contraction in the Sahel and SE Asia

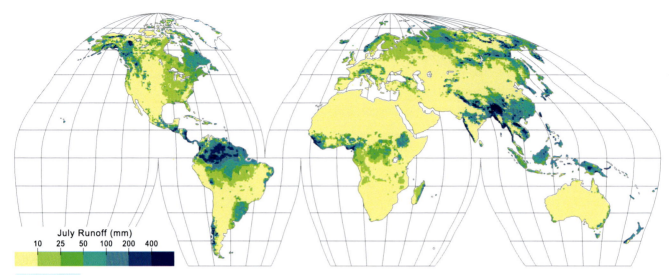

Figure 11.28 July runoff. Central Africa dries, but runoff returns to the Sahel, monsoon runoff appears in SE Asia, and there is more runoff in high latitudes

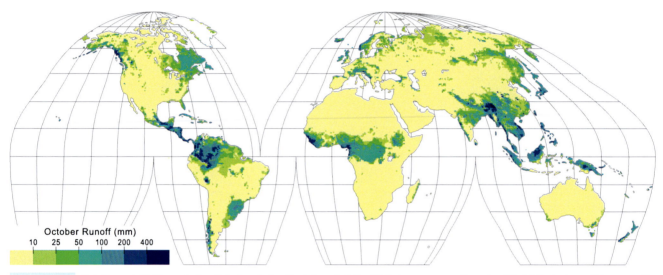

Figure 11.29 October runoff. Rivers in the Sahel are fuller, the tail-end of the SE Asian monsoon flows is still visible, but runoff is generally reduced in both hemispheres

is at a low ebb in the Sahel and south east Asia. In July, Africa is generally drier, but runoff returns to the Sahel with the ITCZ. Southeast Asia experiences runoff from the monsoon rains and there is more riverflow in higher latitudes in the northern hemisphere as depression tracks are pushed polewards by the northward migration of the westerlies and the subtropical high pressure belt. By October, the Sahel rivers are fuller from the West African monsoon, the tail-end of the south east Asian monsoon flows is still visible, but in both hemispheres as a whole runoff is generally below average as the autumn equinox signals the change of seasons.

The seasonality of riverflows can be a major problem for water resources management: it has been the prime

stimulus for dam-building for millennia. The greatest contrasts between the seasons are found on the desert margins, but these 'margins' can extend well beyond into some regions not normally regarded as semi-arid (Figure 11.30). Seasonal contrasts are high throughout most of Australia. Critical contrasts also occur in the western USA and Canada and over much of the Russian Federation, two of the most important 'bread basket' regions of the world.

Unfortunately, riverflow records are generally too short to enable a meaningful analysis at the global scale of inter-year variability or the risk of extreme events. These analyses are only possible for a few countries, principally in Western Europe and North America. The chal-

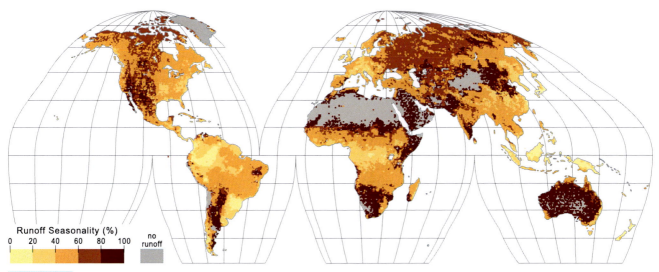

Figure 11.30 The seasonality of river runoff

lenge of improving data collection is a major issue and is discussed in Chapter 18 Improved monitoring and data management.

However, data collated by the Global Runoff Centre in Koblenz give an interesting picture of the geographical distribution of river discharges into the ocean, calculated for units of 5° square (Figure 11.31). These reflect the dominance of the great rivers of the world, like the Amazon, Orinoco and La Plata in South America, the Congo and Niger in Africa, and the Ganges, Irrawaddy, Yangtze and Mekong in Asia, and the Mississippi in

North America. But there are also substantial 'second tier' discharges from many smaller equatorial and tropical rivers, the rivers of the mid-latitude rainforests of British Columbia, and rivers fed by meltwaters in Canada, Alaska and arctic Russia. The other notable feature of this map is the areas of internal drainage – rivers, often seasonal, that naturally do not reach the sea, as opposed to rivers like the Hwang Ho that frequently fail to reach the sea due to over-exploitation by man. Many of these 'internal sinks' have little or no surface water, but when and where rivers do exist they feed salt lakes or 'seas', like the Aral Sea, Utah's Great Salt Lake or Lake Eyre

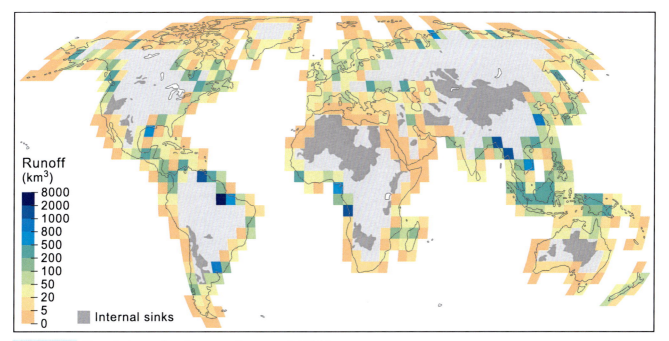

Figure 11.31 River discharges into the ocean. (Courtesy of GRDC.)

in Australia. Salt Lake City is a shining example of successful human habitation in this restrictive environment. The Aral Sea is the exact opposite, where over-exploitation of water resources has created the worst human and environmental disaster in history (see *Aral Sea – lost and regained?* in Chapter 15). Improvements in the technology and cost of desalination are critical for continued survival in these environments (see *Desalination* in Chapter 16).

Finally – a few geopolitical observations. Water resources are not only essential for human life and food production, they are also an important foundation for economic development. Looking at the maps of riverflow reveals that all the so-called 'Bric' countries (Brazil, Russia, India and China), which economists predict will be the powerhouses of economic development in the coming decades, are blessed with substantial water resources centred on some of the world's greatest rivers. Brazil and China have already begun major hydropower developments on their rivers. China is diverting more riverwater from the south towards Beijing. But the blessing comes with a caveat: many are not in total control of their major rivers. China and India need to have control the sources of their rivers in the Himalayas and Tibetan Plateau. Hence retaining control of Tibet is vital to China, if only human rights can be better respected and the Dalai Lama's compromise of autonomy, not independence, be put into effect to the satisfaction of both parties. To India's chagrin, Tibet also puts China in control of the Brahmaputra. Conversely, India claims a sizeable section of the headwaters of the Indus River, which is the lifeblood of Pakistan, and this is an element in the long-running dispute over control of Kashmir.

Brazil has the greatest river of all. Yet it too lacks control of the headwaters, and future developments on the headwaters in Peru, Colombia and Ecuador could reduce resources. But the Amazon is different from the Ganges or Brahmaputra in so far as it is also sustained by substantial rainfall in the rainforests of the central basin, so the mountain headwaters are less significant. Even so, the economic centre of the country lies well to the southeast of the Amazon basin, and its second river, the Paraná, crosses into Paraguay and Argentina, discharging into the La Plata estuary between Argentina and Uruguay (see the peak discharge south of the Amazon in Figure 11.31). Both countries have exerted some political constraints on development in Brazil, although Brazil has the upper hand as it controls the headwaters.

And Russia? In the breakup of the Soviet Union in 1991, Russia retained control of all its major rivers, bar the upper reaches of the large Irtysh tributary of the Ob, which went to Kazakhstan. Arguably, therefore, Russia is the most favoured among the Bric countries to be able to develop its substantial water resources free from international interference – especially the seven high yielding rivers draining over 100 km^3 a year each into the Arctic Ocean (shown in Figure 11.31). The big drawback is that all but two of the large arctic rivers are in remote Siberia, and only the Volga (entering the Caspian Sea) serves the Russian heartland.

China controls all the headwaters of its major rivers, and some others (like the Mekong), but the river that serves its main agricultural region, the Hwang Ho, is failing. Both China and India have major plans to redistribute riverwaters across the country to help water-stressed regions. Russia abandoned similar plans during the 1980s, but what of the future? Many of the old arguments for and against disappeared with Communism, but new ones may be appearing, like climate change, which is likely to adversely affect the agricultural south of the country, and industrial growth, which could benefit from hydropower. Climate change is also likely to add to China's reasons for diverting more riverwater from the south to the increasingly parched north. Many of these issues will be discussed fully in Chapter 7 Dams and diversions.

Figure 11.31 also emphasizes America's current and potential future dependence on Canadian water resources. Apart from the Mississippi, all its major rivers have headwaters in Canada. In contrast, Canada owns the Mackenzie and the rivers of James Bay in their entirety. Canada has exploited James Bay for hydropower, which it exports to the USA (see *James Bay Hydro – redesigning a landscape* in Chapter 7). It has thus far tried to save the Mackenzie as an ecological preserve, but its grip may be slipping (see *The Mackenzie Basin: gained and lost?* in Chapter 5). Climate change is likely to emphasize the disparity between Canada and the contiguous states of America, and is an added consideration behind America's pressing for the full inclusion of water resources in the North American Free Trade Agreement.

The root problem for many countries, not just the Brics, is that the present division of the land into discrete countries pays little attention to the natural boundaries between river basins. The issues surrounding 'transboundary rivers' are analysed in greater detail in the section on *Transboundary waters* in Chapter 9. The same applies to transboundary groundwater, which is becoming an increasingly exploited source of water (see Chapter 12 Shrinking freshwater stores).

Regional variants of the water cycle

The maps of precipitation, evapotranspiration and runoff clearly indicate vast differences in the nature of the water cycle in different regions of the world. It is because of this that 'one-size-fits-all' solutions to water problems can often be dangerous. Measurements of rainfall alone are not sufficient to determine available water resources. Seasonal and inter-annual variability in all the components of the cycle mean that lengthy records are needed in order to determine the range and frequency of extremes, and so estimate risks. Errors in the design of dams and flood control structures, such as reservoirs not filling to capacity or floods larger than expected, have been caused by lack of hydrometric records or misjudging the type of river regimes.

Figure 11.32 gives a selection of generic annual water balances from around the world. Although the highest runoff in these few illustrative examples is generated in the Amazon basin, the highest proportion of rainfall translating into runoff is found in Japan, owing to lower evaporation losses there. Higher precipitation in the Himalayas is partly negated by high evaporation losses, and the greatest groundwater recharge potentials in these samples occur in the Amazon basin, eastern USA and Japan. The paucity of groundwater recharge in the two desert regions represented is a graphic illustration of the problems facing exploitation of groundwater resources in Australia and Saudi Arabia. There are, of course, countless other variants.

Until recently, global views of water balances were limited to mean annual statistics. Since the establishment of global data centres as repositories of data from around the world during the latter part of the twentieth century, however, it has become possible to look in more detail at the global pattern of river regimes. This theme is taken up in Chapter 18 Improved monitoring and data management.

Clouds and rain-forming processes

Scientific understanding of the processes by which rain and snow are created is still developing. And we are beginning to appreciate the importance of understanding these processes for a wide range of activities, not only weather forecasting and artificial rainmaking, but also for the study of future climate change.

The first significant advance in the science began with studies in laboratory cloud chambers by Scottish physicists John Aitken and Charles Wilson in the 1890s. They discovered that microscopic nuclei are necessary to cause water vapour to condense at normal atmospheric vapour concentrations: in pure, clean air with no microscopic dust particles, Wilson measured vapour contents seven times greater than can theoretically be held in air at a given temperature before condensation occurred. The most effective of these nuclei are now known as Aitken nuclei (Table 11.1). Some of the most efficient nuclei are hygroscopic, that is, they absorb moisture at vapour contents below the level where the air is 'saturated'. Common salt is an excellent example, which is capable of scavenging moisture from the air even when vapour content is only half saturation level: technically, at relative humidities as low as 50 per cent. Research has shown that many materials can act as condensation nuclei if the dust is the size classified as Aitken nuclei, including many types of clay minerals. The best nuclei have the same hexagonal crystal structure as ice or a related structure, especially at sub-zero temperatures, when they may be called freezing nuclei. Others can become more efficient nuclei once they have become coated in water.

The second major advance was made in the 1930s by the British-born Swedish meteorologist Tor Bergeron, who

Table 11.1 Types of condensation nuclei

Type of nucleus		Effectiveness
By condensation properties		
Non-wettable and insoluble		Useless
Wettable, insoluble and non-hygroscopic		Some value because they impersonate a water droplet once surface is wetted
Hygroscopic		**Most efficient**
By size (radius in micrometres)		
More than 10 μm		Ineffective as quickly drop out of the air
1–10 μm	Giant	No use unless hygroscopic
0.1–1.0 μm	Large	More effective
0.01–0.1 μm	Aitken	**Most effective**
Less than 0.001 μm		High levels of supersaturation of air needed before effective

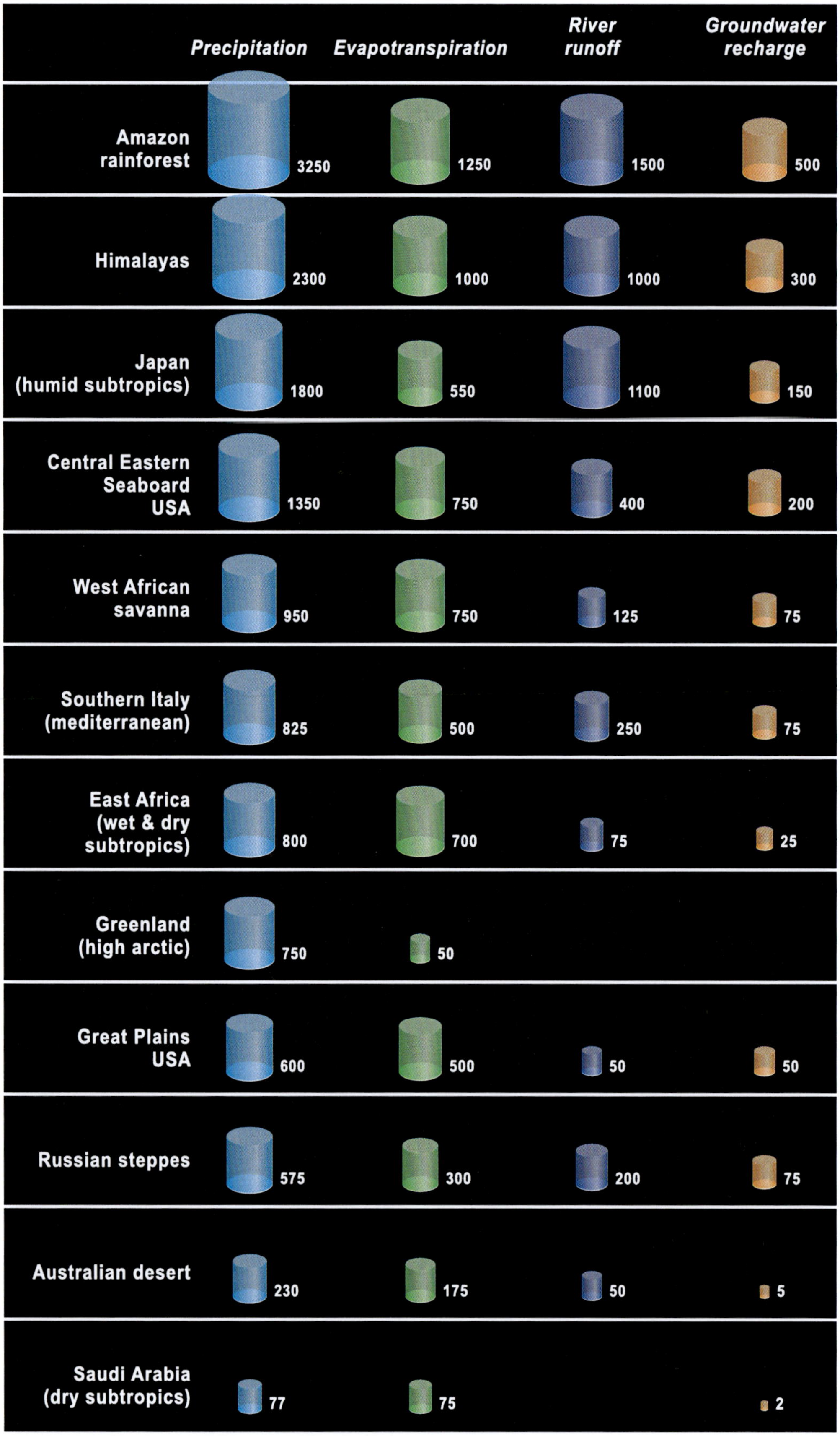

	Precipitation	Evapotranspiration	River runoff	Groundwater recharge
Amazon rainforest	3250	1250	1500	500
Himalayas	2300	1000	1000	300
Japan (humid subtropics)	1800	550	1100	150
Central Eastern Seaboard USA	1350	750	400	200
West African savanna	950	750	125	75
Southern Italy (mediterranean)	825	500	250	75
East Africa (wet & dry subtropics)	800	700	75	25
Greenland (high arctic)	750	50		
Great Plains USA	600	500	50	50
Russian steppes	575	300	200	75
Australian desert	230	175	50	5
Saudi Arabia (dry subtropics)	77	75		2

Figure 11.32 Generic water budgets for a variety of climatic regions. Millimetres per year

discovered that ice crystals are extremely efficient collectors of water vapour and suggested that rainfall may often be snowfall that has melted as it passes through warmer air nearer the ground. The heart of the theory is that at sub-zero air temperatures ice crystals will attract water vapour at lower atmospheric vapour concentrations than will super-cooled cloud droplets, and they will continue to scavenge moisture until the droplets are gone and 'vapour pressure' is too low. Walter Findeisen elaborated on the theory and hence it is more correctly known as the Bergeron-Findeisen theory.

The theory largely displaced the idea that raindrops are formed by condensation creating cloud droplets and then by multiple collisions between these droplets. Most rainmaking activities are still based on this premise, but since the 1990s significant advances have been made using alternative techniques based on the displaced theory of condensation and coalescence (see *Making Rain* in Chapter 17). Recent research is providing fascinating additional insights, particularly on the sources of condensation nuclei.

Most fascinating of all is the discovery of the role of biological ice nucleators. It has been known for 40 years that bacteria can be excellent nucleators, hence their use in artificial snowmaking (see *Snowmaking for leisure* in Chapter 17), but there was no information on how common they are in the natural atmosphere. Now, work published by Christner *et al* (2008) provides more detail on natural bacterial plant pathogens in the atmosphere. The authors estimate that these are widely dispersed in the atmosphere and that they may have an important role in ice formation. The bacteria were found to be abundant in fresh snow and ubiquitous in precipitation worldwide, with the highest concentrations in samples from well-vegetated regions like Montana and France and the lowest in Antarctica. They can operate at higher temperatures than many earth materials, at just –2°C, and are at their best between –4 and –7°C. The activity of most nuclei is mediated by proteins or proteinaceous compounds. Lysozyme decreases activity and they can be completely deactivated by heat. In contrast, heat does not deactivate mineral nuclei, but John Mason propounded a theory of 'training' whereby once clay minerals that have been activated by lower temperatures in high clouds like *cirrus* they can subsequently operate at higher temperatures: thus, he proposed that ice crystals falling from *cirrus* clouds may evaporate leaving trained nuclei to pepper the lower clouds and make up for the lack of first time nuclei there.

This recent research has important implications. The first is that plants play an important role in rain formation: more plants, more rain – another self-sustaining process for James Lovelock's Gaia world. Second, the agents are pathogens from infected plants, so could spreading infection or introducing infected plants help alleviate drought? Third, it has substantial implications for climate modelling in deciphering the biological interactions. It has also been found that extra nuclei are produced on the seashore when the common large brown seaweed called kelp is exposed to the sun between tides; the seaweed releases iodide, which forms condensation nuclei. Smoke from forest fires also provides nuclei. Dense smoke can inhibit rain formation by providing too many nuclei, but smoke that reaches higher altitudes can stimulate intense rainfall and hail formation. Fears have been expressed that increased fire risk due to drought in Amazonia might further reduce rainfall. Over the last half century, Clean Air Acts in most developed countries have reduced the particulate pollution that used to contribute to heavier rainfall over or downwind of urban-industrial areas (Jones, 1997). Is this now being replaced by forest fires like those in the tropical rainforests of Indonesia and Amazonia, and will the high concentrations of particles reduce rather than increase rainfall?

Current research at CERN also suggests that the processes that create aerosols and condensation and ice nuclei from gases might be more varied than previously supposed. It has long been realized that sunlight may act as a catalyst in 'photochemical reactions', for example, converting sulphur dioxide into sulphate particles, which may then act as cloud nuclei. But recent research by Svensmark and others at CERN suggests that bombardment by cosmic particles may have a similar effect (Svensmark *et al*, 2007). Svensmark postulates that electrons supplied by cosmic radiation may stimulate a kind of fast-breeder reaction, so that only a few electrons are needed to start a series of particle clusters that may form condensation nuclei. As Svensmark puts it: 'You can think of an electron as a teacher organizing several teams of children for a game. First one team, then another, and so on. In previous theories of cluster growth, each electron was supposed to remain with just one cluster – as if you needed a teacher for every team. The catalytic behaviour of the electrons is much more efficient.'

Kirkby (2009) notes that estimates of the importance of these ion-induced aerosols vary widely, ranging from

10–20 per cent of all aerosols over the land to more than 80 per cent. But even the lower estimate would mean that fluctuations in cosmic radiation reaching Earth could have a significant effect on global cloud cover and potentially on precipitation as well. The addition of extra cloud condensation nuclei over the land could create denser clouds with more but smaller droplets, as more nuclei compete for a fixed amount of water vapour. This will alter the thermal and radiational properties of the clouds. It could also suppress precipitation and extend the life of the clouds. The net result, however, will depend very much on local atmospheric properties, especially the pre-existing concentration of nuclei. Kirkby suggests they may be relatively even more important at high altitudes over the oceans where there are fewer 'background' aerosols from surface sources and lower concentrations of trace gases (see *Making clouds* in Chapter 17).

Implications for the geography of rain and snow

The processes outlined in these theories have paramount consequences for the geography of rainfall and global water resources, and indeed for their history. The first of these is that aerosols, dust particles suspended in the air, are the root cause of rainfall. Without them, there would be hardly any rain. Second, they play a significant role in causing a disproportionate amount of that rain to fall on the land because more dust is derived from the land. At around five million nuclei per litre of air, condensation nuclei are estimated to be five times more common over land than over the sea. This is supplemented by two other properties of the land – relief and a low thermal capacity. Hills and mountains cause the air to cool as it is forced over them, and sunshine warms up the land faster than the sea, causing convective currents that lift and cool the air.

We have to thank this tripartite combination for all our freshwater resources – the 40,000 km³ of extra precipitation over the land that feeds the rivers every year.

The third consequence is historical and it is down to Bergeron. The fact that most rainfall, at least in mid to high latitudes, is melted snow crystals means that a relatively small reduction in the temperature of the lower air could allow that snow to reach the ground. The factors that cause ice ages and interglacials are then amplified by this change of phase in the precipitation, because snow cools the climate by reflecting more solar radiation back to space. It thus creates the environment for its own continued existence on the ground, until the external warming processes once again cross a threshold, the melting snow exposes the ground and reflectivity suddenly drops from 80–90 per cent to 10–20 per cent. The ground and the air warm rapidly and the days are numbered for the rest of the snow. This is happening on a global scale and is the primary reason why the polar regions are warming much faster than the global average and will continue to do so for many decades. This sudden change also happens every year on a smaller, river catchment scale in regions of annual snowcover and causes the steep rise in river levels in meltwater floods.

The ice sheets and glaciers are only partly within the presently active water cycle. Most of the ice falls into what the Russian hydrologist M.I. Lvovich (1970) called a 'deadlock arc', a relatively inactive part of the system, rather like deep groundwater and the ocean depths. Meltwaters and icebergs are in the 'here and now'. The amount of water added to the ocean by calving icebergs is roughly equal to one-twentieth of annual riverflow. But the bulk of ice on land is stored for millennia before returning to active exchange.

Nevertheless, it is supremely important to the system. The growth and decay of land ice has played an integral part in climate change over the last million years and has had a major effect on the amount of freshwater in the system. At the peak of the last glaciation 18,000 years ago, around 64 million km³ of ice were stored on land. Now there are only 24 million km³ – the amount of meltwater generated by the retreat of the ice would be sufficient to keep all the rivers of the world flowing at present rates for nearly a thousand years.

As current global temperatures rise, the remaining ice is once again becoming a more active component of the water cycle. And as liquid water resources are under stress in some parts of the world, so there is increased interest in exploiting these frozen assets.

Conclusions

Over the last half century, we have learnt how extremely variable the world's climate is and how fragile the water resources are in many parts of the world. We saw in Chapter 10 how relatively small changes in the factors that control the weather can have major effects upon it. The water cycle is an integral part of that system and it shifts and changes endlessly at a vast range of scales.

Riverflow is probably the most sensitive of all, because it stands at the end of the line, gathering together all the variability in its driving forces, from precipitation and evaporation to changes in land surface properties. Yet rivers are the foundation of water resources and understanding what drives changes in riverflow is key to improving their management.

Discussion points

- Select and study an area of the world where climatic variability causes severe problems for water management.

- Investigate the effects of changes in the proportion of precipitation falling as rain or snow on river regimes and water management.

- Study the role of ocean currents in patterns of evaporation and precipitation.

Further reading

An excellent introduction to climatic process is given by:

Barry, R.G. and Chorley, R.J. 2010. *Atmosphere, Weather and Climate*. 9th edition, Routledge, Abingdon.

Sir John Mason's classic texts *The Physics of Clouds* and the easier *Clouds, Rain and Rainmaking* give a fascinating insight into the physical processes.

Precipitation processes are also covered in Jones, *Global Hydrology* (1997), Chapter 2.

12 Shrinking freshwater stores

The two great stores of freshwater, frozen resources and groundwater, are both suffering a steep decline. The icecaps and glaciers are suffering mainly because of climate change. Groundwater reserves are suffering mainly as a result of human exploitation. The loss of land ice is receiving wide media attention, while groundwater hardly gets a mention. Land ice may still hold the largest store of freshwater (Figure 12.1), but while ice masses are a vital source of water in some countries in mid- and low latitude, most especially around the Himalayas and Tibetan plateau, the Rockies and Andes, large-scale exploitation of arctic and antarctic resources is still merely a long-term possibility (see *Exploiting the cryosphere* in Chapter 14). In contrast, groundwater is a vital resource for many more people throughout the world. As surface water supplies diminish, so people and companies are increasingly turning to groundwater.

Shrinking land ice

'Glaciers and ice caps are indeed key indicators and unique demonstration objects of ongoing climate change. Their shrinkage and, in many cases, even complete disappearance leaves no doubt about the fact that the climate is changing at a global scale and at a fast if not accelerating rate … Glacier shrinkage may lead to the deglaciation of large parts of many mountain ranges in the coming decades.'

Wilfried Haeberli, Director
World Glacier Monitoring Service, 2009

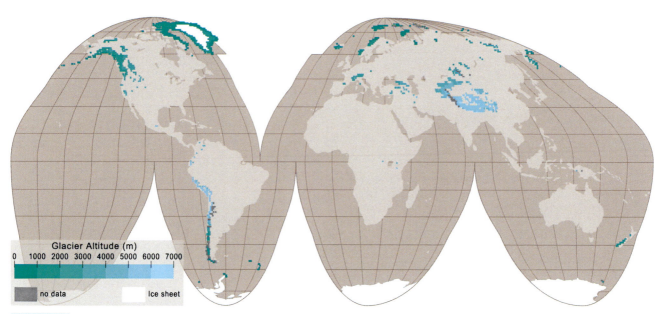

Glacier Altitude (m)
0 1000 2000 3000 4000 5000 6000 7000

no data   Ice sheet

Figure 12.1 World distribution of glaciers and ice sheets. Average altitudes shown are very close to the equilibrium line (After Cogley)

Evidence is accruing which suggests that ice sheets and glaciers are melting at an accelerating pace. Data collated by the World Glacier Monitoring Service in Zurich indicates that the world's glaciers thinned by an average of 74 cm in 2007 and have been melting twice as fast in the 2000s as in the previous 20 years. Virtually all mountain glaciers are melting (Figure 12.2), from the Alps to the Rockies, the Sierras and the Andes. Perhaps the most critical area of all is the Himalayas and the Tibet-Qinghai Plateau, where temperatures are rising faster than elsewhere in South Asia. The glaciers and ice caps there supply essential water to support irrigated agriculture and drinking water for over a billion people through major rivers like the Brahmaputra and Yangtze.

The polar ice sheets are also melting. Although these are not currently significant sources of water supply for humanity, they will have numerous indirect effects on established water resources through the effects of melting on weather patterns and the ocean (see Chapter 10 The threat of global warming).

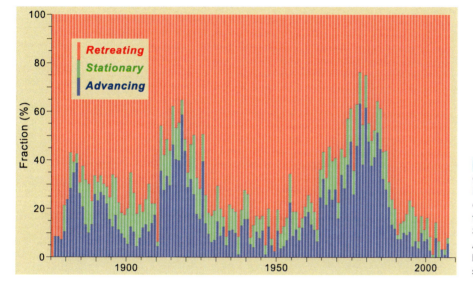

Figure 12.2 Fluctuations in world glaciers since the nineteenth century. Data from the Swiss Glacier Monitor compiled by Dr Andreas Bauder of the Glaciological Commission of the Swiss Academy of Natural Sciences. About 70 glaciers were monitored at the beginning of the twentieth century and slightly over 100 for the last 50 years

Glacier mass balances and recent changes

J. Graham Cogley

Measurements of mass balance are difficult and expensive. Direct measurements of the mass balance have become possible only recently for the two largest ice bodies, the Antarctic Ice Sheet and the Greenland Ice Sheet. The first credible estimates of ice-sheet mass balance were based partly on modelling and partly on measurements with poor resolution. For example, measurements of the accumulation rate are at best climatological, that is, they are averages over several decades, but this problem has now been alleviated by the development of regional climate models capable of simulating the measurements with acceptable accuracy. There are few *in situ* measurements of the melting rate in Greenland, although it can reliably be assumed to be nearly negligible in Antarctica. The rate of loss by calving is also very difficult to measure, but satellite remote sensing technology now yields a clearer picture of rates of loss of land ice (that is, of flow across the *grounding line*, which separates land ice from the floating ice of ice shelves). Most importantly GRACE, the gravimetric satellite mission launched in 2002 and still operational in 2010, has transformed our understanding of the net mass balance, that is, the sum of gains and losses.

Table 12.1 gives the current best estimates for the mass balance of the two ice sheets. Recent measurements make it clear not only that both ice sheets are losing mass to the ocean but that their rates of loss are accelerating.

Table 12.1 Glacier size and mass balances

	Area (10⁶ km²)	Volume[a] (10⁶ km³)	Balance[a] (mm a⁻¹)	Balance[a] (km³ a⁻¹)	Balance effect on sea level (mm a⁻¹)
Small glaciers[b]	0.741	0.241	−690±98	−511±72	1.41±0.20
Greenland Ice Sheet[c]	1.710	2.900	−134±19	−230±33	0.63±0.09
Antarctic Ice Sheet[c,d]	12.144	24.700	−12±6	−143±73	0.40±0.20
Ice shelves[e]	1.591	0.700	−700?	−1110?	0.00

Key:
a: All volumes are water-equivalent (not ice-equivalent). 1 mm water-equivalent a⁻¹ is equal to 1 kg m⁻² a⁻¹. 1km³ of (fresh) water has a mass of 1 Gt (10⁹ t).
b: All glaciers other than the Antarctic and Greenland Ice Sheets. Balance is an average for 2000–2005 based on *in situ* and geodetic measurements, and spatial interpolation; for 1960–1990 the equivalent figure is −173±73 mm a⁻¹, or 0.37 mm a⁻¹ of sea-level equivalent.
c: Balances are from analysis of trends in GRACE gravity fields for 2002–2009.
d: Grounded ice only, excluding ice rises.
e: Antarctica only; including ice rises. The mass balance of the ice shelves is very uncertain. The weight of ice shelves is supported by the ocean and therefore their balance has negligible impact on sea level.

On glaciers smaller than the ice sheets, mass-balance measurements are more numerous, but in a global population of as many as 300,000 glaciers there are *in situ* measurements for only about 350, and geodetic measurements (based on repeated mapping) for another 350. Most *in situ* measurement records are short, and most geodetic measurements have poor time resolution. In order to obtain a comprehensive regional or global estimate of change it is necessary to extrapolate to the many unmeasured glaciers. Uncertainty due to incomplete temporal and spatial coverage is illustrated in Figure 12.3. These uncertainties do not obscure the general picture, which is that the world's small glaciers had slightly negative balance during the 1960s and have been losing mass at an accelerating rate since then. There is some evidence that mass balances were also negative before the 1960s. This is consistent with more abundant, but indirect evidence that in general glacier terminuses have been retreating since at least the nineteenth century. (Terminus retreat, however, is not necessarily the same thing as mass loss.)

Figure 12.3 also makes a different point of significance in the water resources context. No assumptions about temperature are made in measurements of glacier mass balance. For monitoring global environmental change, glaciers and weather stations are therefore independent sources of information. The Figure shows that the two sources are consistent and well-correlated in their depiction of global change since the 1960s.

Modelling of the evolution of glacier mass balance during the twenty-first century suggests that plausible warming trends will bring about substantial reductions

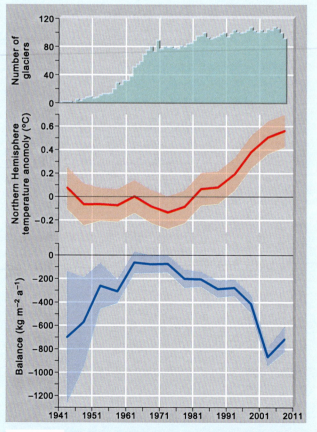

Figure 12.3 Trends in the mass balance of northern hemisphere glaciers compared with changes in average air temperatures (The bands around the lines indicate the level of uncertainty. The top graph shows the number of glaciers monitored.)

in the volume of small glaciers. However, the smaller the glacier the greater is the proportional predicted shrinkage, essentially because smaller glaciers are thinner. Most of the world's ice is in large ice bodies in unpopulated regions, but the smaller, more vulnerable glaciers are disproportionately concentrated in densely peopled drainage basins. Those glaciers which are in present use as water resources are therefore relatively more sensitive to climatic change, or at least to atmospheric change.

Predicting the evolution of the ice sheets is more difficult. Models of ice-sheet dynamics are less well developed than models of the climate, and in particular they are not yet able to simulate interactions with the ocean. This means that they cannot be used to assess 'unpleasant surprises' due for example to the incursion of warmer seawater beneath the ice shelves. Recent measurements in Greenland and West Antarctica have shown that such incursions are indeed responsible for accelerated basal melting beneath the ice shelves. This in turn is largely responsible for accelerated flow across the grounding line, and hence for the recently-detected acceleration towards faster mass loss from the ice sheets.

The Greenland ice sheet

The rate of melting in the Greenland ice sheet has accelerated rapidly during the last 20 years. In the mid-1990s, it was losing nearly 100 km^3 a year. The period 2006–8 was the fastest of all, with warm summers causing an annual loss of 273 km^3 a year, which amounts to an extra 0.75 mm in sea level rise (van den Broeke *et al*, 2009). In 2006, Greenland lost 1 km^3 of water equivalent every 40 days, equal to a whole year's water consumption for Los Angeles. It lost a record amount in 2007. The Jakobshavn Glacier in West Greenland has retreated nearly 50 km since 1851 (www.nsidc, 2009). According to Hanna *et al* (2008), five of the top nine years of melting in Greenland since 1958 have occurred since 2000. About half of the loss of ice is due to increased melting and half to more iceberg production, as the outlet glaciers have speeded up in part because more meltwater has lubricated the flow. By lowering cameras into surface sinkholes called 'moulins', Konrad Steffen of the University of Colorado has found that the ice sheet is riddled with tunnels. Surface meltwaters drain into the moulins and tunnels and lubricate the ice flow. The Greenland ice sheet has lost 1500 km^3 since 2000, equal to a 5 mm rise in sea level. If this rate continues as expected, it would add 400 mm to sea level by 2100. According to the IPCC (2007), if it were ever to melt completely, the global ocean would rise by 7 m – compared with 57–65 m if the Antarctic ice sheet melted or just 0.15–0.37 m if all the world's small glaciers melted. However, evidence presented by Jonathan Bamber of Bristol University to the 2009 Copenhagen Climate Conference suggests that the temperature 'tipping point' beyond which the Greenland ice sheet will melt irreversibly is twice as high as estimated by the IPCC (2007) report: a mean global temperature rise of 6°C, not 3°C, based on a new model which considers more of the physics. The IPCC estimated that the 3°C tipping point could be reached within decades. However, Bamber notes that Greenland retained an ice sheet half the current size throughout the Eemian Interglacial period 125,000 years ago when temperatures reached 5°C above the present.

Predictions are complicated by the fact that temperature and melting are only part of the equation. Snowfall rates can increase or decrease, offsetting meltwater losses or adding to their effect. Warmer seas and lack of sea ice around Greenland have increased snowfall. Hanna *et al* (2008) estimate that increased winter snowfall in recent years has offset about 80 per cent of the summer melting. This is delaying the ice sheet's demise. Nevertheless, melting of the marginal glaciers in Greenland is exposing more land and making mineral exploration easier. The first gold and silver mines were opened in 2009, helping to pave the way for Greenland's independence from Denmark; oil exploration is likely to follow. All of which may also have an effect on melt rates by increasing pollution. Indeed, airborne pollution has been increasing melt rates for some time. Hansen *et al* (2005) identify soot or black carbon deposits as a factor in arctic snowmelt by lowering the albedo of the snow. Flanner *et al* (2007) calculate that the recent effect on melting has been three times greater than that of greenhouse gases.

The latest predictions for the Greenland icecap use a combination of satellite observations, a regional atmospheric model, and observations of small changes in the Earth's gravity field: gravity lessens as the ice sheet melts. But the trend is complicated by climatic variability. Ice loss on two of the largest glaciers in Greenland doubled in one year in 2004 and then returned almost back to previous levels in 2006 (Howat *et al*, 2007). Van den Broeke *et al* (2009) report that the ice sheet would have lost

249

twice as much as observed since 1996 if snowfall had not increased at almost the same rate as melting increased up to about 2005, and if a substantial part of the meltwater had not refrozen in the cold snowpack. Before accurate satellite altimetry, this 'superimposed ice' used to make it difficult for glaciologists to estimate changes in glacier mass balances because it blurs the boundary between the area of net gain on the upper glacier, the 'accumulation zone', and the area of net loss on the lower glacier, the 'ablation zone', which loses its snowcover in summer. This is why the 'snowline', which used to be thought of as the dividing line between these two zones, is not necessarily a good indicator of the extent of the ablation zone. Glaciologists speak of the 'equilibrium line', which is the true dividing line between ablation and accumulation, and the best indicator of changes in the growth or decay of a glacier. Unfortunately, it is more difficult to observe because there is less visual contrast between old ice and superimposed ice than between ice and snow. This was a problem for the first attempts to catalogue the state of the world's glaciers in the 1970s, which were based largely on analysis of aerial photographs and maps. Accurate altimetry overcomes this problem, but many of the old glacier records in the archives only recorded snowline levels, which may be a poor indicator of the state of the glaciers' 'mass balance'. Even now, there are many glaciers worldwide that lack accurate records of the equilibrium line or the mass balance in the World Glacier Inventory (NSIDC, 2009).

Mountain glaciers

The World Glacier Inventory now holds detailed measurements for about half of all ice bodies and one third of

Glacier melt in the Italian Alps

Claudio Cassardo

The Alpine glaciers constitute an important source of water for irrigation, drinking and energy production. There are over 4000 glacial units distributed over the entire Alpine chain, of which 807 covering 22 per cent of the total surface area of all the Alpine glaciers are in the Italian Alps (Haeberli *et al*, 1989). According to the Italian glacier inventory compiled by the Italian Glaciological Committee in 1959–61, at that time there were 745 glaciers and 93 iced snowfields, with an overall surface area of about 525 km². A second inventory in 1989 recorded 29 fewer units and the area covered by glaciers had fallen to 482 km², showing a marked decrease of 43 km² (8.1 per cent) in less than 30 years (Biancotti and Motta, 2002).

A notable example of recent change can be seen at the Ghiacciaio dei Forni in higher Valtellina in Lombardy, which is the largest valley glacier unit in Italy (13 km²). Between 1929 and 1998, the width of the ablation tongue reduced from 1.43 km² to just 0.29 km² and the average thickness of the glacier in the ablation zone fell by 73 m, an average annual loss of about 1 m (Merli *et al*, 2001). Pictures taken at the end of the nineteenth century and at the start of the twentieth century show that practically all the Alpine glaciers show a similar trend.

On the southern side of the Alps, the Lys Glacier on Mount Rosa has the longest series of observations in Italy starting in 1812 (Cassardo *et al*, 2006). From 1860 to 2006, the glacier snout retreated about 1600 m (Figure 12.4). Where only 150 years ago an ice tongue about 150 m deep was present, there is now a larch forest: this landscape change follows an increase of 1°C in average temperature. Figure 12.5 shows that the measurements, though discontinuous, capture the main features of the end of the Little Ice Age well. There was a dramatic retreat after 1860, due to the negative mass balance of the glacier, mainly caused by the reduction in snowfall at high altitudes. The snout is still retreating, interrupted only by two modest advances in 1913–21 and 1973–85.

A glacier database called GLAD has recently been created by the Centro Elettrotecnico Sperimentale Italiano (CESI) and the University of Milan, which contains data on frontal movements for all Italian glaciers (Apadula *et al*, 2001). The glaciers monitored show a series of phases: general advance before the 1920s, intensive retreat from the 1920s to the end of the 1960s, followed by slight advance up to the mid-1980s, and then a strong withdrawal since 1985. In 2005, 94 per cent of glaciers were

Figure 12.4 Lys glacier, Mount Rosa, seen from Aosta Valley, in 1868 (photo: Monterin archive) and in 2007 (photo © Italian Meteorological Society)

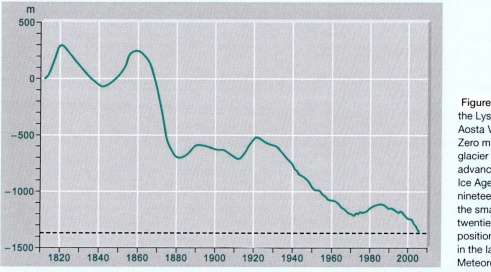

Figure 12.5 Frontal movement on the Lys Glacier, Mount Rosa, in the Aosta Valley.
Zero marks the altitude of the glacier front in 1812. Note the advances at the end of the Little Ice Age at the beginning of the nineteenth century, as well as the small advances during the twentieth century. The present position is probably the lowest in the last 500 years. (Italian Meteorological Society.)

Figure 12.6 Indren glacier, Mount Rosa, Aosta Valley, in 1920 (old postcard) and in 2000 (photo © Italian Meteorological Society)

251

Figure 12.7 Pré de Bar glacier in the Ferret Valley, Mount Blanc, Aosta Valley, in 1897 (photo: Druetti, *Glaciologic Bullettin*) and in 2000 (photo © Italian Meteorological Society)

retreating, some very rapidly. In one year, the Money Glacier in the Gran Paradiso chain retreated 119 m and the Scerscen Glacier in Valtellina 118 m (Citterio *et al*, 2007). These dynamics, which generally agree with measurements of the glacier mass balances, are in quite good agreement with the secular climatic trend, which has witnessed a general increase of the global mean temperature with short periods of slight decrease during the 1920s and between the 1960s and the 1980s (Pinna, 1996).

A study of the correlation between all the Italian glaciers (Cogley and Adams, 1998) shows that some adjacent glaciers can behave differently, whereas those some distance away with similar topographic characteristics can react in the same way. Each glacier's response to climate change is controlled by its altitude, orientation and geometric properties. All the Alpine glaciers are now showing a consistent correlation between retreat and negative mass balance (Diolaiuti, 2001), in remote areas as well as closer to the urban-industrial centres where pollution could play a role. This means that remote monitoring of glacier frontal variations can give a reliable indication of glacier health. And the widespread retreat can be considered a clear signal of a warmer climate.

the total area of land ice in the world, covering 240,000 km² out of a total of 445,000 km². But of the 100,000 or so glaciers listed only 230 have mass balance records, and records of length are available for just 1800 glaciers, respectively amounting to 3400 mass balance and 36,000 length records collated over the 40 years or so since the Inventory began. This is no mean achievement considering the harsh environments involved and limited funding in the early days. But, unfortunately, length is an even poorer measure of a glacier's health than snowline, because advance or retreat of the glacier snout may reflect change in the mass balance that occurred some time back and have only just worked their way down to the end of the glacier. Changes in the length of a glacier can also be affected by sudden 'surging'. Glaciers that suddenly surge forwards, some regularly, some only once in a while, perhaps because of a build-up of meltwater within the glacier or due to earthquakes, are excluded

from detailed analysis of trends. The same applies to glaciers covered by an excessive amount of rock debris, which insulates the ice from changes in air temperature.

Nevertheless, current evidence shows a great preponderance of thinning and retreating glaciers (Figures 12.2 and 12.3). Average annual melting rates in mountain glaciers accelerated in the last two decades of the twentieth century and have doubled since 2000: a pattern very similar to the Greenland ice sheet. Melt rates in 2006 were nearly twice those in the previous high in 1998. Glaciers in the Alps are estimated to have lost 50 per cent of their ice over the last 50 years, more than a third of which has been in the since 1990 (see later in this chapter, 'Glacier melt in the Italian Alps' by Claudio Cassardo, above). Two-thirds of alpine ski resorts could be snow-free by 2050 and the Alps could be totally ice-free by the end of the century (hence the popularity of snowmaking – see

Glacier retreat in the Central Andes of Argentina

Lydia E. Espizua, Gabriela I. Maldonado and Pierre Pitte

Glaciers in the Central Cordillera of the Andes in Argentina show marked retreat since the late nineteenth century (Figure 12.8). The glaciers provide 'natural regulation' for the rivers draining from the mountains. This is critical for maintaining water supply in the semi-arid regions east of the mountains, where precipitation is limited by the Andes, which act as a barrier for most of the humidity coming from the Pacific Ocean on the westerly winds. Loss of the glaciers would be a severe blow for agriculture, viticulture and many cities in the region.

The fluctuations of the Las Vacas, Güssfeldt, El Azufre and El Peñón glaciers in Mendoza Province have been studied from the late nineteenth century to 2007, through historical records, topographic maps, aerial photographs (from 1963 and 1974), recent images from the Landsat satellite (Figures 12.9 and 12.10), and measurements in the field using Global Positioning (GPS). Satellite remote sensing is a useful tool for studying glacier fluctuations because the areas are often inaccessible.

The Las Vacas and Güssfeldt glaciers are large glaciers (between 5 and 10 km long) and El Azufre and El Peñón are medium-sized glaciers (between 1 and 5 km long), all flowing toward the east. Figures 12.8 to 12.10 show that there is a clear, overall retreating trend from 1894/6 to 2007. The Las Vacas glacier showed a marked retreat

from 1896 to 1963, and a lesser one between 1963 and 1974. The glacier advanced in 1974–2003, then remained

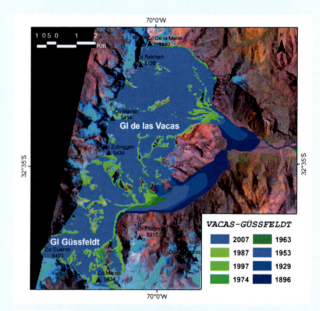

Figure 12.9 The progressive retreat of two Andean glaciers since the late nineteenth century

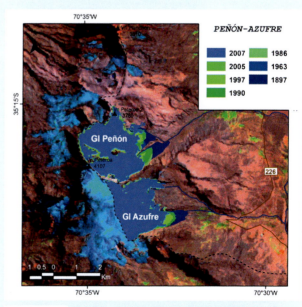

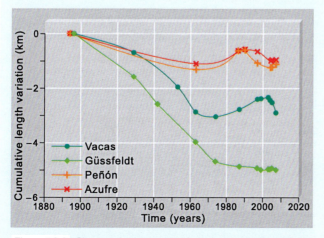

Figure 12.8 Changes in the length of glaciers in the Argentinian Andes since 1896

Figure 12.10 The retreat of the El Azufre and El Peñón glaciers since 1896

stationary, and has retreated slightly since then. The Güssfeldt glacier retreated dramatically between 1896 and 1987, continued retreating slightly between 1987 and 1999 and remained stationary in 2004 and 2007. Large glaciers like these give strong, smoothed signals of trends with a delay of several decades.

The El Azufre and El Peñón glaciers also underwent a general retreat over the period (Figures 12.8 and 12.10).

Both display simultaneous fluctuations with very marked retreats (1894/1897–1963 and 1990–2004), advances (1963–1986), and a near stationary period (1986–1990). The glaciers advanced slightly during 2004–2007.

The advance of Las Vacas, El Azufre and El Peñón glaciers during the 1980s and 1990s could be related with the strong ENSO events of 1982–1983, 1986–1987 and 1997–1998.

Making snow in Chapter 17). The Pyrenees are almost there already. Pyrenean glaciers lost 90 per cent of their ice last century and a report by the Spanish Environment Ministry in 2009 predicts they will be ice-free within decades. This follows a long-term trend. Despite periodic advances, Europe has lost a fifth of its glaciers since 1850.

There are, however, exceptions in northern Europe. Many glaciers have been growing recently in Norway and Iceland as a result of higher snowfalls. Coastal glaciers in Norway thickened in 2007. Glaciers have also been growing in New Zealand for similar climatic reasons. These are recent trends that fit in with the expected effects of global warming. But in the longer term, Iceland has lost so much ice over the last 100 years that it appears to have triggered some of the volcanic activity. The Vatnajökull ice cap has lost ten per cent of its ice since 1890, shedding 435 km³ between 1890 and 2003, causing the land to rise by 25 m a year due to isostasy, which has resulted in an increase in magma production and seismic activity (Pagli and Sigmundsson, 2008).

Glaciers at low latitudes are especially vulnerable. Glaciers throughout Africa may soon disappear. Mount Kilimanjaro has lost 85 per cent of its ice since 1912. Only two large icefields and isolated patches covering 1.9 km² are now left out of the 12 km² a century ago. It could be totally gone by 2020. Thompson *et al* (2009) suggest that the loss of ice is due to a combination of factors that are unique in over 11,700 years: rising temperatures, which are rising faster at high altitudes in the tropics, reduced cloud cover allowing more solar radiation in, drier air and variable snowfall, and finally, a positive feedback as the reduced ice cover lowers the albedo and warms the ground. Some of these effects may be due to extensive human modification of the landscape and vegetation around the mountain in recent decades. The glaciers in the Rwenzori Mountains in Central Africa are also declining. And Mount Kenya has already lost over

90 per cent of its largest glacier, which is the main water supply for many Kenyans who will soon have to rely solely on the increasingly infrequent rains.

The UK Met Office predicts that glacial retreats could cause a major decline in agricultural productivity in the surrounding regions as the rivers dry up. The situation is worse in the tropics than at higher latitudes. Whereas in higher latitudes glacial decay is almost exclusively by melting, which at least enhances riverflow for a while, in the high altitude glaciers of the tropics a substantial amount is being lost by sublimation directly into the atmosphere. Sublimation requires much more energy than melting; enough to propel water through two phase changes in one go, consuming the latent heats of both melting and evaporation (2.83 million Joules per kg, against just 0.335 million for melting on its own). So sublimation is only significant in climates with high inputs of solar energy, which is unfortunate for tropical water resources. Thompson *et al* (2009) find that mass loss on Kilimanjaro is dominated by sublimation from the surface with only a relatively small amount of meltwater.

Considerable concern centres on the glaciers of the Himalayas, Hindu Kush and Tibetan Plateau; some of it rather exaggerated, as in the IPCC (2007) report. They contain the largest body of ice outside the poles with some 36,000 glaciers, hence it is sometimes called 'the third pole' or the 'the water tower of Asia'. Ice covered around 50,000 km² in the late nineteenth century, but has shrunk by 30 per cent since. Over 80 per cent of the glaciers are now retreating and the area under ice has shrunk 4.5 per cent since 1990. Average snowlines have risen to 5300 m above sea level, some 700 m higher than in the 1970s. Even so, a few glaciers in the Karakoram Mountains are advancing.

Unlike the polar ice sheets, the Himalayan glaciers are currently a major source of public water supply. They serve 1.3 billion people in 17 countries via the melt-

Figure 12.11 The retreat of the Rhone Glacier in the Swiss Alps over 200 years: 1856, 2000, and projected for 2050. Note the glacial lake on its surface, signs of which are already appearing. (Simulated images courtesy of the Division of Glaciology at ETH. Zurich)

waters, which feed the Indus, Ganges, Brahmaputra, Salween, Mekong, Yangtze and Hwang. The Hwang Ho supplies 50 big cities and a fifth of China's population. In recent years it has stopped flowing for nearly two-thirds of the year, largely because of over-exploitation from agriculture, but also because of reduced precipitation and higher evaporation in its middle to lower reaches and drying out of wetlands in its headwaters. The glacier area around the headwaters of the Yangtze has decreased 60 per cent in the last 40 years. The Jianggendiru Glacier, which is the main glacier supplying the Yangtze, has been decaying rapidly since 1970. Riverflows in the headwaters of the Mekong are also said to be lower.

There are a number of reasons for this. One is a sharp decline in snowfall over much of the region: parts of Tibet receive 30–80 per cent less precipitation than in the mid-twentieth century. Another is rising temperatures. Temperatures in Tibet have risen 1°C between 1980 and 2010, which is twice the global rate: another case of greater change at higher altitudes. Temperatures in China as a whole have only risen 0.4°C in the last 100 years. But in Tibet there is a major additional reason for glacial decline: black carbon or soot deposits on the ice, which lower the albedo. The amount of soot on the glaciers has risen dramatically since the 1990s, reflecting the economic growth of China and India and their reliance on coal. Without the soot, some suggest that Himalayan glaciers could possibly recover up to a quarter of last century's losses. Problems are also exacerbated by shifts in the pattern of the annual monsoon, which is tending to become weaker and more variable.

There is a lack of detailed mass balance studies of glaciers in the region due to the remote and inhospitable locations combined with a lack of research prioritization and finance. There is now an urgent need to improve glacier monitoring throughout the region. In 2009, The Energy and Resources Institute (TERI) in India made a renewed effort, installing modern, intensive instrumentation on three Himalayan glaciers to be used as benchmarks.

One effect of melting glaciers is a proliferation of meltwater lakes. Nepal now has over 2000 glacial lakes, which are growing, behind glacier end moraines. These moraine barriers are formed by deposits of rock debris left by the retreating glaciers, and they are at risk of bursting as the waters build up, especially if they are affected by earthquakes. Some 20 lakes are now in a dangerous state in Nepal. The rising temperature is also melting the permafrost, causing land subsidence. The situation is similar in the Caucasus Mountains on the border of Russia and Georgia, where glaciation is on a much smaller scale (1600 km²): glaciers are retreating, glacial lakes expanding and the threat of lake outburst floods, like those in the hot summer of 2006, is increasing (Shahgedanova *et al*, 2009).

Glaciers are rapidly disappearing in the Rocky Mountains. Glacier National Park, Montana, could lose its *raison d'être* by 2030. In 1850 the Park had 150 glaciers. It currently has just 26 and they are melting fast. The US Geological Survey began documenting their retreat in its Repeat Photography Project in 1997, returning to sites photographed earlier in the century, and estimates all could be gone by mid-century. Like the Himalayas, soot is a major factor in combination with rising temperatures. In America, much of the soot is from automobiles as well as industry. The same factors are affecting snow-cover in the Sierra Nevadas and adding to California's water shortages. A snowpack that was only 60 per cent of the long-term average contributed to water rationing in the summer of 2009 by reducing the meltwaters feeding the headwaters of the Sacramento and San Joaquin rivers in Central Valley. There are now fears the Spanish epithet 'nevada' might become a misnomer as the Sierras could be completely snow-free before the end of the century.

Ice is melting exceptionally fast in the Andes from the tropics to mid-latitudes, threatening drinking water supplies and hydropower schemes. Until 1992, Bolivia had its own winter Olympics team that used the world's highest ski run on the Chacaltaya Glacier at 5300 m OD. This is no longer the case. The glacier has lost more than 80 per cent of its ice in about the last 20 years. Evidence of retreating glaciers in the Argentinian Andes is presented in the article 'Glacier retreat in the Central Andes of Argentina', above.

Antarctica

The Antarctic ice sheet holds the largest freshwater store on the surface of the Earth and contains 87 per cent of all frozen water (Table 12.1). It is currently only used as a water resource by small scientific missions. There is, nevertheless, every reason to avoid overexploitation for water, just as for oil and minerals. However, there is the possibility for safely exploiting natural wastage, like icebergs or meltwater from beneath the ice shelves (see *Exploiting the cryosphere* in Chapter 14). Whether this might also apply to the hundreds or thousands of lakes that exist beneath the ice sheet on land is an open question.

Fortunately, the ice sheet appears to be less vulnerable than the Greenland ice sheet partly because it is ten times thicker and partly because it is surrounded by the cold Circumpolar Ocean Current and is not threatened by a major warm ocean current like the North Atlantic Drift. However, recent shifts in wind and currents may be warming the water beneath some floating ice shelves. Rignot and Jacobs (2002) estimated that 10 m of shelf ice can be melted for every 0.1°C rise in temperature above freezing point. Moreover, air temperatures on the Antarctic Peninsula, which is the only part of the continent to protrude outside the Antarctic Circle, have risen 3°C in the last 50 years. The Peninsula has shown the highest rate of increased losses, rising from 25 km³ in 1996 to over 60 km³ now. A number of ice shelves have broken away. The Larsen Ice Shelf has completely disintegrated: the 1500 km² Larsen A shelf broke up into numerous small icebergs in 1995, and in 2002 the 3250 km² Larsen B shelf broke up in large chunks, the main part an iceberg measuring 70 by 25 km. Although this was a floating shelf, Larsen B was fed by glaciers and over the following 18 months the glaciers speeded up eightfold. The Wilkins Ice Shelf began breaking up on the west side of the Peninsula in 1998 and even more in 2008.

Over the Antarctic continent as a whole, ice melt has doubled between 2000 and 2010, losing another 80 km³ and raising the total annual loss to around 196 km³. The only relatively stable area is East Antarctica, where annual losses of 4 km³ have remained virtually unchanged over the same period. Satellite radar shows that melting in some areas has increased 140 per cent since 1996 (Rignot *et al*, 2008). Melting is particularly active on the Pacific side of the continent. West Antarctica lost 136 km³ in 2006, up 49 km³ since 1996. All of the eight large ice shelves in Antarctica outside the Peninsula are glacier-fed, which means that if they disintegrate the glaciers backed up behind them will begin to flow faster. The Ross shelf has already begun to break up: in 2000 iceberg B-15 calved off, and iceberg C-19 in 2002, returning the shelf to the size mapped by Captain Scott in 1911. However, there is some doubt as to whether this is due to warming like the Larsen. Like mountain glaciers, the progress of the glaciers feeding the ice shelves depends to varying extent upon past patterns of accumulation and ablation. Some changes occurring now may be a response to changes that occurred 12,000 years ago, and some fluctuations may be part of a natural cycle whereby the shelves grow to the point where they become unstable and break up under their own weight.

Two big questions hang over predictions of what might happen in Antarctica: whether snowfall will increase to buffer losses, and whether some of the large ice shelves will float away and allow the glaciers discharging into them to rush forward, eventually causing the West Antarctic ice sheet to drain away. The argument that snowfall will increase, as it is doing around Greenland, was advanced in the 1990s when early modelling suggested that while Greenland will be a net contributor to sea level rise, Antarctica would offset this by growing, leaving small glaciers and thermal expansion as the main causes of sea level rise. In practice, this was never likely to be true for two reasons:

1   Most importantly, climatically the Antarctic High Pressure belt consists of descending cold air that blows out from the centre of the continent, repelling invading air masses. Its location is fixed by the negative radiation budget in the atmosphere over the South Pole and by the super-cold ice surfaces beneath, and warm air incursions and especially convective cloud formation are therefore strictly limited to the very margins of the land.

2   Whereas an established warm current enters the Arctic Ocean between Greenland and Norway and the recent retreat of sea ice in the arctic is allowing that ocean to warm up, increasing evaporation, atmospheric moisture and snowfall, no such situation exists in and around Antarctica. Satellite radar measurements presented by Rignot *et al* (2008) seem to confirm the view that snowfall is not increasing in Antarctica, and many glaciologists are now beginning to question the older theory.

The fear that the break up of ice shelves might lead to the demise of the whole West Antarctic Ice Sheet also seems to be exaggerated. Original support for the idea came from a suggestion by geologists that it may have collapsed during the last interglacial, but even this is in doubt. However, there is still much to learn about the mechanics of ice shelves: to what extent they may be grounded, to what extent sea level rise may flex and weaken their attachment to the ice sheet or warmer water cause melting underneath the shelves, and how quickly the feeder glaciers might flow into the sea once the shelves no longer hold them back. Only recently, the view that the melting of ice shelves will not add to sea level rise in itself has been challenged on the basis of

the different densities of saltwater and freshwater (see Chapter 10 The threat of global warming). It is even possible that parts of some shelves may be held above normal floating level by their attachment to land ice, i.e. they are not totally floating.

Shrinking groundwater resources

Groundwater is generally considered to be the second largest store of freshwater, after the ice sheets and glaciers (see Figure 11.1). It tends to be more conveniently located closer to the majority of the world population than most ice resources, but the distribution of exploitable volumes depends on suitable rocks as well as the present (or past) climatic water balance (Figure 12.12). Moreover, because it is underground and out-of-sight, and so difficult to assess the amount available, it is often ignored and undervalued. This has led to overexploitation and ecological damage. It is also partly to blame for a tendency among hydrological scientists to overlook it and simply assume that it is a fixed storage year to year. The IPCC (2007) report had little to say about it. Groundwater is very much the Cinderella of hydrology.

Nearly half of the world population drink groundwater. In 2000, groundwater provided over 18 per cent of all water withdrawals and 48 per cent of drinking water (WWAP, 2009). In many regions most drinking water comes from groundwater. In many parts of Europe and Russia up to 80 per cent of drinking water is groundwater, and even more in the Middle East and North Africa. Many cities depend heavily on groundwater supplies and much of the world's agriculture would not exist without it, particularly in North Africa and the Middle East. Increasing demand for water and exhaustion of surface water supplies are forcing all sectors to seek more groundwater.

Yet groundwater is less well understood and monitored than surface flows. Rocks containing sufficient water to be exploited in bulk, known as 'aquifers', underlie nearly half of the total continental land area, excluding Antarctica: 30 per cent is underlain by relatively large, homogeneous and reasonably readily exploitable aquifers and a further 19 per cent is underlain by less easily exploitable

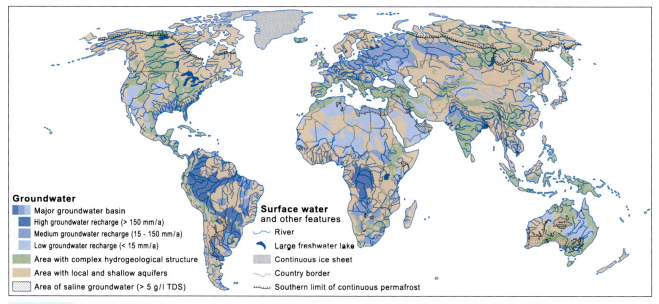

Figure 12.12 Groundwater resources of the world, based on WHYMAP

complex geological structures. In the remaining 50 per cent of the continental land area there are local, patchy and relatively small aquifers, largely in near-surface sand and gravel deposits. In total, these aquifers probably hold around 30 per cent of the Earth's freshwater supplies, although estimates vary widely.

Exploitation of groundwater is often hampered by the difficulties of assessing the level of usable supplies. A good aquifer should possess three key properties: 1) it should be large; 2) it should have a moderate to high drainable porosity, in order to hold enough *extractable* water in the voids; and 3) it should have sufficient permeability to allow easy abstraction of the water. The porosity of the rocks determines the amount of water that can be held, but the amount that is extractable is determined by the storativity or specific yield, which has to be determined from pumping experiments in wells. Excellent numerical models like MODFLOW can be used to predict yields at new sites and the impact extracting groundwater will have on groundwater levels in the surrounding area, but they depend on good field data, especially on rock types and geological structures, and require a higher degree of geological information than is generally readily available. Few rocks are uniform. Water is likely to drain faster through joints and faults in the rocks than through the porous mass of rock. Such discontinuities invalidate the normal assumption of homogeneous porosity made in computer models.

An important distinction must be made between 'deep' groundwater and 'active' groundwater. Groundwater stored in deep aquifers has a very slow turnover rate running into millennia and is generally not actively renewed within the present-day water cycle. Because of its long residence in the rocks, it may contain high concentrations of dissolved minerals that make it less suitable for human use. Despite the fact that this groundwater is not presently being recharged, it is being used in some of the more arid parts of the world, notably in North Africa and the Middle East: in essence it is being 'mined' (see Sonia Thomas's article 'Libya's great man-made river' later in the chapter).

Water held in shallow aquifers tends to be actively renewed in the present-day water cycle, as it is recharged each year by rain and meltwater that seeps into the ground. This groundwater remains underground for relatively short periods, it sustains springs and rivers during dry periods and it is the main source of groundwater for human use in most of the world. But even within this 'active' realm of groundwater there is a wide spectrum of activity, from daily response in which groundwater is forced out into surface waters by heavy rain percolating down through the rocks to groundwater that takes decades or even centuries to return to the active water cycle.

An important consequence of the slow rate of turnover in groundwater is that it is more prone to suffer long-term pollution. Rivers, by contrast, may be polluted more quickly but they tend to recover more quickly as well. Nitrate pollution from agricultural fertilizers is a current worry in many European aquifers (see *Pollution from agriculture* in Chapter 5).

The map in Figure 12.13 shows that substantial resources exist across the well-populated regions of north-central Europe and the eastern United States, but the most extensive and actively recharging aquifers lie in areas with low population in South America and west-central Africa, notably beneath the tropical rain forests.

Dr Petra Döll and colleagues at Frankfurt University and Dr Martina Flörke at Kassel University have modelled recharge rates using a new version of their WaterGap Global Hydrological Model. Figure 12.13 illustrates the results. Taken together, the two maps confirm the view that eastern North America and Northern Europe are generally well supplied with groundwater recharge and, with the notable exception of the low porosity ancient 'shield' lands of northern Canada and Scandinavia, have the aquifers to store it. In contrast, Southern China, Southeast Asia and especially Indonesia have the climatic potential but lack the vital geological substrate for extensive resources.

The WaterGap map shows the strong dependence of groundwater resources upon current precipitation patterns and water surpluses (see Chapter 11, Figures 11.6 and 11.24).

Comparing this map with the global runoff map (Figure 11.25) also suggests that surface and subsurface resources tend to have broadly similar distributions. It is only by resorting to groundwater mining that the constraints imposed by present-day precipitation patterns may be overcome, short of importing water or desalination. However, the detailed pattern is complicated by the way rivers collect and direct surface resources. It is very clear from Figure 12.13 that river basins and groundwater basins rarely coincide. Rivers may drain water surpluses in one direction and the aquifers in another: river basins generally lead to the sea, groundwater basins are defined by rock structures. This has important implications for integrated water resources management: the units of management need to be carefully delimited with these incongruities in mind (see *Integrated Water Resources Management* in Chapter 20).

Groundwater overdraft

Overexploitation or 'overdraft' is the worst and most visible problem affecting present-day groundwater resources. This arises when groundwater is extracted faster than it is replenished. In most cases of overdraft, this is a feature of recent decades, as human demand, especially for irrigated agriculture, has exceeded the natural rate of recharge. Its expression in the landscape is often seen as falling crop yields or dying vegetation. In the most extreme case, when the groundwater that is being exploited is ancient and is not being recharged at all in the current climate, overexploitation is a form of mining: it is consuming a non-replaceable resource (see the section headed 'Mining groundwater', later in this chapter). As with all things climatic, there is a complete spectrum of cases between fully recharging to non-recharging aquifers.

Most of the western USA is suffering moderate to high levels of overdraft (Figure 12.14). Falling water tables in

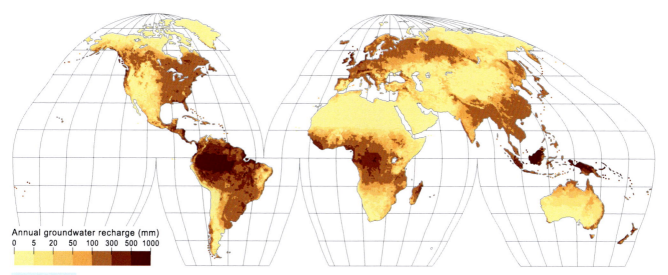

Annual groundwater recharge (mm)

0   5   20   50   100   300   500   1000

Figure 12.13 Annual rates of groundwater recharge, as modelled by WaterGap

the Ogallala aquifer from Nebraska to Texas have been a major concern for some time, with levels falling by a metre a year mainly due to irrigated agriculture. There are fears that agriculture in the region could be forced into rapid decline in coming decades. The problem began 100 years ago, when farmers first settled the High Plains and were encouraged to rely on groundwater to support irrigation schemes.

China's bread basket in the North China Plain is in a similar situation and is the focus of the grandest river diversion project currently in progress (see *China's national strategic plan: South to North transfer* in Chapter 7), extensive rainmaking activity (see *Making rain* in Chapter 17) and attempts by the government to get farmers to use less water (see Chapter 13 Cutting demand). Parts of southern India are worse with water tables falling 30 m in ten years, and this is a major reason for plans to interlink rivers around the subcontinent (see *Dams and diversions in India* in Chapter 7). Many major cities have also been overdrawing their groundwater. In Beijing, the water table has been falling by 2 m a year and

most of the old wells in the city have been dry for years. A side effect of falling water tables beneath Tokyo and Bangkok is land subsidence.

Mining groundwater

North Africa and the Middle East have significant groundwater resources, but they are generally non-recharging owing to lack of rainfall in the region. Many of these aquifers are storing 'fossil' water some 8000 or more years old that collected during the wetter Pluvial Period around the end of the last ice age. There has been very little recharge since. In essence, exploiting fossil groundwater is unsustainable; it is mining a finite resource. Unlike overdraft in much of the USA, reducing extraction would not allow the resource to regenerate: it would not return even if exploitation were halted completely. It is therefore has a limited lifespan. But the effects on the landscape where the groundwater is extracted may be far less than in regions where groundwater is being actively recharged but over-exploited,

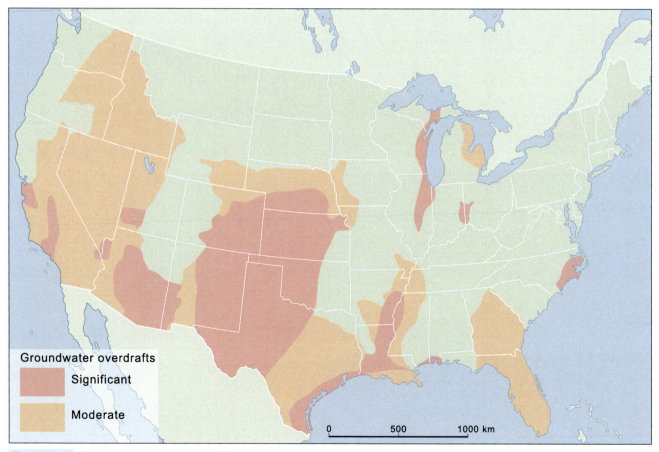

Figure 12.14 Groundwater overdraft in the USA

Groundwater overdrafts

- Significant
- Moderate

0   500   1000 km

because the aquifers lie outside the current ecological system and what vegetation and wildlife there is on the ground does not rely upon it. Groundwater oases in the desert are the exception, where groundwater rises to the surface, and mining is killing many North African oases.

Groundwater mining is widespread in the Middle East, especially in Iraq and the Arabian Peninsula. The groundwater is mostly used locally or transported over relatively short distances, but Libya's Great Man-made River is an example of groundwater diversion on a grand scale (see article below 'Libya's great man-made river'), and Algeria now has a plan for groundwater diversion to rival Libya.

How long these fossil groundwaters might be sustained is unknown; perhaps another 50 or 100 years. The assessment by Salem and Pallas (2001) suggests that current extraction represents only 0.01 per cent of the estimated total recoverable freshwater volume stored in the Nubian Sandstone Aquifer System, which is more encouraging, but it does rely on a number of assumptions.

Libya's great man-made river: unsustainability on a grand scale

Sonia Thomas

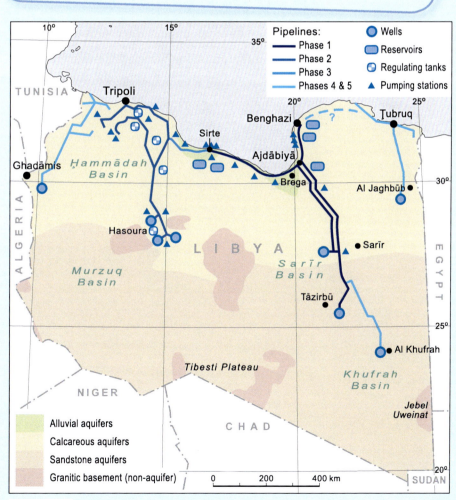

Figure 12.15 Libya's great man-made river

Libya's massive pipeline network, dubbed the 'Great Man-made River' by Colonel Gaddafi, is the largest groundwater piping scheme in the world (Figure 12.15). It was begun in 1985 as the basis for developing irrigated agriculture in the Mediterranean coastlands. The scheme is extracting water from 'fossil' reserves which accumulated during the period of wetter climate that followed the last glaciation: the North African Pluvial period. The groundwater is hardly being recharged under present-day conditions, so it is essentially being 'mined' and will eventually run out, like oil reserves. There are concerns about the environmental impact of the scheme and even stronger concerns among neighbouring countries that the scheme will deplete their own groundwater reserves.

Libya as a whole suffers from a dryland climate, which provides insufficient rainfall for productive agriculture. The sole exception is a small area of the Djebel Akhdar in the north-west of the country, which receives more than 400 mm of rainfall a year. The coastal region, known as the Maghreb, is the most favourable part of Libya for agriculture and home to three-quarters of the population (six million inhabitants). By completion in 2010, the scheme is planned to transfer 6.5 million m³ of water per day, or two billion m³ per year, to irrigate the coastal region, for a total investment of $30 billion. This involves 3850 km of pipeline, 4 m in diameter. The scheme should irrigate between 135,000 and 150,000 ha and increase the cultivated area of Libya by 50 per cent, as its main goal is to provide food self-sufficiency.

The aquifers being exploited cover some 250,000 km² straddling the borders of Egypt, Libya, Sudan and Chad. Egyptian scientists have expressed fears that the groundwater pumping could even reduce flow in the Nile by inducing 'influent' seepage from the river into the aquifer. The Libyan part of the reserves is estimated at between 20,000 and 120,000 billion m³ of water and is divided in four large basins: Sarir and Kufra, in the east and Murzurq and Hamadah, in the west.

The first phase, comprising the eastern or Cyrenaic branch was completed in 1991. Tripoli was served in 1997 in the second phase. The third phase of the project is currently under development and will connect the Cyrenaic and Tripolitan branches. The latter phases will include extra collecting sites in the region of Kufra and in the extreme east and west of the country.

Between 1970 and 1990, the water level of the aquifers fell by 60 m, resulting in the drying up of wells in oases and increased pumping costs. Moreover, while the cost of this scheme was competitive with the cost of seawater desalination 20 years ago, the situation has recently shifted in favour of seawater desalination (less than $0.55/m³ versus more than $0.83/m³).

Paradoxically, Libya is now likely to produce more water than it needs for home consumption and is running its 'river' at a tenth of its capacity. Whereas Gaddafi expelled many Egyptian families in the 1970s, Libya now needs Egyptian farmers to run its agriculture and the two countries have become closer once again. However, should geologists present the proof that mining water in Libya has a direct effect on the Nile, the Egyptian army could try to put an end to the scheme. There may be a standing commission between Sudan and Libya for integrating economic activity, but no agreement has yet been signed between Sudan, Egypt, Chad and Libya for sharing groundwater.

What is a sustainable rate of groundwater abstraction?

The question is not easily answered, partly because of lack of data on the essential properties of the aquifers, but more importantly because of the natural variability of the climate. One year's rainfall is rarely the same as the next, and apparently random fluctuations can create sequences of wet years or drought years. Short records can produce erroneous estimates of long-term resources. The combination of climatic cycles and fluctuations makes estimating average recharge rates very difficult (see Chapter 11 The restless water cycle). Indeed, there is no such thing as a stable average. In theory, resources might be exploitable at a rate above the rate of recharge for a while 'knowing' that statistically rainfall will return to the average or above in the coming years. But this is a dangerous strategy, especially in view of current evidence of systematic climate change.

One of the biggest problems is determining whether increasing droughts are part of a short-term cycle that will correct itself or a long-term trend. Much of the western USA and eastern Australia have been suffering prolonged droughts in recent years that have had severe impacts on groundwater resources.

This is a topic where improved scientific understanding of climatic processes and better monitoring of resources will yield valuable results.

We should value groundwater resources more

There is very little information available on the wealth creation resulting from groundwater supplies. Groundwater that is used for irrigation does not generally require treatment and normally only costs a few cents per cubic metre at most. Groundwater that is treated may cost up to €2 and bottled water can cost €1000 or more per cubic metre. Professor Wilhelm Struckmeier, head of the WHYMAP programme, attempts to assess the value of groundwater to the world economy by assigning it a notional global value of €0.5 per cubic metre, based on

Table 12.2 Volume and value of groundwater production compared with other major natural resources

Resource	Annual Production (million tonnes)	Total Value (million €)
Groundwater	> 600,000	300,000*
Sand and gravel	18,000	90,000
Coal	3640	101,900
Oil	3560	812,300
Lignite	882	12,300
Iron	662	16,400
Rock salt	213	4500
Gypsum	105	1500
Mineral and table water	89	22,000
Phosphate	44	3000

* at notional value of €0.5/m³.

Source: Struckmeier *et al* (2005)

average European values which typically range between €0.8 and €1.4 (Table 12.2). The result is revealing.

Of all the major underground resources, groundwater tops the list in terms of the annual quantity used and is second only to oil in terms of monetary value. Even so, this 'value' is still only a crude indicator of its value to humanity. It is an estimate of what might be charged for it, not of its value to life or as the basis for economic activity. Nevertheless, even the value assigned here is higher than the cost in many parts of the world, including regions where groundwater is being over-exploited. Raising tariffs in some of these regions could be a valuable aid to controlling consumption.

Conclusions

The evidence for shrinking glaciers and ice sheets is now overwhelming. There are also strong grounds for concern over groundwater levels, which are falling at the fastest rates for a century. There is special concern for the role of human exploitation in accelerating the shrinkage, especially the mining of groundwaters that are receiving little or no topping up in the present climate.

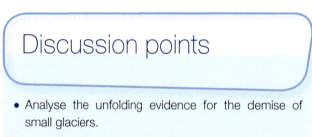

Discussion points

- Analyse the unfolding evidence for the demise of small glaciers.
- Critically review the evidence for and against further growth of Antarctic ice.
- Consider the prospects for halting continued falls in groundwater levels.

Further reading

Two particularly good internet sources of information on glaciers are:

The World Glacier Monitoring Service (WGMS) website at www.geo.unizh.ch/wgms/ provides data free of charge online, including a meta-data file browser facility in Google Earth.

National Snow and Ice Data Center (NSIDC), 1999 updated 2009. *World glacier inventory*. World Glacier Monitoring Service and National Snow and Ice Data Center/World Data Center for Glaciology, Boulder, Colorado. Access: www.nsidc.org.

Part 3 Towards sustainability

13 Cutting demand

Reducing the amount of water consumed is clearly a logical means of alleviating or even avoiding water shortages. It is a way of reducing processing costs for industries. It is also a means of increasing water security for companies. Both were motivations for British industry in the 1970s, especially car manufacturers, following water cut-offs during the severe drought of 1976. During recent decades it has become a widely held policy for governments and private water companies, aimed primarily at reducing the costs of infrastructure and water processing. More recently still, it has become an environmental issue. Campaign groups like Friends of the Earth argue that we have a moral duty to interfere as little as possible with nature. Later in this chapter, Dr David Brooks presents their views and describes some of the 'soft path' solutions being tested in 'Water Soft Paths: the route to sustainable water security'. The director of the Pacific Institute, Dr Peter Gleick (2003), says a transition is under way towards a soft path that complements centralized infrastructure with lower cost, decentralized community-scale systems employing open decision making, water markets and equitable pricing, as well as efficient technologies and environmental protection.

Demand management

Managing demand is the first step in reducing the volume of public water supply needed. Introducing water meters, reducing losses through leaky pipes and running media campaigns to alert the public to the need for frugality are now standard in most developed countries. Responsibility for change has to be shared between water suppliers and water users, and the route to reducing demand is clearly part coercion, part persuasion. Customers can be persuaded to conserve water, but not all the 'demand' comes from the customer. The demand as seen from the supplier's end includes losses en route to the customer. Fixing these leaks is primarily the responsibility of suppliers, and it can have dramatic effects. Solving leakage losses in southern Britain could more than counter the reduction in riverflows due to global warming by mid-century.

Leakage reduction

Levels of leakage from distribution pipes are unacceptably high in many countries. The 4th Annual Global Water Leakage Summit, hosted by the International Water Association in London in January 2010, reviewed the continuing problems and compared solutions. It was attended by delegates from such disparate countries as Gulf states, the USA, Malaysia, Australia, Canada, China, Uganda, Brazil and the EU. China reported that new pilot studies combining leakage reduction and metering had reduced losses of 'non-revenue water' from 39.2 per cent to 31.3 per cent in three years. Qatar launched its own leakage control programme in 2009.

Germany is in the lead. Intense activity to replace old pipes over the last two decades has reduced German losses to single figures, often below five per cent. Denmark is close, having reduced losses from ten per cent to nearly six per cent. In the early 1990s, Britain was losing nearly 25 per cent, more in some areas. A National Leakage Control Initiative was set up to promote methods of detecting, measuring and controlling leaks. Acoustic loggers were introduced and new methods of relining old pipes by injecting polymers into the water. Pressure management was introduced: higher water pressure increases the chance of a burst and raises the level of leakage in pre-existing holes and cracks. Faced with a dense network of old, leaky iron mains, London opted to invest in a new 2.44 m diameter ring main around the city to take some of the pressure off the old 10–90 cm diameter mains. Better monitoring of flows using flowmeters and more rapid response to major leaks were introduced by the water industry, in part as an alternative to the high cost of replacing old mains. This focuses on management of leakage rather than eradicating it.

Ofwat requires all water companies to report leakage statistics annually and it sets agreed targets for each company. Even so, leakage levels continue to run high. Daily losses in England and Wales are still around 200 litres per person. In 2007, Severn Trent Water was fined by the Serious Fraud Office for misreporting leaks in

2000–2002. Ofwat considered fining Thames Water in 2006 for failing to meet the targets for three years running. It was losing over 900 million litres a day, a quarter of all the losses in the UK, despite raising charges and increasing profits by 31 per cent. When Thames Water proposed to develop a desalination plant in the Thames estuary, the London mayor at the time, Ken Livingstone, objected and insisted they control leakage, educate the public and employ more water-saving devices first. The decision was overturned by a new mayor in 2008 on the basis of a new environmental programme agreement including accelerated mains replacement and a leakage action plan. It met the target in 2008.

Metering supplies

Very significant reductions in consumption can be achieved by metering. At the extreme end, an experiment by a Japanese engineer in his own household in the late 1970s led him to estimate that domestic consumption in Tokyo could be reduced by about 40 per cent, which would reduce total consumption in the city from 910 million m^3 to 800 million m^3 a year, obviating the need for extra strategic imports. An official experiment at Normanton, England, in the 1990s showed that water bills for metered homes were 29 per cent lower than for unmetered homes. However, there was no charge for installing the meters and water tariffs were modified during the trial. A three-year experiment on the Isle of Wight involving universal metering of domestic users reported a 20 per cent saving in 1992, but there were suggestions that the community structure on the island may not be representative of the country as a whole. A more reliable comparison between households with and without meters in England and Wales in the late 2000s shows that metered homes use 13 per cent less.

Germany was a leader in introducing water meters for both domestic and commercial users and this has played a significant part in holding demand virtually static for two decades. Its introduction in Britain has been slower and less universal. Part of the problem in the UK may be a propensity to value individual freedom, as perhaps expressed to an extent by Lord Randolph Churchill back in 1884 when he denounced a bill promoted by City of London proposing the introduction of meters as 'the wildest Socialist doctrine'. Churchill's view is, of course, wildly outdated.

In the National River Authority's 1994 'Strategy', it states that 'water companies must be required to achieve economic levels of leakage and metering before new abstraction licences are granted for strategic developments' – strategic developments meaning new infrastructure such as dams and interbasin transfers. South East Water proved the point when it calculated that installing meters in the mid-Kent region at a cost of £30–40 million is a cheaper and better solution to local water shortages, which were aggravated by the 1988–93 drought, than spending £70 million on the proposed Broad Oak Reservoir, which was the subject of vociferous agricultural and environmental objections. During the severe drought in south east England in 2006, Ofwat, the government regulator, called for compulsory metering in the driest areas of south east England, and Folkestone and Dover Water Company became the first company to be granted the right to introduce compulsory metering. The company aims to install meters in 90 per cent of homes (50,000 households) by 2015. Nevertheless, as predicted, demand continues to increase. It appears that Broad Oak is back in consultation with a planned commissioning date of 2024.

Most new properties built in England since 1991 have meters. Some companies, like Thames Water, now install meters automatically when there is a change of owner, and they can insist on compulsorily metering if customers are using excessive amounts of water. The UK Environment Agency wants 75 per cent of homes in England and Wales to be metered by 2025 and 100 per cent by 2035. The EA Water Resources Strategy document (2009) says meters should be compulsory for homes in areas of water shortage, like south east England, within six years. It proposes that water companies should aim to reduce daily consumption from 148 litres per person to 130 within 20 years and be rewarded for reducing the amount of water supplied. However, the National Campaign for Water Justice is concerned about the impact on poor people.

Water tariffs can also be adjusted to encourage frugality, either permanently or as an emergency measure, as in the case of the San Francisco water crisis during 1976–8 (see *Debt and transfiguration* in Chapter 4). Conceivably, there could also be a role for 'smart meters' in the future, similar to those now being introduced by electricity companies. These tell the customer how much they are using and spending in real time, and can recognize different tariff periods. It is also possible that such meters could be updated by information sent along power lines.

Recycling and reusing grey water

The distinction between recycling and reuse is a moot point. Strictly, the word 'recycling' implies returning to use again, but it is commonly used to include water that is simply put to onward use. Recycling may involve treatment before reuse. Simple reuse may not. Indeed, reuse without extra treatment has been practised since time immemorial. It is only really since modern central-ized public water supply and sewage disposal systems were introduced that reuse has become an anathema. We are now realizing how wasteful that is.

Recycling can make a valuable contribution towards reducing demand. All the grey water discharged from the house, from washing clothes, dishes and people, can be reused for processes that do not require water of drink-ing water standard, like flushing the toilet or watering the garden. This could amount to 50–80 per cent of all domestic wastewater. Estimates in Australia suggest that recycling can reduce residential consumption by 40–50 per cent (Vigneswaran and Sundaravadivel, 2004). The rest is mostly black water from toilets, which needs spe-cial treatment (see *Reducing black water* later in this chapter). There is therefore considerable potential for reducing both water consumption and wastewater pro-duction. In addition, vegetable oils and other organic matter in the wastewater may provide useful nutrients for the garden, while soaps are a traditional treatment for aphids. Care is nevertheless needed in matching the type of grey water to the use to which it is applied: some grey water may contain toxins and pathogens.

The US Environmental Protection Agency (EPA) issues guidelines for water reuse. These distinguish between direct potable reuse, which is recycling for drinking water, and indirect potable reuse where wastewater is discharged into surface water bodies before reclamation. Direct reuse is not generally accepted in the USA, but it is successful in Namibia. The German Association for Rainwater Harvesting and Water Recycling also pro-duces a regulatory guide aimed at manufacturers, and urban planners as well as individuals.

One of the biggest obstacles to domestic adoption of recycling is plumbing. Most plumbing systems, both domestic and industrial, were designed when water was more plentiful and the prime objective was to pro-vide a clean, safe water supply. Dual plumbing systems, termed 'dual reticulation', with one system for drinking water and one for grey water, are more expensive to fit than normal single systems and even more expensive

to retrofit in existing buildings. The UK Environment Agency 2009 report proposes that all new housing and industrial properties in England should be fitted with separate supplies of drinking and grey water.

It is most affordable where there is a significant eco-nomic return or an environmental imperative. Larger scale applications are possible in industry, agricultural and landscape irrigation, groundwater recharge, or even at municipal level.

At the time of writing, only a few towns with a dire shortage of water, like Windhoek in Namibia, oper-ate recycling on an urban scale. But the extension of recycling on this scale is more a question of current eco-nomics and degree of water shortage than of available technology. Windhoek's scheme began when the city was approaching the limits of conventional water supplies in the 1960s. Trials began in 1968 and were so successful it became fully operational in 1982. It recycles domestic wastewater and avoids industrial and other potentially toxic sources. The water is blended with fresh water before reuse. In the 1990s, Dubai City adopted a system for irrigating its public parks and golf courses based on partially reclaimed urban wastewater blended with fresh desalinated water. Recycling has been used quite widely in Japan at least since 1951. The Shinjuku dis-trict of Tokyo is a large urban redevelopment that uses recycling.

In America, Chanute, Kansas, was an early, emergency adopter in 1956 after its prime supply, the Nishu River, ran dry following a five-year drought. The authorities built a dam below the sewage outfall to collect the wastewater. The town populace survived with no notable ill effects for five months on water that was retreated and recycled up to 15 times. Altamonte Springs, near Orlando, Flor-ida, adopted a permanent recycling scheme after concern over pollution of the local lake and groundwater. Their scheme involves extensive re-treatment before recycling and the city of 45,000 was retrofitted with a new pipe network. The Denver Potable Reuse Demonstration Project was begun in 1984, but despite its title it is not used for drinking.

Israel is one of the most efficient water users in the world. More than a quarter of its water is reused and it has plans to recycle 80 per cent. Israel hopes to recycle 430 million m^3 annually from 2010.

Figure 13.1 shows an unusual case in Japan where no extra plumbing is required: it is reuse rather than recycling in the strict sense.

Figure 13.1 Reusing artesian groundwater in Japan. A typical domestic case in the Kurobe alluvial fan area of Toyama Prefecture. (Courtesy of Kurobe City Office and K. Mori)

Water-saving technology

Considerable advances have been made in water-saving technology in industry and in domestic appliances over the last two decades. Electric washing machines were one of the main causes of rising domestic water demand in developed economies during the latter part of the twentieth century. The average UK household still uses nearly 21 litres a day to wash clothes – 13 per cent of total consumption. In response to public criticism and governmental pressure, manufacturers of domestic and commercial washing machines have been steadily reducing the quantities of water and energy requirements. Miele have been in the forefront of water-saving technologies. Even so, the pressure group Waterwise says that the use of domestic washing machines in Britain has risen by nearly a quarter in the last 15 years, which partially counter-balances the increased efficiency. The Environment Agency's 2009 Strategy proposes that water-saving domestic equipment be given lower value added tax (VAT) rating to encourage uptake.

Waterless washing

An entirely new approach has recently been developed at Leeds University that raises the prospect of virtually waterless washing. It uses barely a cupful of water per wash, offering a 90 per cent saving on water and 30 per cent saving on energy, and is being promoted by the spinoff company Xeros. It is based on nylon polymer beads that absorb dirt and stains in a humid environment of water vapour and can be reused up to a hundred times. Further development is needed before it is ready for use in domestic machines, but the commercial company GreenEarth Cleaning has signed up to sell the technology in North America.

Flow limiters

The most basic flow restriction is cutting off supplies. It is the weapon of last resort used against customers who fail to pay their bills (see *Debt and transfiguration* in Chapter 4). It is also a frequent and often irregular bugbear for thousands of families in India, more usually because of shortage of electricity to power the pumps than shortage of water.

Some 20 per cent of domestic water use is through taps and part of Germany's success in containing demand is due to mechanical flow restrictors installed by manufacturers as standard in all new equipment. These consist of extra metal filters in taps and showers that can be easily removed where legislation allows. The British firm, Bristan is marketing a range of 'eco-smart' products that includes eco-click taps – a cartridge gradually restricts flow once flow exceeds half capacity, but this can be overridden by clicking the button – and spray taps that aerate the water to reduce the volume used. Bristan claims that eco-click reduces use by 32 per cent and aerating taps by 72 per cent.

Inducements to adoption

The cost of replacing old technology is often a big disincentive for households. The same applies to energy saving, and governments, as in the UK, are responding to the commitments made on greenhouse gas emissions by subsidizing energy-saving measures and encouraging electricity and gas suppliers to do likewise. A strong case is emerging for similar policies to support water saving.

During the multi-year drought in 2006, the Queensland government passed legislation requiring all new houses to have water-saving infrastructure from 2007. The state government set predetermined water saving targets for councils, but left it up to them how this is achieved, for example, by rainwater harvesting, recycling or flow restriction.

Reducing black water

Flushing toilets can account for up to 30 per cent of household water use. They have made a vital contribution to improved public health, especially in urban areas, but the battle is now on to find less profligate systems. The problem is not so much the quantity of water consumed as the fact that it is generally water that has been treated to drinking water standards: sourcing the water from rainwater can reduce the demand on public water supply dramatically (see *Rainwater harvesting* in Chapter 14).

Significant savings can be made simply by reducing the volume of toilet cisterns. Whereas older cisterns may have used 9–12 litres a flush, more modern ones use 7–8 litres, and many now use less than 4 litres: Bristan have a 2.6 litre cistern. After the '400-year' drought of 1976, East Anglia encouraged the installation of 4.5 litre cisterns, half the size of the typical English cistern, and a third of many Scottish cisterns. Early attempts to reduce capacity by putting a brick in the cistern have been succeeded by products manufactured for the purpose, like 'hippos' and other cistern bags, the 'Interflush' system that flushes only so long as the handle is held down, or the now common dual flush levers for selecting long or short flushes. These systems can save more than half the water, but is there scope for an intelligent toilet that can work out exactly how much water to use?

Using no water at all is a logical extension of standard approaches to reducing black water and will be discussed under the section headed *Soft path solutions*, later in this chapter.

The role of media campaigns: persuasion versus coercion

All water management agencies have their own systems for reducing demand during emergencies. These begin subtly by alerting the public to the dangers and gradually escalate from banning garden hoses and topping up the swimming pool to more draconian restrictions policed with the threat of fines.

However, coercion often encourages covert avoidance. Following 18 months of low rainfall in south east England, by August 2006 13 million people were affected by official watering bans. Unfortunately, it also encouraged 'stealth watering'; an estimated 750,000 people crept out to water at night to circumvent the bans. A poll suggested that more people were prepared to break the bans to water the garden than to wash the car or top up the pool. Water companies encouraged neighbours to act as 'water vigilantes' and report transgressors. As a result, Thames Water issued over 2000 warnings and the Three Valleys water company referred people to the magistrates' court with a possible £1000 fine if they ignored three warnings.

Despite bans and advice issued by Queensland authorities during the drought years of the late 2000s, the temptation to cheat is as strong in Australia as England. Many gardeners in SE Queensland responded to the watering bans by sinking their own boreholes to avoid the restrictions. Private boreholes are largely uncontrolled. Toowoomba City has about 650 boreholes, mostly private with no licensing or metering requirements, and there have been moves to bring these more under government control and make bore water usable in gardens only by bucket or can. Ironically, the city council itself began sinking more boreholes in 2006 to ease pressure on its three dams, which were predicted to run dry by 2008 without further rain.

With all bans and restrictions there is always the need to protect the vulnerable. The first act of Queensland's new Water Commission in 2006 was to exempt 200,000 people over the age of 70 and disabled people from level 3 water restrictions, allowing use of a garden hose for one hour a week. Watering of new gardens was permitted for 14 days, but only by bucket. But all exemptions are nullified if deepening drought requires level 4 restrictions.

In a privatized water industry there is also the problem that restricting water consumption also reduces income and profits. The case of the 1978 San Francisco drought orders discussed in *Debt and transfiguration* in Chapter 4 is a graphic illustration.

Drought emergencies can never be totally eliminated, but there is often room for improvements in prediction and management, and in conditioning customers to long-term reductions in water consumption. Improving prediction and forecasting is covered in Chapter 19 Improving prediction and risk assessment.

Public education is the ultimate key to sustainable, long-term reductions. People in developed countries have grown up with abundant supplies and taken them for granted. Most developed nations now have active water conservation campaigns aimed at permanently reducing per capita demand, in many cases supplemented by inducements like free meters and by pointing out the money that can be saved. The US EPA has its Water Sense website. The UK Environment Agency runs its 'Save Water' website, and the theme is taken up by the individual water companies. Thames Water runs the 'Wise Up to Water' campaign. Australia runs a national 'Waterwatch' educational programme. South East Queensland Water has its 'Water for Ever' campaign (www.waterforever.com.au).

It is not just adults that need to be exposed to information. Attitudes are formed at an early age and investment in the young will last longer. The US Geological Survey runs 'Water Science for Schools' with offices in every state and downloadable podcasts. In Melbourne, Melbourne Water and city West Water provide teaching resources for teachers and educational tours of water treatment and sewage plants. Sydney Water Corporation has published a 'Water Cycle Fun Book' since the 1990s. Public exhibitions like the permanent Delta Expo in the Netherlands, and especially Zaragoza's grand International Water Expo in 2008, mix fun with sound facts.

Figure 13.2 Billboards were deployed in the streets of Brisbane during the severe drought in 2006. Even city buses carried messages

More efficient irrigation

Irrigated agriculture is the biggest single water user in the world and clearly there is considerable scope for greater efficiency. Irrigating by sprinkler, or better still drip irrigation from perforated hoses that deliver the water directly to the roots of the crops, are far more efficient than irrigating by open ditches or flooding. Irrigation in most of the former Soviet states in the CIS is notoriously

inefficient, typically consuming around 25 times more water per hectare than in France (Jones, 1997). India is one of the biggest users, with an annual intake of more than 300 km^3, twice that of the USA, and returning only 20 per cent, half the USA. Israel has the most efficient irrigation in Asia, which consumes only half the amount used in Iraq.

The differences are partly down to technology, but there is also a large element due to lack of understanding. Peasant farmers often over-irrigate, either with too much or too frequent irrigation, sometimes through lack of understanding and following the traditional approach, sometimes because they do not have the support of information on soil moisture levels, crop requirements, evaporation losses and weather forecasts. In Chapter 2, Abdullah Almisnid shows how greater understanding and access to better information in the company-run farms in Saudi Arabia results in markedly lower water consumption. Farmers in countries like the USA are not only well-educated, but they also have access to more technical information, such as the US Department of Agriculture's Soil Conservation Service evapotranspiration calculations, which are also used to allocate water in large irrigation schemes.

There is considerable scope for water-saving technologies to be combined with more government-funded information for farmers, both in terms of training and advice and in real time management of irrigation schedules, especially in developing countries.

Some technologies that can improve irrigation in developing countries can be quite simple and cheap. The Indian International Development Enterprises (IDEI), winners of the London Ashden Award for outstanding and innovative inventions in sustainable energy in 2006, has been promoting a suite of 'affordable drip irrigation technologies' to accompany its human-powered treadle pumps (see *Elephant pumps, play pumps, treadle pumps and rope & washer pumps* in Chapter 14). All are supplied in kit form for private or small-scale commercial use, with storage units mounted a metre above the ground to provide gravity feeding through a number of perforated flexible waterpipes. They start with the bucket kit for kitchen gardens up to 20 m^2 in area, which has proved very popular with women growing the family vegetables. The family nutrition kit extends this to 40 m^2 using a 20 litre plastic bag instead of the bucket. The drum kit is designed for small commercial enterprises and uses a pre-assembled 200 litre drum attached to five or more rows of pipes feeding between 100 and 1000 m². Larger customized systems are available that can serve up to 2 ha.

Hydroponics and aeroponics

Hydroponics is not a new concept, but it has been gradually gaining ground in recent years and there are now many manufacturers promoting the equipment. As a growth method that does not rely on soil, it is potentially useful for agriculture and horticulture in hostile environments where soils are unsuitable, and Nasa has shown interest in possible space applications. From the water resources viewpoint there are two strong advantages in the approaches: hydroponic systems usually involve recycling the water and aeroponic systems go further by relying on aerosol water droplets rather than circulating liquid water. Standard hydroponic systems grow the plants in an essentially inert medium, like clay balls or rock wool, and circulate aerated, nutrient-laden water around the roots. Aeration prevents anoxic conditions developing that can kill the plants through waterlogging. Semi-hydroponic systems use less water by relying on capillary attraction to raise water from an underlying reservoir through tiny pores in the growing medium. Finally, aeroponics use even less water by circulating a mist through the growth medium: tests suggest 65 per cent less water than in standard hydroponics.

There are disadvantages. The systems are essentially greenhouse methods, so scale is limited. But the controlled environment means it is particularly useful in desert and semi-desert climates. There is also the danger of the environment favouring pathogens, though less so with aeroponics. But computerized control systems can ensure that watering is kept at exactly the right level for the crop and without wastage.

Soft path solutions

Standard demand management is concerned with saving water, whereas the concept of 'soft paths' that has been developed over the last decade also considers social justice and equality and environmental protection. The concept includes low-cost community-scale systems, decentralized and open decision making, water markets, equitable pricing, efficient technology and environmental protection. The exact division between demand management and soft paths might be debated, but the important point is that both need to work together in an integrated way. Distinctions are to some extent academic – soft path solutions are really offering new and innovative extensions to established demand management, as outlined by David Brooks in the article 'Water Soft Paths', below.

Dry sanitation

Chemical and composting toilets save even more than low-flush toilets. Originally developed for camping and mobile homes, these are now finding favour in more settled environments. Chemical toilets are less environmentally friendly and require regular, professional servicing. Biological or composting toilets can be cleared by the owners and the waste used as compost. Envirolet VF (vacuum flush) of Scandinavia recommends emptying about four times a year for continuous use. Depending on local rules, the waste may be spread on the land or it may have to be put in the regular rubbish bins. However, installation costs can be a disincentive: Envirolet's cost around $4000 for vacuum-flush models, which cut down on smells, and $2000 for non-vacuum models. Envirolet's Low Water Remote Systems connect to composting units and their Waterless Remote Systems, include a 'mulcherator', which works occasionally to aid composting. In standard toilets, up to 90 per cent of all toilet waste is water, but Envirolet's unique six-way aeration and evaporation process eliminates the liquid, leaving ten per cent as dry compost.

Using a composting toilet means that all household wastewater is greywater, and therefore recyclable.

Xeriscaping

Lawn sprinklers are a major user of potable water supplies in America. There and in many other countries there have been campaigns in recent years to reduce water use in the garden. Interesting moves are also afoot to reduce the use of public water supplies for irrigating golf courses (see *Saltwater irrigation of leisure landscapes* in Chapter 16).

In Britain, the Environment Agency publishes a guide to low-water gardening, and most water companies offer their own advice. The severe drought in south east England in 2006 stimulated renewed interest in dry gardening. There has been a large rise in the number of 'drought exhibitors' at the Royal Horticultural Society's annual Chelsea Flower Show, one recommending using old carpets to reduce evaporation losses. However, there is one

Water soft paths: the route to sustainable water security

David Brooks, Friends of the Earth Canada

The key to a fresh approach for water management – one that seeks sustainable water security – lies in shifting policy from expanding supply to moderating demand. A full shift requires a two-step process: first, from supply management to demand management; then, from demand management to what is called a 'soft path'.

Water demand management seeks primarily *economic efficiency*, better ways to achieve the same service with less water. Water soft paths (WSP) accept the importance of greater water efficiency, but go further by searching for changes in water use habits and water management institutions to achieve a *triple bottom line*: economic efficiency, social equity and ecological sustainability. WSP accomplishes this by making two significant changes to the current methodology: 1) it changes the conception of 'water' from an end to a means; and 2) it inverts the 'standard' analytical process that projects the present status forward to one that *works backwards* from a desired future to find a range of acceptable policies that can help guide our efforts to reach that goal.

Demand management asks the question 'How': How can we get more from each drop of water? Cost-effective savings typically reach 30 to 40 per cent of current use. Water soft paths ask the question 'Why': Why should we use water for this task at all? Why, for example, do we use water to carry away our waste? Demand management would urge low-flow toilets; soft paths promote waterless or composting systems for homes, and on-site methods of waste treatment for larger buildings. Using WSP, potential savings reach 80 to 90 per cent of current use.

Modelling studies on the efficacy of water soft paths undertaken by Friends of the Earth Canada, the first such studies anywhere in the world, explore what water soft path policies might achieve at three scales:

- **Watershed:** The Annapolis Valley in Nova Scotia has a maritime climate but only ten per cent of the rain falls in the summer when nearly half of withdrawals occur – mainly for golf courses and agriculture. Sustainable limits are already being exceeded nearly every third year. Setting a desired goal of keeping summer withdrawals below available water in drought years, the study showed that even widespread adoption of conventional demand management measures could not cut overdrafts to less than one year in five. Adding soft path methods including, for example, using wastewater on golf courses, rooftop rainwater harvesting for homes, and high-efficiency irrigation on farms could reduce summer use to half its current level and almost assure sustainable supplies.

- **Urban:** Urban water use, which is dominated by residential, commercial and institutional sectors, was modelled in a generic community that expects to grow in population by 50 per cent over the next 40 years but that seeks to keep water use to current levels. Conventional projections show water use growing by 50 per cent but demand management for enhanced efficiency appliances, low-volume sprinklers etc., can cut the growth in half. Adding such measures as extensive xeriscaping, dry sanitation, and policies directed at behaviour and lifestyle choices are sufficient to keep future water use well below today's level.

- **Provincial:** Ontario is the largest and most heavily industrialized province of Canada. It has a continental climate. Results for the residential, commercial and institutional sectors were similar to those for the urban model described above, with future water use little more than half current use despite growth in population. Cutting a projected doubling of industrial water use was more difficult. Demand management focusing on leak reduction and recycling could cut the growth in half, and soft path measures focusing on reuse and project adjustment could cut it in half again. Getting industrial water use below current levels requires redesign of existing plants to adopt the best available technologies as well as some change in the industrial mix, especially for the more water-intensive sectors.

The results of these initial studies of the application of water soft path analysis should be seen as indicative rather than definitive. Nevertheless, they do show that the goal of sustainable water security is within our grasp.

problem for drought resistant gardens in the UK: cold, wet winters to drown and freeze them!

Queensland has a more consistent need to conserve water, especially in the urbanized south east. Queensland offers advice on designing drought resistant gardens: planting species with low water requirements like *oleander, poinsettia, plumbago, brunfelsia, leptospermum* and *hibiscus*; using mulches and water-retaining crystals, and supplementing supplies with grey water and rainwater harvesting. Official literature points out that leaky hoses can cut water use to ten per cent of the original, and gardening shops in Brisbane advertise 'get your grey water hoses here'. Even so, hundreds of jobs have been lost in Queensland's nursery and garden industry as a result of the drought in recent years.

Water markets

The idea of water markets is controversial. It is a version of the concept of permit trading introduced by the US EPA to combat air pollution and now incorporated into measures to reduce greenhouse gas emissions. Each user or company is issued with a government permit to emit certain levels of pollutant, or in the case of water to access a certain amount of water. If they over-achieve, they can trade the residual amount on the permit with someone who wishes to use more than their own permit allows.

Water markets have operated since the 1970s in California, where owners of historical water rights, usually farmers, sell unwanted water to other farmers, towns or industry. We discuss this in detail in the section on *National law, the public commons and private markets* in Chapter 20, where it is also covered by lawyer Joseph Dellapenna.

Wasting food is wasting water

Despite the expansion of waste recycling in many developed countries, large amounts of food waste are still being consigned to landfill. Not only does this potentially increase methane emissions and pollute groundwater, it is also a waste of the water used in food production, and the portion that goes down the sink adds to treatment costs.

Britain is the largest dumper of household waste in Europe: 27 million tonnes a year, of which up to 8.3 million is food waste, which amounts to around a third of all food bought. Food waste from all sources in Britain, including supermarkets, restaurants and processing plants, amounts to 18.4 million tonnes pa. A sizeable portion of food and drink is thrown down the sink, around 1.8m tonnes in 2009, into the sewers and water treatment plants. This includes an amazing £470 million worth of wine, £190 million of fizzy drinks, £120 million of beer and cider, £100 million of tea and £35 million of bottled water. Fat deposits in sewers due to restaurant waste have caused problems in London.

The water implications of this wastage are hardly ever considered, yet reducing this waste must be seen as a priority for water conservation globally. In Britain's case, most of this waste is virtual water imported from around the world, including vegetables from southern Spain and fruit from Israel – both regions of water stress. In a very full analysis of the problem, Stuart (2009) also argues that reducing this waste could help in lifting the poorest nations out of malnutrition.

There are also numerous ways in which the waste may be reused to reduce water treatment costs and improve environmental disposal. It can be composted or used for energy or fodder. Restaurants could use dehydrating machines to reduce the volume and the dry waste converted into fuel pellets. The water removed could then be used on the vegetable garden. Anaerobic digester plants may produce useable fertilizer and biogas from the waste. The Wrap campaign 'Love food, hate waste' aims to reduce UK food waste by a quarter of a million tonnes by 2011, through raising awareness and recommending methods of reusing the waste, such as freezing waste wine in ice cubes, using it in cooking or decanting it into smaller bottles (www.wrap.org.uk). Unfortunately, the situation in the EU was made worse in 2001 when the response to the foot-and-mouth outbreak was to ban pig swill. Stuart argues that a temporary ban followed by clear guidelines on sterilizing the waste was all that was needed.

However, most of the water issues can only be addressed by cultural change, by creating less waste in the first place. The governments in Japan, Korea and Taiwan have already made some progress by making it mandatory for food businesses to reduce waste by up to two-thirds (Stuart, 2009).

Changing diets and lifestyle

There is no doubt that the burgeoning middle classes of China and India are going to consume more food per capita and that more of that food will be meat. This is likely to have a major impact on water resources. Rice production already consumes 21 per cent of the total water used to grow crops globally. On average, rice harvested from a paddy field consumes 2300 m³/tonne and after it is processed into white rice for food this rises to 3400. In some regions of the world like India and Brazil the average is even higher – 4254 and 4600 m³/tonne respectively for white rice. In contrast, super-efficient Japan manages to produce it for just 1822.

The shift to eating more meat has an even greater impact. According to the calculations of Hoekstra and colleagues, beef requires 15,500 m³/tonne (see Table 8.1). But is meat really that wasteful? Sheep, goats and cattle on mountain pastures use grass grown from green water only, at no net cost to traditional water resources. Even in lowland fields, when there is no irrigation and the animals are fed solely on grass with no extra feed involved, there is minimal blue water cost apart from possible drinking water. The grass would grow anyway. Only where animals are fed by extra fodder is there likely to be significant water costs. This argues for less intensive production and a return to traditional methods of rearing.

One key message from the UN International Year of the Potato (2008) was that potatoes use less water than most other foods. The world average for growing a tonne of potatoes is only 250 m³ of water, an average saving of more than three million litres per tonne over white rice – and potatoes do not require processing to make them edible. While cereals, including rice, are calculated to use 50 per cent of all the water consumed by crops globally, potatoes consume just one per cent. In part, the difference is due to the differing volumes of crop grown, but it is also due to the lower water requirements of potatoes compared with rice, especially when rice is grown in flood-irrigated paddy fields.

There is a serious message here. The potato has the potential to supply the world population with the necessary carbohydrates while consuming a tiny fraction of the water needed to grow grains. But can diets be changed?

Containing population growth

Without question, population growth is the greatest contributor to the water crisis. It is also the most intractable. The UN International Conference on Population and Development held in Cairo in 1994 generated heated debate, both during the conference and afterwards, mostly relating to methods of reining in population growth. Demands from women's groups clashed with religious interests over family planning, and totalitarian China interpreted 'reproductive health' as meaning a right to abortion and the USA and others disagreed. It was seen by many as a lost opportunity, ironically set in a country with one of the fastest growing populations in the world. Although a resolution to improve access to family planning was accepted and forms part of the 20-year Programme of Action of the UN Population Fund (UNFPA), the programme is directed more towards improving health and wellbeing for women and children. There is no prospect of containing population growth in the foreseeable future.

Both India and China have attempted to contain their populations. China introduced its one child policy in the late 1970s, and relaxed it slightly in 1984 to allow rural mothers to have a second if the first is a girl. China now has a population of 1.3 billion and it is calculated that without the draconian measures it would be 1.7 billion, just under a 25 per cent reduction. Such a policy would not be possible in a democracy like India. India's population is increasing at 1.6 per cent p.a. and is expected to overtake China in 20 years. However, some Indian states have encouraged vasectomies with incentives, like fast-tracking gun licences. In 2009, the Indian Health Minister called for a redoubling of efforts to bring electricity to all rural areas, because he believes that late night television could cut population growth by 80 per cent. There are also reports of unauthorized sterilization of untouchables.

The only real hope of slowing population growth is a matter of economics and wellbeing: a populace that is richer and better fed is less likely to have a high growth rate. Japan is a prime example, with a declining birth rate. Germany is another. Although the USA's growth rate remains around 0.9 per cent p.a., it has been falling almost continually since the 1960s. Most West European countries would have a declining population were it not for immigration, which is increasing partly because the expansion of the EU is allowing free movement to migrants from Eastern Europe, and partly because of illegal migrants from disadvantaged regions in Africa and Asia.

Immigration is becoming a water resources issue. Immigration is one reason why the UK's population is rising. It accounted for around one million or approximately half of the population increase during the first decade of the twenty-first century. From being fairly stable at around 56 million in the 1970s and 1980s, the population began to rise significantly around the turn of the millennium and is now 61 million and set to rise to 71 million by 2031 or possibly 75.4 by 2050, according to the Optimum Population Trust. If this transpires, it would represent a 16 per cent cut in per capita water resources. The shock could be absorbed nationally, but south east England, where most of the increase will be, is already parlously short of water (Rodda, 2006). The OPT estimates are, however, based on a simple extrapolation of recent trends – see the critique of this type of approach in Chapter 19 Improving prediction and risk assessment.

Hydropolis – the water sensitive city

'I believe we need nothing short of a revolution in thinking about Australia's urban water challenge. Two assumptions have dominated water infrastructure in our cities. The first is that water should be used only once. The second is that storm water should be carried away to rivers and oceans as quickly as possible. Neither assumption is suited to a world in which we should judge water by its quality and not by its history.'

John Howard, Australian Prime Minister, 17 July 2006, to all state premiers, setting out minimum criteria for projects to deliver a 'genuinely transformative impact on water management'.

Engineers, planners and landscape architects meeting in Perth, Western Australia, in 1989 to discuss the way forward in urban planning coined the phrase 'water sensitive urban design'. Perth has continued to host a number of international conferences on the theme, especially since 2000 (http://www.keynotewa.com/wsud09/). Yet Prime Minister John Howard was still calling for a fundamental rethink of urban planning at the height of the 2006 drought – and coincidentally the year Perth hosted the Hydropolis conference. Howard emphasized the need to 'drought-proof' Australian cities by recycling, desalination and using rainfall and floodwaters.

Cities have continued to expand at increasing rates worldwide over the last 20 years with very little new thinking about sustainability. With an estimated 70 per cent of world population concentrated in cities that occupy just two per cent of the world's land surface by 2050, the issue is becoming urgent. Dubai City emerged as one of the worst planned cities in 2008 when its beaches were covered in illegally dumped sewage. It transpires that the city's rapid growth in the 1990s and 2000s was based largely on septic tanks rather than a centralized sewage system. Tankers making daily collections from thousands of septic tanks faced ten-hour queues at the emirate's only treatment plant in the desert at Al-Awir. Many tankers dumped their loads in the storm drains that run straight to the sea. Since storms are rare, the raw sewage festered for months before being washed to the sea. The media dubbed it 'Poo-bai'.

Las Vegas is a prime example of a desert city in trouble; a city of two million inhabitants and all-year-round tourists that has been expanding by 50,000 a year since the 1990s. The city depends upon the Colorado River for 90 per cent of its water supply, but the water level in Lake Mead, the reservoir formed by the Hoover Dam, fell 37 m between 1999 and 2009. Although the city uses a prolific amount of water, in its fountains as well as hotels and casinos, it actually uses only three per cent of the two per cent of the discharge of the Colorado that is allocated to the State of Nevada. The problem is drought. The Colorado is suffering from a prolonged sequence of drought years. Water engineers estimate that water restrictions will be needed by 2011, the hydropower scheme will shut down by 2013 and the city will be in grave danger by 2017. A plan to pipe water from aquifers in eastern Nevada is not due to come on stream till 2018.

Eco-cities – building the ideal

Ironically, Dubai's neighbouring emirate, Abu Dhabi, is planning one of the first eco-cities, Masdar City, due for completion in 2016. Planned to house 50,000, Masdar will be zero-carbon, use solar power to desalinate 8000 m^3 of drinking water a day, recycle water and collect dew. Sewage and other organic waste will be used as fuel, and wastewater will be processed and used to irrigate green spaces. Planners calculate Masdar will need only a quarter of the energy of a normal city and will have no waste output. Without its advanced water treatment and recycling system, the city would consume 20,000 m^3 of water a day. Where desalinated water is the primary source, the saving is especially cost-effective.

With more than 300 million people likely to be living in its cities by 2030, China has grand plans to follow suit with eco-cities, although there have been some initial problems and delays. Huangbaiyu in the northern province of Liaoning is reportedly languishing with problems of inferior construction, lack of regard for the existing populace and the high cost of the housing. The showcase eco-city of Dongtan near Shanghai was due to have Phase One open for the 2010 Shanghai International Expo, but has hit delays, partly through the economic downturn and partly through controversy. It is planned to house half a million by 2040. It is to be built on marshland considered as ecologically important for wildlife and migratory birds. But it now looks set to proceed, as is the latest development near Tianjin in NE China, where the plans call for daily domestic consumption to be held below 120 litres per person and over half of this to come from rainwater harvesting and recycled grey water. China is planning a number of other eco-cities, including Rizhao, Wanzhuang and Nanjing. These are currently being planned in collaboration with foreign engineering firms like Arup and McKinsey. At Wanzhuang, Arup is planning a water capture and storage scheme to provide irrigation water for local farms. In turn, these international companies are applying the lessons around the world. Arup is currently engaged in designing similar projects in the Netherlands, USA, Chile, Azerbaijan and Helsinki.

In the USA, the Clinton Climate Initiative has selected 16 sites for eco-cities, of which one is Destiny, Florida, south of Orlando. Building is scheduled to start at Destiny in 2011 and take up to 50 years to create a city of 166 km², the size of Washington DC. It will be built on swampland and old cattle ranches. Water management will be based on rainwater harvesting, reusing grey water for the garden and using local brackish water where possible. Ecologically sensitive areas are to be mapped and designated as nature sanctuaries, and construction work confined to disturbed land. Golf courses will be planted with less thirsty grasses. The scheme is intended to be largely self-funding, with some additional support from the founder of the Subway sandwich chain, but the housing market has not been strong since the subprime crash: Florida is second only to Nevada in the number of home repossessions and many buildings have been abandoned through fears of hurricane damage.

Hydrogenic city 2020 is a more futuristic and slightly quirky vision for Los Angeles. The scheme was a finalist in the 2009 Working Public Architecture competition run by the Department of Architecture and Urban Design at the University of California Los Angeles. The designers point out that 41 per cent of southern California's water supplies are now vulnerable to reductions in snowmelt from the Sierra Nevadas and in flow in the Colorado River, and that LA gets 85 per cent of its water from these two sources. The plan is to replace the four present centralized wastewater treatment plants with a decentralized network of water reclamation units using eco-energy and linked to community life (www.hydrogeniccity.com). After treatment, water would be released into wetlands, shallow channels, swimming pools and more curious 'mist platforms', water towers used to mount solar panels, urban beaches and 'aquatic parking lots'. The mist platforms are simply public art using vaporized water to create a chest-high fog. It is based on the creation by the French landscape architect Michel Corajoud, which opened on the quayside in Bordeaux in 2006. He also created 'water mirrors' – thin sheets of water across the pavement. The idea of aquatic parking probably comes from this – underground car parks lit by light filtered through thin sheets of water. The message is that water art can be used to enliven urban living as well as using up surplus water and maybe purifying it. It is the twenty-first-century equivalent of the fountains of Rome, with a more ecological twist.

Other eco-cities are planned near San Francisco, London, Berlin and Warsaw. Britain has unveiled plans for ten new eco-cities, but retrofitting is currently to the fore. Arup is also involved in retrofitting schemes. The former mayor of London, Ken Livingstone, was enthused by eco-city methods after visiting Dongtan and proposed introducing them in London. The C40 Cities Network was established in London in 2005 and aims to foster retrofitting and recycling in partnership with the Clinton Foundation. As a result, buildings in the East London Albert Basin project are being retrofitted. The Thames Institute for Sustainability, a group comprising commercial and academic interests, is also retrofitting 100 houses in Dartford with water efficient systems and planning 'green roof' trials in London (see the section 'Green roofs', below).

New approaches to urban drainage and water sustainability

Recent approaches to urban drainage and water sustainability are taking a more holistic view of water management, considering not just the environment and flooding, but also rainwater harvesting and grey water reuse, and using the drainage and flood water as a resource.

One of the side effects of efficient urban storm drainage, which has traditionally drained stormwater directly into local rivers, is its contribution to flooding downstream and occasionally in the city itself. The expanse of impermeable and low-permeability surfaces results is swifter runoff velocities, flood peaks perhaps two to eight times higher and double the total stormwater volume (Jones, 1997). Lvovich and Chernishov (1977) estimated that every one per cent increase in urban cover is amplified to become a two to four per cent increase in runoff. Observations at Harlow New Town in England found that some summer stormflows were 12 times larger than in the rural area (Hollis, 1979) and at Skelmersdale New Town in Lancashire stormflows with shorter return periods showed marked increases in magnitude and frequency (Knight 1979).

In Britain, the Chartered Institution of Water and Environmental Managers established its Urban Drainage Group in 2009, incorporating the former Wastewater Planning Users Group, in response to concern over a spate of floods in the late 2000s and the Pitt review of the floods in summer 2007. The Institution emphasizes the need to slow the progress of water to the sea so as to reduce flooding and provide greater opportunity for its use as a resource. The CIWEM is concerned to develop innovative technologies, codes of practice and technical guides, like the 'Integrated Urban Drainage Modelling Guide', and to redress the skills shortage by helping train new practitioners. The CIWEM called for statutory requirements for all parties with flood management responsibilities to cooperate and share data. It also supported the cancellation of automatic rights for new urban developments to be connected to existing sewers and the new requirement for Surface Water Management Plans in the government's 2010 Floods and Water Management Act.

The potential effects of urban drainage on runoff were recognized in Ontario in the 1970s with the introduction of policies to ensure that all new urban developments must be designed to have 'zero hydrological impact' (Marsalek, 1977). At the Canadian Inland Waters Directorate, Marsalek used the US EPA Storm Water Management Model (SWMM) to design detention/retention ponds. Measurements of riverflow are often not available prior to urban development, so Marsalek used SWMM to simulate both the pre-urban and post-expansion responses, first calibrating the model on the current urban situation and then changing the parameters that would be affected. The model was then used to calculate the volumes of detention storage that must be provided by artificial ponds in order to maintain peak discharges at or below pre-development levels. The ponds can be incorporated as a scenic feature in urban parks.

America adopted the Low Impact Development (LID) policy and Best Management Practice for stormwater (EPA, 2000). The last decade has seen growing interest in Sustainable Urban Drainage Systems (SUDS) in the UK, or, more generally, Sustainable Drainage Systems (SuDS), which also aim to develop new methods for reducing the environmental impact of artificial drainage. One of these schemes designed to alleviate flooding and water pollution has been successful at Ravenswood, Ipswich. In the mid-2000s, the Ministry of Defence initiated Project Aquatrine for its properties combining rainwater harvesting, grey water use, metering, sustainable drainage systems using reed ponds, swales and soakaways to control peak runoff, and tests for green roofing.

In 2002, the Office of the Deputy Prime Minister predicted that over four million new households would be needed in the UK by 2016. These would have to be built at increasingly high density on both urban 'brownfield' and rural 'greenfield' sites. At the same time, the government initiated a consultation programme within the water industry on sustainable water management as part of its Foresight project. The national consultation highlighted the need for stable investment, management organized around river basins and an unequivocal commitment to SUDS and leakage reduction, and called for ways of improving customer participation. There was a recognition that a wide range of technological approaches are needed, including stormwater detention and attenuation, rainwater harvesting and groundwater recharge, and real-time control. Butler and Dixon (2002) point out that a significant barrier to adopting sustainable approaches is that they are often hard to justify in standard financial terms, because the issue of direct and indirect costs and benefits associated with them has not yet been fully addressed (Foxon *et al*, 2000). The full implications of new practices also need to be explored. The 'obvious' approach of reducing water consumption can also reduce the amount of water available to transport waste products and sewage. The UK Construction Industry Research and Information Association (CIRIA) established the Water

Cycle Management for New Developments (WaND) project (2003–2008) to address these issues. Fu *et al* (2009, 2010) report some results.

Ottenpohl *et al* (1997) report a pilot scheme in a small settlement of 300 inhabitants on the outskirts of Lübeck in Germany, which has separate systems for excreta, grey water and stormwater. Excreta are collected by a vacuum system and treated anaerobically in combination with organic household waste. The digested sludge is used as agricultural fertilizer during the growing season, and the biogas for heat and power generation. The grey water is treated in standard aerated sand filters or constructed wetlands, and disposed of in the stormwater handling system. Stormwater is collected for reuse and excess is either held in retention ponds or spread to infiltrate the soil. The capital costs were about 50 per cent greater than for a traditional system, but the operating costs are only half to two-thirds as much as standard wastewater treatment costs. It is, however, a system for small communities of under 1000. Above that, sludge transport costs escalate and vacuum systems are less appropriate.

Another pioneering experiment in urban flood control has been running since 2004 in the town of Chevilly-Larue in the French Department of Val-de-Marne. The town council gave homeowners €1000 each to install rainwater harvesting and large collection tanks in the gardens. Permits allow recycling of 500 L/m² of ground area from a house of 100 m², or 50 m³ of water. The scheme was extended in 2007 to over 220 buildings, which represent around 40 per cent of the town and which calculations suggest should be sufficient to suppress most floods. But the Environmental Protection Service says it will only stop flooding if the tanks are empty when it rains – so the EPS is also issuing warnings via mobile phones and the internet in the worst storms to warn interested parties to switch on the washing machine as well!

The other approach to reducing urban runoff is to increase infiltration into the soil, not just in soakaways, but also through permeable hard surfaces – leaving permeable gaps between paving slabs or, more inventively, using permeable materials. In Britain, the Construction Industry Research and Information Association (CIRIA) is promoting Sureset resin-bonded paving, which is permeable, for driveways and paths (www.ciria.com/suds/).

Green roofs

Green roofs are another approach to sustainable urban drainage. The idea is that they absorb, retain and detain rainwater, thus slowing and reducing the runoff. Part of the effect is due to increased evapotranspiration. They also increase insulation, reducing energy costs for both heating and air conditioning, as well as reducing greenhouse gas emissions.

The European Federation of Green Roof Associations is helping to promote the technique, as is Livingroofs.com. Ministry of Defence tests in the UK show *Sedum* species to be particularly effective. Tests with turf roofs on Victorian semi-detached houses in Camberwell, south east London in 2002 showed the roofs retained 50 per cent of the rainfall. The Fusionopolis building in Singapore has a 1.4 km-long coil of vegetation on the roof to collect rainwater, assist cooling and improve biodiversity in the city. London, Sheffield, Bristol, Birmingham and Manchester all started studying and developing policies on green roofing in 2008. Green roofs are being fitted to 40 per cent of the 11,000 new houses at the Barking Riverside Development in East London. But the green roof boom has yet to arrive. There is likely to be more understandable resistance to this than to rainwater harvesting. For ecologists, an additional attraction is that it is likely to encourage its own fauna as well – but this might not be to everyone's liking!

Conclusions

Curbing the expansion of demand that has typified the past century is clearly a prime aim for water management now. Population growth, urbanization and increased affluence have increased demand well beyond the levels envisaged when the bases for standard water supply and sanitation systems were laid down in developed countries during the nineteenth century.

Containing demand must be a combination of persuasion and coercion – neither on its own is likely to be successful in meeting targets. There are plenty of technologies available already that will assist in the quest. But two aspects seem paramount: that people need to be thoroughly informed and understand the limits and the aims, and that change should be affordable.

That said, there is still scope for increasing resources, which is the topic of the next chapter.

Discussion topics

- Review the water-saving measures in your home area.
- Explore the prospects for new water-saving technology.
- To what extent is food a water resource problem?

- Compare approaches to designing eco-cities.
- Consider the role of education in saving water.

Further reading

The theory and application of water soft paths, including the models used, are available in a book edited by David B. Brooks, Oliver M. Brandes and Stephen Gurman and published by Earthscan: *Making the Most of the Water We Have: The Soft Path Approach to Water Management* (2009). A summary version by Brandes and Brooks, *The Soft Path for Water in a Nutshell* (2007), is available from Friends of the Earth Canada in Ottawa, Canada, and from the POLIS Project for Ecological Governance at the University of Victoria, Victoria, Canada.

See Gleick, P. 2003: Global freshwater resources: soft path solutions for the 21st century. *Science* 302 (5650), 1524–1528. doi 10.1126/science.1089967.

The ebook *Encyclopedia of Life Support Systems* offers interesting details of recycling, especially in:

Vigneswaran, S. and Sundaravadivel, M. 2004. 'Recycle and reuse of domestic wastewater.' In: Saravanamuthu, V. (ed.) *Encyclopedia of Life Support Systems*, EOLSS Publishers, Oxford, UK, under auspices of Unesco, ebooks at www.eolss.net/ebooks/.

The issue of waste is thoroughly explored in:

Stuart, T. 2009. *Waste: Uncovering the Global Food Scandal.* Penguin.

14 Increasing supplies

'The only way to predict the future is to invent it.'

Vinod Khosla,
co-founder Sun Microsystems,
Silicon Valley, California

The time-honoured alternative to economizing on water usage is to increase supplies. Dams have been the prime instrument and they will continue to serve a purpose. Large-scale strategic imports through river diversions was a concept of the late twentieth century that nearly died until adopted now by the emerging economies of China and India. But innovative thinking has already produced a number of alternatives, some more viable than others, many with less environmental impact.

Indeed, there is a cogent argument that new engineering solutions to the water shortage will multiply in coming years. Over the last half century advances in science and engineering have doubled every 20 years. In 2008, software expert Ray Kurzweil told the American Association for the Advancement of Science meeting in Boston that technical advances over the next 50 years could be 32 times more than in the last. Not everyone totally agrees, but no one can disagree that one very sustainable way that supplies might be increased is to improve wastewater treatment and reduce pollution. And inventions in this field are burgeoning, as we will discuss in the next chapter.

Inventions will include modifications of existing strategies that are less environmentally disruptive and more cost-effective. They are also bound to include grand ideas like 'Red to Dead', the idea put forward by Sir Norman Foster's architectural team to solve the problems of the Jordan Valley by bringing water from the Gulf of Aqaba to the Dead Sea (see *Israel and Palestine* in Chapter 9). This may be too grand and expensive to implement, but brainstorming has its merits.

Desalination and rainmaking are now well established. The use of seawater is being extended in other innovative ways, as is weather modification, even to fog harvesting. These approaches now deserve separate chapters. In this chapter we will look at some less well-developed approaches and consider their potential.

New sources of surface and groundwater

'We have to look at any possible alternative, including towing icebergs from the Arctic and seeding rain clouds ... tankers from Scotland and Norway ... It would be an extraordinary thing to do, but we will have to look at extraordinary measures if we are in an extraordinary situation.'

Richard Aylard, Director,
Thames Water, during 2006 drought

Imports by sea – water bags, drogues and cigars

The bulk transfer of water by sea has generally been done by tankers, often surplus oil tankers. Malta and Gibraltar have benefited from this method in the past. Israel recently established this as a standby option from Turkey, potentially importing up to 400 million m³/year. It can be precarious and vulnerable, and desalination is tending to replace it. But more recently a much more flexible system using floating bags has been developed. In addition to providing a transport medium, dragged behind an ocean-going tug, these can provide long-term or short-term storage as a reservoir anchored in the sea or moored in a harbour and linked to shore by pipeline. The system might obviate the need to build reservoirs on land and so avoid the likely protests and planning restrictions. It can also act as a seasonal supply during periods of high demand, and then be removed. Or it could provide emergency supplies in a severe drought or following a disaster.

The method was pioneered in the 1980s by British university researchers and the Aquarius company was set up in 1992, manufacturing the bags in Southampton. These so-called 'drogues' or 'cigars' are made of thick polyurethane with an anti-tear and anti-UV coating. The material is compatible with drinking water standards and is approved by the American National Sanitation Foundation. Aquarius has been delivering potable water to the Greek Islands using tugs or large fishing boats since 1997.

The first generation drogues were 64 m long and held 720,000 litres of water. Aquarius now offers a 2,000,000 litre bag as well. Buoyancy chambers hold them upright, they float with 95 per cent of the bag underwater like an iceberg, so warning markers are needed at sea, and they automatically fold up as the water is pumped out. The technique is successful and economical and seems likely to become more widely deployed in the future. There are proposals for even larger water bags.

Water bags were first considered seriously by English water companies during the 1997 drought, when it was proposed to transport water from Scotland or Norway to southern counties if the drought persisted into 1998. In the event, they were not required. Thames Water seriously considered the technique again in 2006.

Ocean springs

An unknown amount of groundwater drains directly into the sea. The initial attempt to calculate the amount undertaken in 1970 by R.L. Nace of the US Geological Survey, based on estimated permeability of coastal rocks suggested that it is insignificant in global terms, at only 7000 m^3/s (221 km^3/year): barely one per cent of global riverflow. He called this seepage into the sea 'runout'. Nace was trying to balance estimates of global riverflow derived from discharge records, which gave him about 22,075 km^3/year, with the higher amount of runoff suggested by meteorological records of around 29,139 km^3/year: a shortfall of about 25 per cent. The accuracy of these calculations was drawn into question by the UNESCO (1978) climatic estimate of total riverflow at 47,000 km^3/year, more than double Nace's original figure. This suggests that riverflow gauges are picking up somewhere between 7000 and 25,000 km^3/

year less than the climate is providing. There is no doubt that riverflow records are inadequate and underestimate riverflows globally, and it is possible the climatic estimates are over-estimates, but these figures do suggest that large amounts are also 'going to waste', and this can only mean underground, because evaporation losses are included in the climatic calculations.

A recalculation by Speidel and Agnew (1988) supports this view. Their estimate places runout at around 12,000 km^3/year, which is globally significant: equivalent to a quarter to a third of net atmospheric transport from the ocean to the land. A recent investigation by Burnett *et al* (2006) reviewing a variety of methods for quantifying submarine groundwater discharge corroborates this view. They conclude that the process is essentially ubiquitous in coastal areas.

In the words of Dragoni and Sukhija (2008): 'All this water is lost to the sea, often of acceptable quality, and research to improve the measurement and recovery of it should be strongly enhanced'.

The practical problem is that if this discharge is largely diffuse, it will be virtually impossible to harness. However, we do know that this is not always the case. There are submarine springs. Karstic areas are the most promising. There is a very large spring off the northeast coast of Florida near St Augustine, which discharges 40 m^3/s (1.26 km^3/year) and has already had a proposal to harness it. A survey in 1995 identified 15 submarine springs off the Florida coast, mainly on the Gulf coast between Wakulla and Lee. Freshwater resurgence has been identified up to 120 km offshore in Florida. During the water shortage in Sevastopol in the Crimea in 1993, Ukrainian scientists investigated the possibility of tapping some of the local submarine freshwater springs. One on Cape Aiya was discharging 30,000 m^3/day (1 km^3/year) of low salinity (5.5 per mille) water (Kondratev *et al*, 1999). Submarine springs are common in the Mediterranean, for example, along the coast of Croatia, especially near the Velebit Mountains, Crete and the south of France. A spring off Montpellier yields 50 L/s. Of all the areas so far investigated, Florida seems the most promising in terms of total yields and stability of discharge, particularly the typical deep-seated springs called 'Floridan springs', which are fed from a vast aquifer, and maintain a steady discharge even through droughts, in contrast to the shallow springs.

Conceivably, the spring water might be collected in flexible pods or piped onshore, but the technology is yet to be developed.

Groundwater reservoirs

Artificially recharging groundwater sources has been practised for a while, but its use has increased markedly in recent years. Aquifers can make very efficient reservoirs, as they lose little or nothing through evaporation. This makes them preferable to surface reservoirs in warm climates. Where and when there is spare storage capacity, it can be used to store surplus surface water until it is needed. It is generally a cheap and cost-effective method, requiring little in the way of structural building and maintenance. It may be used to improve groundwater quality by injecting high-quality surface water, or it can be used to filter and purify low-quality surface water, as used during the clean-up of water from the River Rhine in Germany.

It can also be used as an efficient means of water transport, pumping water in one borehole and abstracting from another elsewhere. It can be especially useful when demand is very seasonal. The Algarve is a good example of a warm, dry climate combined with highly seasonal demand from tourism (see the article 'Aquifer storage and recovery' by Ferreira and Oliveira, below). The method has been used in Israel as an efficient means of storing and transferring water at least since the 1970s.

Artificial storage and recovery schemes have proliferated in the USA over the last quarter century. The Everglades Restoration scheme in southern Florida is injecting high-quality surface water, which would normally discharge into the sea during the wet season, into more than 300 boreholes in order to restore the Upper Floridan Aquifer that has become brackish through over-pumping.

It has also been used in England. Edworthy and Downing (1979) made the case for its application in Britain. Under the North London Artificial Recharge Scheme in the chalk and Tertiary sandstone of the Lea Valley, over 20 boreholes are used to inject water during periods of riverwater surplus and the same boreholes are reused to abstract the groundwater when riverflows are low in summer. Tests in a confined chalk aquifer in Dorset were less successful (Hiscock, 2005). Hiscock also gives details of the Great Ouse Groundwater Scheme developed in the 1970s and its subsequent incorporation in the Ely Ouse–Essex Water Transfer Scheme. The original groundwater scheme was designed to top up the River Ouse during the summer with riverwater stored in the chalk aquifer from the winter: an example of river regulation using a groundwater reservoir instead of the more usual surface reservoir, and so avoiding loss of valuable farmland in the area. The Ouse–Essex transfer scheme now takes surplus water from the Great Ouse from a point downstream where it would otherwise just drain to the sea, and diverts it over the topographic divide to supplement flows in the Rivers Stour and Blackwater to serve the growing population in Essex. This has reduced the need for new surface reservoirs in Essex.

Under the Thames Groundwater Scheme introduced in the 1970s, water was pumped out of the chalk aquifer beneath the River Lambourn in Berkshire and thence into the River Kennet to top up levels in the Thames. But it proved of limited use and was environmentally unsustainable, so artificial recharge was subsequently introduced to restore groundwater levels and sustain the river through summer.

A case of an ecological application is presented by Boeye et al (1995) in Belgium. They observed that groundwater recharge from a canal reduced the acidity of a wetland transforming a bog into a rich fen. Aquifers also filter the water, producing better quality water, and artificial recharge can be used as a clean-up strategy for riverwater (see Chapter 15 Cleaning up and protecting the aquatic environment).

A variety of techniques are used, from wells and boreholes to recharge pits and trenches, and permeable surfaces. Where the aquifer is shallow and 'unconfined', i.e. there is no impermeable layer above it, permeable material like sand may be used to create infiltration areas. Water may then be spread artificially over the area or rainwater harvesting may be used to concentrate rainwater in the infiltration area. A similar approach may be adopted within a river channel where the riverbed is very permeable. Dams can be installed to retain the riverwater and increase infiltration. Archaeologists have discovered that this technique was in use in Baluchistan and Kutch in India in the third millennium BC. It is still used in the Middle East in desert *wadis*, valleys where water only flows during the occasional storm. A slightly more complicated modern approach is to induce riverwater to infiltrate the banks. The method involves drilling boreholes parallel to the river on the floodplain and pumping groundwater out. This creates a hydraulic gradient away from the river and causes the river to seep into its banks and bed.

Exploiting the cryosphere

Although ice and snow contain the vast bulk of the freshwater on Earth, only about 0.5 per cent of it is

Aquifer storage and recovery

João Paulo Lobo Ferreira and Luis Oliveira

The technique of artificial recharge of groundwater is used in many parts of the world with several aims, e.g. storing water in aquifers to support future water needs during droughts, as protection against pollution or even for the restoration of groundwater quality.

Aquifer Storage and Recovery (ASR) is a technique with a double purpose, injecting water in a well when surplus water is available and withdrawing it from the same well when the water is needed. The advantage of using the same well is that no extra wells, facilities or water treatment plants are needed: the same treatment plant can be used to treat the water injected and the water extracted. An alternative is Aquifer Storage Transfer and Recovery (ASTR) plants, where the injection and withdrawal wells are different, adding a travel distance parameter to the recovery. This may be used to filter and purify the water.

A large number of Aquifer Storage and Recovery plants are working around the world and many more projects are under construction. There are several techniques for artificially recharging aquifers. The choice of the best technique depends on several factors, such as the natural characteristics of the aquifer, proximity to the water source or the objective of the artificial recharge. *Infiltration basins* for artificial recharge are based on increasing *natural* infiltration to the aquifer by making the surface more permeable. This is the easiest and oldest technique of artificial recharge. Infiltration basins are the preferred method when large open areas are available. When the area available is limited or the water table is really deep, the alternative is to inject water directly into the aquifer using wells.

Artificial recharge in the Querença-Silves aquifer, Algarve, Portugal

Water supply in the Algarve faces two problems: low rainfall and a highly seasonal population. Its permanent population is just 405,380, but it receives over six million tourists a year, mainly in summer, attracted by the beaches and golf courses. Average precipitation is only around 500 mm pa, surface water is limited and the topography is not well suited to large dams. However, the region is rich in aquifer systems. Until about the turn of the century, almost all the water supply in the Algarve came from groundwater. Some 17 systems have been identified, of which the Querença-Silves system is the largest, covering 318 km². The aquifer is mainly composed by karstified Lower Jurassic dolomites with high water productivity.

Artificial recharge of the Querença-Silves aquifer system was listed as a priority in the Regional Land Use Programme of the Algarve (PROT-AL). ASTR could use surface water from the Arade-Funcho group of dams and, in future, also from the Odelouca reservoir, which is under construction.

easily accessible. Nevertheless, the annual melting of glaciers and seasonal snowcover already provides valuable support for water resources in many parts of the world. Natural meltwaters supply more than a third of all the water used in the world for irrigation and are a major source of water supply in 28 countries, most notably in the countries reliant on the rivers of the Himalayas. Many deserts are irrigated largely from meltwater: the Thar from the Indus, the Ulan Buh from the Hwang Ho, and the Kyzl-Kum from the Amu Darya. Typically, 50–80 per cent of flows in Russian rivers are from snowmelt. Meltwaters provide 90 per cent of the flow in the Colorado, 50 per cent in the Mississippi and 40 per cent of riverflow in the Canadian prairies. Fifteen per cent of flow in the Columbia River, Canada, comes from the meltwaters of the Columbia Icefield and sustains numerous hydropower plants. The James Bay hydropower schemes rely almost entirely on snowmelt: 45 per cent of the annual precipitation is snow. Glacial meltwaters also support hydropower in Switzerland and Norway.

It is very tempting to try to enhance meltwater supplies. It is even more tempting to go for the big polar stores.

Controlling snowmelt

By not infiltrating into the ground or running away like rain, snow provides a natural reservoir. It is super-efficient because evaporation losses are minimal and when it does melt it is often over frozen ground, so infiltration losses are small. Why not improve on nature by making a deeper snowpack and controlling the time of melting?

Snow reservoirs

Minimal structural work is needed to trap drifting and blowing snow with permeable barriers. Barrow, Alaska, has been supplied from a snow reservoir created by snow fences since the early 1970s: a 4 m deep pack is all that is needed. Even more environmentally attractive is to plant designer forests. The USDA in Wyoming and the University of Arizona School of Renewable Natural Resources have carried out extensive research into the best tree species and spacing not only to create permeable windbreaks, but also to protect the snowpack from wind deflation and to slow down melting to avoid snowmelt floods. Mixed conifer forest offers the ideal high initial storage followed by slow release. Observations in Russia have shown that the melt period lasts only 5–20 days on the open plain, but 20–30 days under forest. Relatively low-density or 'honeycomb' forests located in low-insolation, high-altitude environments provide optimal conditions, with a maximum 'storage-duration index'. A honeycomb structure allows the maximum accumulation over a larger area, but the trees must also be dense enough to provide sufficient shade to spread the snowmelt, and avoid a snowmelt flood that cannot be harnessed.

Stubble management

Even less effort is needed for the method of 'stubble management' proposed for the Canadian prairies by Steppuhn (1981). He calculated that if crop stubble is left standing over winter to trap and retain just 30 cm of snow across the prairies and northern Great Plains, it would provide 213.5 billion m³ of meltwater for agriculture, comparable to the entire annual discharge of the Great Lakes into the St Lawrence River, which would increase soil moisture levels at the beginning of the growing season.

Artificial avalanches

Another novel form of snow manipulation has been tried using explosives. Artificial avalanching is sometimes used to reduce the risk of unexpected avalanches. But this has been applied to water management in the Colorado Rockies to create deeper, denser bodies of snow which melt more slowly, extending the melt period and reducing the risk of floods (Obled and Harder, 1979).

Quite apart from the dangers of using explosives in this instance, some ecologists are concerned about the possible effect of deeper snowpacks on wildlife. However, studies by the Bureau of Reclamation in Colorado and Wyoming suggest that reducing the stress on wildlife caused by floods and droughts might outweigh effects like restricting access to food beneath the snow or hindering movement.

Artificial icemelt

After Canada published one of the first national surveys of available water resources in 1981, there were suggestions that one possible solution to the predicted water shortage might be to induce more melting from some of its 100,000 glaciers. Many of these glaciers are well placed to feed meltwaters into the prairie provinces, which have been suffering from frequent water shortages for decades. Wisely, Environment Canada's scientists pointed out that this would be a very short-term strategy, in effect a form of mining, and that it would be well nigh impossible to 'turn the tap off' and stop the glaciers and icefields melting completely.

Nevertheless, Russian scientists did propose just such a scheme as a possible solution to the Aral Sea crisis. Krenke and Kravchenko (1996) calculate that glaciers provide 15 per cent of annual runoff and 40 per cent of summer runoff in the basin and estimate that inducement could increase runoff by a much-needed 4.12 km³, but that this is likely to be reduced to 3.93 km³ in a double-CO_2 scenario because of reduced snowfall and higher equilibrium lines.

The only proven method of inducing glacier melt is to lower the albedo over some part of the ablation zone by spraying the surface with dark material, like black carbon or more rarely, black plastic discs. Removing this material once it has induced sufficient meltwater is fundamentally impractical: the 'tap' remains on unless and until the meltwaters can flush the material away.

Inducing freshwater icemelt

Freshwater ice is the ice that forms on lakes and rivers. Induced melting is usually undertaken to reduce flooding. Ice-jam floods are a common problem in Canada and Russia. Ice cover on the Great Lakes can also be a flood hazard when meltwaters flow down the rivers and

285

hit a solid cover of lake ice. The problems may be relieved by spreading black carbon, by cutting leads around the edge of lake ice or by designing channels to eliminate jamming points and trapping ice floes in booms, even fixing steel cutters in the channel to slice up the ice floes into smaller blocks, as at Belleville, Canada.

Icetec – harnessing icebergs

The idea of harnessing icebergs was proposed in the early 1970s. One estimate suggests that a trillion cubic metres of water a year could be harvestable, equivalent to half the annual discharge of the Mississippi. Weeks and Campbell (1973) expected it would become an operational procedure by the mid-1980s, but it has never been applied. At the time, about 5000 icebergs were issuing from Antarctica each year, containing a total of 1000 km^3 of ice. Yields have increased in recent years (see *Shrinking land ice* in Chapter 12). Despite lack of development and considerable technical doubts, 'icetec' is still occasionally raised as an option. Even Thames Water considered it in 2006, before opting for desalination.

Antarctic icebergs are considered potentially more useful than icebergs from the arctic because they tend to be larger and more manoeuvrable. Most calve from ice shelves and are flat or tabular rather than the arctic bergs that calve from valley glaciers and are more pyramidal in shape. Greenland icebergs also get trapped by the shallow lips of the arctic fjords and can melt away slowly there for up to a decade. Antarctic icebergs can be over 200 m thick and contain 250 million tonnes, against just a few million for a Greenland berg. They contain thousands of years of snowfall. Suitable icebergs might be identified from satellites, marked by radio beacons and allowed to float until they reach a location where they can be harnessed by ocean-going tugs. The 'superberg' that broke away from around the Mertz Glacier in East Antarctica in March 2010 contained the equivalent of 20 per cent of the world's annual water use.

Antarctic icebergs could be a useful resource for some of the water scarce regions of the southern hemisphere. But there are three major handicaps. Their slow movement in ocean currents means that they lose a lot of their useful mass while drifting. Once harnessed, towing speeds are limited because the drag increases with the square of the velocity. And their large draught of perhaps 200 m or more means they can only approach land where the ocean is exceptionally deep. Although not best placed in terms of calving sites, ocean currents

or deep water, Australia could well benefit from icetec in future. Because the icebergs are so large, it has even been suggested that India might benefit. Prince Mohammad al-Faisal hosted a conference in 1977 in which he proposed that within three years Antarctic icebergs would be towed to Saudi Arabia by his new company, Iceberg Transport International. The icebergs would be wrapped in sailcloth and plastic to reduce melting and calculations suggested this might be achieved at a cost of 50 cents per cubic metre – less than the 80 cents cost of desalination at the time. It did not materialize and improvements in desalination technology are making it increasingly unlikely that icebergs can usefully be dragged across the equator.

Better placed would be Chile or Namibia. The cold Humboldt Current flows from Antarctica along the deep Peru-Chile Trench close to the Chilean shore and extra water could assist irrigation in the Atacama Desert. The coast of southern Africa is similarly favoured by the cold Benguela Current and the Namib Desert could benefit. Even so, the icebergs will need to be broken up by explosives to make them small enough to tow inshore.

One suggested treatment involves generating electricity at the same time to reduce the net cost. The smaller pieces of iceberg would be towed inshore and left to melt in an artificial lagoon impounded by a flexible floating barrier. The meltwater-seawater mix would be piped off and purified, while ammonia would be piped through the cold lagoon and back to the land where it gasifies in the warm air and drives gas turbines (Figure 14.1). Calculations in the 1970s suggested that a million tonne berg could fuel a 100 MW station for three years, supplying electricity worth over £100 million and water worth about £75 million. Another estimate suggested that a typical delivery could irrigate 1600 km^2 and cost only 1–2 cents for 10 m^3 of water in regions where 8 cents was the norm.

In environmental terms, icebergs are a natural discharge and using them would not harm the ice sheet in the way induced melting harms glaciers. Although Antarctica is protected by the Antarctic Treaty – renewed in 1991 – it would be most cost-effective to delay harvesting until they have floated outside Antarctic waters. The main concern relates to the destination rather than the source. It is whether it is possible to water the deserts to such an extent without inducing salinization and waterlogging or whether it is sound to encourage towns to grow there.

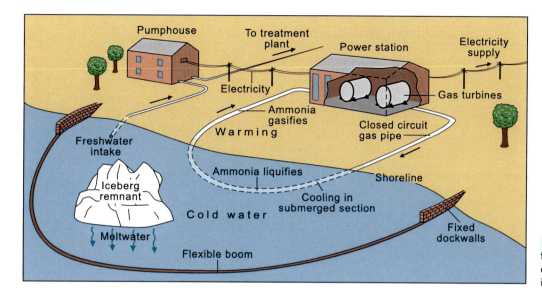

Figure 14.1 A scheme for creating freshwater and electricity from melting icebergs

Figure 14.2 An Iron Age urban rainwater harvesting system using street drains to feed a heated washhouse (right) at the Celtic oppidum of Citânia de Briteiros in northern Portugal

Meltwater from ice shelves

An alternative source is natural meltwater from the ice shelves. This has a more limited capability and is more suited to public supply than irrigation. But the George VI Ice Shelf is estimated to yield 40 km³ a year that forms a freshwater pool beneath the ice. This could be tapped using supertankers or rubber collection pods.

Rainwater harvesting

Rainwater harvesting is among the most ancient of methods, beginning with capturing overland flow in agricultural fields and developing into systems to supply small towns by the Romans and ancient Celts. The Roman theatre at Douga in North Africa was designed to harvest rainwater. Three-quarters of the farmland of the Hopi Indians in the Mojave Desert was fed by runoff harvested from a wide area of hillslope and directed to the crops. Gilbertson *et al* (1994) describe the low stone walls or 'bunds', terraces and spillways used in ancient Libya to collect storm runoff. Similar systems are still in use in many parts of the world.

The methods have largely been replaced in developed countries, with the exception of water tubs attached to existing roof guttering and drainpipes collecting water for the garden. This is changing. Harvesting systems are

Figure 14.3 A rainwater harvesting system built by the ancient Romans to serve a small settlement near Bujaraloz in northern Spain

now being actively promoted as water-saving devices in modern buildings. German manufacturers are in the forefront of developing the technology. And non-profit organizations, like the FBR at Darmstadt, promote it and encourage adoption through conferences and advertising media (see the article 'Rainwater harvesting in Europe', by Glyn Hyett, below). France and Germany offer tax advantages for installing rainwater harvesting. In Britain, the Environment Agency Wales set up a test bed and showcase for rainwater harvesting technology at Crosshands in the early 2000s. All toilet flushing and vehicle/plant washing water is derived from the roof, with mains backup in dry periods. Tests were successful and attracted the attention of a number of organizations including British Airways, local colleges and the Welsh Health Estates. A number of prominent new buildings, and whole new towns, have now been planned *ab initio* to incorporate rainwater harvesting. London's Millennium Dome collects rainwater to flush its 900 toilets. One unfortunate case of eco-minimalism has been working against wider adoption in some areas of Britain: namely, the high carbon footprint of large plastic tanks imported from Germany!

Nevertheless, equipment sales in Britain doubled 2004–2008 and the value of the industry rose from £500,000 to £10 million. The 2007 government code for sustainable homes made it mandatory for all new publicly financed buildings to reduce water use from 150 to 105 litres per person per day and suggested that this is most easily achieved with the help of harvesting rain. In May 2008, the government extended the same code to privately financed social housing. The same is likely to extend to all new housing within a few years. Small domestic systems cost £2000 and can save 40 per cent of mains water use. The British manufacturer Stormsaver claims that commercial customers can reduce mains use by 80 per cent, saving up to £10,000 per annum on water bills. Many stores run by Marks and Spencer, B&Q and the supermarkets Tesco, Asda and Sainsbury's already use rainwater for flushing toilets and washing floors.

Queensland is encouraging rainwater harvesting as part of its drive to cut domestic demand. Under the 2006 'new house sustainable living legislation', local councils were given discretionary powers to require rainwater tanks to be fitted in new housing estates. Take-up was slow because some councils felt the water shortages were due to the State's lack of planning and would make new homes more expensive. They argued that setting targets would be better than making rainwater tanks mandatory. Nevertheless, some private developers like Clarendon Homes offered 3000 litre rainwater tanks free.

During the extremes of the drought in Queensland in 2005–6, rainwater tanks were in heavy demand. Between the introduction of level 2 water restrictions in late 2005 and July 2006 around 8600 new rainwater tanks costing seven million Australian dollars were installed in Brisbane alone. One factor was a council rebate scheme that initially favoured smaller tanks. The scheme was modified in summer 2006, clipping a third off the $Aus 750 rebate for tanks under 3000 litres and raising the rebate to $750 for tanks over 5000 litres, which caused a scramble for small tanks before the change came into force. The boom was also higher in Brisbane than in the surrounding areas because it offered higher rebates.

Rainwater harvesting is also being promoted as a flood protection device (see section in Chapter 13 headed *Hydropolis – the water sensitive city*).

Rainwater harvesting in Europe

Glyn Hyett

Rainwater harvesting is an increasingly accepted and often legally required technology for new buildings across NW Europe. In all European countries there is an increasing awareness that catching rain from the roof makes sense from an economic and environmental point of view. The drive for a sustainable society requires technologies like rainwater harvesting, which may be capable of replacing as much as 50 per cent of the mains water used in homes. The costs of collecting, treating and distributing high quality mains water are ever increasing in NW Europe. The core market for modern rainwater harvesting installations is in Germany, from where much of the designs, technology and principles have originated.

Typically, rainwater is filtered before being stored underground. The rainwater is directed into the bottom of the tank, allowing the freshwater to reach the base of the tank. Sedimentation while in the tank is a vital cleansing process. Particles less dense than water float, and all tanks are allowed to overflow at least once a year. Water is abstracted by electrical pumps, usually submersible pumps. These pump on demand to points of use on a pressure switch or to a header tank. Systems are usually connected to the mains systems for automated top-up in the event of prolonged drought.

The economics of using rainwater vary from country to country, and the lengthy payback periods in some countries (15 years in the UK) often deter private investors. In some countries State Regulations require rainwater harvesting to be incorporated in all new buildings, and often planning authorities push to have rainwater harvesting included where it is not mandatory. The drive to sustainable drainage systems often means that rainwater harvesting is adopted as the first line of defence against future flooding risks.

The benefits of rainwater harvesting are wide ranging, for the rainwater user and the economy as a whole. The benefits include saving often scarce water resources, reducing the costs and energy use associated with water treatment, reducing the runoff to sewers or watercourses, saving on metered water supplies for homeowners, enhancement of building values, and assisting the future sale of buildings. Rainwater is seen as ideal for washing machines, for flushing toilets, and especially for using in gardens and general washing-down purposes. Gardeners, in particular in urban areas with hard water supplies, find rainwater more beneficial for plants.

Detailed information is available from the UK Rainwater Harvesting Association at **www.ukrha.org**, the European Rainwater Catchment Systems Association (**www.ercsa.eu**), the International Rainwater Harvesting Systems Association (**www.eng.warwick.ac.uk/ircsa/**) and the German FBR (**www.fbr.de**).

New rain-harvesting methods are also being introduced in developing countries, notably by charities. CAFOD has installed more than 50 domestic rainwater harvesting systems in Bangladesh. In Africa, the Benevolent Institute of Development, a Christian Aid partner, is promoting rainwater harvesting as part of a package to alleviate food shortages in Kenya. The Institute helps poor farmers to conserve water by digging ditches alongside crops to hold the rain, along with terracing to reduce soil erosion and planting grass between the terraces to bind the soil. It is also promoting multi-storey gardens that optimize on water and space by planting crops on top of soil-filled sacks and through slits in the sides, all fed by harvested rainwater directed into a layer of stones in the centre whence the water trickles through each layer of the sack.

Rainwater harvesting has found a renewed potential for irrigating paddy fields in eastern India. Srivastava *et al* (2004) also claim that these paddies reduce the flow of nutrients to the rivers and are useful as a tool to restore and manage riverwater quality.

Rainwater harvesting is one of a number of solutions that Professor Nurit Kliot considers for small islands in her article 'Water sustainability on small tropical and subtropical islands', in Chapter 16.

Harvesting fog and dew

If and when they think of tapping the atmosphere for more water, managers normally think of 'weather modification', which effectively means rainmaking (see Chapter 17 Controlling the weather). But there is water in fog and dew that could be harvested without modifying the weather, albeit in small amounts. Nevertheless, small amounts might be crucial in a desert and the technology involved could really be said to be human scale. The term 'occult precipitation' is often applied to fog and dew as it goes largely unmeasured.

Nature uses it already. Some ecosystems survive partially or entirely without rain by harvesting fog. Many mountain ecosystems where the mountains rise through the common cloud layers collect undocumented amounts of extra water through interception and fog drip. In the Namib Desert, *Stenocara* beetles collect fog droplets by raising their forewings (Parker and Lawrence, 2001; Harries-Rees, 2005).

Minute inspection of the beetles' forewings has led researchers at Massachusetts Institute of Technology to develop a material that emulates it. The wings consist of hydrophilic wax-coated zones surrounded by non-waxy hydrophobic areas that direct the water towards the hydrophilic areas, increasing the amount of water in these limited areas so that it becomes enough to drain down the forewing. The MIT researchers believe it can be produced cheaply at a commercial scale and used as a coating to collect water on tents or buildings in arid environments. It could also be used as a fog-free nanocoating or self-cleaning surface for glass, so reducing the need to waste water on window cleaning.

Fog is now being collected artificially in various parts of the world. FogQuest is a Canadian non-profit organization that promotes fog and rain harvesting in developing countries. It has installed equipment in the coastal desert of northern Chile and in the mountains of Haiti. It has also advised users in Nepal and the Dominican Republic. Professor Jana Olivier reports on her experiments in her article 'Fog harvesting in South Africa' (below). Similar fog collectors installed at the highest meteorological station in the Velebit Mountains (1594 m) in Croatia have yielded up to 27 L/day/m² of collector on rainy days and up to 19.1 L/day/m² on non-rain days – so they also act as rainwater harvesters, but even the lower figure is potentially useful for a family (Mileta *et al*, 2006).

Dew is also an undervalued resource. Atmospheric humidity is often remarkably high in arid and semiarid regions. Atmospheric moisture levels in the Rajputana desert, NW India, are comparable to humid tropical forests, the problem is the lack of rain-forming mechanisms (Bryson and Murray, 1977). Palestinian farmers have traditionally watered their orchards by 'dew farming': rocks are piled around the trees, these cool radiatively at night causing condensation. Measurements by Nagel (1956) on Table Mountain, Cape Town, using a receptor covered by twigs indicated total annual receipts of 3300 mm compared with just 2000 mm measured by a conventional rain gauge.

Attempts are now being made to create modern dew-collecting systems. Mileta *et al* (2006) describe dew condensers installed on specially designed roofs on the Croatian coast in 2003. The roofs are inclined at 30° and coated with polythene sheeting embedded with titanium oxide and barium sulphate (barytes) to make it hydrophilic. The tests suggest the approach has potential in the Mediterranean during the hot, dry season when there is little cloud and radiative cooling is at maximum. A curious approach is described by Rajvanshi (1981) in India, which does not appear to have found favour. The author admitted it was not economically feasible compared with desalination by reverse osmosis, even though it produced 643 m³ in a day. It involves pumping cold (5°C) seawater up from a depth of 500 m and passing it through a heat exchanger onshore, where condensation occurs.

Low-energy solutions: human and animal power

Ancient solutions to raising water based on human or animal power have served humanity well for thousands of years. Archaic they may be, but they are still serving many communities in developing countries, which is more than may be said for some fossil-fuel powered systems of the twentieth century, as the mechanisms fail and the prices of fuel and repairs escalate.

The way forward for many peasant farmers is to improve power efficiency and reduce dependence on outside sources. In this, many charitable organizations are playing a major role in education and equipment development.

The most ancient human-powered device for raising water was the *shaduf,* a pivoted wooden pole with a

Fog harvesting in South Africa

Jana Olivier

South Africa is an arid country with only 35 per cent of its surface areas receiving more than 500 mm rain per year (South African Department of Water Affairs and Forestry (DWAF), 1986). Surface water is scarce and groundwater may be inadequate, inaccessible or of low quality. Even in mountainous areas with high rainfall, water often runs off quickly and is not accessible to rural communities. It is estimated that rural women spend up to five hours per day collecting water for household needs. According to the DWAF, groundwater is the sole source of water for 280 South African towns, five million cattle, 26 million small stock and about 90 per cent of people living in rural areas (Pretorius, 2005). In 2005, around nine million South Africans did not have access to sufficient supplies of clean water (Vermeulen, 2005). This makes water scarcity one of the most crucial environmental problems facing this country and it is imperative that alternative water augmenting strategies be sought. Fog water harvesting is one such possibility.

Fog occurs most often along the arid West Coast and in the mountainous areas of the country. Advection sea fog, formed when moist South Atlantic Ocean air is advected over the cold Benguela Current upwelling region, is the most frequent source of fog along the West Coast, while orographic fog and surface clouds frequently shroud the mountains of the Eastern Escarpment and the Western Cape.

Fog is the major source of water for many plants and animals living in arid regions. Numerous algae, lichens, succulents such as *Trianthema hereroensis* and grasses (*Stepagrostis sabulicola*) of the Namib desert in Namibia are adapted to using the moisture from fog. The latter flower and produce seed throughout the year and are thus a source of food and shelter for many dune animals (Seely, 1987). However, one of the most unusual adaptations to the use of fog water is the *Onymacris unguicularis* beetle. During foggy conditions, it climbs to the top of a sand dune where the fog is thickest, turns into the wind, lowers its head and straightens out its hind legs. The back of the beetle serves as a condensation surface for the fog droplets that glide downwards and hang suspended from its mouthparts. It has been estimated that a beetle can take up an amount of fog water that is equivalent to 40 per cent of its body weight in one morning (Seely, 1987).

The feasibility of using fog harvesting to augment domestic water supplies in South Africa was investigated during the late 1960s and again in 1995. During the latter study, pilot 1m² fog collectors were erected in various parts of South Africa and the water collection rates monitored over a three-year period. These were found to be a function of elevation, wind speed, moisture content of the air and site characteristics. Along the West Coast yields ranged from 1 to 5 L/m²/day while in the mountainous regions, they exceeded 10 L/m²/day at elevations greater than 1700 m. Fog contributed around 90 per cent of the water harvested at the West Coast sites and 35 per cent in the mountains of Limpopo, Mpumalanga and the Western Cape (Olivier and van Heerden, 1999).

Large fog water collection systems have been erected to supply communities with water. Each system comprises of a water collection screen and one or more storage tanks. The screen consists of three long wooden poles mounted 9 m apart. Steel cables stretch horizontally between the poles and anchor the structure. Two sections of 9 m by 4 m shade cloth netting (30 per cent) are draped over the top cable and secured to the middle

Figure 14.4 A fog water collection screen in South Africa

and lower cables and to the poles on either side. This forms a fog collection screen of around 70 m². A gutter is attached to the lower ends of the screen. During foggy and wet conditions, droplets are blown against the screen and are deposited on it. As the drops become larger, they trickle downwards and drip into the gutter. From there the water is channelled through a sand filter to a pipe that leads to 10,000 litre tanks located down slope. When the top tank is full, the overflow is channelled to the next tank further down the hill, and so on. This system was specifically designed to be used in poverty-stricken rural areas, to be as cost effective as possible, to use material that is readily available, and to be suitable for use in areas without electricity. With sufficient funding, more sophisticated and higher-yielding systems can be used.

Seven large fog water collection systems have been erected to date – one at a school in the Soutpansberg Mountains in Limpopo, one at a small West Coast village and five at schools in the Eastern Cape. The first two comprise the experimental sites where water yields and quality are monitored on a continuous basis. Daily water collection rates range from zero on a sunny day to more than 4000 litres during wet and foggy conditions (Olivier, 2004). The water was found to be of an exceptionally high quality (Olivier and van Heerden, 2002). Sufficient water is collected to fulfil all the water requirements of the schools and local communities. At some, vegetable gardens have been established using harvested water.

Fog water collection has several advantages. It is relatively cheap, can be used in large areas of the country where water shortages are acute, it is environmentally friendly since it does not influence the amount of water available to adjacent vegetation, it is clean, and it does not require an external power source. It also has numerous applications. However, it is only viable in those areas where fog occurs frequently (>70 days per annum) and persists for some time. Since the amount of water harvested is limited and the supply erratic, it cannot replace other sources of water. Nevertheless, it could be used on small scale to alleviate local water shortages and to decrease pressure on existing water sources.

bucket on one end and a counterweight on the other using the principle of the lever. It was invented in Pharaonic times in ancient Egypt in about 1500 BC for lifting water from rivers and canals into irrigation ditches. It can still be found in use today. This was supplemented around 500 BC by the Archimedes screw, thought to have been invented by the ancient Greek scientist, which comprises a broad-threaded screw inside an enclosing pipe, which could be turned by hand. The water wheel was a slightly later development from the Egyptian Ptolemaic period in the third or fourth century BC, which used oxen to turn a horizontal wheel geared to a vertical wheel with buckets attached. Again, these are still used. The idea of turning a horizontal wheel to raise water is embodied in twenty-first-century play pumps (see later in this chapter), with children power replacing the oxen, which may now be too expensive or in short supply.

Sand pumps or 'sand-abstraction pumps' are another adaptation of the millennia-old strategy of digging for water in dried-up riverbeds, common throughout Africa. When the rivers cease to flow in the dry season, there is often a remnant of groundwater left beneath the riverbed. Christian Aid is now financing sand pumps, which do the job without the digging and also provide cleaner water because it is filtered by the sand. Sand pumps are now widespread in Zimbabwe where they are promoted by the Dabane Trust, for example, on the Shashane River, which becomes just a parched strip of sand in the dry season. Women operate simple hand pumps that can provide enough water for domestic use and to irrigate vegetable patches, which provide a much-needed food supplement.

Elephant pumps, play pumps, treadle pumps and rope and washer pumps

These human-scale inventions are beginning to revolutionize access to clean water in Africa and southern Asia. Play pumps were invented by British entrepreneur Trevor Field and were quickly endorsed by Nelson Mandela in 2005 as a solution for South African townships (Figure 14.5). Elephant pump wells have been constructed in a number of countries under the aegis of the British charity Pump Aid, starting in the late 1990s in Zimbabwe, where 25 per cent of the rural population have no clean water and diarrhoea is the third biggest killer of the under-5s (Figure 14.6). They are now found in thousands of villages.

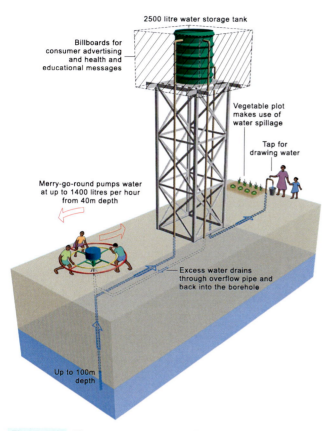

2500 litre water storage tank

Billboards for consumer advertising and health and educational messages

Vegetable plot makes use of water spillage

Tap for drawing water

Merry-go-round pumps water at up to 1400 litres per hour from 40m depth

Excess water drains through overflow pipe and back into the borehole

Up to 100m depth

Figure 14.5 Play pumps are capable of delivering 1400 litres an hour from a depth of 40 metres with the merry-go-round operating at just 16 revolutions a minute. It is effective up to 100 m. (Based on www.playpumps.org)

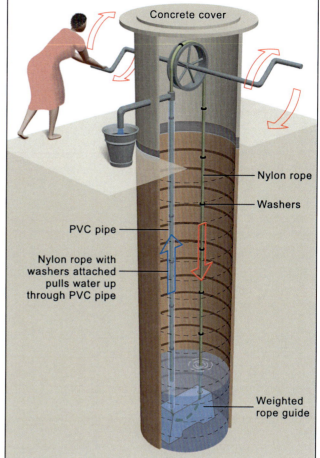

Concrete cover

Nylon rope

Washers

PVC pipe

Nylon rope with washers attached pulls water up through PVC pipe

Weighted rope guide

Figure 14.6 Elephant wells

Treadle pumps were developed in Bangladesh in the early 1980s. International Development Enterprises of India played a major part in developing the technology in the 1990s and claims a benefit-cost ratio of 5. Treadle pumps are now widely used, for example, in Cambodia, Bangladesh, Zimbabwe, Niger, Kenya, Zambia and latterly in Nepal. Many are built locally by non-professionals, often with the aid of charity workers, but there is also a wide variety of readymade treadle pumps on the market using a range of designs (Figure 14.7). The original pumps, based on suction, have now been supplemented by pressure pumps that can raise the water above themselves and be used to pump water uphill over distances of half a kilometre or so. The amount of water raised depends on the depth of water and on the weight or strength of the one or two operators.

Whereas the treadle pump is currently limited to raising water from a maximum depth of 7 m, rope and washer pumps are designed to raise water from up to 20 m depth. These are hand operated and consist of a washer or piston attached to a polypropylene rope, which sucks water up a 19 mm diameter pipe. Women farmers can easily raise up to 20 litres a minute from a well or river for drinking or irrigation.

These inventions tackle major problems in a sustainable way. The traditional method of lowering a bucket into an open wellhead is inefficient and dangerous: bird faeces and dead animals can foul the water, lowering buckets is slow and tedious and risks knocking debris in. The high-tech alternative has been piston pumps, but these can be too efficient, draining other wells by extracting too much, and once they break down they are expensive to repair and require expert servicing. Elephant pumps are more reliable and easily repaired: in Malawi over 95 per cent are estimated to be in operation at any one time compared with only 50 per cent for piston pumps.

All of the inventions are founded on the principles of low cost and ease of building and maintenance, using cheap local materials and unskilled labour. There is no

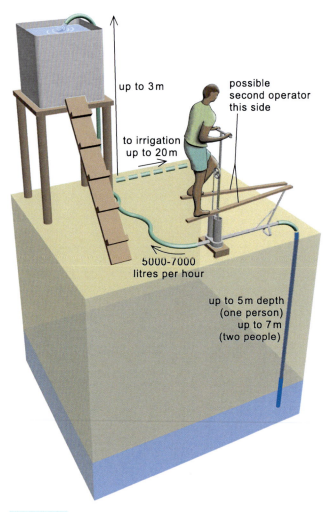

up to 3 m

possible
second operator
this side

to irrigation
up to 20 m

5000–7000
litres per hour

up to 5 m depth
(one person)
up to 7 m
(two people)

Figure 14.7 A treadle pump

need for expensive imported fuel. They aim to provide reliable, clean supplies and prevent the sort of contamination of groundwater that was so devastating in the 2008–9 cholera epidemic in Zimbabwe. Elephant wells achieve this by the simple expedient of a raised edge and concrete covers. The Zimbabwean epidemic, which official (under)estimates suggest killed nearly 4000 and infected over 60,000, was largely caused by raw sewage being washed into the old, hole-in-the-ground wells by rainwater, especially during the monsoon. Play pumps have the added dimension of using children-power in a fun way on the merry-go-round. The storage tank can also be used to raise funds and educate people by acting as a four-sided billboard for commercial advertising and health information.

Women are major beneficiaries, especially young girls. Using the old method of lowering a bucket or tin into the wells could take up to six hours to collect the fam-

ily's daily water requirement – with the pumps it takes only two hours, which means that girls are free to attend school. All the pumps are easily operated by women. In India, women have reported increased confidence and a sense of ownership as many have been able to become independent farmers rather than labourers.

Treadle pumps are commonly employed by small farmers to irrigate crops from groundwater or streams. Farmers can extend the growing season and even cultivate all year round, as well as possibly growing new varieties of crop. In Malawi, Christian Aid's partner CARD reports that treadle pumps have doubled family harvests. In India, family nutrition levels have improved and family earnings raised by at least $100 a year, according to International Development Enterprises. This is in a country where small farmers may have savings of less than $15.

The environment can also benefit, as the human-powered pumps yield less water than diesel-powered pumps, typically 0.6–1.25 litres a second from treadle pumps, which means that less fertilizer and pesticide is washed away into groundwater or streams.

Carbon offsets and treadle pumps

Human-powered pumps have also found support from the British carbon offset charity Climate Care, as used by the Prince of Wales and the UK Prime Minister David Cameron on international flights. Climate Care promotes treadle pumps and claims that one treadle pump saves 0.65 tonnes of CO_2 a year that would have been produced if the farmer had used a diesel pump, although a study in Uttar Pradesh where the farmers hire the diesel pumps suggested that savings are considerably less, no doubt partly because they only hire when absolutely necessary. The pumps are distributed as part of the International Development Enterprises IDE-1 programme funded by governments and charities (see *The role of charities* in Chapter 21). Climate Care is introducing the pumps in Chhattisgarh and West Bengal in India. The treadle pumps also free farmers from the tyranny of unreliable electricity supplies that besets electrical pumps in parts of India, where supplies may operate for only an hour or so at a time, or not at all. It allows crops to be irrigated when needed, the pumps are less likely to deplete the water table because their sucking power is limited, and most importantly they help farming families to stay on the land rather than swelling the drift to the cities.

Water beyond Earth

As human beings embrace the prospect of extended space travel and possibly even colonization of other planets, finding or making water is a critical issue. In his book *Biospheres: metamorphosis of Planet Earth*, biologist Dorion Sagan sees humanity's colonization of space as an inevitable extension of natural biological processes. Just as life emerged from the ocean in encapsulating bodies of various types, each designed to contain the water-based fluids that supported life beneath the waves, so spacecraft and space stations will act as the shells to transport the 'seeds' of life to even harsher environments.

The stuff of science fiction? Perhaps. But Nasa has plans to make it happen and Russia, China, Japan and perhaps India appear to have similar designs. Nasa started its Constellation project in 2006, its first manned space flight programme since the 1970s, and in 2009 launched Ares 1-X, its prototype replacement for the ageing Space Shuttle, which is due to carry crew to the International Space Station by 2014. Nasa produced plans to establish a manned base on the Moon, the first off-Earth colony, by 2024 and for this to act as a launch pad for the onward 400 million km, multi-year journey to Mars by 2037. But this was put on hold in 2010 as a result of the recession, and Nasa is now likely to use robots for the foreseeable future. However, this is unlikely to mean the end of America's manned missions to the Moon, as President Obama has expressed a wish to engage commercial interests. Although lack of money will delay the timetable, international politics mean that the USA is unlikely to abandon the plans entirely and the recent discoveries of water on the Moon provide a convenient boost. International collaboration could be the way forward. Nasa plans to collaborate with the European Space Agency (ESA) in launching a robotic exploration of Mars by 2023.

Russia plans a cosmonaut landing on the Moon by the 2030s. China plans a taikonaut landing perhaps as early as 2020 and Japan is following a similar timetable. Currently, each plan separate bases. India is following, aiming to put a man in orbit by 2016. In contrast, Google's Lunar X-Prize, to be awarded in 2011, has encouraged commercial organizations to devise robotic explorations of the Moon. These will overcome the lack of water, but they are generally seen as only a stepping-stone to the more expensive manned exploration.

Finding a suitable source of water will be key to the success of the manned missions, not only for drinking, but also to rehydrate food and possibly even irrigate crops, although recycling will also have a large role. There is also the prospect of manufacturing water as well as finding it.

Water in the solar system

Recent discoveries indicate that water is more common in the universe than previously thought. The discovery of water on the Moon from Nasa equipment onboard the Indian Space Research Organization's orbiting probe Chandrayaan-1 caused great excitement when it was announced in September 2009. It confirmed the indications of a 'water signature' picked up by the Cassini probe in 1999 en route for Saturn, and by the Deep Impact Probe before it crash-landed in November 2008. However, the amount of water is tiny and the results, which are based on the absorption of light in the 3 μm wavelength, could equally apply to hydroxyl molecules (HO). Evidence suggests both HO and H_2O coexist. Chandrayaan-1 found water in fine films adsorbed or bound on the surface of Moon dust. An initial estimate suggests there could be up to 1 litre of water per cubic metre of soil (strictly called 'regolith' as there is no evidence of organic material as in a true soil). The regolith is actually drier than any desert on Earth and the water is not immediately useable. There is speculation that it might eventually be possible to concentrate this tiny quantity (1 in 1000 parts by volume) to provide a useable resource for resident astronauts. But because the water film appears to be concentrated only in the top few millimetres of dust, this could require heating 730 m^2 of regolith to 200°C in order to get just a glass of water. There are therefore two problems to overcome: scraping enough topsoil together and then separating the water from the minerals and condensing it. Enough heat might be generated by solar panels. There is then the possibility of applying electrolysis using solar-powered electricity to split the water into hydrogen, for fuel, and oxygen, for breathing.

The evidence from Nasa's Moon Mineralogy Mapper on Chandrayaan-1 suggests that the water may still be forming on the Moon. The Moon is continually sprayed with hydrogen by the solar wind, a stream of charged particles emitted by the Sun. Because the Moon surface lacks the protection afforded on Earth by its magnetic field and atmosphere, the positively charged hydrogen ions reach the ground unhindered, where they could bond with oxygen bound in the rock particles. The water molecules are only on the surface because the ions cannot penetrate deeper. Although water was found all across the Moon, there is more water in the polar regions, probably because they are colder, there is less evaporation and the water clings to the surface materials. It is even suggested that water created outside the polar regions might migrate towards the poles, perhaps as vapour evaporated by sunshine at lower latitudes condensing or sublimating near the poles. Small amounts of water even exist at the Moon equator, where temperatures reach over 100°C, perhaps because the natural production of water replaces any evaporated. Previous speculation that ice may have existed for billions of years in the Moon's polar regions was largely based on indirect evidence from several probes of abundant hydrogen there and temperatures of −238°C (Hand, 2009). However, recent re-analyses of Moon rock collected on the manned missions of the 1960s and 1970s also indicates that volcanic eruptions delivered water molecules to the Moon's surface some three billion years ago.

The month after the results from Chandrayaan-1 were announced, Nasa fired its Centaur rocket at the Moon to look not just for water but also for substances that might allow water to be manufactured. The upper part of the rocket struck the Cabeus crater near the dark south pole of the Moon at a speed of nearly 10,000 km per hour to create a huge dust plume. The plume was to be analysed by the Lunar Crater Observation and Sensing Satellite before it too crashed four minutes later, but no plume was visible, perhaps because the rocket hit solid rock. The LCROSS was looking for hydrocarbons and hydrated minerals as well as water and ice. Hydrocarbons are already being exploited to create hydrogen for fuel cells, the product of combustion being water, but this does require a source of oxygen as well. So the 'hydrogen economy' solution to earthly energy needs could also produce a by-product that will support space travel.

Astronomers ostensibly interested in signs of extraterrestrial life are already finding more evidence of water elsewhere in the solar system. There is evidence of traces of water vapour in the Martian atmosphere and pictures of the legs of the Phoenix Mars Lander's legs published in March 2009 appear to show water droplets forming and falling off. The dry channel networks identified on Mars by the earlier Mariner 9 and Viking probes show that liquid water once flowed there, perhaps when it was warmer and before the low gravity force, roughly a tenth

of Earth's, allowed too much greenhouse gas to escape, or else when ice was melted by volcanic eruptions or meteor impacts. It is now clear that the polar ice caps on Mars are composed of water, not CO_2 as once supposed. However, its lifespan may be limited if the warming trend reported by Fenton et al (2007) continues: Mars has warmed by half a degree since 1970 due to stronger winds causing dust storms that trap heat, and the authors suggest the southern ice cap could disappear, perhaps in 500 years.

Jupiter's moon Europa also has ice-covered oceans, which appear to contain more water than all the Earth's oceans. Evidence collected from Nasa's Cassini probe between 2005 and 2009 suggests that high velocity spurts erupting from geysers on Saturn's moon Enceladus contain ice and water, some of it about as salty as Earth's oceans, together with organic compounds. Saturn's E-ring is composed of ice crystals from these water jets. Saturn itself appears to be composed largely of hydrogen with a small core of ice and rock. Nasa's Spitzer Space Telescope made another exciting discovery in October 2009: a previously unknown, very sparse ring around Saturn some 13 million km from the planet, which also contains ice. The ice may come from the moon Phoebe due to impacts. The ring's dust and ice are slowly succumbing to Saturn's gravity and depositing a super-fine rain on the inner moons, like Iapetus, and even possibly on Saturn itself. There is speculation that Earth may also receive cosmic ice particles. If so, they would sublimate in the ionosphere, but they could be adding minute amounts of water to the atmosphere. Whether it survives as water and the hydrogen and oxygen remain bonded is an open question (Verbiscer et al, 2009).

The discovery of water on the surface of an asteroid called 24 Themis, orbiting 480 million km from the Sun, has renewed speculation that most of Earth's water, barring the small amounts of 'juvenile' water produced in volcanic eruptions, may have been brought here by asteroids (Campins et al, 2010). The physical properties of asteroid water match Earth's better than does the other possible source, comets. It has long been speculated that Earth would have become so hot when the impact that tore the Moon out of the Earth occurred 4.5 billion years ago that all water would have evaporated, and then must have been slowly replenished from space. Campins and colleagues speculate that the water on 24 Themis is either being sublimated on the surface or that 'impact gardening' is raising it from inside the rock.

In yet another milestone revelation, in August 2009 Nasa released the first evidence of extraterrestrial life, the protein glycine, found in material collected from the tail of a comet by the Stardust probe. It adds weight to the theory that life arrived on Earth from space, as proposed by Fred Hoyle and Chandra Wickramasinghe (1993). Moreover, the presence of basic proteins together with the discovery of numerous sources of water suggest that higher life forms may indeed exist in space which have the same architecture as on Earth, without resort to postulating alien life systems, e.g. based on silica.

The wider universe

Beyond our solar system, astronomers have identified over 300 planets orbiting distant stars in our galaxy that may have water and support life. The Kepler Spacecraft was launched in March 2009 to seek out these 'exoplanets' in the so-called 'Goldilocks Zone', where planets are close enough to their star for water to be liquid, not too close so it evaporates away nor so far away it freezes, and where temperatures are neither too hot nor too cold for life. These planets will then be targeted by another generation of spacecraft looking for oxygen and water, not with a view to colonization – these planets are far too distant for any currently conceivable means of transport – but to try to answer the old question of whether intelligent life may reside there.

Charbonneau *et al* (2009) report a super-Earth (GJ121b), nearly three times bigger than Earth and seven times heavier, 40 light years away in the constellation Ophiuchus, that is believed to be half to three-quarters water, largely ice.

Water for space travel now

Meanwhile, in the real and present world, the Italian Società Metropolitana Acque Torino (SMAT) Research Centre is producing purified Earth water for space travel. In order to minimize processing costs, SMAT selected well water and spring water from sources near Turin that most closely meet the chemical, bacteriological and physical standards specified by Russia and America prior to treatment, and eschewed taking water from the River Po, which is normally used for the public water supply together with well and spring water. The process is complicated by having to meet different water standards for the Russian cosmonauts and American astronauts (Table 14.1). The water was launched with the ATV module Jules Verne to supply the International Space Station in March 2008.

Figure 14.8 Water processing facility for the International Space Station, SMAT, Turin

Table 14.1 Water quality requirements for the International Space Station

	Russian	American
Source	Groundwater with natural mineral content complying with	Spring water
Mineral content requirements	Moderate mineralization (TDS* 100 to 1000 mg/L) Must contain calcium, magnesium and fluoride within specified limits	Minimal mineralization (TDS<100 mg/L)
Disinfection and purification	Silver ions and sodium fluoride, followed by microfiltration	Iodine, followed by microfiltration
Total organic carbon (TOC)	Not specified	<0.5 mg/L
Volatile organic carbon (VOC) and semi-volatile organic carbon (SVOC)	Not specified	Complying with US EPA limits
Toxic metals	Stringent control	Stringent control
Microbiological safety	Stringent control	Stringent control
Materials in contact with water	Stringent control: all stainless steel in dedicated clean zone	Stringent control: all stainless steel in dedicated clean zone

* Total dissolved solids

297

require a limited degree of desalination. Certain aspects of exploiting snow and ice also have prospects, though this is an area where serious environmental harm could be done. Human-scale methods like rainwater harvesting offer considerable potential for decentralized water supply, and low-cost/low-maintenance techniques such as play pumps offer valuable solutions for developing countries. Further, large amounts of water are lost to human use because of pollution – reclaiming this water will be considered in the next chapter.

Finally, water is already being prepared for space travel. Will sustainable supplies be found or manufactured in space in coming years?

Figure 14.9 Bottled cosmonaut water produced at SMAT, Turin

Conclusions

There are numerous ways to increase water supplies. Some appear to have considerable potential. Others may be of value as domestic sources, comparable to home energy generation, or in special circumstances such as desert environments. Long-distance water transport is an established option and is sound so long as it is carrying water from a region of water surplus, but it is more expensive than transferring virtual water. However, importing virtual water can raise issues of security of supplies, and of environmental and socio-economic impacts in the source countries (World Water Council, 2004). It is viable for Saudi Arabia, which has very little natural freshwater within its borders. However, China and India have opted to develop long-distance transfers from regions of water surplus within their own borders.

Submarine freshwater seems to show promise and demands serious investigation, even though it may

Discussion points

- Explore the possibilities of making greater use of ice and snow, or any of the other sources discussed in this chapter.
- Analyse the environmental impacts of any of the techniques discussed in this chapter.
- Consider solutions for developing countries.
- Are there yet more sources that might be tapped?
- Consider the pros and cons of increasing effective supplies by importing more virtual water compared with long-distance transfers of real water.
- Explore the prospects for finding and making water in space and follow the emerging news.

Further reading

There are few texts that cover the field as a whole, but useful insights into new developments can be found by signing up to the European Water Partnership's Water News. This is a free weekly service available at www.ewp.eu and www.european-waternews.com.

Cleaning up and protecting the aquatic environment

15

Cleaning up drinking water sources and the safe treatment of sewage and industrial and agricultural wastewaters are priorities for maintaining the health of humans and the environment, and preventing further reduction in the available resources. Restoring the physical environment of rivers and lakes is also important for wildlife and human recreation.

We devote this chapter to methods and case studies, but legislation also plays a crucial role in this. Legislation is covered in the wider ethical context in Chapter 20, but here we look specifically at clean-up legislation.

Legislation for clean-up and protection

The USA led the field of environmental protection in 1970 with its National Environmental Policy Act (NEPA) and the setting up of the Environmental Protection Agency. But American legislation specifically relating to water goes back at least to the Water Pollution Control Act of 1948 and the Clean Waters Restoration Act of 1966. The subsequent Clean Water Act (1977) and its 1987 extension from point sources, like industrial discharge pipes, to diffuse or nonpoint sources, like agriculture, form the present-day backbone of American legislation. This requires States to set local water quality standards for their rivers, while the EPA sets national effluent standards for industries and municipalities, and both authorities then adjust the permitted effluents to the local water quality requirements.

Experience in Colorado raised an important issue: should industries and landowners be held responsible for pollution caused by a previous occupier? Should regulators aim to retain the total pollution within general ecological or health limits, or should they merely control the additional pollution above local background levels? Monitoring subsequent to the Colorado authorities setting the water quality standards revealed that many streams already had background levels that exceeded the limits, due to centuries of accumulated mining spoil. The State

reviewed the standards in light of the mining companies' difficulties. EU legislation has taken a somewhat different view, so that occupiers are now held responsible for pollution on their land even if they did not cause it.

There has been a marked proliferation in environmental legislation in the last 20 years, with Environmental Protection Acts, for example, in the UK (1990), the Netherlands (1993) and Canada (1999). The Netherlands is a world leader in polluted land reclamation and groundwater protection. Its Soil Protection Act (1994) is among the most demanding – and sometimes too expensive to comply with totally, as illustrated by the case of the Kralingen clean-up, described below. The European Union has produced a raft of legislation since the 1970s, for example, giving quality targets for drinking water (1980, 1998), and urban wastewater treatment (1991). The European Water Framework Directive (2000) is a landmark in assessment and protection of the aquatic environment.

The European Water Framework Directive

The Water Framework Directive marks the pinnacle of EU legislation to clean up and protect the water environment and to ensure sustainable water resources for the future. It brings many aspects under central, legislative control for the first time and it improves on earlier, piecemeal legislation in the light of experience, improved knowledge and with the overriding aim of integrating all aspects, from surface to subsurface, chemical to ecological, and restoring water bodies damaged by human interference. The WFD also sets deadlines for achieving its aims. Some previous Directives remain at its foundation – Urban Waste Water, Nitrates, the 1996 Directive for Integrated Pollution and Prevention Control, which deals with chemical pollution, and the 1998 Drinking Water Directive – but seven others have been repealed and replaced, for example, on the frequency of sampling and exchange of information.

Minimum standards of pollution are redefined for many substances in surface waters, especially for very toxic substances, but the legislators decided in general not to set minimum chemical standards for groundwater, as the aim should be that groundwater is not polluted at all and setting limits might suggest that a little is permissible. However, limits are in place for nitrates, pesticides and biocides. The approach taken for groundwater is prevention of all direct discharges into groundwater and to reverse any previous pollution. The other arm of protection is to prevent overexploitation and downdraft by setting limits to amount of abstraction.

The Directive requires all member states to bring all water bodies into 'good ecological status' and 'good chemical status' by 2015, and includes measures to ensure that all member states interpret the rules in the same way – which has been a problem with some other EU legislation. However, through the 'reciprocity' principle each country can define 'good status' as it applies to their rivers, especially for ecological status. Defining good chemical status is relatively easy, but ecological status is more difficult because the ecological diversity across the Union means that specific indicator species may not serve everywhere. The stated aim is therefore that all water bodies should show only slight anthropogenic effects.

Management is based on the river basin, crossing administrative and political boundaries, following UN principles and the experience of effective international cooperation in the Rhine, Scheldt and Maas basins. All river basins must have River Basin Management Plans that are updated every six years and must include deadlines for meeting the requirements for drinking and bathing waters and for ecology and habitat preservation, plus an economic analysis of water use. These plans are developed in consultation with the public and other stakeholders for two express reasons: balancing of interests and enforceability – on the basis that participation improves willingness to accept enforcement. Integrated management of surface and subsurface resources is also a core principle. This involves strict controls on discharges from urban, industrial and agricultural sources.

Rather than opting to follow one or other of the previous approaches to pollution used by member states (either concentrating on controlling sources or focusing on the actual river quality of the riverwater – as measured by quality objectives) the WFD combines both approaches. Focusing on wastewater discharges alone can allow cumulative levels of pollution from different sources to mount up to levels that are seriously detrimental to the environment. Equally, the riverwater quality standards alone can underestimate the ecological impact of individual pollutants. The WFD approach proceeds by first requiring all technical means be used to reduce pollution at source, then determine the substances needing priority attention and devising cost effective measures to control them.

The economic analysis is also seen as a crucial element and a means by which pricing of water may be used to restrict usage and conserve resources. The principle had been applied previously in a few countries, but it now applies to all. In the UK, the regulator Ofwat allowed water charges to rise by 12 per cent in 2005 in order to enable water companies to finance the measures required by EU legislation, particularly the WFD between 2005 and 2010. Wessex Water estimates that the WFD could force the UK water industry to invest £500–600 million p.a. United Utilities is undertaking one such scheme to 'rewet' parts of Ribble valley in the Peak District after nearly 40 years of farmers draining the moorland. Wessex Water is encouraging farmers to limit the spraying of pesticides on crops to save the company building a new purification plant. One of the most pressing issues to rectify is sewage overflows from combined drainage systems and other system overloads (see Chapter 5 Pollution and water-related disease). Climate change will make the issue even more urgent. When London received one month's rain in two hours on 3 August 2004, Thames Water sent 600,000 tonnes of untreated sewage into the Thames, killing thousands of fish. This has happened on a number of occasions since 2000 on a smaller scale and has led to a proposal to build a supersewer to a processing plant in East London, but it would require funds of £1.7 billion.

The Framework does allow some derogations from the standards, notably for vital flood protection, public water supply, navigation and power generation. However, these are limited and it has to be proved that all mitigating measures have been taken. For navigation and power, three tests must be applied: that no alternative is possible, such as land transport instead of river transport, that the alternatives are prohibitively costly, or that the alternatives would result in a worse environmental impact.

Clean-up in the Netherlands

The Netherlands suffers from being at the end of the line for two of the great rivers of Europe, the Rhine and the Meuse (Maas). Worse, both rivers drain major industrial regions that released serious pollution into the rivers from the nineteenth century right up to the late twentieth century, when European Union legislation combined with the demise of some old industries began the clean-up. Worse still, in their attempt to ensure that the devastating destruction of the North Sea floods in 1953, which killed 1835 people and flooded 200,000 ha, the Dutch authorities began constructing the dams of the Delta Project in the 1960s just when pollution in the Rhine was reaching its peak. The result is extensive deposits of heavily polluted sediments in the lakes created by the dams (Figure 15.1).

Cleaning up the Delta

As the flood protection dams were built they immediately began to hold back water from the Rhine and Meuse, causing the fine, polluted sediments to be deposited behind them. River engineering work to improve navigation upstream added to it. Pollution in the Meuse continued longer than on the Rhine (see the section headed *The Maaswerken* later in this chapter). The Hollands Diep, where a major distributary of the Rhine meets the Meuse, has collected around 30 per cent of all the sediment from these rivers and contains 40 million m³ of moderate to severely polluted sediment (Figure 15.1).

Clean-up is a long-term project. Most likely, the sediments will never be totally cleaned up, but the riverwater certainly will. About 45 million m³ of sediment have to be dredged annually, largely to aid navigation. Several million cubic metres are strongly contaminated with heavy metals and toxic organic substances. They used to be used to build dikes, new land or dumped at sea, but none of these is now permitted for anything other than the least polluted sediment (class 1), of which about 10 million m³ are disposed of annually in the old ways. All sediments have been surveyed in situ and classified in order to reduce the amount of sediment that needs to be dredged. Around 12 million m³ of moderate to severely polluted sediments containing heavy metals, oil and toxic organic compounds is removed annually, but it has not been dumped in the North Sea since 1985. For many years during the clean-up period, these sediments were comparable to the level of polluted sediments coming down the rivers. The most severely polluted sediments, often containing PCBs, are being stored in a leak-proof container at the Parrot's Beak near Europoort, but storage capacity there could run out in less than a decade. The hope is that an affordable process will be found for treating the material by then.

Meanwhile, progress focuses on two areas: ensuring that the incoming riverwater is clean and not causing unnecessary disturbance in the remaining polluted sediments. The Rhine is now considerably cleaner as a result of international collaboration. The International Commission for the Protection of the Rhine, officially incorporated as an intergovernmental organization in 1963, became involved with controlling pollution in 1983. However, the defining event came in 1986 with

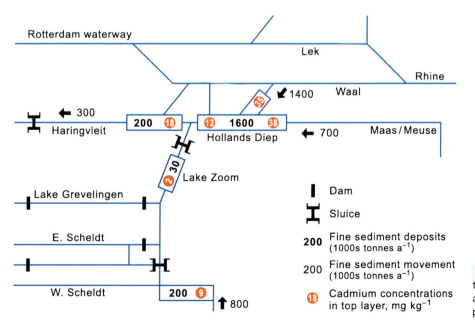

Figure 15.1 Pollution trapped behind the Delta dams. Cadmium is selected as a good indicator of industrial pollutants

301

a devastating fire at the Sandoz chemical factory near Basle. Chemicals drained into the Rhine, some of the most critical coming from the firefighting operation itself, and killed wildlife in the river all the way to the North Sea. The event brought Switzerland into the Commission. As the headwater user, Switzerland had long been reluctant to join. The event added urgency to discussions already in progress, which soon led to the 1987 International Rhine Action Plan. Industrial effluents are now strictly controlled. Calamity basins have been installed, especially at German and Dutch factories, and wildlife and chemical pollutants are regularly monitored. Special riverboats pump wastewater directly from shipping and the largest industrial wastewater plant in the world, the 'tower biology' plant, has operated in the industrial complex around Europoort for the past two decades.

Disturbance of the sediments is minimized by three approaches: controlling the velocity of flow in waterways, sanitizing and protecting floodplain deposits (see the section headed *The Maaswerken* later in this chapter), and restoring storage on the floodplains to contain floodwaters and reduce their ability to erode (see the section headed *Soft solutions to flooding* later in this chapter). One proposal has been to cover the most polluted sediment in the Hollands Diep with a protective layer of clean sediment to prevent its re-entrainment.

Current plans aim to increase six-fold the area of unpolluted deposits in the Haringvliet and to reduce the area of very severely polluted sediment from 16 km² to just 1 km² by 2035.

Over the years, as the Delta Project developed between building of the first dam on the Gravelingen in 1965 and the completion of the Eastern Scheldt barrage in 1986, environmental awareness was growing. From solid dams, the engineers progressed to barriers only lowered in times of flood risk. This was partly to serve shipping, but also to reduce the effect on tidal ranges and the impact on the ecology, particularly on the Eastern Scheldt. During the 1990s, ecological concern grew further and plans were set in motion to alter the operating rules of the dam sluices to reconnect the estuaries and the sea as much as possible. This time, the evolution was down to software, and computer models were used to balance increasing flow exchanges against the prospect of increasing erosion and the remobilization of polluted sediments. The process was set in motion in 1990 by the National Policy Plan for Water Management, which proposed that the operation of the Delta dams could

be improved to reduce the risk of pollution, as well as improving the ecology.

Cleaning up groundwater in the Kralingen

Rehabilitation of a city housing estate in the Kralingen district of Rotterdam was undertaken between 1994 and 2000. It was the first big test of the Dutch Soil Protection Act. Although the clean-up was a response to residents' complaints of headaches and nausea – which turned out to be due to vaporizing hydrocarbons left from the town gas site that the housing was built over – the site was also built over an aquifer that was designated as an emergency source of drinking water for the city when the quality of riverwater in the Rhine fell below safe, treatable standards through accidents or during drought.

Surveys of the 11 ha site showed that it was polluted by PAHs such as naphthalene, mineral oil, cyanide and volatile aromatics, and that they had reached the top of the aquifer and were threatening groundwater abstraction points half a kilometre downstream. New applications for groundwater abstraction were refused, but full rehabilitation according to the rules of the Soil Protection Act was too expensive at around £200 million. Plan B+ was adopted, which was half as costly and aimed to remove 97 per cent of the sources of groundwater pollution and 87 per cent of the pollution already in the groundwater (Figure 15.2).

The clean-up involved removing 270,000 m³ of polluted soil, of which 54,000 m³ was purified, returned and relaid over an impermeable layer of clay (National Research Council, 1994). A layer of gravel laid beneath this acts as a filter draining polluted water from the Holocene clay-peat deposits towards the new pumps. It is estimated it will take 100 years for them to clear the pollution because of the low permeability of that layer. In general, that low permeability seems to have provided some protection for the main water supply aquifer below, the Pleistocene sands. However, drilling did reveal pollution had already reached the top of the aquifer – whether naturally, or as some supposed, at least partly as an unfortunate consequence of the survey drilling. Pumps installed in this layer are expected to have to operate until 2035 (that is, for about 35 years since installation).

The residents suffered six years of chaos, some apartment blocks were raised on stilts while soil was removed and pile drivers shored them up. The headaches may have gone, but the hum of the pumps will last for years.

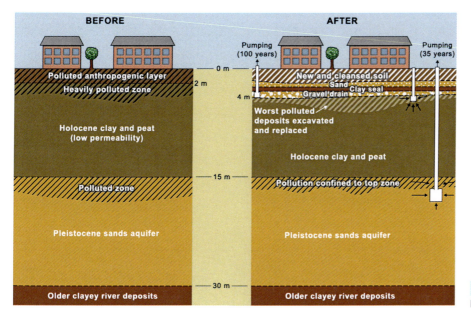

Figure 15.2 Groundwater clean-up project in Rotterdam lasting a century

The Maaswerken

One of the largest clean-up projects has been undertaken on the River Meuse (Maas). The water pollution comes from two sources: sediments and soils in the riverbank and floodplain that contain pollution from now-defunct industries in Belgium (which are remobilized by erosion during floods) and from modern pollution, mainly from poor wastewater treatment in Belgium especially from the city of Liège, which only recently constructed a modern water treatment plant. The strategy is to remove the polluted material from the path of eroding flood-waters, burying it in clay bunds at the top side of the floodplain, and to reduce bank erosion and over-bank floods by deepening and widening the river channel and creating backwaters and wetlands to contain floodwater. The latter also act as an ecological haven. To help fund the project some of the sand and gravel is sold commercially. Dr C.J.J. Schouten and colleagues take up the story in the article below, 'Cleaning up contaminated sediments on the Meuse floodplain'.

Cleaning up contaminated sediments on the Meuse floodplain

C.J.J. Schouten, W. Mosch, H.R.A. van Hout and L.A. Dam
Karnel Environmental Services, The Netherlands

Flood risk in the lower Meuse, or Maas, has increased over the years as a result of changes in land and water management practices along with canalization of the upper Meuse and its tributaries in Belgium and France. The dramatic floods of 1993 and 1995 prompted the Dutch Government to develop the Delta Plan for Large Rivers. Under this plan the Dutch State Water Authority and the Province of Limburg launched the Maaswerken project in April 1997. The aims of the Maaswerken were to reduce flooding, to improve navigation, to create new riparian wetlands in the lower Meuse and to extract gravel and sand (Figure 15.3). Measurements to reduce the flooding included the building of embankments along the part of the River Meuse that was not protected by dikes,

303

in combination with river channel enlargement. The cost of the project increased dramatically in areas where the floodplain sediments along the Meuse were highly polluted with heavy metals and organic contaminants.

Investigations during the last 20 years have revealed the presence of strongly polluted sediments throughout the Dutch floodplain of the Meuse and its tributaries. During floods, significant amounts of bedload and suspended load are transported and deposited. Fresh sediments in the Meuse floodplains originate from three district sources: harbour mud, mixture of fine loamy sediments from upstream areas and local floodplain deposits eroded from nearby soils. Investigations of the riverbanks showed that the strongly polluted sediments date back to the period after the industrial revolution in the Meuse basin (early nineteenth century) and consist of the heavy metals, like zinc, cadmium and lead, and organic contaminants, like PAHs (polyaromatic hydrocarbons) and PCBs (polychlorinated biphenyls). Some of the original major industrial and mining sources of contamination disappeared

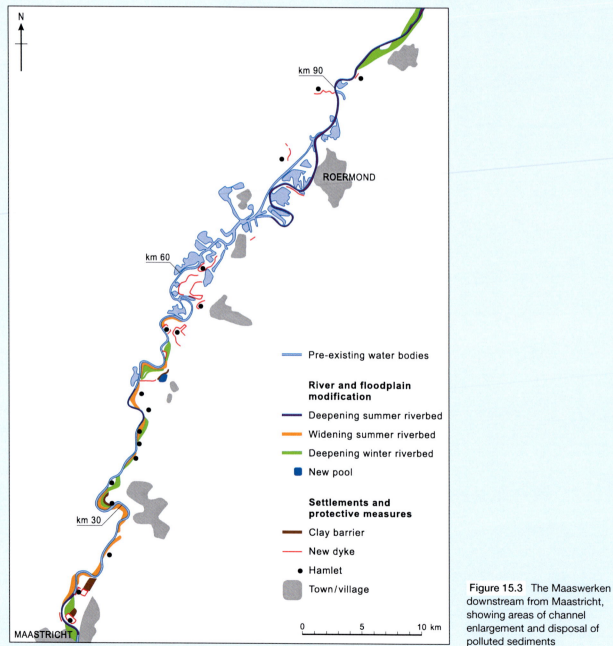

Figure 15.3 The Maaswerken downstream from Maastricht, showing areas of channel enlargement and disposal of polluted sediments

years ago, but their pollution is still moving through the river system as a result of erosion and re-sedimentation. Sediment sampling surveys of freshly deposited sediments were carried out after floods in 1993, 1995, 2002 and 2003. From these surveys it became clear that the recent flood deposits are much less polluted upstream of Liège than downstream. The flood deposits of the Meuse are eventually carried to the harbours and waterways in the lower parts of the Netherlands and combine with Rhine sediments to form strongly polluted harbour mud.

In order to control costs while maintaining the original river management goals and reducing the risk to human health and the environment, a new policy of dynamic soil management was developed within the Maaswerken project. The policy makers developed dynamic soil management, a risk-based remedial approach combined with an accurate soil quality distribution map of the polluted sediments, to target and remove contamination that represents the highest risk and to control costs of cleaning-up operations.

Treating unsafe water – human scale

An increasing number of products and systems are being developed for decentralized and even personal water treatment. Numerous proprietary home-filter systems are available in developed countries. For the affluent traveller, there is the SteriPEN: a pocket-sized, battery-powered pen that emits sterilizing ultraviolet light. But at nearly £85, it is beyond the means of the needy in the developing world.

Cost and ease of obtaining the necessary chemicals or filtering materials are crucial for any deployment in developing countries: expensive polymeric membranes, molecular sieves or ceramic filters are not generally applicable. A Danish company designed the LifeStraw specifically for the developing world. It is a plastic straw containing filters with activated charcoal and a chamber filled with iodine. It removes bacteria and disinfects the water. Manufactured in China, it can treat up to 700 litres and lasts 6–12 months. The straw is marketed at £3.50, but WaterAid points out that even this is beyond the means of most in developing countries. WaterAid says the real problem for many is the long walk to get water and that it costs just £15 per person to provide the full package of water, sanitation and hygiene education.

A number of home construction designs have been developed to remove arsenic. One developed under the aegis of the United Nations University in Japan uses two 35-litre buckets, one above the other (Ali *et al*, 2001). Water is poured into the top bucket and the active ingredients, ferric chloride and potassium permanganate, added. This is stirred and allowed to settle, before draining to the lower bucket, where a cloth strainer and sand filter out the flocs.

Similar systems can be used to filter out pathogens using even more common materials, such as clay, straw, peat, sand and activated charcoal (active carbon). Tony Flynn at Australian National University has invented a very simple system, in which coffee grounds are mixed with clay and animal manure and fired on the ground to create a filter.

Filtration systems can be powered by human energy too. Again, a number of designs have been experimented with. The winner of Google's 2007 'Innovate or Die' competition was 'Aquaduct', a bicycle designed by Californian students that uses a peristaltic pump powered by pedalling to force water through a filter. It can work while stationary or travelling. Aid workers in Africa have expressed interest in the design. It is, however, a rather more costly device than the one designed by the Società Metropolitana Acque Torino Research Centre in Figure 15.4.

Figure 15.4 Domestic water filtration by pedal-power. A system developed and tested at SMAT, Turin

However, most of the problems with unsafe water in Third-World countries come not so much from lack of facilities as from established 'bad practice' and lack of understanding. Open defecation close to water sources is a common example. Unicef's very successful Community-led Total Sanitation (CLTS) programme proves the point that it is often more about changing behaviour than infrastructure. Community level solutions tend to work better and open defecation is declining as a result.

Even so, low-cost infrastructure helps. Pump Aid has also been promoting sustainable, low-cost 'Elephant loos' in Malawi. The toilets comprise a concrete-rimmed hole with two small depressions for the feet, dubbed the elephant's ears. The real work is done by the trunk – a duct that drains urine into a straw pit to make fertilizer. Dry excrement rots faster. A plastic water container is provided for handwashing.

The cleaning bug

After remaining more or less static for over a century, clean-up science is suddenly expanding in a variety of ways, from anaerobic digesters for organic waste to genetically engineered bacteria to deal with heavy metals or oil spills.

The science of using natural bacteria to degrade and sanitize biodegradable waste began in the mid-nineteenth century, when Sir Edward Frankland showed that the sand filter beds installed by James Simpson to purify London drinking water at the Chelsea Water Company in 1828 were more than simply physical filters: bacteria in the beds made them biological filters as well. The gelatinous bacterial film that grows on the sand removes almost all solids, although further chemical precipitation treatment is still needed to remove solutes. It was a while, however, before sand filter beds were also applied to wastewater treatment in special 'sewage farms'. Joseph Bazalgette's great sewer system for London, which did so much to combat cholera outbreaks in the city after its completion in 1870, still disposed of untreated sewage directly into the tidal reaches of the lower Thames: treatment that was no more advanced than 5000 years ago in the Neolithic village of Skara Brae in the Orkneys, where domestic lavatories were piped directly to the sea. Indeed, this system of sea disposal persisted in many coastal towns until Britain was forced to accept EU legislation in the final decades of the twentieth century.

After 50 million litres of oil were spilt on the Alaskan coast by the Exxon Valdez in 1989, the search for a bacterial solution to oil spills intensified. General Electric researched the use of *Pseudomonas* bacteria, which contain an enzyme that breaks down hydrocarbons and can thrive in both salt and freshwater. Under the right conditions with sufficient nitrogen and phosphorus they multiply and metabolize oil quickly. Recent research at the Helmholtz Centre for Infection Research (formerly the German Research Centre for Biotechnology) in Braunschweig, Germany, has focused on *Alcanivorax borkumensis,* which lives almost exclusively on hydrocarbons.

Reed beds and compact wetlands

Decentralized sewage systems are making a small comeback. According to one interesting 'what if' calculation, Singer (1973) concluded that if the population of the USA were perfectly distributed according to the available water resources, 250 million people could be accommodated without requiring any municipal sewage treatment. Because of the high cost of building and running standard tertiary level treatment plants, even municipalities are now exploring cheaper alternatives. Secondary treatment removes organic pollutants, commonly by bacterial biodegradation in trickling filter beds, and tertiary treatment removes nitrates and phosphates. The tertiary stage can be bypassed by directing effluent from secondary treatment into holding ponds, where algal blooms, duckweed and water hyacinth scavenge the nitrates and phosphates. The plants can be harvested and used for animal or even human consumption.

The trouble with individual septic tanks is that they do not always filter out or destroy harmful bacteria. Many septic tanks round Brisbane and the Queensland Gold Coast perform badly because they are set in sandy soil. Septic tanks perform best in well-structured, moderately permeable clay soils. All new septic tanks in Queensland are now fitted with filters to remove bacteria from the effluent and some retro-fitting is taking place on old ones.

Basle, Istanbul and Stockholm are among the municipalities that have successfully installed wetland processing systems. Switzerland saves $64 million a year using forest wetlands as a source of drinking water, which requires little or no treatment. The UN Economic Commission for Europe extols the value of forests and wetlands as both sources for public water supply and for wastewater treatment and points out that many European cities are following this path (UNECE, 2004). It estimates that

China's forests are worth three times more as water filters and flood controls than their value as wood. Trials suggest that wetlands can remove 20–60 per cent of heavy metals, trap 80–90 per cent of sediments and capture 70–90 per cent of nitrates, while alluvial forests can reduce nitrates by 90 per cent and phosphates by 50 per cent.

However, one story that is widely repeated – by *Nature* (1998), *The Economist* (2005), the UNECE (2004) and the World Wildlife Fund (wwf.panda.org) – relating to New York City opting either to create a wetland to treat wastewater at a four-fold saving compared to a conventional facility, or to buy land in the Catskill catchment area and restore it to pristine conditions to effect savings of up to eight-fold compared with a water filtration plant over ten years, turns out to be entirely false (Sagoff, 2002). Sagoff, nevertheless, says the 'parable' is worth repeating because most of us would rather believe that nature knows best.

In reality, compact wetlands, constructed wetlands or reed beds are now well established as an effective method for treating sewage from small communities, like that at the Centre for Alternative Technology (CAT) in Wales (Figure 15.5) or on the Prince of Wales's estate at Highgrove. They use a combination of physical filters and plants that have high transpiration rates to reduce the volume of effluent and host microorganisms that decompose organic material. They have also proven effective in treating effluent from abandoned mines in South Wales by capturing metal pollutants. Most solids are filtered out in the first stage, then composted and used as fertilizer, while the reeds and other plants are selected to provide high uptake of both nutrients and pathogens. The CAT system combines two established approaches, vertical filters and horizontal filters to get the best from both. The processes are described fully in Figure 15.6.

Anaerobic digesters

These use anaerobic bacteria rather than the aerobic bacteria used in standard filter beds, and break down food and farm waste into fertilizer and gas that can be burnt for heat or to generate electricity. They are widely used in Germany. In 2009, Environment Minister Jane Kennedy announced UK government hopes for 1000 digesters in the UK by 2020. The 1.73 million tonnes of sewage sludge treated by the British water industry each year could be digested and produce enough electricity to power two million homes. A significant portion of the 90 million tonnes of agricultural waste, including slurry and manure, along with the 12 million tonnes of food waste that goes to landfill, could be similarly treated, reducing the potential hazards for groundwater and streams.

Biobricks

The concept of trapping pollutants inside a 'brick' has proved effective but costly. (They are not to be confused

Figure 15.5 Two reed-bed sewage purification systems trialed at the Centre for Alternative Technology, Wales. Above: The bin in the foreground filters large material for composting. Water then trickles through the lush vegetation where microbes break down contaminants and the greenhouse increases evaporation to reduce wastewater volumes. Right: view of the system illustrated in Figure 15.6, looking up the cascade from Point 5: slow horizontal filter bed in foreground, rapid vertical filter bed behind

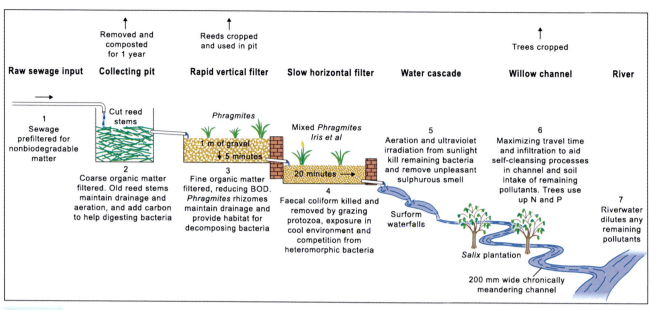

Figure 15.6 Reed beds at the UK Centre for Alternative Technology, combining vertical and horizontal filters

with 'BioBricks', which are composed of sawdust or straw used in domestic fires.) The polluted sludge should contain an amount of organic material mixed in with sediments – sand, silt and clay. It is heated to 1000°C until the silica fuses into a glassy material like an igneous rock. Proponents claim this will fix the pollutant for at least 1000 years, and it can be used to make decorative building materials, such as a false marble floor covering. Washington Suburban Sanitary Commission built a complete building with biobricks made by combining sewage sludge with clay and slate. Lin and Weng (2001) describe tests in Taiwan on urban sewage. Tests in Australia produced bricks that are 15 per cent lighter and stronger than standard materials. Tests in the Netherlands using heavily polluted sludge dredged from waterways in the Rhine Delta proved successful but uneconomic because of the high energy requirements. Concern for carbon footprints may well be the deathknell of the process, but the Netherlands might have to fall back on it as the only way to safely treat the millions of tonnes of sludge dredged from its waterways and stored in the watertight pond at the Parrot's Beak near Europoort, called the 'slufter', for which there is no currently feasible treatment. The slufter is rapidly approaching full capacity.

Developing approaches

Nasa has refined an old invention that could be placed in a river and act as a sponge to absorb heavy metals

from the water. It is a silica gel called aerogel, or more popularly 'frozen smoke', and is literally a sponge with millions of tiny pores that is light enough to float. The prototype was invented in the 1930s but was too brittle. Nasa fitted it into the Stardust space probe in 1999 and established the Aspen Aerogel company in 2002 to develop and market it.

At the Queensland University of Technology, research on substances that break down pollutants under ultraviolet light or sunlight, called photocatalysts, has focused on titanium dioxide. Unlike normal photocatalysts, titanium dioxide can be easily filtered out of the water. It is claimed this could be commercially viable by 2010 for breaking down a wide range of pollutants, such as pesticides, oil, organic wastes, bacteria, and even viruses and moulds.

Oculus of California has developed a water that disinfects. The product, called Microcyn, is water that has been 'superoxidized' with oxychlorine ions. The ions pierce the walls of the microbes. Human cells are unaffected because the cell walls are too strong. It is being trialled in the EU and UK.

Geotechnical engineering is expanding into biology in developing the use of bacteria in 'bioremediation'. Research at the Cambridge Department of Engineering is experimenting with bacteria to decompose hazardous chemicals in soils or bind them to calcium carbonate so they cannot be washed into groundwater or streams.

At the time of writing, bioremediation has focused on techniques that depend on the pollutant making contact with the cleaning agent. A new field of possibilities is opening up with the use of heterotrophic plasmodial slime mould, which actively seeks out its food like an animal: the cleaning agent initiates the contact. These are amoeba-like cells with a dendritic network of tubular structures called pseudopodia that help it move to the food and thus digest the pollutant.

Synbio

Synthetic biological products or 'synbio' are a new approach to cleaning up. If there is no readily available bug to metabolize the pollution, then why not engineer one?

The 'inventor' J. Craig Venter is building new species at his Institute in Rockville, Maryland, by genetic engineering. He is aiming to produce new bacteria to eat oil or heavy metals, create fuel by breaking down cellulose to ethanol or from slurry pits, or algae that create fuel by soaking up more carbon dioxide. He is already marketing a cheap anti-malarial drug produced by inserting a gene into a strain of yeast (Lartigue et al, 2009).

Will these products be environmentally safe? Will they become useful tools for terrorists?

Green chemistry

This is a new science devoted to developing environmentally sound chemicals for industry and better ways of cleaning up. It has been promoted in recent years by the Canadian Green Chemistry Network and the American Chemical Society. Green chemistry is a totally new approach, using catalytic reactions in water or air, requiring less energy than established chemistry and producing no hazardous waste. For example, it uses enzymes to do the chemical processing at room temperature; these are large molecules that can easily be filtered out once their work is done. Enzymes may be engineered to do other jobs. In response to EU legislation requiring imported products to meet EU environmental standards, Canadian green chemists are also hoping to find an alternative to the large quantities of chlorine used for bleaching in the pulp and paper industry, that are currently a major environmental problem in Canadian rivers.

So-called 'downstream' applications include using natural proteins and enzymes to treat waste, for example to degrade plasticizers (see the section headed *Endo-*

crine disrupters – a sea of oestrogens in Chapter 5) or to degrade PCBs into non-toxic substances. Research at McGill University is aiming to develop a 'green' type of plasticizer as an alternative.

Robots in charge

Robotics offer tremendous potential for undertaking tedious and repetitive work without tiring or lowering standards of accuracy (King *et al*, 2009). They can help in both the development and the application of new approaches. They can search for genes in bacteria to speed bioengineering and they can carry out routine analysis and surveillance. There is no doubt that robots will become essential tools – and workmates – in the near future. In the process, they will upgrade present-day standards of surveillance and clean-up, and maybe even reduce the costs.

At the meeting of the American Association for the Advancement of Science in Boston, 2008, 'futurologist' Ray Kurzweil predicted that nano-robots would be developed to do jobs in the environment. Kurzweil noted that science and engineering advances have recently doubled every 20 years and he predicts that over the next 50 years they will advance 32 times more than in the last. Among his other predictions are new systems of desalination that consume less energy and smaller-scale technologies for water purification.

Restoring and protecting the aquatic environment

Water sustainability is about more than supplying present and future human needs. The health of wildlife and the maintenance of biodiversity require us to consider the ecological demand for water in any development of water resources for human society. Indeed, we now have the technology to go further and to restore, rehabilitate or even create aquatic ecosystems from new for their own sake, without recourse to their direct value to human society. In practice, there can be many indirect advantages for humans in considering ecological requirements, ranging from the technical value of preserving the self-cleansing ability of biochemical exchange processes within water bodies to the purely aesthetic well-being created by living among healthy water courses. As with humans, a healthy wildlife needs both sufficient water and water of sound quality.

In many ways, the methodology for determining the optimum and minimum quantity of water needed to maintain healthy aquatic ecosystems is still very much in the developmental stage. This is partly because such considerations are a very recent development, but it is also because of the inherent complexity of diverse ecosystems. This complexity is created by the wide environmental tolerances of many species, their adaptability and the complex feedback systems by which different environmental factors, like levels of light, water, oxygen or nutrients, or the frequency of events, like flooding, desiccation or flow velocities, affect metabolism and community structures. One of the few attempts to quantify ecological requirements is described below by Professor Xia, who is a specialist in this emerging field.

Determining optimum discharges for ecological water use

Xia Jun

Despite some recent progress, the issue of determining the correct balance between ecological and socio-economic water use has been largely ignored in water resources allocation. Sound allocation in the future depends on establishing viable scientific criteria for determining local ecological water demand. This is of special concern in China, where a rapidly expanding economy poses a growing threat to nature.

We used the Hai River, which flows into the sea southwest of Beijing, some 200 km north of the Yellow River, as a case study to develop a methodology. The river basin suffers from frequent droughts and is generally short of water. At the same time, it is an important industrial area, which accounted for 15 per cent of national GDP in 2000, with a population of 126 million, 10 per cent of the national population, and a population density of 397 per km². The average water resource per head is just 305 m³, compared with 2388 m³ in the Yangtze basin, 2342 m³ for China nationally and even 749 m³ for the water-stressed Yellow River basin.

The ecological problems in the basin include:

1 A major reduction in water supplies from the mountains since the 1950s.

2 Reduced discharge into the sea with serious siltation in the river mouth: discharge in the 1990s averaged just 6.85 billion m³, compared with 24.1 billion m³ in the 1950s.

3 Reduction in wetland areas, drying out of lakes and falling water tables due to over-exploitation. In the 1950s wetlands covered 9000 km², whereas now wetlands and reservoirs combined only cover 3852 km².

Assessing the river's ecological water requirement

Any assessment is partly determined by the natural requirements of river ecology, including both the biotic communities and the river morphology, and partly by society's expectations or 'ecological objectives'. As society becomes more affluent and more knowledgeable, these ecological objectives are becoming more demanding. The natural ecological water requirements also vary in time and space according to life cycles, seasons and fluctuations in the weather. Figure 15.7 illustrates the hypothetical range

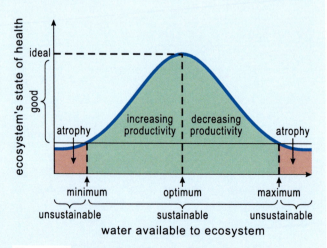

Figure 15.7 Hypothetical range of water availability for a healthy ecology

for a healthy ecology, with an optimum state lying somewhere between upper and lower threshold limits for river discharge. The lower threshold is the minimum discharge needed to maintain the current constitution and functioning of the ecosystem. When the volume of water exceeds the upper limit, too much water may restrict healthy development. In present-day China, however, water resources are very tight and there is no problem with overabundance of water for ecosystems. We have therefore not considered the maximum limit here.

There are three key ecosystems in the Hai river basin that need consideration: the river itself, the wetlands and the estuary. We calculated the requirements of each using the 'Montana method' developed in the USA by Tennant (1976), which is widely used.

1 **The river system.** In the river itself, the assessment is complicated by the rise in demand for public water supply from the rivers for municipal and industrial purposes since the 1970s and for agriculture since the 1960s, all of which are now accelerating and causing widespread deterioration of ecosystems (Table 15.1). The late 1960s marked the turning point from a satisfactory state to a deteriorated state.

Ecologically, the ideal would be to return to the water availability of the early 1970s, defining this as the 'optimum' and that of the late 1970s as the minimum state. If, however, ecosystem requirements were to be fully met, socio-economic development would be restricted. Striking the correct balance between these conflicting demands has become an urgent area for research. The water available for river ecology is equal to total discharge minus the portion of domestic, industrial and agricultural water abstracted from the river (i.e. excluding the portion abstracted from groundwater). We need to find an acceptable balance by first defining reasonable ecological objectives. We define sustainable ecological water use as lying between the minimal and optimal water levels.

According to the Montana method, the minimum ecological water requirement for a river is 10 per cent of mean annual discharge and the optimum is 40 per cent. On the Hai, mean annual discharge is 22 billion m^3, so the minimum level is 2.2 billion and the optimum 8.8 billion m^3. This is called the 'threshold interval'.

2 **The wetlands and lakes.** The total ecological requirement of the wetlands is equal to the total water stored in the wetlands plus the amount lost by evaporation and drainage out of the wetlands. Inputs must balance these outputs in order to maintain water levels at the depth required by plants and animals and covering sufficient area to maintain populations. For the Hai, the minimum goal would be to protect an important wetland at Baiyangdian and to improve three others at Tuanpowa, Dalangdian and Qianqinwa. The water needed for this is calculated at 966 million m^3. The optimal goal would add improvements at a further six wetlands (Ningjinpo, Dongdian, Qingdianwa, Xiqilihai, Dahuangpuwa and Enxianwa), increasing the wetland area to 1068 km^2, compared with only 471 km^2 in the late 1970s. This would require 2.956 billion m^3 of water. The threshold interval is therefore 0.966 to 2.956 billion m^3.

3 **The estuary.** The estuary had no real problems before the early 1970s, when the minimum flow into the sea during a drought was 2.5 billion m^3. The upper limit was 11.6 billion m^3 (Figure 15.8). The requirements of the river, wetlands and estuary ecology overlap. The

Table 15.1 Features of the key ecosystems in the Hai river basin

Ecosystem	Feature	Current state	State circa 1970
River	Channel dry all year	4000 km	<1000 km
Wetlands and lakes	Total area	122 km²	1500 km²
Estuary	Discharge in drought year	1200 million m³	2500 million m³

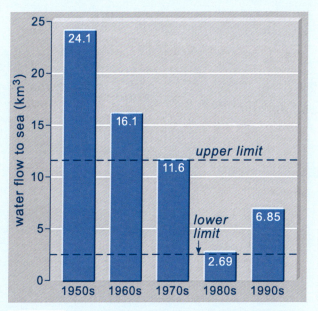

Figure 15.8 Changing discharge of the River Hai, North China

basic ecological requirement of the rivers should be within the threshold value interval for water flow into the sea. The maximum value of the two (11.6) is therefore chosen as the water requirement for river and estuary. To this must be added the requirement of the wetlands and lakes, giving a total requirement of between 3.47 and 14.56 billion m³ (Figure 15.9).

During the drought years of 1999 and 2002, exploitation rates were over 1000 per cent and in-stream discharge was mainly discharged wastewater. Ecosystems were badly damaged. In the less severe droughts of 2000 and 2001, small amounts were left for ecological use (16 per cent and 6 per cent respectively). 2000 was close to the minimum threshold for the ecology. Since 2002, minimum ecological requirements have hardly been met for four consecutive years. While ecosystems often have the ability to self-regulate and regenerate after an extreme event, prolonged year-on-year events can lead to permanent degeneration. Therefore alternative plans are needed. In a wet year, minimum ecological demand may be only 12 per cent of total water resources, but in a dry year this can rise to 35 per cent. We therefore propose that this figure of 35 per cent should be adopted as the minimum allocation for the ecosystem in severe dry years. For optimal conditions the proportions are 50 per cent for wet years and 148 per cent for severe dry year. Since the latter is impossible to supply, there has to be a trade-off between ecology and human requirements. We decided that 50 per cent of mean annual runoff should be the highest volume for the most suitable ecological water supply, which places the range between a minimum of 35 per cent and a maximum of 66 per cent.

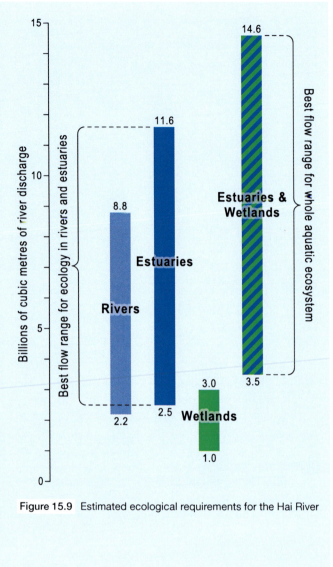

Figure 15.9 Estimated ecological requirements for the Hai River

The Aral Sea – lost and regained?

In April 2009, the five Central Asian states participating in the International Fund for Saving the Aral Sea held a water summit in the Kazakh capital Almaty to coordinate international efforts to restore and conserve water resources in the region. The Aral Sea basin has been the victim of the worst hydrological disaster ever created by water resource management. Most of the region is semi-arid, yet populations have been growing and the countries – Kazakhstan, Kyrgyzstan, Tajikistan, Turkmenistan and Uzbekistan – now occupy five of the top seven places in the league table of per capita water consumption. Water is not only in short supply, much of it is heavily polluted and has had awful effects on public health, livelihoods and wildlife.

In the middle of last century the Aral Sea was the fourth largest inland sea in the world. As the main feeder rivers, the Amu and Syr, dried up between 1960 and 2000, the Sea lost 70 per cent of its volume and the water surface contracted from 68,000 km² to 23,000 km². Sea level fell by up to 20 m and the coastline retreated 130 km, exposing approximately four million hectares of environmentally hazardous land (Dukhovny *et al*, 2009). The exposed seabed contains not only salt, but also large quantities of pesticides and herbicides washed in from the irrigated cotton fields.

Excessive water use in irrigation was the main reason why the feeder rivers dried up, but Normatov and Petrov (2009) point out that huge hydropower reservoirs also contributed. By the 1990s, the total irrigated area in the Central Asian Republics reached 8.8 million ha and hydropower accounted for almost a third of the total electricity generating capacity of 37.8 million kW. In 1961, the Sea was fed by up to 100 km³ of riverwater, roughly a tenth of its total volume, every year. By 1975, only 7–11 km³ were reaching the Sea: nothing from the Syr between 1974 and 1986. The Amu failed to reach the Sea for half the 1980s. At the same time, the Sea was losing 60 km³ a year in evaporation. The Sea split in two in the late 1980s and in three in the early 2000s (Figure 15.10).

Concerted efforts are now being made to restore the rivers feeding the Sea and to conserve at least one part of the Sea, but there are huge technical, political and economic hurdles. The problems began with Soviet plans to irrigate large areas of the basin principally to grow cotton, but also to increase employment and food production. The two major crops are both prolific consumers of water. Cotton uses 10–15,000 m³/ha and rice 25–55,000 m³/ha: over 10,000 m³ of water are needed to produce one tonne of rice. This involved massive water diversions from the main feeder rivers. The original Rakitin and Litvak Plan aimed to increase the irrigated area to 10 million ha by 1990, effectively doubling the total irrigated area within the USSR, and eventually increasing it to 23.5 million ha. Another early plan to provide supplementary water by diverting the north-flowing headwaters of the Ob-Irtysh basin was discarded in the 1980s. That was part of the grand reversal of the arctic rivers project that was halted by President Gorbachev in 1986 (see *US and Soviet diversion plans* in Chapter 7). Even so, the benefits of this extra supply would have been short-lived. Now climate change is exacerbating the problems as the glaciers in the Hindu Kush, Tien Shan and Pamirs begin to melt and will eventually supply less water to the Amu and Syr.

The breakup of the Soviet Union in 1991 also left the new independent states with major economic and political problems: the initial collapse of the rouble, subsequent lack of funds, lack of a coordinating body, interstate rivalry, states putting their newfound national interests ahead of neighbours' and no clear and enforceable international law to guide the way (see Chapter 20). It also left the states without the means to process the cotton crop: following Marxist spatial and political theory, the USSR separated crop production and cloth production, so that the crucial, value-added processing facilities were retained within Russia proper.

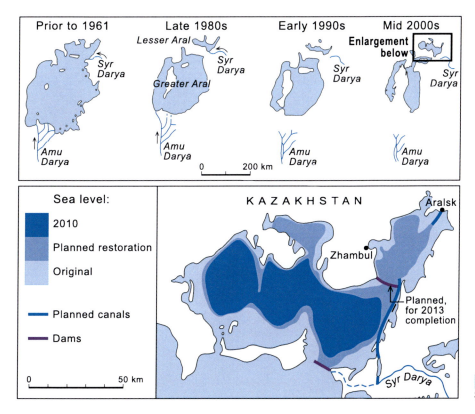

Figure 15.10 The breakup of the Aral Sea

Desertification, salinization and pollution

By 2003, the world's newest desert, the Aralkum, had been created, covering 40,000 km² of the former seabed and heavily laden with dried salt and with fertilizers, pesticides and herbicides carried in the return water from the irrigated fields. Large amounts of these contaminants also entered the groundwater. In addition to the usual story of diminishing returns from irrigation due to waterlogging and salinization, dust storms have caused 'secondary salinization'. Intensified in part by the loss of trees and the Sea's ameliorating influence on temperatures, dust storms have become nearly three times more frequent over the past 25 years, carrying 40–150,000 tonnes of salt onto the land for distances of up to 70 km every year. Salinization and waterlogging have caused falling productivity. Productivity per hectare has fallen by at least 50 per cent since the Second World War (Kamalov, 2009). Despite expansion of the irrigated area, the gross yield of cotton is not increasing in Kazakhstan and Kirgizia. In the 1980s, Karakalpakstan around the old delta of the Amu produced over a million tonnes of cotton, rice, wheat and vegetables; now it produces less than 300,000 tonnes.

In order to sustain the cotton monoculture, applications of mineral fertilizers increased six-fold and chemical sprays increased up to 50 kg/ha, 25 times the norm in the CIS. Banned chemicals like hexachloran and the defoliant Agent Orange are present in near-fatal concentrations in the lower reaches of the Amu and Syr. Root-crops contain 12 times the permitted level of DDT.

The result has been a gross deterioration in public health and decimation of fisheries and wildlife.

Health and livelihood

Population in the Amu and Syr basins trebled to 30 million, putting further pressure on water resources, and has remained high despite some emigration from the worst affected towns. Eighty per cent of the population does not receive a glass of clean water a day. People drink the polluted water, eat polluted food and inhale polluted air. Unsurprisingly, life expectancy has fallen from 64 to 51 years in the regions bordering the old Sea and 75 per cent of the population suffers from illness. Infant mortality is among the highest in the world. Dysentry is rife. The 1980s saw a tremendous increase in disease: cardiovascular disease increased 1.6 times, cancer of the oesophagus increased ten-fold, tuberculo-

sis doubled, the incidence of deformed babies increased dramatically, and general mortality rose 15-fold. In the Kzyl-Orda region east of the Sea, gastric typhoid increased 20-fold in just five years. In Karakalpakstan the risk of contracting paratyphoid is 23 times higher than in the rest of the CIS. Karakalpakstan now has the highest incidence of oesophagal cancer in the world, and has exceptional rates of other cancers, TB, liver and kidney disease, allergies and immunological and reproductive disorders.

The impact on wildlife and the falling production rates have had a drastic effect on employment and livelihoods, further raising levels of poverty and ill health.

Impact on fisheries and wildlife

Around 80 per cent of wildlife species have died out. Fisheries have suffered as the Sea's salinity has increased from 10 per cent to 31 per cent and rivers and seawater have accumulated a 'cocktail' of poisons. Bream, sturgeon, chub and pike-perch have all disappeared. The Aralkum is now a ships' graveyard of disused fishing boats rotting in the sand.

Many people made a living from the reed beds that used to cover a million hectares in the river deltas, but these have dried out and reed production has reduced from 3–4 tonnes/ha to 7–130 kg/ha. Muskrat fur used to be an important export: a quarter of the total production in the USSR. Now, barely 1000 rats survive since the demise of the woodlands and wetlands.

'Priority should be given to water conservation at all levels of the water hierarchy and end-users that will promote increases in the volumes of environmental flows. This will allow more water for use in the lowlands and permit complex protection measures to be organized by a combination of developing wetlands, concentration on water delivery to the Western Bowl and the development of afforestation of the ecologically unstable zones.'

V.A. Dukhovny et al, 2009.
The Interstate Commission for Water Coordination in Central Asia, Tashkent, Uzbekistan, on measures to conserve water in the Aral Sea basin

Solutions ahead

Any rehabilitation is bound to be expensive and current plans for recreating the Sea are limited to a very small part of the original area. One official estimate for reclamation in the 1990s suggested that total costs could reach 60 billion roubles or more, eight times the cost of the Chernobyl disaster. An official report in 1994 recognized that it is impossible to restore the Sea to its original volume and biological status. The Aral Sea Commission proposed creating two small, protected 'sub-seas' around the mouths of the Amu and Syr to restore fisheries and the commercial exploitation of musquash for fur, to allow horticulture to be re-established and to stimulate tourism. Another plan proposed that salt tolerant tree species be planted as a 'green barrier' to stabilize the seabed. Over the past two decades, about 225,000 ha of exposed seabed have been artificially reforested and a comparable area has regenerated naturally. Solutions are discussed in more detail in the article below.

The problems facing the Aral Sea basin demonstrate the need for an integrated, multi-pronged approach with international collaboration to tackle economic and social problems along with environmental rehabilitation. The big questions now are whether the money is going to be available, whether proper feasibility and EIA studies will be carried out, and most of all whether the new independent states can work together. Local NGOs and people-led collaboration campaigns are calling for more government involvement, hence the importance of the 2009 Almaty Water Summit. A key element of political collaboration is the need to establish binding and egalitarian agreements between upstream and downstream countries on water rights and responsibilities (see *Transboundary waters* in Chapter 9). There is also hopeful evidence of wider international aid: the World Bank has already shown interest and the WMO's first satellite surveillance system has been deployed in the region (see the section headed 'Satellites' in Chapter 18).

Historical footnote

The current disaster is man-made, but historically nature has had much greater impacts on riverflows and the size of the Sea, only without the poison. Around 5000 BC, swelled by meltwaters after the last glaciation, the Sea covered an area some three times that in 1950. Yet in 300 AD it was broken into two lakes, similar in total area to that in 1990, which were linked together and had an outlet to the Caspian Sea and were therefore freshwater. As recently as 1500 AD, it consisted of three saltwater lakes a little larger than at present.

Some possible solutions for the Aral Sea basin

Numerous proposals have been made. Many are highly original, perhaps to the point of being questionable.

1 **Withdraw land from irrigation:** could save 15–20 km³ of water a year. Taking a million hectares out of rice production and importing rice would save 3 km³ of water. Simply reducing the high rejection rate (over one-third) for cotton cloth could save 10–15 km³.

2 **Improving primitive irrigation systems:** could save 40–70 km³ pa. Most are only 55–67 per cent efficient.

3 **Improve management of drainage waters:** 46 km³ of return waters currently drain into rivers, lakes and groundwater, adding to salinization. This could be diverted directly to the Aral Sea or barren closed depressions, or else desalinated and reused.

4 **Re-establish soil fertility:** this may be done partly by crop rotation and replacing crops with a high water demand, especially cotton, by vegetables and grapes. But desalinizing the soils is far more difficult. The only possible method is flushing with large quantities of freshwater, but this risks raising water tables and bringing with it yet more salts dissolved from the evaporite rocks beneath, which were laid down millions of years ago under an ancient ocean.

5 **Improving the quality of drinking water:** while most projects concentrate on the environment, recent efforts are focusing on improving human health by dealing with the polluted drinking water. In 2007, 3000 households were issued with water treatment filter systems and nine mini-purification plants were produced. These use newly developed ion-exchange sorbents based on polyacrylonitrile cloth with immobilized silver nanoparticles (Khaydarov *et al*, 2009).

6 **Conserve and preserve the Aral Sea:** 35 km³ of extra water are urgently needed simply to stop further reduction in sea level. Schemes must determine the ecological requirements, including containment of salt blowout, and the technical limitations; several dozen schemes have been proposed. A detailed ground survey in 2005–2007 mapped and classified the unstable areas, and identified areas for priority reclamation. This delimited more than half a million hectares to be protected, including nearly 120,000 ha at high and very high risk of erosion (Dukhovny *et al*, 2009).

These measures include augmenting local water resources, as in the following examples:

• A newly constructed system of waterways from the Amu to Adjibai Bay now feeds the Western bowl.

• **Inter-basin transfers:** these include proposals to revive parts of the defunct Soviet plan to reverse the arctic rivers.

  – Transfer from the Caspian Sea to the Aral – but Caspian water is saltier.

  – Transfer from the Volga – but this would reduce the input to the Caspian, and necessitate dams that would affect migratory fish.

  – Transfer from Siberian rivers – this seems more feasible, but fears have often been expressed that this would be seen as a panacea and slow down vitally necessary reforms in agriculture and water management. Even so, one estimate suggests that perhaps no more than 3–12 km³ of this would finally reach the Aral. However, the Mayor of Moscow, Yuri M. Luzhkov, revived the proposal in the early 2000s and the presidents of the Central Asian Republics held informal talks with Russia and China over the proposal in 2006.

• **Rainmaking:** before the breakup of the USSR, the State Committee for Hydrometeorology calculated that 25 km³ could be created in Central Asia, but at a cost exceeding 100 million roubles.

• **Intensifying glacier melting:** several cubic kilometres might be gained by artificially inducing icemelt, but at the expense of runoff later (Krenke and Kravchenko, 1996). Glaciers have been melting naturally: over the twentieth century the headwater glaciers have lost 12–15 km³ of ice and just in the last 40 years 14 small glaciers have disappeared entirely (Normatov, 2009).

• **River regulation:** proposed by Lvovich in 1980, this is seen by many as the most promising. It might allow up to 30 km³ to be retained and released in drier years. It would help irrigated agriculture, but not have much effect on the Aral Sea itself, since only the temporal distribution rather than the total discharge is affected. Evaporation losses would also increase.

• **Exploiting groundwater:** groundwater resources are several times greater than surface waters. Up to 10 km³ of extra groundwater could be used without causing a reduction in riverflow, and even brackish water might be used. Ultimately, however, the danger of adding to salinization is ever-present and this is likely to be a short-term and ecologically unsound solution, since much of the basin is underlain by ancient evaporite rocks.

• **Political solutions:** the mountain countries are trying to make downstream states pay for water delivered from upstream. Kamalov (2009) says they should pay for security of supply not the water itself and that this should oblige upstream states to provide measures of annual resources by monitoring rivers and glaciers, and to clean up the riverbeds, preserve the forests and prevent floods and droughts. They should also pay compensation when they fail. Nazruzov (2009) points to disagreements between states over hydropower developments upstream, for example, in Kyrgystan. Compensation arrangements have now been established.

The Mesopotamian marshlands – lost and regained?

Whereas the Aral Sea disaster was more of an accidental by-product of agricultural development, albeit founded on faulty Soviet philosophy, the second great disaster, the draining of the Tigris-Euphrates Marshes, was the result of a malicious act of genocide.

The restoration of the marshlands of southern Iraq promises to be a good example of international collaboration. In his article 'Restoring Iraq's Mesopotamian marshlands' Richard Zuccolo, who was part of the restoration research team, outlines progress. By 2008, half the marshes had been reflooded, in part unofficially by the returning Arabs themselves, who broke down the earth dams and dykes. Fish are returning. The ecological disaster was caused by Saddam Hussein's drainage programme (see *War against the Marsh Arabs* in Chapter 9). But restoration will be hampered by changes in the river regimes caused by Turkey's GAP project and Syria's dams. The more than 30 dams on the Tigris and Euphra-tes have halved river discharges. There is no longer a spring surge from snowmelt in the Turkish mountains, which used to bring freshwater into the marshes when the fish were spawning and the reeds, so valued as a source of income and building materials by the inhabitants, were beginning to regrow. The water is dirty, saline and undrinkable. And land disputes are proliferating as people return.

Restoring Iraq's Mesopotamian marshlands

Richard Zuccolo

Until the mid-1970s, the marshlands around the junction of the Tigris and Euphrates covered an area of about 20,000 km² and were characterized by a dense network of interconnected channels, mudflats and lakes (see Figure 9.1). The 25 years of Saddam Hussein's policy on agricultural development and the construction of drainage and water diversion systems, progressively desiccated these wetlands and transformed them into a desert-like area.

By 2002 the marshlands had been reduced to less than ten per cent of their original area and by 2003 the majority of the marshlands were wastelands. The impact on the local population and wildlife were devastating and it has been estimated that more than 500,000 marshland dwellers, the Marsh Arabs or Ma'dan marsh dwellers, were displaced during the 1990s. In addition, there has been a marked degradation of water quality in the mainstreams of the Tigris and Euphrates due to saline drainage from irrigation schemes and dams retaining the soil sediment and contaminants.

With the liberation of Iraq, the decimated communities of marsh dwellers and Iraq's Ministry of Water Resources have begun to return water to portions of the marshlands. In 2004, Iraq's Ministry of Water Resources declared that the restoration of the marshlands was its highest priority. In the same year, about 30 per cent of the marshlands were reflooded and some areas have experienced

Figure 15.11 Satellite image of the Marshlands in 2004. This shows the reflooding and the re-establishment of vegetation resulting from the direct actions of the marsh dwellers and Iraq's Ministry of Water Resources. (Landsat EROS imagery courtesy of NASA and USGS)

rapid re-establishment of native flora and fauna. However, the effects of reflooding on marshland restoration remain unknown or not well understood (Lawler, 2005; Richardson *et al*, 2005).

There is a critical need for long-term coordinated scientific and technical initiatives to guide the process of water resources governance and marshland restoration and to avoid irreversible deterioration of the ecosystem. Water resources planning regulating the development of hydraulic connections between wetlands, rivers and the ROPME Sea Area is critical for the long-term rehabilitation and health of the Mesopotamian marshlands. Planning for the marshlands must address critical issues of water availability and allocation water quality, land-use schemes and environmental conditions.

During the last decade, an international consortium of agencies from the USA, Canada, the UK, Italy, the Netherlands, Japan, the World Bank and UNEP, coordinated by Iraq's Ministry of Water Resources, Environment, Agriculture and Public Works have addressed these issues. This produced a water resources Master Plan, linking science to the policy and planning process, identifying solutions and prioritizing water management activities, whereby at least 50 per cent of marshlands were to be restored by 2010. Research indicates that enough water is available to restore the marshlands. Most importantly, the water resources Master Plan aims to ensure that the voice of marsh dweller communities is heard and incorporated into the decisions about the restoration of the marshland ecosystem and their social and economic future, through collaboration with regional ministerial offices.

However, the restoration of the marshlands must be part of an overall national restoration strategy and plan with a common vision clear to all ministries involved in the efforts. In fact, the water resources optimization policy will have to be coordinated in order to meet water demand from different sectors: domestic consumption, industrial and agricultural uses, and environmental rehabilitation.

Improving farming practices and landscape management

More efficient irrigation may be the most obvious way of reducing the huge global demand from agriculture, but farming has a much wider role in altering and therefore potentially protecting many water-related aspects of the landscape. Better management of the soil can help reduce flooding and lessen the need for irrigation. It can reduce water pollution from fertilizers, pesticides and sediment. It could also be a valuable means of combating climate change.

The answer lies in the compost

Good soil management is an essential part of water economy. Maintaining a good structure of soil aggregates improves infiltration and drainage, and reduces waterlogging and wastage of water through surface runoff, with its potential for soil erosion. Maintaining an optimal content of organic matter not only contributes to aggregate stability, but also improves water-holding capacity. Burying compost deep in the soil can also aid long-term carbon sequestration and so combat global warming.

At his Soil Association organic farm, Blaencamel, in Wales, Peter Segger has reduced water use in his tomato houses by 40 per cent compared with conventional methods by using his own compost and zero watering July to November. Research at Sekem Experimental Farm in Egypt found that organic farming employing compost reduced water use by 50 per cent compared with neighbouring farms operating similar but non-composting systems. Segger cites equally dramatic results from composting trials at the Volcani Institute in Israel and the Fibl Institute in Switzerland, and points out that reductions in water use of 20–60 per cent have been achieved using organic wastes, especially compost, in a range of European experiments.

No till

During the 1980s, the University of Leuven in Belgium carried out experiments which replaced traditional ploughing with no tillage using mulching and seed drilling. The soil at the university's Huldenberg station is developed on silty loess deposits and is very prone to erosion, especially when the fields are ploughed and left bare. Grain production fell by about 20 per cent, but soil erosion was dramatically reduced. At the time, there was concern about the EU 'grain mountains' resulting from overproduction fuelled by the Common Agricultural Policy, so reduced productivity was regarded as less of a problem compared with the huge benefit from reduced erosion, conserving vital soil and decreasing flash floods and siltation in the rivers. Loessial soils are particularly vulnerable to gullying and subsurface piping erosion (Jones, 2010). Faulkner (2006) estimates that 260,000 km² within the European Union are at risk from serious pipe erosion.

The UK's Department for Environment, Food and Rural Affairs (Defra) is showing renewed interest in 'no till or low till' (Defra, 2009). The report favours 'precision farming' using machines that only turn over a shallow surface layer, and satellite-based Global Positioning Systems that allow farmers to focus pesticides or fertilizers only where needed. Defra's official encouragement to British farmers is aimed primarily at conserving the store of carbon in the soil, an anti-greenhouse measure, by reducing exposure of the organic material in the soil to oxidation – British soils are estimated to hold ten billion tonnes of carbon, some 57 times the UK's total annual emissions of greenhouse gases.

The drive would also benefit water resources: enhancing the water retention in soils, reducing irrigation requirements and the amount of pesticide and fertilizer spread over the fields. Predictably, the Soil Association advocates full organic farming practices. Although these undoubtedly have merit, widespread adoption is unlikely.

Better management of the mountains and uplands

Mountains and uplands tend to be the source of most surface water resources. In general, they are also the best quality waters. In the UK, more than two-thirds of the drinking water comes from the 40 per cent of the country that lies above 250 m. Yet in the lead-up to the implementation of the European Water Framework Directive, three-quarters of the rivers and lakes in the English Lake District were judged to be in an 'unfavourable state', due in part to intensified agriculture.

Preserve the bogs

Preservation of peat bogs is also being promoted as a measure to fight climate change. Half of the soil carbon in Britain is actually stored in untilled peat bogs, hence a second Defra (2009) initiative to protect these as well. A Royal Society for the Protection of Birds report notes that many peat bogs in Scotland are suffering severe erosion caused by artificial drainage and conifer plantations (RSPB, 2007). The RSPB point out that peatlands in England and Wales are losing 381,000 tonnes of carbon a year because of erosion, whereas properly maintained they could sink 41,000 tonnes p.a.

Protection of peat bogs is often advanced as a means of reducing flood risk by virtue of their supposed 'sponge' properties. With this in mind, a major plan was initiated in 2009 to fell 17,000 conifers on the North York Moors to double the size of the May Moss bog to prevent flooding in nearby towns like Pickering, which was swamped in 2007. Whether the scheme is successful or not remains to be seen. Conifer forests can be effective anti-flood devices themselves, because of the large

Figure 15.12 Soil erosion on loessial soils near Leuven, Belgium

319

water losses through evapotranspiration. Where they have contributed to flooding it has been due to ditching and drainage rather than the trees themselves (see Chapter 6). Whether the extra water is mopped up by an expanded bog is an open question. In practice, the sponge-like properties of bogs are highly variable. By their very nature, bogs are wet most of the time and so may have little capacity to hold more water. The drier bogs often have their own internal drainage systems of natural pipes and tunnels, which channel rainwater into the streams with little or no delay (Jones, 2010). The real danger comes from accelerated erosion of the bogs, either through 'bog bursts', when the build-up of water pressure causes a landslide, or through destruction of the bog by fluvial erosion, as on Kinderscout in the Pennines. Water retention is then dramatically reduced and sediment may cause problems downstream. Some of the problems in the Pennines have been put down up air pollution in the nineteenth and early to mid twentieth centuries, depositing acid aerosols that killed surface vegetation. Inept use of fire to manage the vegetation has also exposed peat surfaces to erosion, as well as contributing to the discoloration of drinking water. The most recent threat is from climate change, which is blamed for drying out some bogs.

Combating climate change and saving water

Soil is the largest store of carbon after the ocean. The world's soils hold three times as much as the atmosphere and more than four times as much as vegetation. It is therefore vital to preserve soils and do everything to enhance a greater uptake of carbon. Intensive cereal production has markedly increased the loss of soil carbon during the last 50 years. Reverting to pasture-fed rather than grain-fed livestock would help redress this. So too, according to the Soil Association (2009), would reverting to more perennial plant crops for food and fodder. The current top ten food crops are annuals that require large inputs of oil-based nitrate fertilizers, pesticides and tractor fuel to get them going, and in many cases also the irrigation used to sustain them. Perennials have established root systems that are better at utilizing water and nutrients from deeper in the soil. Most grazing land is not irrigated, while many grain crops are.

The advantages of grazing animals would, however, be significantly reduced if there are greater numbers of ruminants producing more methane. And as emerging nations like China and India demand more meat in their diet, this is likely to be the case.

Soft solutions to flooding

'10,000 River Commissions … cannot tame the lawless river, cannot curb it or confine it, cannot say to it, Go here or Go there, and make it obey; cannot save a shore which it has sentenced; cannot bar its path with an obstruction which it will not tear down.'

Mark Twain, *Life on the Mississippi* (1883)

Mark Twain wrote those words before the US Corps of Engineers took control of the Mississippi. From the 1930s, the Corps built over 200 flood control reservoirs, 3200 km of earth levées or dikes, and thousands of stone wing-dikes designed to force the flow into the middle of the channel and scour it to maintain a 2.75 m deep route for navigation. They also cut off numerous meanders, so that by the end of straightening the river in 1942 it was 240 km shorter. This is just one example of many around the world. Some engineers in Japan have rued the day they decided to accept Western engineering philosophy.

Confining and straightening the rivers, cutting them off from their natural floodplains and increasing the channel gradient, leads to more rapid flows with larger volumes. It can be valuable around a city, but it makes floods worse downstream.

The events of June and July 1993 on the Mississippi caused the philosophy to be questioned. It was the worst flood in the river's recorded history. Levées burst, thousands were made homeless, entire towns were made uninhabitable, water treatment plants were flooded. The town of Grafton, Illinois, has been rebuilt on an entirely new site well away from the river. In truth, it *was* a severe weather event and following eight months of unusually high rainfall and heavy winter snowfall in the Rockies, the catchments were saturated. Many other mid-western rivers continued in spate till September. The Corps pointed out that society pays for protection against a certain flood risk – say the once in a hundred years event – and this can always be exceeded. It is a cost-benefit balance (but see how the '100-years event' is calculated in Chapter 19). Early warning systems have been considerably improved since under NOAA's '1995–2005 Strategic Plan', with doppler radar and interactive processing at River Forecast Centers.

Nevertheless, critics pointed to the lack of floodplain storage – the draining of the wetlands, the cutting off of the floodplains for agriculture and cities. Engineers modelling the section of river around St Louis, Missouri, proved that it was only the failure of a levée that saved the city from the flood.

Action began to restore the natural flood storage areas along the Missouri River in Nebraska. Wetlands are being recreated in the Mississippi headwaters. After the flood, more money was put into the national Wetlands Reserve Program, which pays farmers to recreate wetlands. The programme already existed – established to reverse the trend that had drained 475,000 km^2, half the nation's total, over the last two centuries. But progress had been slow and by 1994 barely 125,000 ha had been restored. The National Wildlife Federation also renewed its call for an end to the National Flood Insurance Program, which they believe encourages building on floodplains, although the false security given by the levées is probably more to blame. Meanwhile, the Corps has largely abandoned the old policy of engineering meander shortcuts.

Europe is following suit. Under the European Water Framework there is now collaboration across EU state boundaries. In the Netherlands, the Rijkswaterstaat and provincial governments are encouraging the opening up of the 'forelands' between the summer and winter dikes by breaching the summer dikes to allow floodwaters in. The government is buying back the forelands, or ending leases on land it owns, and getting wildlife organizations to manage them. In some cases, like the De Blauwe Kamer near Wageningen, backwater ponds have actually been excavated to increase floodplain storage. Between 1990 and 2010, some 2500 ha of farmland in South Holland adjacent to waterways and lakes were turned into wildlife refuges. In some cases, even land reclaimed from the sea in the polders is being abandoned: the de Dood and Selena polders are now wetlands, part of the Flevoland polder was never completely reclaimed and became the Oostvaardersplassen wetland nature reserve in 1968, and there are plans for more wetlands along the Eastern Scheldt. The Danube has been taken as a model for European rivers, largely because little development took place under the communist regimes. Restoration is also taking place on the Danube downstream of Vienna,

where the floodplain has been naturalized and a national park created. An EU research project led by Professor Keith Richards of Cambridge is exploring opportunities to restore floodplains to their natural role as water meadows or wet woodlands termed 'alluvial forests'.

Britain too is planning to restore wetlands, mainly for ecological reasons, but they will also be large areas for flood relief. In 2008, Wetland Vision began mapping out areas best suited to restore in collaboration with Natural England and Wildlife Trusts. The project aims to recreate thousands of hectares of new wetland over the next 50 years. The model for this is the Great Fen Project in Cambridgeshire where 3400 ha of wetland have been created by blocking ditches. Over the last 400 years, 99 per cent of the Fens have been lost to agriculture. Now, the largest reed bed creation scheme in Europe has been undertaken on the Great Ouse Wetland, using former gravel and sand quarries to create lakes. New wetlands are being created on the River Avon downstream from Stratford as part of a combined response to the floods of August 2007 that flooded most of Tewksbury, except for its medieval cathedral. Clay levées are being built around the town of Pershore and the site where the clay is being dug out is to become a flood-relief wetland. Some 25,000 tonnes of clay are being transported by barge – the first to use the canalized river for 40 years – saving 3000 truckloads. Other artificial wetlands on the Avon are now ten years old. This contrasts with a rather different solution on the River Severn, which joins the Avon at Tewksbury. There the town of Bewdley is now protected by removable steel barriers – unfortunately in the first flood test, the panels arrived too late, delayed by traffic!

Many other possible soft or green approaches to flood alleviation are being explored; some new, some reviving old traditions. Flood-proofing the ground floors of buildings, zoning floodplains for less sensitive uses like playing fields, grading levées and floodwalls according to the sensitivity of specific human activities. Around the town of Yarm in Cleveland, the town is given protection against the 40-year flood, but the farmland upstream is only protected against the ten-year event. This type of prioritization is being widely used. Control of runoff from urban areas may help, even rainwater harvesting and turf roofs.

Early warning is still one of the most effective anti-flood strategies, as no amount of physical protection can ever protect against every flood. Early warning systems range from the highly sophisticated Mississippi system to simple loudspeakers and mobile phones in Africa. Christian Aid is funding mobile phones for flood alerts in Malawi to complement an educational and tree-planting programme with the help of EU support.

Wild rivers and protected landscapes

The debacle at Hetch Hetchy in the newly created Yosemite National Park 100 years ago spawned the modern environmental movement (see *Dam protests*). However, it was not till the Wild and Scenic Rivers Act of 1968 that the USA established a national system for protecting rivers. The system classifies free-flowing rivers as wild, scenic or recreational and assigns appropriate levels of protection to each. Both wild and scenic rivers must be free from impoundments or diversions. A scenic river may be upgraded to a 'wild river' by cutting off road access and removing human constructions just as 'wilderness areas' are created within national parks. Recreational rivers have good access, have some constructional developments and may have been dammed or diverted in the past.

There are now over 31,000 km of river that are protected on 252 rivers in 39 states, and the list is growing. However, as the government points out, this only amounts to around 0.25 per cent of national rivers. Compare this with the 75,000 large dams in the USA that affect flow over 1.5 million km of river, amounting to 17 per cent of US rivers. In other words, the length of river protected, from the basic level of recreational upwards, amounts to less than one per cent of the length of river that is severely modified by large dams alone, disregarding small dams and other engineering works.

It is a small but welcome move, and it has to be said that America is leading the way. This is largely through pressure groups like the Sierra Club and American Rivers, which was founded in 1973 to 'protect our remaining natural heritage, undo the damage of the past and create a healthy future for our rivers and future generations' (www.americanrivers.org).

Academics and wildlife organizations in Britain began to follow the same route in earnest in the 1990s, with books like Cosgrove and Petts (1990), Calow and Petts (1992), Boon *et al* (1992) and Harper and Ferguson (1995), extolling the ecological value of restoring forested river corridors, and the Royal Society for the Protection of Birds *New Rivers and Wildlife Handbook*

(Ward *et al*, 1994). The EU Water Framework offers some protection, but America appears to be in the lead with its championship of truly 'wild' rivers.

Artificial floods

For 60 hours beginning on 4 March 2008, over 1100 m³/sec was deliberately released from the Glen Canyon hydropower dam on the Colorado River into the Grand Canyon. The release was designed to restore and sustain the ecology of the river and its floodplain. One of the key targets was the humpback chub, a rare fish that became almost extinct after the dam was opened in 1963, trapping 98 per cent of the river sediment and absorbing the floods that used to build the sandbars where the chub breed.

Yet there are serious questions about the effectiveness of the remedy. An official of the Grand Canyon Trust called the Glen Canyon release a charade and a glamorous event mounted for the media. Ecologists point out that the temperature and clarity of the water released from the dam is very different from the natural events – cool, clear water where it used to be warm and muddy – and that flooding needs to be repeated every year to sustain the effects. But the Bureau of Reclamation does not plan any more flood releases in the next five years, and one powerful reason is that the dam provides electricity for 650,000 people and the Colorado is dangerously short of water: less snowmelt from the Rockies caused discharge to fall to an 85-year low in 2008 and the Lake Powell reservoir is suffering from severe drought.

A number of African countries have introduced similar procedures, notably Senegal, South Africa and Cameroon, but aimed more at reviving the livelihoods of local people living on the floodplains than for wildlife.

Don't forget the canals

The value of conserving and improving natural water bodies for the benefit of nature is now widely accepted. Hydropower is seen by many as a key 'green' alternative energy source. But ecologists have often denigrated river navigation for destroying natural rivers with dredging, straightening, dikes and locks. However, the indirect value for the environment of expanding water transport systems is generally overlooked, and it is possible to create more natural riverscapes while maintaining efficient water transport.

Water transport is a more sustainable and greener alternative to fuel-hungry road, air and even rail. The value of navigable rivers and canals for the wider environment needs to be recognized, whether this is in reducing greenhouse gas emissions, or traffic congestion and damage to roads, or even reducing the need to build new roads. Yet the canals have always had other roles, like water supply and flood relief, and they have always altered natural riverflows, withdrawing water, releasing water or simply harvesting rainwater within their own feeder reservoirs. Canals are now enjoying a renaissance and an extension of their roles. They have an important part to play in sustaining a wide range of water services for both humans and wildlife, from leisure to nature conservation.

Canals and canalized rivers provided the transport system that was the backbone of the Industrial Revolution in Britain and much of Western Europe. They fell into a gradual decline, as they began to be superseded by faster rail and road systems: the first modern canal was dug in Britain in the 1770s and the railways arrived in the 1820s. The nineteenth-century network of canals and canalized rivers in Britain was in sad decline during much of the twentieth century, and from the 1960s many were filled in.

Britain now has just 3500 of its original 8000 km of canals and canalized rivers, but these still carry nearly a quarter of all freight moved in tonne-kilometres. Government support for reviving and expanding water freight has been half-hearted, notwithstanding its legislation on carbon emission targets. The large leisure lobby – fishing and pleasure craft – is now also an obstacle to expanding freight. Nevertheless, a major restoration programme was begun in 1999, 320 km of new or restored canals have been added and the canals are now busier than when they were built over 200 years ago: more than 60,000 boats and canoes and about half a million tonnes of freight ply Britain's canals annually. More recently, a completely new canal has been proposed for dual leisure and freight use in Bedfordshire. British Waterways plans to spend £160,000 between 2008 and 2012 on restoration and maintenance projects.

Transporting goods by water rather than road typically produces 80 per cent fewer carbon emissions: the benefits for climate change and national carbon emission targets are clear. Heavy, bulky, low-value items are particularly suited to water transport. Cemex barges carry 20 truckloads of aggregate each up the River Severn in England, equal to 34,000 lorry journeys a year for 1/1000th of the fuel consumption. Successful experiments on the Grand Union Canal in London recently saved 45,000 truckloads of sand from road transport, and the system is being employed in constructing the 2012 London Olympic site. Waste is also a suitable cargo. Fifteen per cent of London's waste is already transported on the River Thames, saving 100,000 lorry journeys a year. A new wharf has now been built on the Grand Union to feed a recycling plant in west London processing 400,000 tonnes of waste a year. In 2007, supermarket chain Tesco became the first major British retailer to transport by canal, using the Manchester Ship Canal to ship bulk wine from Liverpool docks to Manchester. It is a revival of the old system whereby freight was transferred to smaller boats for transport inland. Ninety-five per cent of freight by volume is still carried to and from Britain by sea – 573 million tonnes in 2004. This is largely transferred to road, but a proposal has been made recently for 'feedering' this onto small boats again.

The main limitation of water transport is its slowness. It is not suited to every type of good, particularly high-value, low-bulk luxury or perishable goods. But it is eminently suited to transporting bulky and heavy materials that are in constant demand. For these, the principle of the conveyor belt makes its tardiness irrelevant: with a continual stream of barges, goods are offloaded at one end the instant goods are loaded on at the other.

Britain's canals are estimated to contribute £500 million to GDP. A substantial part of this revenue comes from leisure activities, which dominate water use, but the public health value of those activities remains largely unquantifiable. Many urban canals have been restored as linear parks and recreation areas as part of urban regeneration schemes, for example in Birmingham and London. Half the population of Great Britain live within 8 km of a canal or canalized river and over 11 million people visit the canals annually for boating, fishing, nature watching, walking and cycling.

More quantifiably, Bristol gets 50–60 per cent of public water supply from the Gloucester and Sharpness Canal. Grey water is taken for cooling purposes directly from the Forth and Clyde Canal in Scotland and the Grand Union Canal in London, with substantial savings in carbon emissions.

A 1994 report by the National Rivers Authority even considered the possibility of using existing canals as part of an interbasin transfer scheme to bring water from the underused Kielder Dam in Northumbria to London. A full-scale national water grid has been mooted, which would use the canal and river networks to deliver water to the water deficit regions of SE England from the water surplus regions of north west Britain. This is currently rated as unnecessary, and possibly too costly and inefficient, given losses through leakage and evaporation, pumping and maintenance costs and pollution potential.

British Waterways estimates that it provides £67,000 worth of flood relief a year by lowering water levels to capture surplus riverflows. Gloucester was saved during the catastrophic floods on the River Severn in 2007 by pumping water from the river into a canal. As flood frequency increases with global warming, this role is likely to become increasingly important.

There are also proposals to generate hydropower from canal locks and reservoirs at up to 50 sites in Britain. These could generate 100 MW, emission free, saving 100,000 tonnes of CO_2 emissions a year.

Elsewhere in Europe, canals and canalized rivers remain active. In Germany, 64 per cent of internal freight transport is by water, against barely one per cent in the UK. The bulk of this is carried on the Rhine and its tributaries, which comprise one of the most canalized river networks in the world. Canals and canalized rivers have remained extremely important in the Low Countries. They form vital links in flood control and water supply management in the Rhine Delta Project, as well as acting as transport arteries and a means of regulating water quality. Locks are opened and closed in response to incoming river discharges, pre-eminently on the Rhine at Lobith, and tide levels. The Scheldt-Rhine canal was only completed in 1975. Details of the complex water management system are presented in the discussion on the environmental protection and rehabilitation in the Rhine Delta in Jones (1997). During severe droughts, the canals and canalized waterways enable water managers to divert water to where it is most needed.

France had 4800 km of canal and over 6000 km of navigable rivers at the beginning of the twentieth century, but between 1970 and 2000 the network gradually declined as freight moved to road and rail and the canals began to collapse. Towards the end of this period tourism began to revive interest. Today the Canal du Midi in southern France carries twice the traffic it did in the heyday of industrial trade, now mainly for leisure. It too has had a flood control role since it opened in 1681, eliminating the annual floods that devastated large areas in the Aquitaine and Carcassonne Gap. It also contributed to major land-use change by opening up markets for grain and creating 'the bread-basket of the south'.

There is a strong argument for not overplaying the restoration of river environments, 'reverting to nature', to the detriment of river transport. Of course, commercial interests will largely see to this, at least on the bigger rivers. But a more integrated and planned approach is called for. The direct benefits of naturalizing rivers and the indirect environmental effects of expanding water transport need to be considered together. Recent developments on the Meuse and the Rhine indicate the way forward: a symbiosis in which naturalization of the old river channels coexists with improved canals and canalized reaches running alongside. A key issue here, as ever with canals, is: where does the water come from to feed the canals and what does this do to ecology in the old rivers?

Conclusions

Methods for cleaning up and restoring the physical framework of aquatic environments are expanding rapidly. In many parts of the world they are expanding faster than aquatic environments are being polluted or destroyed. Dire predictions made in the 1970s have mercifully not materialized as a result of efforts to reduce effluents and widespread moves to clean up and restore catchments and water bodies. Sadly, this is still not true the world over. The race is being lost in many developing countries. Some legacies remain very challenging and continued vigilance is needed everywhere.

Discussion topics

- Explore new biological approaches to cleaning up pollution.
- To what extent will it be possible to restore the Aral Sea and its tributary rivers?
- Analyse the roles of wetlands and forests in soft flood control.

Further reading

A special section on the Aral Sea with contributions from Central Asian scientists and administrators, much of it available in English for the first time, can be found in:

Jones, J.A.A., Vardanian, T. and Hakopian, C. (eds) 2009. *Threats to global water security*. NATO Science for Peace and Security Series – C, Springer, Dordrecht, 365–400.

See also: Pearce, F. 2006: *When the Rivers Run Dry: what happens when our water runs out*. London, Eden Project Books, Chapter 25.

16 Using seawater

'There is compelling evidence that desalination will have to be adopted as the only acceptable option for potable water provision in some developing countries, particularly those in the Arab world. Reducing the cost of desalinated water, therefore, is becoming a pressing issue … as it is not only dependent on improving desalination technology, but also on improving the management of the conservation and utilization of desalinated water.'

Taysir Dabbagh *et al*, 1993,
Desalination: the neglected option,
Kuwait Fund for Arab Economic Development

The ocean contains over 97 per cent of all water on Earth and is more widely accessible than the polar ice caps. It is a very attractive source to plunder. If only it were drinkable, it would solve future water crises for all nations with direct or indirect access to it. Until recently, desalination has been almost the only method of creating usable water from seawater, but times may be changing. Japanese industry has been using seawater directly in some processes, especially as a coolant, for decades: nearly 30 per cent of industrial requirements were met by seawater as long ago as 1965. An innovative scheme in Helsinki is now using seawater from the Baltic, which drops below 8°C in November to May, to cool a computer data centre and deliver the excess heat to the city's heating system, which pumps hot water to 450,000 people. Britain and America have also been siting nuclear power plants on the coast to use seawater for half a century, in contrast to France, which has preferred inland waterways. The fallacy of France's policy was demonstrated in the summer of 2009 when lack of riverwater along with repair works caused most French nuclear power stations to close down and forced France to import electricity from Britain and other neighbouring countries. It was the second time this had happened since the millennium.

In the last few years the possibility of using untreated seawater for agricultural irrigation has also emerged. Research is progressing with the development of crops for food or biofuel that can tolerate saltwater irrigation. Seawater is being used to irrigate golf courses in America using specially selected salt tolerant grasses. All of these developments will help relieve pressure on hard-pressed freshwater resources.

Desalination

Desalination is an expanding industry (Figures 16.1–16.3). The International Desalination Association announced in 2009 that there are now 14,451 major desalination plants worldwide and another 244 in progress. These have a total installed capacity of 60 million m³ a day or 22 km³ a year, with an extra 9.1 million m³/day due from the plants in progress. Expansion is accelerating: the increase of 6.6 million m³/day between 2008 and 2009 was the largest increase on record for a single year. The British manufacturer, Modern Water, expects the Middle East market alone to increase by £13 billion between 2010 and 2016.

New technologies and changing economics are progressively reducing the main obstacle to its use: cost (Tables 16.1 and 16.2). Multi-flash distillation is the oldest technique, dating back to the 1960s, but this has now been overtaken by reverse osmosis, first introduced in the 1980s. Hybrid plants combining the two approaches now offer to exploit the best of both (Table 16.1). A small minority of plants use multi-effect distillation, also called multi-effect vapour compression or humidification. The costs vary very widely as much depends on the fuel used: present prices vary roughly from 40 to

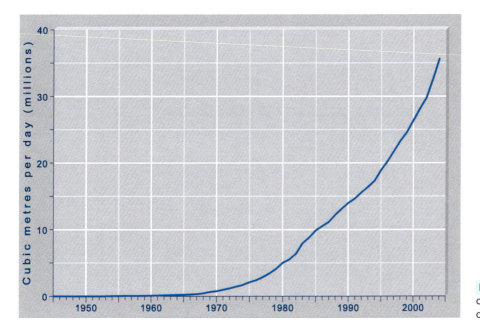

Figure 16.1 Increases in the volume of desalinated water. (Pacific Institute data)

Table 16.1 Methods of desalination

Method	Description	Comments
Distillation methods:	Evaporation produces freshwater	Still widely used but can be capital intensive and expensive to run
Multistage flash distillation	Reduces pressure so boiling occurs at 80–85°C to reduce fuel costs	Popular form since 1960s. Good use of waste heat from other processes, but needs to be large. Produces very pure water
Vapour compression distillation / multi-effect distillation	Pumping steam through mixture of new seawater and recycled seawater (brine) from previous passes through system	Economizes on fuel and wastewater. Low pressure requirement. Very pure water
Glasshouse distillation	Using sunlight to evaporate seawater pool	Cheaper but limited capacity
Reverse osmosis	Water forced through thin filter membranes whose pores allow water molecules through but not larger dissolved salts	Introduced in 1970s. Widely used and increasingly popular; energy efficient. Needs high pressure, but can be large or small. Well suited to brackish water and wastewater. In 1980s developed for seawater. Best where fuel expensive, but high salt content clogs membranes
Refrigeration methods:	Impurities are expelled during ice-formation	Less common
Vacuum freezing	Boiling point is lowered to freezing point at pressure of 400 Pascals (4 mb). Brine boils, seawater freezes. Freshwater from both vapour and ice	
Secondary refrigeration	Evaporating butane gas freezes the salt water	
Chemical methods	Chemical separation of salts	Chemical costs proportional to salt content
Water softeners	(1) sodium carbonate or polyphosphates form insoluble complexes, (2) sodium aluminium silicate uses ion exchange	Mainly for removing excess Ca and Mg from alkaline terrestrial water
Electrodialysis	Separating salt and water electrically in brackish waters	Electricity costs proportional to salt content. Best for small units and brackish water

90 US cents per cubic metre. Recent innovations, such as molecular and nanotechnology have raised efficiency and new energy sources like solar have reduced running costs. 'Forward osmosis' promises another 30 per cent reduction once the filters are perfected.

As ever, adoption of the technology depends on the balance between need and ability to pay the cost. Saudi Arabia, the United Arab Emirates and Kuwait led the way, along with the USA. Most desalination plants are still in the Middle East, especially in oil-rich countries

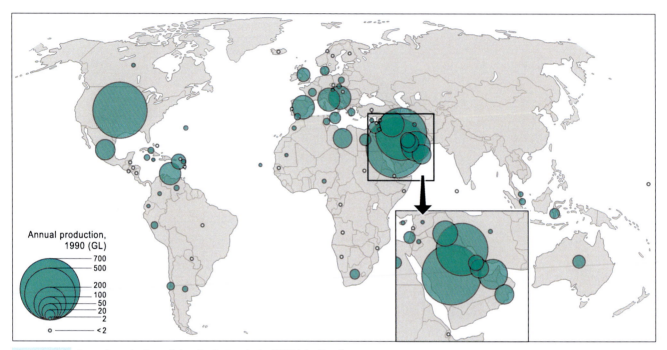

Figure 16.2 Annual production of desalinated water in 1990 (gigalitres)

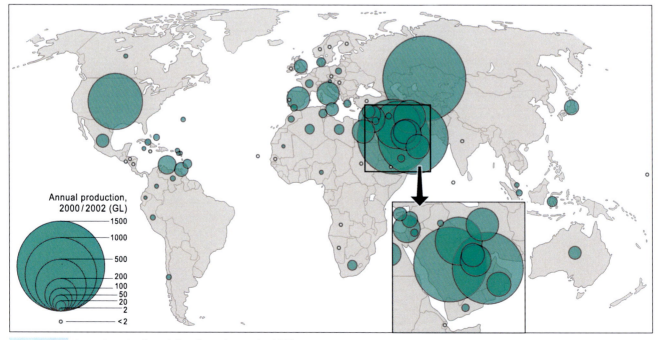

Figure 16.3 Annual production of desalinated water by 2002

Table 16.2 Effects of plant size on desalination costs for oil-fired plant*

Size of plant	Water production millions m³ per year	Cost US$ per m³
Large	50–100	60–65c
Small	5–10	70–80c
Small brackish water plant (TDS salinity 1000–5000 ppm)	5–10	40–50c

*After S. Arlosoroff (pers. comm., 2008), based on use of oil at $100 per barrel, ex plant.

where fuel has been cheap. In combination with the far less environmentally supportable exploitation of fossil groundwater (see *Mining groundwater* in Chapter 12), desalination allows Libya and most of the Arabian Peninsula to consume water at rates equivalent to more than 100 per cent of their conventionally available water resources. In 2002, Saudi Arabia had some 28 plants that produced 2.5 million m³/day, meeting 70 per cent of its drinking water needs. The Saudi capital, Riyadh, is served by the largest desalination plant in the world on the Gulf coast at al-Jubail, which produces nearly 945,000 m³ daily or 0.4 km³ a year. Saudi's Shoaiba 3 project can produce up to 880,000 m³/day. Israel is expanding its desalination facilities in response to potential loss of the West Bank aquifers, the relatively high cost of importing water by ship under its agreement with Turkey – let alone the ruptured diplomatic relations caused by the slaughter of Turkish peace activists by Israeli commandos in 2010 – and the perennial need for security. The plant at Ashkelon has operated since 2005 and can produce up to 165,000 m³/day by reverse osmosis. Israel has a second plant at Palmachin, and in 2010 unveiled the large reverse osmosis plant at Hadera near Tel Aviv, capable of producing 456 million m³ a day, plus an extension of 88,000. Funded by the European Investment Bank, Hadera alone can now provide 20 per cent of the country's needs at a cost of NIS 2.6/m³ ($0.67) – the cheapest to date. Before Hadera was built, desalination provided 10 per cent of Israel's water requirements. Prior to resuming Palestinian peace talks in 2010, Prime Minister Netanyahu also urged speedier progress on large plants at Ashkelon and Sorek (due 2012). Most North African states now have a desalination capacity, making the Middle East–North Africa region the largest producer, outstripping its rival, the rest of Asia, by more than twofold.

A review of the potential for expansion published by the Kuwait Fund for Arab Economic Development back in 1993 noted that the inception of desalination took place in unusual circumstances (Dabbagh *et al*, 1993). Whereas major technological advances usually come from industrial countries to satisfy increasing demand, desalination arrived when they had no need of it. In contrast, desalination has largely expanded out of the Middle East. The oil boom, combined with the economic and demographic explosion, created a market for desalination that grew rapidly, based on energy-intensive technology. Lack of competition between manufacturers, cheap fuel, a captive market, the lack of research outside the main manufacturers and the lack of university training courses in the new procedures, which are substantially different from other water treatment plants, all conspired to limit research into alternative methods that would be more affordable for non-oil producing countries. The hegemony of the major manufacturers was further boosted by security considerations. The Kuwaiti review noted that groundwater resources in the region are unsustainable, which leaves just two solutions to water shortages: desalination and long-distance transfers. Coming immediately after the Iraqi invasion of the country and the first Gulf War, the review concluded that the proposed 3000 km long Peace Pipeline from Turkey would be highly vulnerable to attack as well as requiring consent from a number of independent states – Kuwait would be at the end of the line. This left desalination, combined with recycling, as the only solution.

Desalination is often seen as a technology best suited to the 'islands and drylands'. Certainly, it is the salvation of many islands or effective islands (see box 'Water sustainability on small tropical and subtropical islands'). Malta gets two-thirds of its drinking water from desalination by reverse osmosis: Singapore 10 per cent. Gibraltar cut its dependence on ship tankers and by the mid-1990s had a capacity of over 20,000 m³ a day. It began installation of a new plant in 2008 and the Modern Water company reported in 2009 that the plant, which is piloting its new 'manipulated osmosis' technology, was performing ahead of expectations. Hong Kong was another pioneer of the technology. At the time of its return to China by the UK in 1997, it had a developed capacity for over 183,000 m³ a day, exceeded only by the islands of Qatar, Bahrain and the Netherlands Antilles. Other Caribbean and Atlantic islands heavily dependent on desalination include Antigua, Ascension Island, the Bahamas, Bermuda, Cape Verde, the Dominican Republic, Jamaica, the Turks and Caicos, and the Virgin Islands. Around a third of the total desalination capacity on islands is in

Water sustainability on small tropical and subtropical islands

Nurit Kliot

Islands have long been noted for their unique ecosystems, but they are particularly vulnerable to disturbance and destruction by human activities. There are nearly 2000 'significant' oceanic islands (up to 17,000 km² in size) around the world (IUCN/UNEP, 1991). Because of their isolation and often remoteness, the transfer of resources or commodities needed on the islands is more difficult and expensive. The size of island states has an effect on their sustainable development because resources are often limited, especially water, and they can support only limited population sizes. Many islands rely on narrow resource bases which do not allow the economies to diversify. Water resources therefore need careful management to match water availability to water needs. This can be particularly difficult in tropical and subtropical regions where water consumption is relatively high due to elevated temperature and evapotranspiration, and because of their economic specialization in agriculture and tourism, which need large amounts of water. Singapore, Hong Kong, Bahrain, Malta, Barbados and the Maldives in particular suffer from water scarcity, with total domestic use below the 50 litres per day which is considered the basic human need.

Availability of water resources in small island states

Islands are classified as high or low: high islands are divided into continental and volcanic, and low islands into atolls and raised limestone islands. The most severe water shortage is experienced in the low islands, where there are no rivers and the inhabitants must rely on small groundwater lenses floating on the saltwater, e.g. The Bahamas, Seychelles, Maldives, Niue Nauru, Tuvalu, Marshall Islands, Northern Cook Islands, Tonga and Kiribati. Once the lens is contaminated by saltwater intrusion, the delicate balance between fresh and saltwater may take years to re-establish. Depletion of the freshwater lenses, exceeding their natural recharge, has contaminated freshwater resources in the Maldives and many Pacific islands.

The geology of many small islands, especially granitic and metamorphic rocks with low permeability, limits groundwater yields, e.g. in the Cyclades and Dodecanese. Karstic islands in the Bahamas or the Adriatic also have scarce water resources. In many high and mountainous islands, deforestation and mismanagement of watersheds increases runoff and little water can infiltrate into the ground, as in the Windward Islands, Antigua, Barbuda, St Kitts and Nevis, and Dominica.

Many islands also suffer from low rainfall, like Cape Verde, the Cyclades, Dodecanese, Cyprus, Malta and Barbados. Many others with abundant precipitation also experience water shortages because of high evaporation or strong seasonality, e.g. the Seychelles, St Kitts and Nevis, Antigua, Barbuda, Grenada, Trinidad and Tobago.

Many are particularly vulnerable to the effects of climate change and variability. Rainfall data show considerable decadal-scale fluctuations on many Pacific islands of the order of 200 mm in mean annual rainfall, and 50–100 mm in seasonal rainfall, and there is a growing consensus that global warming may lead to more droughts and floods in the Pacific. Sea level rise will increase intrusion of saltwater in fresh water lenses. El Niño episodes in the last two decades have reduced precipitation by as much as 87 per cent in the western Pacific. In the Indian Ocean, Madagascar, the Comoros, Seychelles and Mauritius are becoming increasingly short of water. Droughts are frequent and desertification is occurring in Mauritius, Madagascar and even the Seychelles. The same is true in the Mediterranean basin, in the Balearics, Malta, Cyprus and Sicily, and overexploitation of groundwater is practised in almost all the islands.

Patterns of demand

Rapid population growth and rising numbers of tourists are placing increasing strain on water resources on many islands. The seasonal demand from tourists can be several times higher than that of locals. In the Seychelles it is estimated that a tourist consumes 3.5 times more water than a local and in the Caribbean tourists consume 5–10 times more than locals. The demand from tourism exceeds water supply in the Windward and Leeward Islands, the Balearics, the Seychelles, Malta, Maldives and most Greek islands. This demand is also spatially uneven, concentrated in just a few tourist hotspots.

Agriculture is traditionally important on most islands and is in severe competition with the domestic and tourism sectors. This is common to many Mediterranean islands, such as the Balearic, Malta and Cyprus. Demands from irrigated agriculture are growing on Barbados and St Lucia and other Caribbean islands. Irrigated agriculture has a very limited role in the Pacific islands, but it consumes about three-quarters of water withdrawals in Madagascar and Mauritius. Water use in industry does not constitute a problem in islands and island states.

Sustainable and unsustainable management of water resources

The main mismanagement practices are: over-depletion of water resources, water contamination, and little or no conservation of water resources. Many of the islands listed in Table 16.3 are water stressed, with less than 500 m³ per capita per year. Barbados, Bahamas, Dominica, Grenada, St Kitts and Nevis, St Lucia, Cape Verde, Maldives, Bahrain and Malta are water scarce. The Solomon Islands, Papua-New Guinea, Fiji, Jamaica, Maldives, Haiti and Madagascar are unable to provide even the min-imal 50 litres per person per day, though some of these islands are rich in water resources.

The main mismanagement practices are over-depletion and pollution. Over-depletion of groundwater resources has caused significant saline intrusion into aquifers in many Mediterranean islands, such as the Balearics, Sicily, Sardinia and Malta, and elsewhere in Malaysia, Maldives, the Balearics, Bahamas, French Polynesia, Cyprus, Guam, Rhodes and Cape Verde. Pollution by untreated or partially treated wastewater is common due to lack of sanitation, even in major tourist islands, like Antigua and Trinidad. There is inadequate sewage disposal infrastructure in most of the Caribbean islands, Madagascar, Mauritius, the Maldives and Seychelles. However, the worst pollution is in the Pacific islands. Only 11 per cent of the population in the Pacific islands is serviced by a reticulated wastewater system. As a result, Antigua and Barbuda have a problem with drinking water pollution, as do Malta and many Pacific islands. Water-borne diseases and bacterial contamination have been detected in Kiribati, Tuvalu, Niue, the Marshall Islands and Micronesia. Industrial pollution is also a problem in many Caribbean islands, Mauritius and Madagascar.

Measures to increase water supply

Rainwater harvesting: Effective rainwater collection is often under-utilized or inoperative, mainly due to cost. Rooftop catchments and cistern storage have been the

Table 16.3 Water availability and water conservation measures on small islands

Country	Water availability per capita m³	Conservation measures					
		Recycled wastewater	Importation	Storages, cisterns & dams	Desalination m³/day	Rainwater harvesting	State of water resources
Bahamas	63	None	yes		11,681,936	yes	scarcity
Barbados	294	Local			40,000	yes	scarcity
Grenada	125	None			(destroyed by hurricanes)	¼ of drinking water	scarcity
Jamaica	3482	None			6094		
Seychelles	150	Some		yes	3900	yes	scarcity
Malta	153	None		yes	72,600		scarcity
Maldives	88.8	None			58,800	yes	scarcity
Singapore	137	Partial	yes		133,380	Treated stormwater	scarcity

basis of domestic water supply on many small islands in the Caribbean and Mediterranean. It is the most important source of freshwater in the Marshall Islands and Tuvalu, and serves as a supplementary source in the Cook Islands, Micronesia, Fiji, Kiribati, Nauru, Niue, Palau, Papua, Samoa, Solomon Islands and Tonga.

Recycled brackish and seawater: Recycled water is used in irrigation in Cape Verde, Cyprus and Bahrain. Seawater and brackish water is used for flushing in some of the Maldives, Kiribati, Marshall Islands, US Virgin Island and Hong Kong.

Desalination: Although expensive, desalination is sometimes the only or the most important source of potable water, as the cases of Nauru, Malta and Antigua show. In all, 60–70 per cent of potable water is produced by desalination. Desalination is the main source of water in the US Virgin Islands, Aruba and the Cayman Islands. Desali-

nation provides more than one-third of supply in Cyprus and Cape Verde, and is prevalent in the Bahamas, Barbados, Bahrain, Seychelles, Singapore and Tuvalu, although many plants are poorly maintained.

Importing water: Water importation is practised in all parts of the world, usually if no other measures to increase supply are available. Nauru used to import water by ship, and this method is still used on small islands in Fiji, and Tonga, and small islands in the Mediterranean.

Conclusions

Many islands suffer from natural shortage of water, either annual or seasonal. But there are many cases of mismanagement, unsustainable usage and pollution. Many islands have begun to adopt desalination and rainwater harvesting, but most are failing to curtail pollution.

the Caribbean, reflecting the tourist trade rather than the permanent population. Even so, at an average per capita capacity of around 5 m³ a year, the Caribbean islands generate little more than a tenth of the per capita capacity in Saudi Arabia.

Nevertheless, many better-watered countries have now opted for the technology. By the mid-1990s, states in Western Europe and Scandinavia had a total installed capacity approaching 1.5 million m³ a day, including 2725 m³/day in Ireland. Climate change and the expansion of desertification onto the northern shores of the Mediterranean are causing many southern European countries to plan major expansions. Spain is planning 20 new plants in its southern coastal region. Europe now has roughly half the desalination capacity of the USA.

California has been so short of water – a problem that began developing soon after the 1922 Colorado Compact – that Los Angeles has opted for an expensive mega-plant jointly fuelled by oil and nuclear power. Australia is also embracing desalination following the recent severe droughts. In 2006, Prime Minister Howard called for 'drought-proofing' of cities by using desalination in combination with other sources. Adelaide now plans to get 25 per cent of public water supply from desalination combined with a doubling of its Mount Bold reservoir.

The UK has changed its mind since the mid-1990s. The National Rivers Authority's 1994 strategic plan ruled it out, yet even in 1996, Britain had a desalination capacity of over 100,000 m³ a day. Anglia Water opened a

£2 million test plant at Felixstowe in 1998. The cost of desalination plus pumping water up to reservoir level to get tap pressure proved to be five times higher than rainwater, but it was judged to be of potential use to top up supplies *in extremis*. Plants are now operating around the south and east coasts from the Scilly Isles to the Isle of Wight, and Southern Water and Anglia Water are developing further plans. The driving factors are increasing water demand and projected climate change: even before the latest influx of migrants from the new EU countries, south east England was projected to be using all its locally available resources by 2021. Climate change estimates suggest that riverflows could be more than 25 per cent lower by mid-century (Pilling and Jones, 1999) and aquifers are already over-exploited across most of the region.

The high price of oil is beginning to alter the fuel of choice even in the Middle East. In 2008, it became more economic for Dubai to sell oil abroad and to start using gas and to import coal from South Africa to feed a new mixed fuel plant. Co-production of electricity and water is another option. Waste heat from electricity production can be used to power desalination; multistage flash distillation is well suited to this, with cost advantages for both products.

Lack of proximity to the sea can be another limitation. But saline and brackish lake water or groundwater can also be desalinated, and at much lower cost than seawater because of the lower salt concentrations: brackish water has a typical salt content of 0.5 per cent (compared with 3.5 per cent in standard seawater) and can therefore be treated more cheaply by chemical 'water softeners' or electrodialysis. Since costs in chemicals or electricity are proportional to the salt content of the water, these methods are best used for water with lower salt content. A clever feature of the proposed London desalination plant is its location in the Thames estuary at Beckton, where salinity is reduced by the river discharge to just one third that of seawater. Operating only on outgoing tides can also reduce processing costs and the clogging of the membranes in reverse osmosis. The Thames Water plant is planned to provide up to 150 million litres a day, enough to support 900,000 of the eight million Londoners. Even so, the plant is expected to operate part-time only, offering support to public supplies about 20 per cent of the time, during peaks in demand, because of the relatively high processing costs, estimated at double that of conventional water processing.

Landlocked countries with desalination plants like the water-poor Central Asian states of Azerbaijan, Kazakhstan, Turkmenistan and Uzbekistan, have been joined by such unlikely users as Switzerland, Austria and the Czech Republic. Indeed, Kazakhstan vies with Saudi Arabia as the leading desalinating nation (see *Aral Sea lost and regained?* in Chapter 15). In these instances the water comes from saline groundwater, rivers, lakes or inland seas. Inland plants are not always successful, however. A large plant completed in Colorado in 1992 was closed down after just eight months of operation because of the $30 million annual operating costs.

Beside the sea or not, however, topography may add to the problems and costs of desalination: that of raising the water from sea level up to storage reservoirs. Whereas traditional engineering can use gravity to transport water supplies from the hills, most desalinated water has to be pumped uphill from the sea incurring more energy costs.

Is desalination environmentally safe?

Three main objections have been raised on grounds of environmental sustainability. Some concern has been expressed about the return of waste brine to the sea and its possible impact on marine life. The policy of using seawater for cooling in coastal nuclear power plants is also not entirely eco-friendly. Along with the huge amounts of water intake – 60 m³ per second at Britain's Dungeness, and twice this at France's Gravelines

station – a tremendous number of fish are attracted by the warmer water and killed, either stuck on the intake gratings or, if they are small enough to pass through into the pipework, by radiation and chlorine. Killings have been equivalent to half the commercial catch for some species: 250 million fish have been killed in a five-hour period at Dungeness. France's freshwater intakes must have a similar, if smaller, effect.

A second issue is the high energy demand, especially where fossil fuels are emitting greenhouse gases. This is an issue that is likely to become less evident as new technologies improve efficiency and as fossil fuel use declines. But it can often be a question of the lesser of two evils and of the immediate need for water versus long-term and externalized paybacks.

This does, however, lead on to the third and perhaps greater issue: should humanity be colonizing water-poor regions in increasing numbers? Clearly, there are important political barriers here, as there are with large dams and river diversions: issues of sovereign aspirations. Until recently, Dubai appeared to be a paragon of a desert economy relying on desalination. That is, until raw sewage began appearing on its beaches in 2008, because it had concentrated on water provision and largely ignored sewage treatment apart from diverting some wastewater to landscape irrigation. The rapid urban expansion of the previous two decades had been achieved without installing a centralized sewer system, and the sewage collection trucks were queuing for so long to get into the one sewage treatment plant that the drivers were tipping their loads into the stormwater drains. A cautionary tale, but one that is probably now unlikely to be a blueprint for future desert city developments.

Reducing costs

The cost of desalination is still seen as a major obstacle to its adoption. Technology is diversifying and efficiency increasing. The grip of a few companies producing highly expensive equipment seems to be reducing as competition increases. There is now greater realization that one size does not fit all, for example, that smaller plants and cheaper methods may serve a useful purpose for small numbers of consumers or that different fuel sources and systems may be more cost-effective in different environments.

A new process invented by Adel Sharif at Surrey University goes a long way to reducing energy costs. It uses chemicals to reduce the need to generate high pressure to push saltwater through the filter membrane in reverse osmosis. It is called 'manipulated osmosis' and the manufacturer, Modern Water, claims it reduces energy costs by 30 per cent. Oman began operating one of these plants early in 2010. It could be the first of many, especially in the Middle East.

There is also greater understanding of the economics to do with traditional ways of increasing water supply: it is not simply a question of the cost of the energy input. Desalination suffered in the recent past from comparison with the costs of traditional water supply methods, which have frequently been underestimated, for example, by omitting the cost of infrastructure like reservoirs and pipelines that were paid for long ago. The matter then becomes very sensitive to the methods used for discounting or calculating depreciation, which vary considerably. The cost of supplying water to 15 million people in Kuwait at 350 litres per person per day via the Peace Pipeline was calculated to vary from $0.735 per m³ to $1.758 depending solely on the interest rates used in the calculation (Dabbagh *et al*, 1993). To this must be added numerous costs like wayleave for countries it passes through, equipment and energy for pumping, etc. Using equivalent discount rates, the Kuwaiti study found that desalination would be at least as economical as the pipeline, without the security problems. The problem is to compare like with like, and how this may be achieved is still being debated, as seen from the debate on the pricing of water at the 2010 World Economic Forum. The social and political advantages or disadvantages are always among the most difficult elements to quantify in any cost-benefit analysis.

Desalination without fossil fuels

Desalination does not have to be a costly option. When the oil runs out there is still the Sun. And sunshine is available to many poorer countries. Nor is there need for expensive and inefficient conversion of solar energy into electricity. A large pool on the Greek island of Patmos, filled with seawater and covered by a glasshouse, doubled the local water supply in the 1970s. The technique replicates the natural cycle of rainwater purification that operates over the oceans every minute of the day, collecting the evaporated and purified water condensing on the glass and running down into the drains. Production was small, just 27 m³ a day, but the technique has proved a valuable aid to development of the tourist industry on the Egyptian Red Sea Riviera in recent years. As climate change reduces water supplies in the southern and eastern Mediterranean and as countries deliberately boost tourism to develop their economies, so the technology is bound to develop and become more widely adopted. There are still maintenance costs and major limitations on the volume of water that can be produced per day, but free fuel is a key economic advantage. It is currently a viable option for small islands and settlements in sunny climes, especially to meet seasonal demand.

Fears of global warming and carbon footprints are opening up new approaches to fuelling desalination. More conventional use of solar power to generate electricity is becoming more viable as technology gets more efficient, especially if need is more important than cost. Massive solar collectors are being designed covering several square kilometres to power multi-effect humidification plants. The United Arab Emirates is currently incorporating a solar-powered desalination plant in its showcase eco-city, Masdar City, due for completion in 2016.

Nuclear desalination has been around for some time. Kazakhstan was a prime adopter 30 years ago and is a major user along with India and Japan. Now the nuclear plant manufacturing industry is targeting the developing world marketing dual electricity and water production plants. Commonly, these use surplus electricity from the nuclear generator to power reverse osmosis. Some multi-flash distillation facilities use waste heat from nuclear plants. The cost of nuclear desalination is currently in the region of 70–90 US cents per m³, compared with around 50 cents from fossil fuel. Japan has ten nuclear dual plants and plans are in progress in South Korea, China, Russia, Egypt, Algeria, Qatar. Jordan is actively considering it to relieve its estimated water deficit of 1.4 million m³/day.

Nuclear power has its environmental problems, not least the unsolved question of what to do with the waste and the ever-present questions of security and nuclear proliferation. Britain's Nirex company failed for decades to find suitable sites for disposal amid concerns about the possible effects on groundwater, and the UK has now begun exporting its waste. In 2010, President Obama declared that the grand Yucca Mountain storage facility would not be made operational, and fears emerged that nuclear waste buried in Germany may be leaking into groundwater.

Among the safer alternative fuels, solar shows the most potential, but interesting developments may be afoot for wind and waves. Perth has a new facility powering reverse osmosis from a wind farm yielding 130,000 m³/day. But an entirely new approach, yet to be made commercially viable, is using wave power.

A new way to desalinate?

Professor Stephen Salter, the engineer who invented the 'Edinburgh duck', a devise to generate electricity from wave movement, has recently extended the idea to desalination by wave action. The system uses the rocking motion to work a water pendulum and a heat exchanger to force the seawater to alternately evaporate and condense (Figure 16.4). The device appears to have potential: no input of artificial energy and low environmental impact. Perhaps the main issue is the quantities of freshwater that can be created, but as with the solar powered desalination experiment on Patmos, it could be a valuable supplement in coastal tourist areas.

Seawater irrigation

At first sight, irrigating with saltwater would seem an anathema. Salt accumulation in the soil normally sounds the death knell for crops. Over the history of irrigation, it is estimated that more than 25 million hectares have been rendered infertile by salinization, which also destroys the soil structure, reducing aeration and causing waterlogging. In Mesopotamia, the 'cradle of civilization', 80 per cent of all the land that has been irrigated is now suffering from salinization and one-third has been totally abandoned. One of Saddam Hussein's excuses for draining the Tigris Marshes in the 1990s (see *War against the Marsh Arabs* in Chapter 9) was to use the water to flush out the salt from abandoned fields, though the feasibility of rehabilitation was questioned by many soil scientists. In recent history, two-thirds of the land reclaimed from the desert by irrigation schemes in Egypt since the construction of the Aswan High Dam are now out of production or marginal due to salinization. Productivity has even fallen on the Nile floodplain below the dam, because the annual flood uscd to flush out the salt. Three-quarters of the gains in food production due to the new irrigation schemes have thus been lost.

Water managers in dry countries often take the view that salinization is an inevitable consequence of all irrigation. But it does not have to be inevitable. Technology has moved on. Coupling irrigation with adequate drainage helps reduce salt build-up: the International Commission for Irrigation and Drainage acknowledges this in its very title. Much of the latest extension of irrigation in the evaporite lands of the Ebro basin is now served by

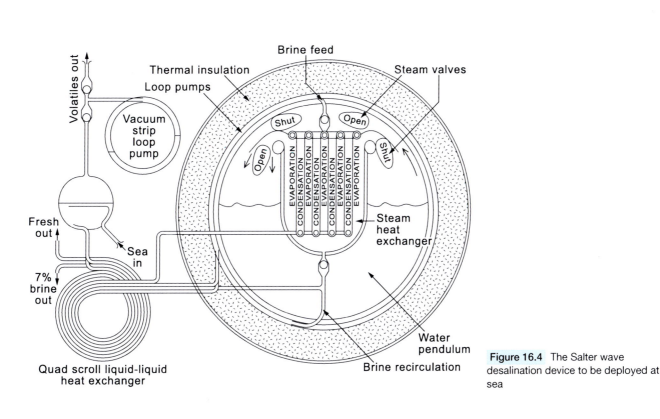

Figure 16.4 The Salter wave desalination device to be deployed at sea

frugal sprinkler systems using low-salinity water piped from the Pyrenean rivers, rather than the old-style flood irrigation, using local riverwater recycled downstream from farm to farm.

When Cyclone Nargis hit Burma on 3 May 2008, it not only killed thousands and destroyed countless villages, it also flooded the rice fields of the Irrawaddy Delta with saltwater immediately ahead of the planting season for the autumn rice crop. The salt is likely to poison the land for some time. As the Delta is responsible for 15 per cent of the national rice crop, the government swiftly began importing salt-resistant seed. And perhaps this points to one solution.

Salt-tolerant food crops

Since the late 1990s, experiments with saltwater irrigation have been undertaken in the USA, Japan, East Africa, Mexico and China. Chinese scientists claim to have grown tomatoes, hot peppers and aubergines successfully and even maintain they taste better – presumably because of the extra salt! Claims have been made in America that seawater irrigation in desert coastal areas could create new farmland equal in area to more than half that currently under freshwater irrigation in North America. If this can be achieved, then the saving in freshwater would be truly considerable.

There are caveats, however. First, it is limited to salt-loving plants, which currently rules out most food crops. The solution has been to use genetic engineering, but this raises a whole new field of controversy. The Soil Association in Britain, pioneer of organic farming since the 1940s and chief certifier of organic foods in the UK, is firmly against GM crops for many reasons, including the fact that genes may be doing more than just the job the engineering is aimed at (www.soilassociation.org). Moreover, experts have suggested that the GM crops produced by China's 'pollen tube' method may not be stable, although Hainan University scientists claim to have produced seed that has been genetically stable over four generations.

The Chinese transferred salt-tolerant genes from mangrove plants and have made further progress with rice, wheat and oil-seed rape, crops of real significance for human sustenance. This could be vital technology for China with its population demanding more food as affluence and numbers rise: agriculture accounts for 70 per cent of national water consumption, but per capita resources are only 20 per cent of the world average and 60 per cent of farmland is short of water, especially in the best arable land in northern China. Shandong University scientists report salt-tolerant wheat producing 26 kg/ha and estimate that salt-resistant crops could produce 150 million tonnes, or 30 per cent of China's current annual food output, from 40 million ha of new land, while the Ministry of Agriculture estimates that irrigating 13 million ha of coastal land could feed 150 million people (Hargrove, 2001). It is also estimated that seawater irrigation could save China 300 billion tonnes of freshwater at a thirtieth of the cost of desalination.

The technology offers an additional means of reclaiming both naturally saline soils and soils destroyed by inappropriate irrigation methods, which in China amount to 20 per cent of the total cultivated area. Unsurprisingly, the Chinese government approved nationwide promotion of the developments soon after the millennium.

Saltwater biofuels

Using seawater to grow biofuels may have significant advantages for both freshwater conservation and food production. The sudden increase in world food prices in 2006–8 was in part due to agricultural land being turned over to biofuel production: in 2007 a quarter of the American maize crop went to produce biofuel. As a consequence, the UK government began to reconsider its own policy and that of the EU of forcing oil companies to include biofuel in petrol as a means of reducing carbon footprints.

Tropical rainforests are being cleared at an accelerated rate in Amazonia, Indonesia and Malaysia by companies that are cashing in on the biofuel bonanza, with as yet unrealized implications for rainfall, let alone biodiversity. Accompanying this, there have been widespread reports of land grabbing and human rights abuses as indigenous peoples have been forced off the land, and small farmers resettled from areas flooded for hydropower reservoirs in Brazil have been displaced for a second time. In Colombia, farmers have been removed at gunpoint.

Large American and European companies are investing in land in Africa to grow oil-rich crops like soya bean, rape and palm are planted for biodiesel, and starchy food and fodder crops like sugar cane, beet, maize and wheat for bioethanol, the petrol substitute.

Seawater irrigation *per se* is just one of a number of possible ways of reducing pressure from biofuels on farmland, food crops, food prices and freshwater resources. Seawater might be used to grow non-agricultural biofuel crops that are naturally salt-tolerant without taking up agricultural land. This is perhaps the best alternative, although it is judged to be some years away from profitable exploitation. Alternatively, second generation biofuels, like cellulosic ethanol, could require less land by using the whole plant not just the seed ears. Biofuels may also be derived from sea kelp, or from elephant grass that does not require any irrigation.

Whether based on freshwater or saltwater, the future of biofuel production itself is in its early stages and doubts have already been expressed about its adequacy as a substitute for fossil fuels. The CEO of the British Royal Society of Chemistry, Dr Richard Pike, wrote in a letter to *The Times* (28 April, 2008) that biofuels are likely to underperform significantly in carbon trading and 'in the extreme represent the junk bonds of the sub-prime carbon market'.

Hydrogen cells could be a greener and more efficient power source, and their only by-product is water. If the hydrogen is produced by electrolysis of water, then the electricity should ideally come from renewable sources. But the hydrogen can also be produced biologically, for example, by feeding algae to microbes.

Saltwater irrigation of leisure landscapes

A prime objection to the stalled Spanish National Hydrological Plan is that a large amount of the water transferred to southern Spain would be used to irrigate golf courses. But the technology is already available to use non-potable water. Dubai has been irrigating its urban parks and golf courses with a mixture of desalinated water and partly reclaimed wastewater for more than a decade. The American Golf Association has explored techniques for using seawater, salty recycled wastewater and brackish water for fairways and greens in some detail. Mike Huck of the USGA Green Division in California says he now views the Pacific Ocean 'as potentially one of the largest irrigation reservoirs in the world'.

In addition to saving freshwater, irrigation with seawater could be maintained throughout a drought or water shortage, and there would be no charges to pay water companies or government agency licence fees. But there are higher construction and maintenance costs: planting

different species of grass, designing adequate drainage, adding fertilizers to correct nutrient balances, adding acids and larger amounts of water to assist leaching, monitoring corrosion and replacing damaged pumping equipment, even the possibility of building reverse osmosis plant into the distribution system to control the salinity.

A key advantage over using saltwater for food crops is that appropriate salt-tolerant plants already exist for landscape design. There is no need for genetic modification. US golf courses have been successfully planted with *Paspalum* grass, naturally found on seashores, which even tolerates low mowing on putting greens. Similarly, there are many species of trees and shrubs that are salt-tolerant, as well as drought tolerant, that can be used in parks.

Conclusions

With desalination and new methods of using seawater directly for irrigation or cooling, need may be more important than cost in determining their adoption. The ocean certainly offers an unlimited source of water. The amount of seawater currently desalinated is equivalent to about 0.5 per cent of all annual freshwater withdrawals and a further tiny fraction is used directly. But technologies are being refined, new approaches are being developed and relative costs are generally falling. The islands and drylands are the main beneficiaries at present and likely to remain so, although adoption is expanding into well-watered regions as demand increases. Desalination offers water security for big cities near the coast like London, Los Angeles and Adelaide. It also means that standard methods for calculating available water resources do not apply.

Discussion topics

- Is desalination the solution to the water crisis? Consider costs and environmental impacts.
- To what extent is using seawater for irrigation a safe solution?
- Could saltwater biofuels release more freshwater for food crops?
- Does the expanding use of seawater spell the end for jobs in standard water resources assessment?

Further reading

It is difficult to recommend a text as desalination is a profitable industry and information is often treated as commercially valuable. It is hardly covered in most standard water resources texts. Direct use of saltwater is even more sparsely covered. However, the following publication from the Pacific Institute is worthwhile:

Cooley, H., Gleick, P.H. and Wolff, G. 2006. *Desalination: with a grain of salt. A California perspective.* A Report of the Pacific Institute for Studies in Development, Environment, and Security, Oakland, California. Available from www.pacinst.org.

17 Controlling the weather

The desire to control the weather is ancient. For most of human history this has amounted to praying for rain. The practice continues today in many faiths, including the Roman Catholic liturgy: Father Angelo Marcandella is credited with ending a three-month drought around Piacenza in northern Italy in 2007, while denying the suggestion of the local mayor that he looked at the weather forecast. At the height of the Australian drought in April 2007, even Australian Prime Minister John Howard urged people to 'literally and without any irony, pray for rain'. With the late arrival of the monsoon in June 2009, the government of the Indian state of Andhra Pradesh ordered all religions to pray for rain. Among the more colourful cases, in the severe drought in Uttar Pradesh in 2002, women were reported to have ploughed the fields naked at night to appease the rain gods.

With the advent of aviation and advancements in meteorology following the First World War, however, a new age of direct, physical attempts at rainmaking began. Early experiments concentrated on a macrophysical approach, which tries to augment convective turbulence and so stimulate collision and coalescence among the cloud droplets. The Soviets tried this approach in the 1930s, using aircraft to dive-bomb or jettison bags of cement. But this approach was not very successful and energy costs are disproportionate. Indeed, the amounts of energy that mankind can muster are small compared with natural rain systems and any approach based on energy alone is unlikely to make much difference.

In an interview in 2008, veteran British weather forecaster Michael Fish supported this argument against weather modification in general. A single thunderstorm is equivalent to 1/10,000th of mankind's total daily energy consumption. A mid-latitude depression contains the energy of 1000 large nuclear bombs. Even so, science took a major step forward in the mid-twentieth century, when the focus shifted to influencing microphysical processes, tipping the balance with minimal input of energy.

Methods of rainmaking have advanced considerably since then, especially in the last 20 years, but there are still substantial questions as to its physical effectiveness and reliability and also its social desirability. There is, however, no question that it can have value in emergency situations, to relieve drought in the Sahel, to limit hail damage or to redirect heavy rain away from sensitive areas. Quite apart from its value to water resources, research is likely to continue to improve the technology because of its perceived value to the military (see Chapter 9 Water, war and terrorism).

Yet rainmaking may not be the most cost-effective means of improving water supply via controlling the water budget. Any budget comprises inputs and outputs. Controlling outputs or 'losses' can be just as important, and there are many methods, some tried and tested for centuries, that aim to control evaporation losses. Many of these are readily applicable to farmers and water managers without access to high tech solutions. In practice, we should be concentrating on controlling the water budget rather than the weather.

> 'I am convinced that by cloud-seeding techniques it will be possible to exercise a considerable control over the weather for the general good of the country.'
>
> Dr Bernard Vonnegut,
> Rainmaking pioneer at General Electric Laboratories, to
> US Senate Committee on Weather Modification, 1949

Making rain

The foundations for scientific rainmaking were laid between the First and the Second World Wars by the Swedish meteorologist Tor Bergeron. The Bergeron–Findeisen process centres on the fact that at subzero temperatures water vapour sublimates preferentially onto ice crystals rather than condensing onto water droplets, and that these ice crystals grow at a faster rate until they are heavy enough to fall (see *Clouds and rain-forming processes* in Chapter 11). It transpires that a very substantial proportion of the world's rainfall depends on this frozen source and that most rain outside the tropics is snowfall that has melted in warmer air nearer the ground. The theory was translated into operational methods of inducing rain immediately after the Second World War by American scientists Bernard Vonnegut, Vincent Schaefer and Irving Langmuir.

Vonnegut chose to kick-start the process by introducing a substance with a similar crystal structure to ice to fool the water vapour, silver iodide (AgI). This has become the leading method, although a variety of other substances, notably lead iodide (PbI) and cupric sulphide (CuS), have been experimented with.

The substances may be introduced into suitable clouds, i.e. clouds that are already part-way to producing rain naturally, by sprinkling from aircraft, by exploding rockets or in air rising from ground-based burners. Ground burners are less used nowadays, but they were quite popular in early experiments. Experiments with aircraft-based seeding by the American Institute of Aerological Research in 1948 in Arizona proved too costly and they noted that the optimum conditions for rainmaking were also the worst conditions for flying – with icing and turbulence. Instead, General Electric favoured ground burners after successful tests around the Roosevelt Dam reservoir. They found the best method was to burn coke covered with AgI, which vaporized and rose in the superheated air currents. Using this method, the Water Resources Development Corporation (WRDC) of Denver claimed to have nearly doubled rainfall in experiments in Nebraska in 1951. Other successful experiments were carried out with AgI vaporized from a large number of heaters near the Oquirrh Mountains in Utah during the 1950s. However, the WRDC did express concern that farmers using large numbers of burners without meteorological knowledge or advice might cause unexpected results. Some environmentalists remain concerned about the pollution, although no actual problems seem to have been identified with silver – lead or copper compounds would be different.

Schaefer and Langmuir tried a different approach. Rather than introducing nuclei, their aim was to cool the clouds below freezing point with dry ice, solid carbon dioxide, which has a temperature of −78°C and volatilizes at normal atmospheric pressures taking the latent heat of volatilization from the surrounding air. Apart from adding a little extra CO_2 to the atmosphere, this method is non-polluting. Unfortunately, it requires more special equipment and handling than AgI.

Many believe the technology has potential for alleviating serious problems, from drought mitigation and hail suppression to flood prevention, hurricane-busting, dousing forest fires, and, less seriously, preventing parades being rained upon. Many US water and hydropower companies regard the 'meteorological option' as a legitimate part of their arsenal for combating drought or catastrophic flooding. With flood prevention, the aim might be to stimulate heavy rain over a less critical area before it reaches a risk sensitive catchment. A similar result may be achieved by 'overseeding', introducing an excessive number of freezing nuclei which then compete for the finite amount of water vapour in the air and produce too many small droplets that do not grow large and heavy enough to fall.

American power companies began experiments in the 1950s. In the autumn of 1951, the hydropower plants of the Bonneville Power Administration were in crisis because of lack of rain and the American Northwest was threatened with 'power brown-outs'. Aluminium smelting plants began planning to move out of the area. The Department of Interior contracted the Water Resources Development Corporation to seed clouds throughout the Columbia River basin and the results seemed to pay off: the day the power cuts were due to begin, the Grand Coulee Dam began spilling water and the risk was averted. Experience in the 1950s led American power companies to conclude that seeding could be used in a dry year to boost hydropower resources sufficiently to avoid having to increase output from less efficient ther-

Figure 17.1 Early rainmaking experiment by the American Institute of Aerological Research using a ground-based burner in 1948

mal power stations. The experience suggested that the technology is best applied to larger basins: as little as two per cent more runoff could cover the costs of seeding in basins over 1000 km², compared with a ten per cent increase needed to offset costs in basins a tenth of the size.

A similar principle applies to agricultural applications: because of the uncertainties surrounding how much will fall when and where, it makes more sense for large ranches in the American West than a small New England farm – where any extra rain could well fall on your neighbour. The Water Resources Development Corporation made some grand claims for the benefits to agriculture (Table 17.1). As so often the case, putting the gains in monetary terms encouraged acceptance of the technology. The claims may now be regarded as somewhat over-enthusiastic.

Table 17.1 Agricultural benefits of rainmaking in Nebraska, according to an enthusiastic report published by the Water Resources Development Corporation in America at the peak of interest in the early 1950s. Estimates from the American Institute of Aerological Research in contemporary dollar values (WRDC, 1951)

Percentage increase in rainfall	Value of extra yields for summer rainmaking (all crops)	Value of extra yields for all year round rainmaking (all crops)
1%	$93,310	$159,100
10%	$933,103	$1,591,088
50%	$4,665,520	$7,955,442

Rainmaking has its military applications as well, and some militarists are hoping for further developments as part of future remote-controlled wars (see Chapter 9 Water, war and terrorism). Recent experiments have been directed at another type of war – fighting climate change (see the section later in this chapter headed *Making clouds*).

Rainmaking might also be used to combat air pollution. The Chinese used it in 2008 to clear the smog that had been plaguing Beijing in the run-up to the Olympic Games. It has emerged that the Soviet Union used the technology to protect Moscow from radioactive pollution in 1986. Immediately after the explosion at the Chernobyl nuclear power station in the Ukraine, the Soviet air force flew sorties over Belarus for two days to try to prevent the toxic cloud reaching Moscow. It would not have been welcome to the population of

present-day Belarus had it been generally known: people in the city of Gomel reported heavy black rain and later suffering from radiation poisoning, but the authorities kept it secret for 20 years. In the event, they seem to have been less than totally successful in raining-out the caesium, as the cloud continued on a northerly airstream to rain heavily on Finland, especially in Lapland, reaching Sweden, where the first alarm was raised, since the USSR had suppressed the news, and eventually infecting Britain with so much caesium that the Ministry of Agriculture instigated a ban on the movement of sheep from some upland farms that lasted well into the 1990s.

In recent years, rainmaking has been used in at least 24 countries, notably the USA, France, Russia, Japan, Australia, Mexico, South Africa, Israel and China. India began experimenting in 1951, but has only used it occasionally. It still has no national policy on rainmaking, but this may be about to change as a result of the havoc caused by the late arrival of the monsoon in 2009. French companies used the technique to alleviate cyclical droughts in former French colonies in the West African Sahel since the 1970s. Israel carried out three cloud-seeding operations in the rainy season in 1961 and succeeded in making flowers bloom around the Dead Sea. Seeding has been most widely used in America, where it has been dubbed 'the meteorological option' for water management to alleviate both droughts and flood hazard. Water companies have hired commercial 'weather modification' companies to make rain in reservoir catchments or to make clouds threatening heavy rain to deposit their load away from sensitive catchments, especially those where floods could affect large populations.

During the proliferation of commercial weather modification companies in the USA in the 1940s and 1950s, up to $5 million were being spent annually on cloud-seeding exercises covering about ten per cent of the USA and many US rainmaking companies claimed increases of several hundred per cent. However, investigations carried out by the US Defense Department under Project Cirrus and the President's Advisory Committee on Weather Control report of 1958 reduced the figure to only 10–15 per cent. Even that amount can be worthwhile in a drought. Nevertheless, interest declined in the 1960s.

It was reborn in the 1970s by fears of global food shortages and the 1976 US National Weather Modification Policy Act empowered the Secretary of Commerce to establish a national policy and to direct research funding. Subsequent research by NOAA in Florida suggested that as much as 20–70 per cent more rainfall can be obtained from thunderclouds. Seeding became commonplace in the southern High Plains where groundwater resources have been in marked decline for decades (see *Mining Groundwater* in Chapter 12).

California has been one of the biggest users of rainmaking contractors. Los Angeles County has signed many contracts with North American Weather Consultants from 1957 to the present day. It is believed that the LA Department of Water and Power frequently seeded clouds in the Eastern Sierra Nevada Mountains, which provide much of LA's water. During the California-wide drought in the summer of 2008, LA initiated a $400,000 seeding contract aimed at achieving a 15 per cent increase in rainfall in the San Gabriel Mountains.

Yet even after well over half a century of trials, the technology is largely unproven and many regard its application as an art more than a science. A number of experiments have shown no increase or even a decrease. Some now claim that even the President's Advisory Committee overestimated the effects. In 2003, a National Academy of Science report concluded that the science is still unproven.

Results very much depend on choosing the right clouds with suitable physical properties (temperatures, temperature structures, vapour content, thermal updraughts, cloud thickness, etc.) and on the stage of development. Flat layer clouds are generally useless. Billowing clouds with strong vertical development are best. Introducing too many nuclei can cause too much competition among them for a finite amount of vapour, resulting in too many crystals so small and light they do not fall. This does, however, offer one approach to rain suppression and yet another route to flood prevention, an alternative to creating rain away from the flood sensitive areas.

The complexity of the processes involved is compounded by the normal lack of sufficient observational data on critical physical parameters within the clouds, like the strength of updraughts, vertical profiles of temperature and humidity, and the concentration of natural condensation nuclei, and ultimately by the difficulty of proving success or failure. No technique works unless the clouds are naturally near to producing rain. Then the question of how much rain would have been produced naturally and how much is man-made presents a big problem. Even the total amount of rainfall in a particular event may be in doubt, as raingauges may not be in the best

place and they may be limited in number; weather radar may help, but these are large, fixed site installations that are rarely to be found in a particular target zones (see *Weather radar* in Chapter 18).

Some sophisticated statistical experimental designs have been developed to test results, such as selecting clouds to be seeded by random numbers or 'target-control regression', which is a test based on the difference in the relationship before and after seeding between rainfall in an area to be targeted by seeding and an unseeded control area. Unfortunately, weather patterns can change from one period to the next and confound the effects. It is also possible to falsify the results by using weather forecasts to select the best occasions for seeding and so bias the results. The best design combines both approaches in the 'target-control crossover' method, whereby target area and control area are randomly interchanged for seeding. This overcomes the problem of shifts in weather patterns and reduces the probability of a biased sample, while still identifying a fixed area that can be instrumented for rainfall measurement (Jones, 1997). In practice, however, most cloud-seeding has been conducted without such rigorous statistical controls, more commonly as a commercial service or an urgent drought-busting exercise.

New developments and warm cloud processes

Although the Bergeron–Findeisen theory continues to be the basis of most cloud-seeding, many are disaffected with the results and the high cost of AgI. The Chinese are currently experimenting with varying combinations of substances, principally using AgI as a freezing nucleus, but combined with dry ice or liquid nitrogen as coolants. In Western Australia, the Northampton pharmacy has been mixing a variety of substances for the last 20 years, including silver nitrate, potassium iodide and silver iodide. Silver nitrate and potassium iodide are mixed in water, allowed to crystallize and then ground into a powder. These are designed to cover the spectrum of cold cloud and warm cloud processes. Farmers have found that cold front clouds offer the best target and when a weather front is approaching they club together to vaporize the substances over a wide area using oxy-torches or else fire them from rockets.

Others have turned to the question of making rain in clouds where temperatures are wholly or largely above freezing. Early 'warm cloud' experiments based on adding large drops to clouds to stimulate growth by collision and coalescence were not very successful. Early experiments run in the Caribbean by the University of Chicago required vast quantities of water: aircraft sprayed nearly 1300 litres/km onto the clouds and still only half the seeded clouds rained.

In contrast, pulverized salt is cheap and effective as a hygroscopic nucleus. Delivery by aircraft in the West Indies and from ground sources in India have both proved effective. Common salt (NaCl) is also a less problematic environmental pollutant than AgI or PbI. The amount of salt used in rainmaking should be much less than occurs naturally in rainfall in coastal districts.

A chance observation in South Africa led to claims that salt injections are the way forward and more reliable and effective than AgI or dry ice. After decades of research on rainmaking and with new evidence from laser monitors showing that the classic approach was ineffective, Dr Graeme Mather was about to give up his life's work, when he noticed heavy downpours in the vicinity of a newly expanded paper mill, and decided to investigate the process. In 1990, Mather and Dr Deon Terblanche of the South African Weather Bureau began a five-year programme of trials, which happened to coincide with a severe drought. Using salts similar to those emitted from the paper mill sprayed from aircraft, they achieved up to 60 per cent higher rainfall, with maximum yields of 140 tonnes of rain just 90 minutes after seeding, compared with only 40 tonnes from the unseeded clouds. The first effects were visible barely 15 minutes after seeding. The experiments were well designed. Because natural variations may be more than 30 times greater than the effects of seeding, 150 seeding experiments were conducted using random selection with a 'double blind': an envelope was opened on the ground to select 'seed or not seed' and another was opened by the pilot to accept or reject the instructions from the ground. Weather Bureau radar and the TITAN software system were used to determine the results.

Mather says that he was lucky to be working in South Africa rather than his native Canada, because he was free from the grip of entrenched scientific opinion, and the tests certainly appear to have convinced an initially doubting international scientific community. NCAR joined the experiments in Mexico in the 1990s spraying a mixture consisting mainly of potassium chloride (KCl) with some NaCl. As vapour condenses on the hygroscopic nuclei, the release of latent heat causes increased updraught, allowing the clouds to grow and increasing collision and coalescence. The severe Mexican drought of 1993–98 proved an ideal time to test the technique. Ranchers were selling cattle and facing ruin. Mexican agricultural authorities collaborated with NCAR in a successful four-year programme seeding convective clouds, particularly *cumulus congestus*, which have naturally strong vertical updraughts of 1–2 m/s and may only drop below freezing at the very crest, if at all. Additional confirmation came from hail-suppression experiments in France (see the section headed *Suppressing hail*, later in this chapter).

In May 2009, as a direct result of the late arrival of the monsoon, the Indian Institute of Tropical Meteorology began a three-year cloud-seeding experiment that covers both cold cloud and warm cloud processes by using KCl and NaCl together with AgI and dry ice. 2009 was devoted to data collection. Random seeding of monsoon clouds began in 2010–11, and computer modelling will proceed in 2011–12. The hope is to produce practical guidelines for seeding monsoon clouds.

The new technology that has developed over the last 20 years, based on hygroscopic nuclei, offers a number of significant advantages. It is more applicable to tropical regions with more warm clouds, hence the reported interest from rice farmers in SE Asia. It uses cheaper materials and is more reliable. And it gives quick results, which is an enormous step forward. Proponents claim that it could be a powerful technique for controlling wildfires. A quick response method should certainly remove some of the problems of making rain fall on specific target areas, which makes it more valuable to water companies and flood control agencies, and could avoid some of the neighbourly litigation problems outlined below.

Finally on the subject of new developments – a mystery. A secret experiment has been underway in Alaska that looks like technology out of the US government's Star Wars II project. At first sight, the High Frequency Active Auroral Research Program (HAARP), with 180 antennae shooting electromagnetic beams to heat up the ionosphere, would seem to have nothing to do with rainmaking. The project began in 1993 ostensibly to improve radio communications. Yet journalists have discovered the following enigmatic statement in patents filed for HAARP: 'weather modification is possible by, for example, altering upper atmosphere wind patterns by constructing one or more plumes of atmospheric particles, which will act as a lens or focusing device.'

Don't rain on my parade

Both Russia and China have embraced the technology to guarantee dry weather during national celebrations. The Russians use it each year to prevent rain disrupting the annual May Day parade in Red Square, by inducing rain to fall away from the centre of Moscow. They dispatch jet aircraft with canisters of seeding material mounted beneath the wings, as during the 1995 VE Day celebrations and the 2006 G8 summit, when President Putin reportedly ordered rain to fall on Finland not Moscow. Some Russian operations around Moscow use cargo aircraft equipped with a variety of seed material, including liquid nitrogen as a coolant and cement as well as AgI. The bags of cement are intended to disperse in the air and add nuclei rather than just create turbulence: clay minerals are natural condensation nuclei and lime also attracts moisture. But things went badly wrong in 2008 when a 25 kg bag failed to open and crashed through the roof of a house.

By far the biggest exercise was undertaken by China to ensure a rainless and smog-free opening to the Olympic Games in August 2008. August is normally a rainy time of year. Far more people (some 40,000) were enlisted into the operation to make it rain away from Beijing, than were involved in forecasting the weather at three-hourly intervals using the new IBM supercomputer. China used three aircraft, deployed 21 sets of four 37 mm anti-aircraft guns at locations, equipped farmers with mortars, and fired 1100 rockets charged with AgI and other chemicals to trigger rain showers in the Fragrant Hills and elsewhere west of the city a day before the opening ceremony. In combination with draconian measures banning traffic and closing factories, the rain appeared to wash the air and help disperse the city smog that had been worrying contestants in the weeks before. Whether this was a true demonstration of the effectiveness of the technology or not, immediately after the supreme effort for the opening, the rain returned for days, drenching outdoor competitors.

The opening of the Games is widely believed to be the first time that China has used seeding for a major event. It used seeding again to ensure a fine day over Tiananmen Square for the 60th anniversary of the Communist state in October 2009. In fact, China has a long history of rainmaking beginning in 1958, especially to combat drought in the North China Plains, which is now one of the biggest threats to national economic growth. China currently has one of the largest, regular programmes in the world, ahead of Russia and Israel. China spent $266 million on it between 1995 and 2003, which it claims created 210 billion m³ of extra rain, equivalent to the annual needs of 400,000 people. Spending doubled over the next five-year period to $500 million in 2003–2008. It employs more than 50,000 people with over 4000 rocket launchers and over 7000 artillery guns. The current five-year plan calls for an extra 50 billion m³ a year of rain to be created.

Making snow

Most rainmaking over the last 60 years has actually been snowmaking. The difference in the final product is controlled by the temperature of the air between the cloud and the ground. Cloud-seeding aimed at producing actual snowfall has been used in the USA to bolster winter snowpacks in order to enhance water supplies during the spring melt. Early in 2009, a long drought in North China was ended by three days of snowfall following AgI seeding with rockets.

Operational use began in America in the early 1970s after the second Project Skywater demonstrated its effectiveness in Nevada. At that time, Howell (1972) estimated that the current technology could increase snowfall by 15 per cent and that this might improve to 25 per cent by 2000. In addition, Howell suggested that overseeding could also be used to reduce hazards caused by excess snowfall, and he estimated that by 2000 the technology would be capable of reducing snowfall in individual storms by up to 30 per cent. He predicted it would be more effective in the eastern USA, especially in New England, the very region where it would have most value to the economy, than in the Rocky Mountains. But Howell's predictions remain unfulfilled, there is still no concerted effort to use the meteorological option, and New England remains prone to heavy winter snowfalls, as in early 2009 at the tail end of the La Niña effect (see *El Niño/Southern Oscillation (ENSO) and similar Pacific oscillations* in Chapter 19). However, the lack of effort is due to many factors, including doubts about the effectiveness and controllability of the technology.

The mayor of Moscow took up the idea in 2009 when he called for cloud-seeding to be used to keep the city snow-free for the whole winter. He claimed that such a snow management programme would only cost a third of the current expenditure on 2500 snowploughs and 24-hour clearance operations by 50,000 workers, and that farmers outside the city would benefit from increased yields. Newspapers reported that people outside the city were more concerned with possible flooding in spring. In fact, Moscow currently deposits the snow cleared from its streets in dumps of up to 100,000 m³. These are often in river channels but the snow melts more slowly because of the bulk, and causes less flood hazard than it would if it melted on the streets and drained into the stormwater sewers.

A classic case of the use of snowmaking to enhance agricultural water resources, which was deemed a waste of effort, is described later in this chapter, in the section headed *Disputes and litigation*. Nevertheless, in that instance failure was more the result of an inept hydrological forecasting model than the cloud-seeding.

Snowmaking for leisure

Not all snowmaking begins in the clouds. The combination of climatic variability, climatic change and the increasing popularity of skiing, snowboarding and children's 'snowparks' has led to greater pressure on winter sports resorts to guarantee snowcover. Increased traffic alone erodes the snowcover. Artificial ice is also made for skiing, as in the ice cascades in the Chisone Valley at Pragelato in Piedmont.

In 1950, three Americans – Art Hunt, Dave Richey and Wayne Pierce – invented the snow cannon. These machines splay fine 'nebulized' water droplets into the cold air, which freeze and fall. Biodegradable, non-toxic nucleating agents, both organic and inorganic, are often added to speed the process. The pumping equipment is expensive to buy and run and the process produces a fine powder of ice crystals rather than the ideal snowflakes, but it has proved its economic worth for commercial ski companies and the machines are now widely used from the Alps to the Rockies. Snow machines were deployed in the Sauze d'Oulx, Sestriere and Bardonecchia basins to support the 2006 Winter Olympics in the Italian province of Piedmont. Snowmaking now keeps many downhill and cross-country ski tracks in the Italian Alps permanently operational to Italian Federation of Winter Sports standards throughout the season. Over 250 km of Piedmontane pistes are currently maintained with artificial snow, 18 per cent of the total length of piste. Forty per cent of the resorts now have snowmaking installations and the system is continuing to expand to meet increasing leisure demand and following a number of years with low snowfall.

Huge amounts of water and electricity are needed. A typical machine can require over 830,000 litres to cover one hectare to a depth of 12 cm and needs a large source such as a river or reservoir. At Sestriere in the Italian Susa Valley, an array of 950 cannons is needed to support 80 km of pistes, even at an altitude of 2000 m. Machines are now mainly computer-controlled and programmed to monitor weather conditions in order to reduce unnecessary waste, and they generally operate at night to reduce evaporative losses and use cheaper electricity.

In the USA, smaller domestic versions are widely available that operate from the fully-treated public water supply and are capable of producing up to 15 cm of snow an hour.

Suppressing hail

Hail can be a major problem for farmers, especially since it tends to occur in summer and autumn towards harvest time, when the land is warm and thermals provide the ideal vertical currents for towering *cumulus* clouds to develop. In medieval times, church bells might be rung or rockets fired to try to stimulate precipitation before the hailstones grew to a dangerous size. In the last 50 years or so, more scientific rainmaking methods have been used from France to the USA and Argentina. Most commonly, the aim is not to prevent hail, but to make hail that is less damaging. However, to some pastoral farmers, like ranchers in the western USA, hail is as good as rain to water the grass, and they may hire rainmaking companies to make it.

Hail can cause enormous damage in France, especially in the valuable vineyards. Half of the grapes in the Chablis Grand Cru area were destroyed by hail in May 1998. In June 2007, 60 per cent of the harvest was destroyed in the Côte Rôtie on the Rhône, and the extreme damage in the Bordeaux region in May 2009 resulted in up to 75–100 per cent losses in Graves. The same month, thunderstorms, heavy rain and hail destroyed a sixth of the hop crop in Hallertau, Germany, the world's largest hop-growing region.

A chance meeting between Graeme Mather and Jean-Francois Berthoumieu of the Climate Association of the Garonne led to a series of experiments in France during the 1990s that proved highly successful. The idea was to use Mather's hygroscopic nuclei approach to cause more condensation in the lower region of the thunder clouds and so starve the higher, colder regions of moisture. Mather reported rain within ten minutes of seeding accompanied by a small amount of hail, resulting in no hail damage and also no need to irrigate. However, there is little evidence that growers are using the technology and there is still the problem for weather forecasters of predicting exactly where hailstorms will occur.

A similar approach was taken in Palm Beach, Florida, in 2001 using an artificial polymer-based hygroscopic powder called Dyn-O-Gel to disperse a thunderstorm. Dyn-O-Gel is extremely hygroscopic, capable of absorbing 1500 times its own weight in moisture. In July that year, a B-57 bomber dropped $40,000 worth of the substance onto the storm and seemed to defuse it successfully – but at what cost. And once again, there is the serious question of polluting the environment to be answered.

Busting Hurricanes

The idea of using cloud-seeding to reduce hurricane damage centres, not on suppressing the heavy rain, but on encouraging release of the latent heat of condensation by stimulating condensation and making rain away from the core region. The intense winds and the towering convective clouds that surround the eye of the storm and deliver most of the rainfall are fuelled by the natural release of latent heat there. The theory is that the amount of heat released in this wall of cloud can be reduced by causing rain to fall away from the core region in the spiralling 'lanes' of clouds that carry the moisture towards the wall.

Hurricanes typically combine intense rainfall with high winds, raised sea levels and coastal 'storm surge' floods. Hurricane-generated floods are a major hazard in many areas on the edge of the tropical world, often accompanied and exacerbated by landslides and mudflows. Hurricanes are currently responsible for $10 billion worth of damage every year in the USA. Evidence seems to be mounting that hurricanes and tropical cyclones are likely to increase in frequency and intensity with global warming and that the trend may already have started (see Chapter 10 The threat of global warming).

In the 1960s and 1970s, experiments in long-distance airborne seeding of hurricanes carried out from Florida by the US National Weather Service, seemed to successfully demonstrate reductions in windspeeds and hurricane intensity a few hours after seeding in a number of storms. For many years afterwards, the US Air Force maintained planes on alert in Florida during the summer hurricane season. The practice was subsequently dropped on grounds of expense and lack of sufficient effectiveness, and at a time when satellite surveillance and improved weather forecasting also made early warning and evacuation procedures more effectual.

Hurricane Katrina's devastation of New Orleans in 2005 changed the government's view: the tens of billions of dollars spent on reconstruction could surely have been better spent on prevention. The research division of the US Department of Homeland Security has been looking again at hurricane busting and investing in trials of a number of new techniques. One project from a team at the Massachusetts Institute of Technology has proposed painting cloud tops black with soot to kill convection by absorbing sunlight at different sides of the storm in order to nudge its track in the desired direction. It began with computer simulations and aims to progress to real-world trials. The American company Space Island Group is planning to launch satellites with mirrors that will aim hot beams at hurricanes to steer their trajectory. The Nasa-funded project hopes to launch its first satellite by 2012.

In 2009, Microsoft's founder Bill Gates, also stepped in joining and financing a team of inventors that applied for patents on a new geoengineering approach. This focuses on the ocean rather than the clouds. Since the critical source of a hurricane's energy is evaporation from oceans where the temperature is above 27°C, the aim of the new approach is to cool the ocean surface to below the critical threshold by pumping the warm surface water deeper down and pumping up cold water from depth. This would be done by giant turbine pumps mounted in two vertical tubes, one for warm water, one for cold, floating on a fleet of barges deployed around the southern coast of the USA and ready to be moved into the hurricane's path. The approach has been given a cautious welcome by some scientists, but there seem to be two crucial issues: can the technology respond quickly enough to reach the critical area of ocean, and what if it merely succeeds in diverting the storm elsewhere (say to Cuba?) – the patents actually refer to 'hurricane suppression' and 'hurricane deflection'. Since hurricanes can shift trajectory naturally and spontaneously – as Katrina did, saving New Orleans from the worst possible damage – chasing them with barges could be rather difficult.

In itself, the idea of suppressing ocean evaporation is not new. Suggestions have been made in the past, prior to modern environmental fears, of spreading oil slicks in the Gulf of Mexico to act like the monomolecular layers on reservoirs. Another suggestion was to use fleets of aircraft to produce interlacing condensation trails (contrails) to create *cirrus* clouds at high altitude that shield the sea surface from the sun. Subsequent studies stimulated by modelling the processes of global warming have since proved that these thin ice clouds have a greater effect in retaining heat emitted from the Earth than cutting out the sunshine; and naturally the aviation fuel expended will add to carbon emissions. Any advantages would be local and at the expense of the wider picture, but creating low level clouds might be more effective (see the next section). In contrast to these 'solutions', the new ocean-pumping approach could have significant environmental advantages.

In his book, *Cool it*, the global warming sceptic, Bjorn Lomborg, suggests that hurricane damage is likely to increase 500 per cent over the next 50 years, not so much because of global warming, but as a result of the attraction of population and urban development to coastal regions. He hazards a guess that if people could be prevented or discouraged from moving into harm's way the increase in damage could be reduced to a mere ten per cent. Better building codes and evacuation plans would also help. Comparison of the impact of hurricanes in the Dominican Republic with that in adjoining Haiti is very instructive: the Dominican Republic has better preparedness and infrastructure than its impoverished neighbour Haiti and its hurricane death rate is one thousandth that of Haiti. But Haiti's problem is more deep-rooted than just poverty. Over the last 50 years, the country's environment has been critically impoverished by forest clearance for firewood and subsistence agriculture, so that the hillsides are bare and unstable and prone to landslides. The heavy rains wash away the topsoil and further impoverish the agriculture.

Making clouds

Two natural or semi-natural phenomena have been the inspiration for engineering solutions to offset global warming. One is the observation from satellite imagery of condensation trails forming clouds across the sea following the routes taken by ships, like the more familiar 'contrails' of aircraft. The condensation forms around sulphates emitted by ships' engines. The second is the natural production of dimethyl sulphide (DMS) by marine plankton, which also stimulates condensation and cloud formation.

Along the lines of the first of these, John Latham of NCAR has seen a potential use for making thin low altitude layer clouds, like *stratocumulus*, thicker and whiter (Latham, 2002). These are not new clouds, merely enhanced clouds, and they will not produce rain, but they might prove a weapon against global warming by reflecting more solar energy back into space, technically, raising the 'albedo' of the Earth. In collaboration with engineering and computing colleagues at Edinburgh University, Latham published a practical guide to how this may be achieved in a special themed issue of the prestigious *Philosophical Transactions of the Royal Society* on 'Geoscale engineering to avert dangerous climate change' (Salter *et al*, 2008).

The aim is to use sea salt rather than sulphate to expand low-level clouds at sea and cool sea surface temperatures (and not annoy land-dwellers). This would be achieved by specially designed 'cloud ships' that suck in seawater and spray micrometre-sized droplets of saltwater into the air through tall funnels. The ships' propulsion and the lifting mechanism would be powered by special wind turbines called Flettner rotors, and electricity generated by passive propellers dragged along under the ship. The droplets of spray would evaporate, leaving hygroscopic salt particles that would be wafted higher up and encourage condensation. The large concentration of salt particles would compete against each other for the fixed amount of water vapour near the cloud base and so create numerous very small cloud droplets, rather than a few larger ones. The smaller the droplet, the more effective they are at reflecting sunlight, thus making the clouds whiter from above. Sites need to be free from higher level clouds that would trap the reflected radiation and reduce the effectiveness. One question not addressed is how fine droplets added to the bottom of the clouds will have much effect on reflection from the top. Increasing cloud thickness alone will help, but vertical turnover is limited in *stratocumulus*, updraughts are typically less than 0.1 m/s, and the clouds may be 300 metres thick, so it could take eight hours for droplets to have much effect on cloud top albedo. Given the minimum windspeed of 8 m/s needed to power the boats, this very crude calculation suggests that optimum results would only be achieved 24 km downwind.

The authors calculated that an increase of just 1.1 per cent in the global albedo would be sufficient to offset the effects of doubling CO_2 concentrations, and thus surprisingly small amounts of spray would be needed to stabilize sea temperatures. They suggest a fleet of 50 remote-controlled ships each costing up to £2 million could negate one year's increment in greenhouse gases. These could be strategically placed to cool seas where hurricanes develop or to extend the area supporting phytoplankton growth. However, the original proposal did not appear to have considered an important downside: that cooling the sea and reducing evaporation could reduce rainfall on the land downwind. This could be a significant detraction and lead to legal challenges similar to those discussed under the section headed *Disputes and litigation*, below. According to their original analysis, the best all-season sites are off the coasts of California, Peru and Namibia, none of which are likely to appreciate reduced rainfall. Salter has since suggested that the ships could be deployed in the Pacific Ocean well away from land.

Other researchers have proposed increasing atmospheric albedo by injecting sulphur aerosols into the upper stratosphere where there is no water vapour, so clouds are not formed, but the aerosols themselves increase the albedo (Caldeira and Wood, 2008; Rasch, 2008). The Royal Society (2009) evaluated these projects and concluded that many deserve to be explored.

When things go wrong

Cloud-seeding disasters

Tom Dunne and the veteran USGS hydrologist Luna B. Leopold began their groundbreaking and generally optimistic book *Water in Environmental Planning* (1978) with the case of the tragic flashflood that swept away much of Rapid City, South Dakota, in 1972, killing 237 people, which followed cloud-seeding in the area. Lawsuits were filed, but it was impossible to prove cloud-seeding was responsible. In 1978, another devastating flood struck in the Big Tunjunga Canyon in the San Gabriel Mountains, California, killing 11 people and causing $43 million of damage. Many residents sued Los Angeles County, claiming that the flood had been aggravated by seeding. Los Angeles won the court case, but stopped seeding till 1991. It stopped again in 2002 when it feared that wildfires had left the land vulnerable to mudslides.

A similar case hung over the disastrous flood in south west England in August 1952 that washed away most of the Devon village of Lynmouth with 90 million m^3 of water, killing 35 people. The flood was estimated to lie above the UK Institution of Civil Engineers' envelope curve relating peak discharge to drainage area, to which all flood protection works were to be designed. It led to a major revision of the curve (ICE, 1960). But fears circulated in secret that the event might not have been entirely natural, as the Ministry of Defence and the Met Office were thought to have been conducting cloud-seeding experiments in the vicinity of the southwest peninsula at the time. After years of official denial, information has been released in the last ten years confirming that experiments were conducted around the time. Witnesses who experienced the Lynmouth flood spoke of a smell of sulphur and rain so hard it hurt their faces. However, a similar flashflood in 2004 at Boscastle in Cornwall seems to confirm that these extreme events can be natural features of the southwest, where intense summertime convective storms, fuelled by evaporation and heat from the warm waters of the North Atlantic, meet the first hills and deposit intense rainfall (reaching 90 mm/hr at Boscastle) directly into very narrow and steep river basins. There was a smell at Boscastle as well, but it was traceable to sewage overflows.

It is possible that the unusual snowfall that hit Beijing and the North China Plain in November 2009 was a cloud-seeding project that went wrong. The Chinese had been seeding to increase water resources in the region, which is suffering long-term water stress. A cold front with a strong updraught increased the effect and produced the heaviest snow that early in the winter since 1987. It caused buildings to collapse and killed 32 people.

A spectacular failure of attempts to reduce hail damage that resulted in a 200 per cent increase is reported from Argentina. The Argentinian technicians were using ground-based burners to introduce AgI into the centre of the updraught zone beneath the convective rainclouds. The aim was to reduce the size of hailstones by premature overseeding, i.e. catching the hail before it had time to circulate and grow by accumulating water within the clouds, by introducing so many competing condensation nuclei that there was not enough water vapour available to allow the individual hailstones to grow larger. The approach was successful in frontal storms, reducing hail by an estimated 70 per cent. Unfortunately, the vertical updraughts are much stronger in convective clouds – up to 30 m/s in *cumulonimbus* compared with 1–2 m/s or less in most frontal clouds – and cloud-top temperatures can be much lower: −50°C against −25°C. These are the prime producers of hail, and the burners merely fuelled the process.

Events in Washington State in the winter of 1976–7 represent yet another form of failure, this time with snowmaking and because of the reasons for initiating the programme rather than with the physics of the process (see next section).

Disputes and litigation

Three inadvertent side-effects of rainmaking have led to a welter of disputes over the years, especially in the USA where environmental legislation is highly developed. The first problem is the uncertainty of where the rain may fall. The second is the lack of control over the type and intensity of precipitation generated. After his experiments in the 1940s, Langmuir claimed that rainmaking could increase rain over most of the US east of New Mexico by three to ten times. When seeding was big business in the USA, Canada and Australia in the 1960s, some dryland farmers threatened to shoot down aeroplanes seeding to reduce hail, because they frequently suppressed rain as well.

The third problem is the fact that rainmaking can rarely be said to actually *make* rain. It is more an exercise in its redistribution. Consequently, in 2004 five cities in Henan Province, China, accused each other of 'cloud theft'. A study of rainfall in the Netherlands showed increased rainfall generated by urban heat and pollution around urban areas, with distinct 'rain shadows' downwind (Yperlaan, 1977), and similar effects are likely to occur in the short-term downwind of artificial rainmaking activities.

The USA has some of the best environmental protection measures in the world. Citizens' rights to the natural water on their land extend to the water in the atmosphere above. Not surprisingly, private litigation and public legislation to control rainmaking activities have become widespread in America. The 1972 Colorado Weather Modification Act remains one of the best of its type. Under the Act, all weather modifiers are licensed annually by the State. If they prove incompetent or troublesome, their licence can be revoked. Roberts and Lansford (1979) describe the case of a public hearing called under the provisions of the Act in San Luis valley, southern Colorado, in the 1970s to settle a dispute between farmers and ranchers over cloud seeding.

Farmers growing top quality barley to sell to a brewery had hired Atmospherics Inc. of California to maintain the optimum 190 mm of rainfall for the main growing season, suppress damaging hail and provide a dry period for ripening and harvest. The brewing company made it a condition for purchasing the barley that it had been grown under weather-controlled conditions. However, the neighbouring ranchers and fodder barley growers were not so fussy about the amount, form or timing of precipitation and they objected to activities that could reduce their own income. The case was finally decided against the beer-barley farmers by expert evidence, which suggested that overseeding, whether deliberate or accidental, could reduce precipitation over thousands of square kilometres in south west USA.

Another classic case of litigation followed a widespread programme of snowmaking in the western USA during the latter part of winter 1976–7. Seeding was commissioned in a number of States, especially Colorado and Washington, after water agencies predicted a 50 per cent shortfall in winter snowfall from the Coast Ranges to the Rockies. Washington State spent $400,000 on seeding in the Cascades to prevent a springtime drought caused by reduced snowmelt, despite threats of legal action by Idaho and Montana to prevent it 'stealing' their water. In the event, the measures were probably largely unnecessary. Certainly this appears to have been so in the Yakima valley, where farmers using riverwater for irrigation were advised by the Bureau of Reclamation to take precautionary measures (Glanz, 1982). Many sunk new wells or even transplanted sensitive crops, like mint, to adjacent valleys. Statisticians and scientists were ambivalent about the claims that snowmaking had been successful. But there was no shortage of water in the spring. This was largely because the drought prediction was based on a faulty water management model which failed to account for the amount of return water reaching the river from the irrigation schemes. The farmers subsequently filed complaints against the Bureau for the costs incurred needlessly.

In spite of the problems, cloud-seeding remains a feasible alternative to interbasin transfer or to changing agricultural systems. Interest was reborn in the 1970s by fears of global food shortages, out of which came the US National Weather Modification Policy Act of 1976, which empowered the Secretary of Commerce to establish a national policy and to direct research funding. Subsequent research by NOAA in Florida suggested that as much as 20–70 per cent more rainfall can be obtained from thunderclouds, and seeding has become common-place in parts of the High Plains where groundwater resources are in marked decline.

Making or dispersing fog

Fog is usually considered a hazard. A variety of fog-dispersal methods were developed to aid aviation during the Second World War. But is there now an argument for *harvesting* fog as a water resource? Professor Jana Olivier discusses this in the box 'Fog harvesting in South Africa' in Chapter 14. Could this approach also be applied to artificial fog? Climatologist Reid Bryson observed that the air over the Rajputana desert in India is as humid as that in tropical rainforests, but the natural precipitation-forming processes are lacking (Bryson and Murray, 1977). Could we assist the natural processes?

To date, most experiments with fog have been aimed at dispersing it, and none have focused on making it. The macrophysical techniques of fog dispersal used in the Second World War were somewhat more success-ful than parallel techniques used for rainmaking, but they were still energy intensive: the 'Fido' system of propane burners along airstrips to raise air tempera-ture and evaporate the fog, or hovering helicopters to mix ground-hugging radiation fog with drier air above. Microphysical approaches seem to have been less used: sprinkling the fog with water or hygroscopic salt parti-cles to make the fog rain out. Some commercial airlines, notably the former Northwestern Airlines in America, have claimed that fog-clearance has been profitable, by allowing more flights. But the methods remain relatively expensive options and no method has yet been found to deal with ice fog at temperatures below −35°C.

No one seems to have thought it worthwhile to *make* fog yet – though a military use might suggest itself. If fog were to be made for water supply, then the techniques outlined for cloud-making could be applied. Like Profes-sor Olivier's fog-collection system, the amount of water gained would be relatively small, but it could be crucial in a desert.

Controlling evaporation losses

Reducing the loss of water by evaporation is a major focus in modern irrigation methods. Evaporation can also be a major drain on resources in reservoirs, espe-cially those in hot, dry climates where large reservoirs are most needed. Less obvious, but potentially just as significant, are losses through transpiration. While tran-spiration from crops is essential for crop growth and generally of economic value, transpiration from forests and wildland vegetation may be regarded as a loss, at least from the point of view of local water resources. Hence there have been moves, especially in the USA, to control this transpiration, either directly by applying antitranspirants or indirectly by changing the vegeta-tion.

There have been two principal approaches to controlling losses from reservoirs: reducing the temperature of the water or spreading a physical barrier across the surface to contain the water vapour. The earliest methods devel-oped in the 1950s used alcohols like hexadecanol and octadecanol, which spread across the water surface in thin 'monomolecular layers'. These long-chain molecules act like a net to hold down the hotter molecules in the water and so hinder evaporation. They were particularly popular in America and Australia, where losses from some reservoirs were excessively high. Unfortunately, such layers are easily broken by waves in winds over 2 m/s. In low wind situations, Australian tests reported savings of up to 50 per cent in lakes of less than 11 km^2 in area, but in the windy environment of Lake Heffner (11 km^2) in Oklahoma, savings of only 9–14 per cent were achieved. By holding the hotter molecules in, the monomolecular layers also raise the water temperature and increase the chances of water vapour eventually breaking through the barrier. This was a major reason why the US Bureau of Reclamation originally overesti-mated the effectiveness of the method by nearly 14 per cent. Even so, calculations suggest that the approach is relatively cost-effective and costs per litre of water saved are comparable to many methods of alternatively increasing water inputs.

The other method of saving water is to increase the reflectivity of the surface to keep the water cool. This may be achieved by floating white wax, rubber or styro-foam blocks on the water. The method became more popular in the 1960s, partly because it is non-polluting and partly because it is less sensitive to wind: the aim is not to create a continuous cover but merely to reduce the average amount of solar radiation entering the lake by reflecting more, i.e. raising the albedo.

The most effective way of all to reduce evaporation losses is to bury the reservoir. This is regularly done in cities, where covers, often turfed, are used primarily as a safeguard against pollution, but on a larger scale water may be pumped into natural rock aquifers to create or enhance a groundwater reservoir. This is currently being planned in the Algarve (see the box 'Aquifer Storage and Recovery' in Chapter 14). The technique offers an extra advantage in that continuous rock strata may be used as a means of transporting water from one area to another, free of charge. Israel has used this to transfer water from the north of the country to the Negev. It has been operated successfully in England where water from East Anglian rivers is pumped into the chalk aquifer (see Groundwater reservoirs in Chapter 14). It works well so long as the aquifer has no significant leaks and it is especially attractive in hot climates, where surface reservoirs are very inefficient.

Methods of controlling evapotranspirational losses on the land follow broadly the same lines: providing physical barriers, controlling the heating or removing surface sources. Irrigation demand can be reduced by conserving moisture with water-retentive materials like compost (see *The answer lies in the compost* in Chapter 15), by insulating the soil with a layer of mulch, such as leaves, wood chips or gravel that have a low thermal conductivity, or raising the albedo with white paper or lime. Dry farming is another traditional technique, especially in drier climates. This uses the soil itself as an insulator by lightly tilling the surface to increase the total porosity, so that the air-filled pores reduce the thermal conductivity. Even monomolecular films can be sprayed onto the soil, but there may be a pollution issue.

Monomolecular films have been sprayed onto vegetation as an antitranspirant. But there is a danger of causing overheating and their effectiveness can be reduced if sprayed by aircraft from above, as stomata tend to be mainly on the underside of the leaves. Spraying with white substances removes this problem, but there remains the big question of whether the environmental pollution is acceptable. This is also the objection to chemical antitranspirants designed to encourage the stomata to close. In commercial aspen forests in Utah, phenyl mercuric acetate has been shown to be an effective agent, delaying the pattern of water use by up to six weeks. However, application at the wrong stage of growth or in the wrong amount could cause irreversible damage to the plants' metabolism. Wider concern could be expressed at the prospect of highly toxic mercury getting into the wildlife food chain.

Removing vegetation with high water demand and replacing it with other species, especially dryland 'xero-phytic' plants, may have a major aesthetic effect on the landscape, but it generally avoids chemical pollution and it can yield substantially more surface water runoff and groundwater recharge. The US Forest Service estimated that manipulating vegetation in the upper Colorado basin could save more water than could be generated by cloud-seeding. The USFS estimated that 1850 million m^3 could be saved this way, whereas the US National Weather Service estimated that 1233.5 million m^3 could be created by cloud-seeding (Stetson, 1980). If these two approaches had been combined in the 1980s, they could have halved the calculated shortfall in supplies to Arizona and California. There is a wealth of experimental data on the hydrological effects of vegetation change, especially with afforestation and deforestation, but there is also a large degree of inaccuracy in predicting the effects in a different location (see Chapter 6 Water, land and wildlife). Current hydrological methods of prediction lack transportability because they are based on statistical extrapolation from individual catchment experiments, not on botanical processes. It is to be hoped that future botanical research will help to improve the technology.

Conclusions

The long-held wish to control the weather is becoming a reality in limited ways. Rainmaking is the best and most commonly used example. It is an established 'meteorological option' for water managers in America and China. The results are often disputed and can occasionally go badly wrong. But cloud-seeding can be used to increase rainfall provided the target area is large enough. It is also regularly used to keep city centres dry during celebrations in Russia and China. Snowmaking is essentially the same technology and is similarly used to increase local water resources. In addition, cloud-seeding may be used as a flood-diverting tactic or to suppress or divert hurricanes.

After some decades in the doldrums, the approach is being actively pursued once more, and new techniques are being tested that seem to promise greater success.

The alternative approach to increasing available resources – reducing *losses* through evaporation and transpiration – also began some 60 years ago. Efforts to reduce evaporation from reservoirs have fared less well than improved agricultural methods.

351

Discussion topics

- Review the effectiveness of recent rainmaking methods.

- How environmentally sound are methods for reducing transpiration?

- Could snowmaking be used as a way of reducing some of the negative effects of global warming on glaciers, mountain water resources and the flow regimes of mountain rivers?

- Explore the prospects of new geoengineering proposals for making clouds to combat global warming.

Further reading

The special themed issue of the *Philosophical Transactions of the Royal Society* on 'Geoscale engineering to avert dangerous climate change' (2008) is a fascinating read.

Sir John Mason's classic is also fascinating:

Mason, B.J. 1962. *Clouds, Rain and Rainmaking*. Cambridge, Cambridge University Press.

18 Improved monitoring and data management

Monitoring and communication lie at the very heart of modern water management, from the assessment of available resources to early warning of disasters. There has been some dismay at the loss of many ground-based monitoring systems over the last 40 years. The number of raingauges has declined on every continent. The number of stations measuring river discharge increased dramatically in North America and Europe during the 1980s, but has shown a slight decline since. Numbers have remained almost static elsewhere, but the total number of river gauging stations in all the rest of the world is only around half the total in Europe and North America. The contrast in the deployment of water quality stations is even starker: the rest of the world has only around a tenth of the number in Europe and North America. The contrast in water quality measurements swelled in the 1980s as Europe and North America introduced environmental legislation, but has levelled off since. Evaporation and evapotranspiration are hardly measured at all, and normally only calculated from data collected at meteorological stations or agricultural research stations (Jones, 1997).

There were many reasons for the decline. The worst reasons have been lack of funds and lack of prioritizing. Many newly independent countries were either short of funds or diverted funds to other uses perceived as more important. The same has been true even in the UK, especially in the 1990s after privatization. Some monitoring networks have been rationalized on a sensible, scientific basis, but these are a very small minority. There are two principal bases for scientific rationalization: statistical weeding out of superfluous stations or replacement by a more advanced technology, such as remote sensing.

Most ground-based networks inherited today have never been designed scientifically. They were located for convenience and generally near major centres of population. This meant that more populated areas have tended to be better sampled than the remote areas, especially the uplands where ironically most riverflow originates. In the UK, only around 1 in 20 raingauges are in the uplands, where most rain falls. Much of the poor spatial coverage of ground stations comes from inheriting the pattern of meteorological stations. When water resources monitoring developed its own priorities the 'pendulum' sometimes swung too far the other way: in order to cover for a lack of understanding of the spatial variability of precipitation some catchment areas were originally over-supplied with raingauges. Statistical rationalization is often based on correlation between gauges: if more than 90 per cent of gauge B's record can be predicted from gauge A, then gauge B is deemed redundant.

Improvements in the instruments used in ground-based observations are continuing, particularly in monitoring snow and riverflow: snow pillows, better wind shielding for snowgauges or electromagnetic monitoring for riverflow. Ironically, however, the best raingauge ever designed, the ground-level gauge produced by the UK Centre for Ecology and Hydrology, which minimizes the effects of wind turbulence as well as minimizing splash-in from the ground, is hardly deployed anywhere outside the CEH itself. This is sad because it means that most rainfall is under-measured, and more so in storms. It is a conservative error as regards water resources – combined with the lack of raingauges in the uplands this means resources could be significantly under-estimated. But it could be a liability for flood prediction.

There is one powerful argument for retaining the old instruments: continuity of record. A 'statistically homogeneous' record is important as the basis for statistical assessment of risks and averages. This is the argument maintained by most meteorological services. But it could also be a biased record, and the degree of bias is not constant; wind speeds and the size and intensity of raindrops affect it. In the Unesco (1978) atlas of the world water balance, the compilers concluded that precipitation was under-measured by 60 per cent in parts of Central Asia, because of the design of the Soviet precipitation gauge, set at 2 m above the ground in order to stay above the snowcover.

An ideal solution would be to run old and new side by side. Indeed, the newest 'gaugeless' technologies, remote sensing by weather radar and satellite, still require some ground gauges to calibrate them.

Automation is clearly the way forward. Automatic stations can maintain 24-hour coverage and operate for weeks or months without human assistance, either recording data on solid state loggers or transmitting it by radio telemetry or telephone landline. 'Intelligent' data loggers are programmed to carry out calculations before transmission or storage, such as summarizing data to reduce storage or performing quality control. Stations or individual gauges can be interrogated automatically by computer, which may then store the data, issue a warning to a flood officer or input it into a hydrological model. The computer used in the pioneering River Dee regulation scheme in North Wales in the 1960s performed all these functions, accessing remote rain and river gauges every ten minutes, inputting data into the then ground-breaking rainfall-runoff model developed by the Institute of Hydrology and the Soil Survey (SSEW) called DISPRIN (Dee Investigation Simulation Program for Regulating Networks) and telephoning officers when it deemed necessary (Jamieson and Wilkinson, 1972). Weather radar (see below) was also tested in the system before deployment nationally. Subsequent experience proved that the telemetric access of 'strategic' gauges gave sufficient warning of critical events, and – rather sadly for keen modellers – the hydrological model was used less frequently. The information provided by the Dee system continues to inform decision making to meet the legal requirement to alleviate the 100-year flood. It

Figure 18.2 Solar power is now widely used to operate remote monitoring stations like this river gauging station on the Rio Gallego in northern Spain

does this by helping managers regulate levels in the four reservoirs and release enough water to maintain minimum levels for public water supplies at seven abstraction points downstream, including the cities of Chester and Liverpool and the Shropshire Union Canal.

Weather radar

Over the last 50 years, weather radar has established itself as a vital tool in monitoring storms. There are now extensive radar networks in Europe and North America. In the USA, new radar installations were a key element in NOAA's 1995–2005 Strategic Plan, following the disastrous 1993 Mississippi floods. The new generation of doppler radar systems that monitor the speed of storm movement is linked to interactive processing and mosaicking of scans from neighbouring radar stations at River Forecast Centers. This information is then used to feed upgraded riverflow models for real-time flood warnings and also used in offline modelling of river response for a range of meteorological situations in order to improve decision-making procedures.

Weather radar has important advantages over the old raingauges. It provides spatial coverage, not just point samples like gauges, and so can identify the location of the centre of the storm (which often falls between gauges). It is particularly useful for convective storms, which often occur where there are no raingauges at all, or for picking up intense convective cells within a weather front. It can also provide immediate, dynamic coverage of changing rainfall intensities and storm migration better than individual automatic gauges. The

Figure 18.1 UK Institute of Hydrology (NERC Centre for Ecology and Hydrology) automatic met station in the Plynlimon Experimental Catchments, which is designed to record rainfall and collect data for calculating evapotranspiration losses

results can be linked to hydrological models and can be valuable for modelling urban stormwater (Cluckie and Rico-Ramirez, 2004; Reichel et al, 2009).

Weather radar can also be linked with satellite coverage in the visible and infrared wavelengths. Satellites essentially record cloud cover and the state of cloud tops, not the precipitation beneath. By associating the reflection and emission of the cloud tops, the 'spectral signature', which indicates the temperature and ice/water content, with rainfall intensities measured beneath the cloud by ground-based radar, it is possible to extrapolate from a radar image that may cover just 100,000 km^2 to a full weather satellite image covering more than two million km^2. This approach overcomes two limitations of weather radar: the area covered is limited by the weaker levels of radiation backscatter further from the station, and also because the radar beam is normally directed upwards at an angle of half a degree or so to overtop any hills or mountains, which means that it measures precipitation progressively higher up the further away from source, eventually rising into the clouds above the precipitation.

Radar is now incorporated into satellites, but for different purposes. Passive microwave sensors detect the natural emissions of microwaves from the Earth. The attenuation of the natural microwave signal can be used to calculate the water content of a snowpack by comparing emissions at a site with and without the snowcover. Active synthetic aperture radar is used to penetrate cloud cover to give a very accurate measure of topography. This is valuable for modelling topography in regions with frequent cloud cover, like tropical rainforests. The result is a 'digital terrain model' (DTM) of 'digital elevation model' (DEM) that can be used as a basis for modelling drainage within a river catchment (Chen et al, 2004). (See Notes.)

Satellites

Satellites have revolutionized observations and communications and their potential is still being realized (Figures 18.3 and 18.4). Communications satellites collect data from ground sensors connected to a transmitter called a 'data collection platform' (DCP). Meteorological satellites incorporate scanners that produce scanned images from a variety of visible and infrared wavelengths which map cloud cover, together with profilers that measure vertical profiles of air temperature and humidity. They have the coarsest 'resolution', with the smallest resolv-

able area in the images typically a few kilometres across. The Japanese Geosynchronous Meteorological Satellite (GMS) series, which operated from 1977 to 2003, was replaced by the broader based Multifunctional Transport Satellite (MTSAT) in 2005. Meteorological satellites can also act as communications satellites. These satellites form part of the WMO's World Weather Watch (WWW) programme. Plans for the latest European meteorological satellite system, Meteosat Third Generation, will comprise six satellites. This is due to be operational by 2016 and provide much improved early warning.

Earth Resources Satellites use multispectral scanners to map surface features, patterns of vegetation, soil moisture, water cover (floods), snowcover, etc. They have much higher resolutions: 30 m for the later Landsats and 10 m for the French SPOT satellites. However, whereas communications and meteorological satellites are typically geostationary or 'geosynchronous', i.e. they travel at the same speed as the Earth so as to maintain a fixed position over the target region, Earth Resources Satellites are orbiting and sun-synchronous. This means that they may not be overhead during part or all of a given flood event. Nasa's Landsat series provides the longest continuous coverage, from 1972 to the present, giving a valuable record of changes in land cover and vegetation. Landsat-7 has been operational since 1999 and Landsat-5 returned to working order in 2010. Landsats typically only return to a specific point once every 18 days. A key feature of Landsat data is that it is available to download free to anyone through a variety of agencies, like Nasa's World Wind project and the University of Maryland Global Land Cover Facility. The free exchange of data is a major step forward in global cooperation.

NOAA (the US National Oceanic and Atmospheric Authority) operates over 10,000 DCPs. The transmissions are picked up by Geostationary Operational Environmental Satellites (GOES) and relayed to ground receiving stations, like that at Wallops, Virginia, under the auspices of NOAA's National Environmental Satellite, Data, and Information Service (NESDIS). Transmissions are either timed or 'random': random transmissions are initiated in an emergency, when some measurement exceeds a preset threshold, for example, in a flood.

The World Hydrological Cycle Observing System (WHYCOS) was initiated by the WMO Division of Water Resources under the directorship of John Rodda in the early 1990s as a way to stem the decline in hydrometric stations, to strengthen the global programme of water

355

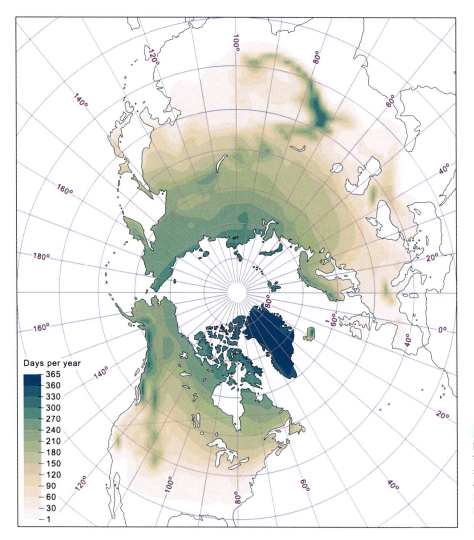

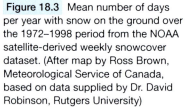

Figure 18.3 Mean number of days per year with snow on the ground over the 1972–1998 period from the NOAA satellite-derived weekly snowcover dataset. (After map by Ross Brown, Meteorological Service of Canada, based on data supplied by Dr. David Robinson, Rutgers University)

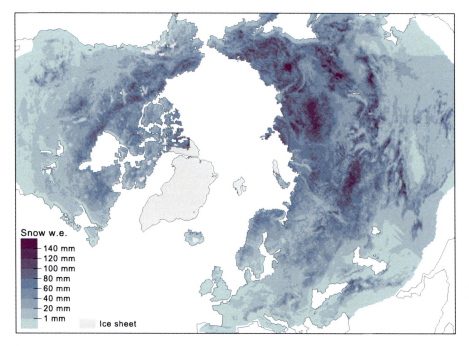

Figure 18.4 The water equivalent of the northern hemisphere snowpack at the end of the accumulation period, based on NOAA data

resource assessment, and to collect timely, if possible real-time, quality-controlled data worldwide (Rodda *et al*, 1993). The scheme is the hydrological equivalent of the WMO's international system of meteorological satellites, the World Weather Watch, although it does not aspire to complete global coverage like WWW (see WHYCOS box). It uses satellite communications to link strategically-selected new and existing stations reporting water levels, discharges, water quality, precipitation and other meteorological measurements via DCPs. It is also promoting free exchange of data.

WHYCOS is more than a satellite data collection system. It also aims to strengthen the technical and institutional capacities of National Hydrological Services and improve their cooperation in the management of shared water resources. The scheme is expanding gradually with the help of international sponsorship (Figure 18.5). The scheme for the Mediterranean was implemented with World Bank funding and uses 31 DCPs. Britain and France are supporting a scheme for the Aral Sea basin, now at an advanced stage, and progress is being made with a number of schemes in Africa. The first phase of the Southern Africa Development Community project (SADC) includes 43 DCPs. A pilot scheme covering 22 states in West and Central Africa (AOC) was set up to help countries that have suffered badly from the severe

Sahelian droughts and where the hydrometric infrastructure, such as it is, has been declining. France has provided funds and the French Research Institute for Development provides scientific and technical support from its Montpellier office. Plans for more complete and localized schemes have been implemented for the Niger and Volta river basins, and are at an advanced stage for Senegal, Lake Chad and IGAD in East Africa, and a scheme for the Congo is in preparation. Plans for the Baltic, Danube, Black Sea, Amazon and La Plata appear to have been delayed. A new scheme for the Arctic to improve the coverage of melting ice bodies in also in preparation.

Many other satellite systems have been implemented to meet specific goals with direct or indirect application to water resource management. Following the first extreme ENSO event on record in 1982–3, which caught the world by surprise with its intensity and the global reach of its effects, Nasa and France's Centre National d'Études Spatiales (CNES) collaborated in a combined buoy and satellite warning system for the Pacific. The TOPEX/Poseidon satellite returns high accuracy measurements of sea surface elevations, which can pick up the beginnings of serious wave migrations just a few centimetres high (see *The ocean* in Chapter 10). This is supplemented by an array of 70 Tropical Atmospheric Ocean buoys measuring surface meteorology and

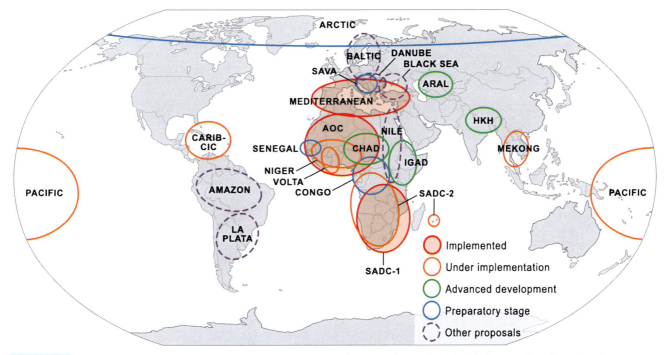

Figure 18.5 Unfolding projects in the WMO World Hydrological Cycle Observing System (WHYCOS). (Sava = River Sava; IGAD = Intergovernmental Authority on Development; SADC = Southern Africa Development Community; AOC = West and Central Africa; HKH = Hindu Kush Himalaya)

The World Hydrological Cycle Observing System

John Rodda

The WMO conceived WHYCOS as a response to the aspirations aired at UNCED in Rio de Janeiro in 1992 and in its Agenda 21, aspirations which reflected the call in the Dublin Principles for an improved knowledge base for the hydrological cycle. These are the hydrological data vital for deciding investment in sustainable water resources development and management, for environmental protection, for studying global change and for other purposes.

WHYCOS was launched in 1993 by WMO with the support of the World Bank and a number of national developmental agencies with the aims to:

- strengthen the technical and institutional capacities of hydrological services to capture and process hydrological data and to meet the needs of their end users for information on the status and trend of water resources

- establish a global network of national hydrological stations to provide data of consistent quality transmitted in real time through the global network of geostationary satellites and the global telecommunications system to national and regional data bases

- promote and facilitate the dissemination and use of these and other water related data using modern information technology.

Initially, WHYCOS was planned as a global network of 1000 river gauging stations linked by data collection platforms (DCPs) through geostationary satellites to a series of regional centres where the data would be gathered, quality controlled, analysed and turned into products for dissemination and use. Each station would record and transmit weather and water quality variables in addition to flow, probably some 15 variables in all, to agreed standards. The idea was to strengthen national networks and the bodies that operate them through a global network, not to supersede them. Attention was first directed to Africa where the decline in networks and capabilities was most severe. Since that time a number of additional HYCOSs have been implemented (Figure 18.5).

profiles of water temperature, since the critical waves of warm water run at 100–150 m depth. The data are relayed to satellite and published on the internet daily.

There are also moves to establish a monitoring scheme for the thermohaline circulation in the North Atlantic. An international team of researchers led from the National Oceanography Centre, Southampton University, set up a prototype scheme to profile temperatures, flows and saltiness in 2004, which they hope will lead to a blueprint for a permanent system (Cunningham *et al*, 2007). The scheme, part of the UK Natural Environment Research Council's RAPID Climate Change programme, includes an array of seabed pressure sensors to calculate flows and uses satellite measurements of wind-driven surface flows at 26°N. Initial results suggest that there are so many frequent changes in fluxes and salt content that only a long-term observational system will provide sufficient understanding of the processes to allow accurate predictions and incorporation in GCMs to improve climate change predictions.

Diminishing polar ice has now become a priority for monitoring. NOAA's NSIDC launched the Global Land Ice Measurements from Space (GLIMS) project in 2005, using the Terra satellite and Landsat Thematic Mapper. In 2010, Cryosat-2 was launched by the European Space Agency (ESA) to improve monitoring of polar snow and ice, especially focusing on the thickness of marine and land ice. The first Cryosat, which failed at launch in 2005, was intended to check whether ice cover is diminishing. By 2010, ESA stated there is 'little doubt' there is a trend and the need now is to gather data on ice thickness, not just areal coverage. This represented a major shift in views in just five years.

Groundwater is the other priority for monitoring. Probably the most exciting development of all is the use of changes in the Earth's gravitational field to plot changes in water distribution. The American Geophysical Union in conjunction with Nasa and the German Aerospace Centre launched the twin satellite GRACE project (Gravity Recovery and Climate Experiment) from the Plesetsk

Cosmodrome in Russia in 2002. Small changes in the gravitational pull on the satellites caused by the mass of water above or below the surface are converted into estimates of the changes in water distribution around the globe. This is the biggest expansion in technology since multispectral scanners first mapped the Earth surface in visible and infrared, or synthetic aperture radar saw through clouds.

GRACE is proving a unique answer to the decline in land-based observing stations and adding more opportunities than have ever been available from traditional monitoring. It is enabling regional and global maps of water and water movement to be produced, covering every aspect from surface flows, soil moisture and snow water equivalent to groundwater. This is giving us a completely new understanding of continental hydroclimatology. The satellites cover the Earth once a month. Nasa's website (www.nasa.gov) displays some of the resulting maps. Plots of the Amazon basin show the monthly shift in water through 2004 and highlight the summer dry season very clearly.

The University of Texas Center for Space Research (www.csr.utexas.edu/grace/), which oversees the project, reports a number of successful applications of the technology to water resources. In addition to estimating changes in large groundwater systems and continental freshwater discharges, researchers have extended the methodology to estimating the component of the water balance that is so poorly monitored by ground networks: evapotranspiration. Global maps of evapotranspiration have been produced by combining the output from the GRACE Land Waters calculations with precipitation and runoff data. Coupled land-atmosphere water balance calculations have also successfully simulated the discharges of the Amazon and Mississippi and offer a new means of monitoring global riverflows.

The most astonishing result to come from GRACE to date is the revelation that groundwater overdraft in California is over three times greater than officially reported by the California Department of Water Resources (Gleick, 2009). GRACE shows that between October 2003 and March 2009 more than 30 km³ of groundwater were pumped up for irrigation. This represents a totally unsustainable rate of over 5.5 km³ per year. It is one of the worst cases of groundwater mining in the world and neither the Department, the Californian government nor the people had any idea of the magnitude of the problem. Nobody actually measures groundwater

use. Recent State legislation calls for limited monitoring, but does not provide for a comprehensive monitoring system or for regulation. It is still a free-for-all: 'Whoever can pump it can have it, to the detriment of everyone else, our wetlands, and runoff into our rivers', says Gleick. At the same time, Nasa reported that GRACE shows massive losses of groundwater in India over a similar period, although these were only half those in just the one State of California.

'California is heading for a catastrophe of huge proportions if the overdraft of groundwater continues at the same rate ... Groundwater levels will drop, the economic and energy costs of pumping will go up, and agricultural production will falter.'

Peter Gleick, Director, The Pacific Institute, after GRACE satellites revealed massive groundwater overdrafts, 2009

International data archive centres

One of the other great advances of the last 50 years has been the establishment of a number of international databases that collate hydrological data from around the world, carry out quality controls, publish progress reports and offer data free of charge to researchers and water managers.

These organizations operate broadly under the UN umbrella, including the UN Environment Programme (UNEP) and the World Meteorological Organization (WMO). Each centre is funded by the host country. UNEP established its water quality database in 1978 as part of its Global Environment Monitoring programme, GEMS. GEMS/Water is based at Environment Canada's National Water Research Institute in Burlington, Ontario (www.gemswater.org). It holds water quality data from over 100 countries and now stores some four million entries (see Chapter 5 *Pollution and water-related disease*). GEMS/Water published its first global assessment as 'Water Quality for Ecosystem and Human Health' in 2007.

The hydrometric, or water quantity, equivalent is held at the Global Runoff Data Centre (GRDC), hosted by

359

the German Federal Institute of Hydrology in Kobenz on behalf of WMO (www.grdc.bafg.de). GRDC holds daily and monthly data for over 7300 river-gauging stations in 56 or more countries, totalling more than 280,000 'station-years' (Figure 18.6). The average record is 38 years long; more than enough for calculating climatic averages. Unfortunately, only 11 countries in Africa send data to GRDC, and there are notable gaps in the Middle East and Central Asian Republics. GRDC offers particular advice to UNEP's Division of Early Warning and Assessment (DEWA), UNEP-GPA (in full the UNEP Global Programme of Action for Protection of the Marine Environment from Land-based Activities!) and UCC-Water (the UNEP Collaborative Centre on Water and Environment – www.ucc-water.org).

The Global Terrestrial Network for River Discharge (GTN-R)

Thomas Maurer, GRDC, Koblenz, Germany

The Global Terrestrial Network for River Discharge (GTN-R, http://gtn-r.bafg.de) is a recently launched project of the Global Runoff Data Centre (GRDC, http://grdc.bafg.de), aiming at improving access to near real-time river discharge data for selected gauging stations around the world, capturing the majority of the freshwater flux into the oceans (Figure 18.6).

The GTN-R activity is a contribution to the Global Terrestrial Network for Hydrology (GTN-H) of the Global Climate Observing System (GCOS) and the World Meteorological Organization (WMO). It has been defined as the GCOS Baseline River Discharge Network and is formally supported by action item T4 of the Implementation Plan for the Global Observing System for Climate in Support

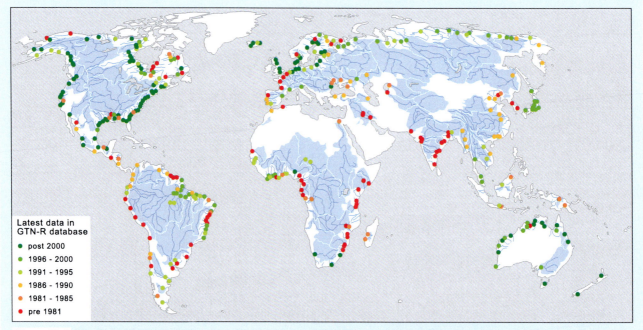

Latest data in GTN-R database
- post 2000
- 1996 - 2000
- 1991 - 1995
- 1986 - 1990
- 1981 - 1985
- pre 1981

Figure 18.6 The Global Terrestrial Network for River Discharge (GTN-R) reporting to the Global Runoff Data Centre. Only the dark green stations are currently active. Note the dates of termination of records and the large gaps (red) in Africa and many other parts of the developing world

of the United Nations Framework Convention on Climate Change (UNFCCC), published by GCOS in October 2004. It will serve an increasing number of purposes and projects in the fields of climate and hydrological research and monitoring.

As a first step, GRDC has identified a priority network of around 380 river discharge reference stations and associated basins as depicted in Figure 18.6. The stations selected so far are the ones initially proposed by GRDC and may be subject to local feedback.

Through the WMO Secretary-General, the GCOS Secretariat approached the Permanent Representatives and the Hydrological Advisers with WMO of the 82 countries involved in April 2005, concerning the 'institutionalized regular provision of daily river discharge data for selected rivers and gauging stations of the GTN-R' together with a country-tailored information package.

As a minimum requirement, it is envisaged that GRDC will receive regular updates for GTN-R stations within one year of their observation. However, as far as possible, countries are requested to provide more frequent and timely updates using the computerized infrastructure of the so-called GRDC Near Real Time River Discharge Monitor (GRDC NRT Monitor). In essence, this is a software for copying and reformatting data, by which the data are automatically downloaded via internet protocols (HTTP or FTP) from servers all over the world (provided by the National Hydrological Services) and redistributed again by uploading the harmonized data in a standard format to one central GRDC file server. Users may then retrieve the data from GRDC for their own applications.

One product of the harmonized data will be an internet map server (IMS) that graphically displays the harmonised NRT-discharge stations in an interactively scaleable world map at a web page (similar to **http://gtn-r.bafg. de/?9931**) displaying current discharge values.

Groundwater is covered by IGRAC, the International Groundwater Resources Assessment Centre, an arm of WMO in The Netherlands, founded as a result of The Hague Declaration (www.igrac.net). IGRAC operates the Global Groundwater Monitoring Network and the Global Groundwater Information System (GGIS) which is an interactive portal. IGRAC has developed an international classification system for aquifers, which is used in the portal to allow searches at a variety of levels of generalization (van der Gun, in press). It operates Global Overview, a web-map giving access to groundwater-related aspects of countries around the world, together with a meta-database listing information on organizations, experts and international projects. It has also hosted ISARM, the Internationally Shared Aquifer Resources Management programme, which has been coordinated by UNESCO, the UN Food and Agriculture Organization (FAO), the UN Economic Commission for Europe (UNECE) and the International Association of Hydrogeologists (IAH). ISARM was founded in 2002 in response to The Hague Declaration on Water Security in the twenty-first century, signed at the Ministerial Conference during the second World Water forum in 2000. It aims to encourage cooperation between countries that share transboundary aquifers, and it organizes regular ISARM conferences. It published the *Atlas of Transboundary Aquifers* in 2009, covering over 200 transboundary aquifers. Related work is also undertaken by the older Federal Institute for Geosciences and Natural Resources in Hannover (www.bgr.bund.de), which hosts WHYMAP (Struckmeier *et al*, 2008) (see section headed *Shrinking groundwater resources* in Chapter 12). Both IGRAC and WHYMAP have addressed issues of identifying and classifying aquifers, and of communicating information succinctly and meaningfully to the general public and policy makers. A major hurdle in this work is to establish a common international standard for data collected under many different national schemes and at different scales.

Precipitation and other climatic data are held in a number of centres. The World Data Centre holds geophysical data from 1957–8, the International Geophysical Year, to the present. Operating under NOAA and the International Council for Science (ICSU) there are now 52 centres in 12 countries, which cover a wide range of environmental and human data in timeframes ranging from seconds to millennia. The WDC for Climate is located at the Max-Planck Institute for Meteorology in Hamburg. The WDC for Glaciology is housed at the Cooperative Institute for Research in Environmental Sciences at Boulder, Colorado, at the Scott Polar Research Institute in Cambridge, and at the Chinese Academy of Sciences Lanzhou branch. The WDC for Meteorology is at the US National Climatic Data Center in Ashville, and in Beijing, and Obninsk in Russia.

361

In 2009, a new NOAA-funded centre was announced that links satellites and climate. The Cooperative Institute for Climate and Satellites is a major extension of the existing Cooperative Institute for Climate Studies at the University of Maryland. It aims to combine satellite observations with climate change modelling and to develop 'climate products' for policy makers and the public, such as long-term drought assessments.

The United Nations World Water Assessment Programme is the summit of international collaboration (see article by Simone Grego, below). It was established in 2000 at the request of governments within the Commission for Sustainable Development. It collates data and case studies from some 30 UN water-related agencies and the international data archives, and publishes reports to accompany the World Water Fora.

The United Nations World Water Assessment Programme

Simone Grego

The United Nations World Water Assessment Programme (WWAP), founded in 2000, is the flagship programme of UN-Water. Housed in UNESCO, the WWAP monitors freshwater issues in order to provide recommendations, develop case studies, enhance assessment capacity at a national level and inform the decision-making process. Its primary product, the World Water Development Report (WWDR), is a periodic, comprehensive review providing an authoritative picture of the state of the world's freshwater resources.

WWAP's general objectives

WWAP's general objective is to influence leaders in government, civil society and private sector, so that their policies and decision making that affect water promote sustainable social and economic development at local, national, regional and global scales. At the same time, WWAP's efforts are aimed at equipping water managers with knowledge, tools and skills, so they may effectively inform and participate in the development of policies and in decision making and they may plan for, develop and manage water resources to meet these objectives.

WWAP's specific objectives

- Monitor, assess and report on the world's freshwater resources and ecosystems, water use and management, and identify critical issues and problems;
- Help countries develop their own assessment capacity;

- Raise awareness on current and imminent/future water related challenges to influence the global water agenda;
- Learn and respond to the needs of decision makers and water resource managers;
- Promote gender and cultural balance;
- Measure progress towards achieving sustainable use of water resources through robust indicators;
- Support anticipatory decision making on the global water system including the identification of alternative futures.

WWAP's main products and activities

WWAP's main product is the World Water Development Report Series (WWDR). The Third Edition of the World Water Development Report (WWDR-3) was launched at the fifth World Water Forum in Istanbul, on 16 March 2009. Coordinated by the World Water Assessment Programme, the *United Nations World Water Development Report 3: Water in a Changing World* is a joint effort of the 26 United Nations agencies and entities that make up UN-Water. The report brings together some of the world's leading experts to analyse the state of the world's freshwater resources: it monitors changes in our water supplies and in how we manage them, and tracks our progress towards achieving international development targets. In addition to the main volume of WWDR-3, WWAP produces a Case Study volume, where the main findings of WWDR are reflected at a local level. Over the life of the programme to date, more than 54 countries have been covered at basin or national level.

In addition to the WWDR-3 and to the Case Studies volume, the WWAP Secretariat produces other publications, divided into Messages Series and Side Publications. The Messages Series (ten published so far) are four-page documents that address a specific target audience with messages extracted from the WWDR-3. The Side Publications (17 published so far) provide more focused, in-depth information and scientific background knowledge on particular areas touched on in the

WWDR-3 and a closer look at some less conventional water sectors.

As a side process, WWAP is also co-coordinating UN-Water efforts for the development of a reliable set of water indicators which will allow the international community to monitor trends in water availability, quality and quantity.

Finally, starting with its fourth phase, WWAP is also initiating a process for the development of World Water Scenarios, which will be an invaluable tool to predict possible futures and to support anticipatory decision making on the global water system.

A climatic classification for hydrology

This is just one illustration of how the international data archives can be used to get a more reliable global overview, based on longer records and global coverage. Figure 18.7 takes the updated climatic classification which Peel and McMahon developed using data from the Global Historical Climatology Network (GHCN), and adds a selection of annual river discharge hydrographs representing most of the major climatic types, using data from the Global Runoff Data Centre archive. The hydrographs have been placed in their approximate latitudinal positions. Peel and McMahon describe their approach in their article 'Updating the Köppen-Geiger climatic classification for hydrology', below.

The hydrographs illustrate the great range in annual discharge and seasonal flow patterns around the world. They represent 20 of the main climatic types from tropical rainforest to snowy subarctic, all plotted at the same scale. The two tropical rainforest rivers, the Lang Suan on the Malay Peninsula of Thailand and the Kawaikoi in Hawaii, together with the Lillooet in the Coast Mountains of British Columbia, far outstrip the others in total annual discharge and also show some of the strongest seasonality. In the Lang Suan this seasonality is due to the monsoon rainfall; on the Lillooet it is due to spring

snowmelt. The Lillooet lies on the eastern edge of the mid-latitude rainforest belt that stretches down the west coast of Canada.

Water management becomes more problematical where annual discharges are lower and the dry season more marked. This begins to show in the wet-and-dry tropics as seen on the Niger, where it is further aggravated by high inter-annual variability and relatively large populations highly dependent on local rain-fed and irrigated agriculture.

The tropical and subtropical desert and savanna (steppe) climates, illustrated here by three South American rivers, and the mid-latitude equivalent represented by the Bad River in South Dakota, have very low annual runoff and suffer long periods with little or no flow. Discharges increase again in the more humid and temperate mid-latitudes, although Mediterranean-type climates still suffer from many months of little or no riverflow, like the Wungong Brook and the Torrens in southern Australia. In contrast, mid-latitude marine west coast climates present a much more stable pattern, e.g. the Latrobe. Further polewards, the polar and subpolar climates show great variety, from the relatively high snowfall combined with rather cool summers in the Vuokski basin in Finland that produce one of the most reliable all-year-round discharges, to the highly seasonal tundra and warm summer climates that produce the meltwater freshets in the Arnaud, Desna and Yana.

363

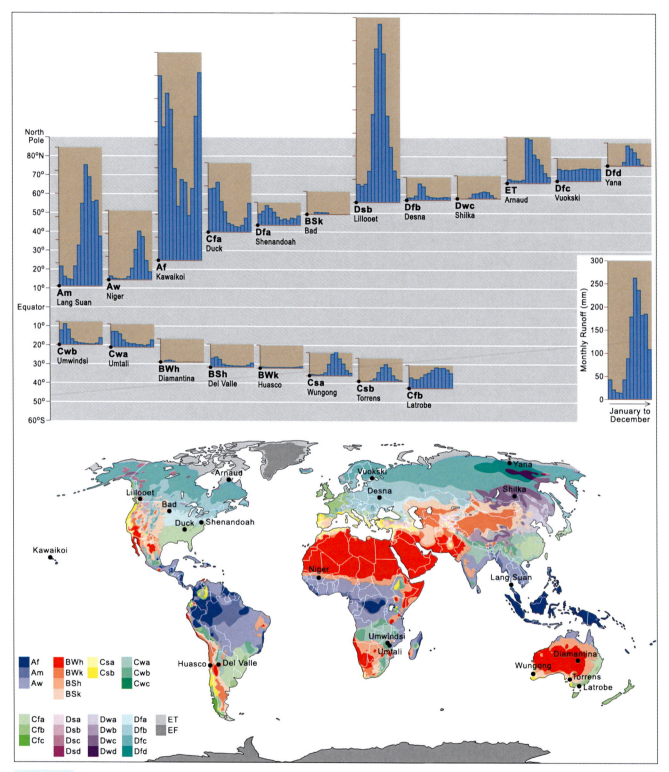

Figure 18.7 Annual hydrographs of rivers representing the main climatic types (above) and the new climatic classification map (below). Details of the climatic notation are given in Table 18.1

Updating the Köppen-Geiger climatic classification for hydrology

Murray C. Peel and Thomas A. McMahon

The most widely known and used climate classification scheme is that initially devised by Wladimir Köppen, first published in 1900, and later updated and revised by Köppen and his co-worker Rudolf Geiger (Köppen, 1936). Following criteria based on annual and monthly temperature and precipitation statistics, the Köppen-Geiger climate classification scheme divides the world into five broad climate types: Tropical, Arid, Temperate, Cold and Polar. These broad climate types can then be further subdivided using additional criteria. During Köppen's lifetime considerable debate about these criteria existed and subsequent to his death the debate continued (see Wilcock, 1968; Essenwanger, 2001). In the present updated world map of the Köppen-Geiger climate classification, we have used criteria that follow Köppen-Geiger, with the exception of the boundary between the temperate (C) and cold (D) climates. Here the temperate/cold climate type divide is made when the temperature of the coldest month is above 0°C, rather than above –3°C as used by Köppen-Geiger (see Wilcox, 1968 and Essenwanger, 2001 for a history of this difference).

Long-term station records of monthly precipitation and monthly temperature were obtained from the Global Historical Climatology Network (GHCN) version 2.0 dataset (Peterson and Vose, 1997). Stations from this dataset with at least 30 observations for each month were used in the analysis (12,399 precipitation and 4846 temperature stations). The variables used in the criteria to determine the Köppen climate type are listed in Table 18.1. The station values were interpolated using a two-dimensional tension

Table 18.1 Description of the updated Köppen climate notation and defining criteria

1st	2nd	3rd	Description	Criteria*	1st	2nd	3rd	Description	Criteria*
A			Tropical	$T_{cold} \geq 18$	C			Temperate (continued)	
	f		– Rainforest	$P_{dry} \geq 60$			b	– – Warm Summer	Not (a) & $T_{mon10} \geq 4$
	m		– Monsoon	Not (Af) & $P_{dry} \geq 100 - MAP / 25$			c	– – Cold Summer	Not (a or b) & $1 \leq T_{mon10} < 4$
	w		– Savanna	Not (Af) & $P_{dry} < 100 - MAP / 25$	D			Cold	$T_{hot} > 10$ & $T_{cold} \leq 0$
B			Arid	$MAP < 10 \times P_{threshold}$		s		– Dry Summer	$P_{sdry} < 40$ & $P_{sdry} < P_{wwet} / 3$
	W		– Desert	$MAP < 5 \times P_{threshold}$		w		– Dry Winter	$P_{wdry} < P_{swet} / 10$
	S		– Steppe	$MAP \geq 5 \times P_{threshold}$		f		– Without dry season	Not (Ds) or (Dw)
		h	– – Hot	$MAT \geq 18$			a	– – Hot Summer	$T_{hot} \geq 22$
		k	– – Cold	$MAT < 18$			b	– – Warm Summer	Not (a) & $T_{mon10} \geq 4$
C			Temperate	$T_{hot} > 10$ & $0 < T_{cold} < 18$			c	– – Cold Summer	Not (a, b or d)
	s		– Dry Summer	$P_{sdry} < 40$ & $P_{sdry} < P_{wwet} / 3$			d	– – Very Cold Winter	Not (a or b) & $T_{cold} < -38$
	w		– Dry Winter	$P_{wdry} < P_{swet} / 10$	E			Polar	$T_{hot} < 10$
	f		– Without dry season	Not (Cs) or (Cw)		T		– Tundra	$T_{hot} > 0$
		a	– – Hot Summer	$T_{hot} \geq 22$		F		– Frost	$T_{hot} \leq 0$

Key: MAP = mean annual precipitation, MAT = mean annual temperature, T_{hot} = temperature of the hottest month, T_{cold} = temperature of the coldest month, T_{mon10} = number of months where the temperature is above 10, P_{dry} = precipitation of the driest month, P_{sdry} = precipitation of the driest month in summer, P_{wdry} = precipitation of the driest month in winter, P_{swet} = precipitation of the wettest month in summer, P_{wwet} = precipitation of the wettest month in winter, $P_{threshold}$ = varies according to the following rules (if 70% of MAP occurs in winter then $P_{threshold} = 2 \times MAT$, if 70% of MAP occurs in summer then $P_{threshold} = 2 \times MAT + 28$, otherwise $P_{threshold} = 2 \times MAT + 14$). > = above, < = below. All precipitation variables are measured in millimetres (mm) and all temperature variables are measured in degrees Celsius (°C). The order in which the criteria were applied was B climates first, followed by A, C, D and E.

spline for each variable onto a 0.1 × 0.1 degree of latitude and longitude grid for each continent. The criteria for determining the Köppen climate type were then applied to the splined variables. Further details about the process followed to make the map and the accuracy of the map can be found in Peel *et al* (2007). This map is based on the climatology at stations over their entire period of record, with each variable individually interpolated and dif-

fers from the recent work of Kottek *et al* (2006), which is based on coarser (0.5 × 0.5°) gridded temperature and precipitation data for the period 1951 to 2000. Although broadly similar to the map of Kottek *et al* (2006), the present map is more detailed due to the finer grid resolution and has some features that may be due to the longer period of record used.

Conclusions

New technology, especially satellite remote sensing and digital recording, is adding new capabilities and helping to offset the decline in manned ground measurement stations. Digital data capture is also making it easier to archive the measurements. The real-time return of data is enabling better tracking of events and helping disaster warning. The data archives are helping us to make better assessments of long-term changes and spatial patterns.

Discussion points

- Follow the developments in satellite remote sensing. (See Further reading.)

- Discuss the view that the provision of free data collection and dissemination is the best form of international aid from western countries to the developing world.

- Explore the ongoing work of the World Water Assessment Programme.

- Study the work of the major international organizations. (See Further reading.)

Further reading

A full coverage of new observational technology is given in:

Chen, Y., Takara, K., Cluckie, I.D. and Hilaire De Smedt, F. (eds) 2004. *GIS and Remote Sensing in Hydrology, Water Resources and Environment.* IAHS Pub.

A more detailed chapter on instrumentation and the design of observational networks can be found in Jones (1997) *Global hydrology: processes, resources and environmental management.* Harlow, UK, Addison Wesley Longman, 399pp.

Up-to-date news on new satellites can be found on the websites of the space agencies: Nasa (www.nasa.gov), the European Space Agency (www.esa.int), the Japan Aerospace Exploration Agency (www.jaxa.jp), the China National Space Administration (www.cnsa.gov.cn), the Indian Space Research Organisation (www.isro.org) and the Russian Federal Space Agency – Roscosmos (www.federalspace.ru).

More information on the GRDC's global river discharge network can be found online at:

Maurer, T. 2005. 'The Global Terrestrial Network for River Discharge (GTN-R) – Near real-time data acquisition and dissemination tool for online river discharge and water level information.' Proc. First International Symposium on Geo-Information for Disaster Management, 21–23 March 2005, Delft, The Netherlands, 18pp. Online at: http://grdc.bafg.de/?8168.

For more information on major organizations, visit their websites, e.g. the World Water Assessment Programme (www.wwap.org), the Global Environment Monitoring Programme, GEMS/Water (www.gemswater.org), the International Groundwater Resources Assessment Centre (www.igrac.net), WHYMAP (www.bgr.bund.de), and the Global Runoff Data Centre (www.grdc.bafg.de).

19 Improving prediction and risk assessment

'There are known knowns – things we know we know. There are known unknowns… things we don't know. But there are also unknown unknowns – things we don't know we don't know.'

Donald Rumsfeld,
former US Defense Secretary

Improving forecasting and prediction is fundamental to sustainability. Weather forecasts and forecasts of river-flows are necessary tools for day-to-day management. Risk assessment is the foundation for designing infrastructure – dams that will hold enough water to sustain a community through the most severe drought, flood protection that will contain the highest riverflows. And predicting longer-term trends and cycles is necessary for forward planning to ensure that systems designed today will be robust enough to continue to be a good investment and serve well into the future.

Great strides have been made over the past century, not only in developing new models and procedures, but also in linking these to the new observational technology discussed in the previous chapter. Developments in computer technology have been critical to both aspects. While meteorological forecasts and climate change predictions are continually advancing with the aid of bigger, faster mainframe computers, hydrological models designed to forecast riverflows have been downscaled to run on laptops that can be deployed in the field – and those laptops can now handle what mainframes did two decades ago.

The UK Met Office's £33 million IBM supercomputer, which went fully operational in 2009, does 125 trillion calculations a second and collates data from satellites, aircraft, ships and robotic buoys, as well as conventional weather stations. The previous grid resolution was 4 km square, the new one is 1.5 km square, which makes it able to cover small convective storms better, like those that caused the exceptional flashfloods in Lynmouth (1952) – which changed the Institution of Civil Engineers' assessment of maximum flood dis-

charges – and in Boscastle (2004) – which drew attention to the hazards of sewage overflows during floods. The Met Office says it could save millions of lives from floods, typhoons and hurricanes. Even so, after being tempted into making more long-term forecasts – of a British barbecue summer in 2009 and a warm winter 2009–10, neither of which materialized – it has ceased to publish seasonal forecasts. And an even faster computer, the Sequoia, is being built at the Lawrence Livermore National Lab in California capable of 20,000 operations a second and due to be operational by 2012, primarily for nuclear weapons testing, but IBM say it could also be used for climate change research (Avro, 2009).

Real-time forecasting

For the last half century, hydrologists have been attempting to understand and model the physical processes by which precipitation translates into riverflow or 'runoff'. The ultimate aim is to produce a computer model that simulates the processes based purely on understanding of those processes and the environmental characteristics of the catchment area (Dawdy and O'Donnell, 1965). A multitude of models have been produced, from the groundbreaking Stanford Watershed Model of the 1960s and its derivatives in America to the Système Internationale Européen (SHE), constantly approaching the 'ideal' of simulating real processes in the real landscape on as fine a resolution in space and time as possible. The latest models are called 'distributed' because they divide the river basin into smaller units, ideally each with its own uniform characteristics.

367

'The ideal model would specify completely the properties of and the processes that occur in all the relevant components of a basin. The specification would be given in terms of physical parameters and would involve all the behavioural relationships within the basin.'

The ideal hydrological model, according to Dawdy and O'Donnell (1965)

We are closer to the ideal. But we are not there yet. And one of the problems is lack of the necessary data on environmental variability within catchments, especially the hydrological properties of the soils. The trouble with this 'ideal' approach is that it generally requires more environmental data than are readily available and it may still be constrained by lack of hydrometric data with which to 'calibrate' it. Remote sensing from aircraft or satellites may help with some aspects, such as vegetation cover and soil wetness, but there are still aspects that require ground survey. To a lesser extent, there are also a few gaps in our understanding of the mechanics of hydrological processes. In practice, however, the complete specification of processes may not be necessary and the art of the modeller is in deciding what is truly relevant for a given application.

The ideal model that is fully 'transportable' so it can be used in river basins where there is no previous hydrometric data is still awaited. Calibration is still the norm. In this regard, the International Association of Hydrological Sciences current PUB project (Prediction in Ungauged Basins 2003–2012) is a laudable and ambitious attempt to fill this critical gap. It is, nevertheless, a difficult task, progress is slow, success is not guaranteed, but hopes must be kept up.

Improving risk assessment

Improving risk assessment is a priority. Assessing the risk of extreme events is one of the most crucial tasks for water managers. It is also one of the most difficult, if not intractable, and inadequate risk assessment lies at the root of most system failures. There is always a trade-off between the cost of planning to protect against a certain size of extreme event and the cost of not doing so. Society sets targets in proportion to the risk and the cost. It might be a flood that happens once every 20 years for a rural river or the once every 100 years flood for a town.

Extreme event analysis

There are two fundamental problems: lack of data and the ever-changing climate. Short periods of measurement and runs of data collected by different methods or at different sites provide an inadequate sample for statistical analysis. How do we estimate the size of events that can be expected only once every 50 or 100 years – the common design events for flood control or dam engineering - when we have records for just 30 years or less? The average length of riverflow record in the UK is only about 20 years and this is much longer than the global average. Rainfall records are generally longer. Official records from the British Rainfall Organization go back to 1860. Records at Cherrapunji, India – the wettest place on Earth – go back to 1851. But even so, most records go back less than 50 years and many stations in Africa have disappeared since the 1970s. Records have never been kept for long in the developing world and they are getting scarcer: a major problem for engineering design and for estimating climate change.

The truth is that almost all 'once every 100 years events' are estimates based on extrapolating from the data according to assumed statistical distributions. All standard 'extreme value' analyses try to fit the existing data to one of a number of known theoretical statistical distributions, and then extrapolate beyond the data to worse cases using that distribution (WMO, 1994). Over the years, standard distributions have progressed from the symmetrical 'Normal' distribution – which often fits precipitation quite well – to distributions that are more and more skewed, with long tails in the upper extremes – which tend to fit river discharges better. Jones (1997) looks at possible reasons for this. But the practical problem is that the further the extrapolation goes above the largest event in the observed data, the greater the uncertainty. This can be compounded by the fact that climate is never stationary. Yet 'stationarity' is the basic assumption in statistical analysis. Statements like 'the 1976 drought in England was a once in 400 years event' are really quite meaningless, for two very basic reasons: measurements are short and the English climate has changed dramatically since the depths of the Little Ice Age. They give science a false appearance of accuracy and may even damage the public's perception of it. Where does this leave Dutch coastal defences, said to be designed for the 4000-year flood? Even the common 100-year flood protection for rivers goes beyond the actual data available.

Even if we took the most extreme event from an actual 100 years of record, it might not be any better guide, because the climate is likely to have changed, and of course it could have really been the 200-year or 400-year event that just happened to occur in our 100-year record. This is a particular problem: if an event that is really a 100-year event got embedded in our 30-year record, then it would be assumed that this is only a 30-year event and extrapolation would give us a gross overestimate of the size of a 100-year event. Actually, the probability that this might happen is surprisingly high: 26 per cent, assuming random occurrence. Conversely, given the same assumption of randomness, there is a 64 per cent chance that any particular 30-year period will not contain a real 30-year event. Very dispiriting. So dispiriting that some water companies have dispensed with this 'extreme event analysis' altogether, perhaps constructing their own 'worst event scenario' to aid planning and management: Welsh Water takes the worst recorded drought and extends it a month or so.

To cap this, practically speaking there is virtually no limit to the potential size of an extreme event. All statistical distributions recognize this with a tail of ever-decreasing probabilities that never quite reaches zero. This fact was indelibly engraved on the memory of more than a generation of hydrologists by Mandelbrot and Wallis (1968) when they dubbed the most extreme events 'Noah floods'.

> 'No amount of observations of white swans can allow the inference that all swans are white, but the observation of a single black swan is sufficient to refute that conclusion.'
>
> John Stuart Mill,
> nineteenth-century philosopher

Black swan events

John Stuart Mill's comment makes the point that predicting from experience has its limits. Everyone thought all swans were white until Australia was opened up and black ones were found. None but the legendary Noah predicted the great Mesopotamian flood – it was outside experience. So were the 9/11 events in New York and

Washington, and the recent world recession (the Great Depression of the 1930s had different causes). Economic models took a knock in 2009. Many economists blame too much faith in economic models for the crisis – the risks were severely underestimated. One Goldman Sachs executive described the crash as 'a 25σ event' (meaning 25 standard deviations beyond the average), i.e. totally beyond the realms of expectation. But was it? Or was it just that the models were wrong, that is, not adequately covering the systems they were designed to mimic? Was there too much modelling of the essentially unmodellable? Too many believed the models and were led badly astray.

Models in physical hydrology and water management are somewhat different in that they do not have to consider human nature. But they too can mislead if key parameters are not correctly set or represented, or they are based on insufficient data. And this also comes down to experience. But there will always be events beyond experience.

In his fascinating book, *The Black Swan: the impact of the highly improbable* (2007), N.N. Taleb, imagines two worlds – Mediocristan, where things are fairly predictable, and Extremistan, where big, unexpected events change everything.

Persistent patterns

Statistical methods assume that the data constitute an 'independent, random sample' – yet another basic requirement that hydrological data sets regularly flout. On the daily level, tomorrow's weather or river discharge is frequently related to today's. At the annual level, wet years and dry years often cluster in runs of similar weather – because of slowly developing oscillations in the atmospheric circulation.

This property is called 'persistence'. It is commonly seen as a problem that can give a biased sample. This was the source of one of the classic and most critical misjudgements in the history of hydrometric assessment: the 1922 Colorado Compact. The Compact apportioned the rights to water from the Colorado River on the basis of records taken during what turned out to be an exceptionally wet period. In the years 1896–1930 the average flow at Lee's Ferry was nearly 21 billion m³, whereas the 1931–1965 average was only 16 billion m³; nearly a 25 per cent reduction. The Compact was drawn up when the ten-year moving

369

mean was at its highest. The record was collected during a period that proved unrepresentative of the rest of the twentieth century. The consequences may have hampered development over a large portion of the south western USA ever since, particularly aggravating California's water shortage.

But persistence can also be turned to advantage. Mandelbrot and Wallis (1968) labelled the process the 'Joseph' effect, after the biblical Joseph who famously advanced his career and the cause of the Jewish people by predicting that the Nile would give Egypt seven good years and seven lean years. Mandelbrot and Wallis noted that just as 'Noah' events 'prove' that extreme events can be very extreme indeed, so 'Joseph' events prove that events can be very persistent indeed.

It was also the Nile that led the British engineer, H.E. Hurst, to formally draw the attention of modern water science to the phenomenon, now called the 'Hurst effect', in the 1950s (Figure 19.1). So-called 'Markov' models incorporate this element of persistence. One of the most important developments in stochastic hydrology in recent years has been the creation of a new family of models known as autoregressive (integrated) moving average (ARIMA or ARMA) models. These extend the Markov approach by combining the dependence on preceding events, the 'autocorrelation', with a smoothed running mean of simulated random oscillations.

Extending the record

The results can be used to extend records artificially in order to better assess probabilities when designing water resource systems. Another modern approach that aims to extend records and improve risk assessment is Time Series Synthesis. This involves decomposing the recorded data into trends, cycles and random elements, and then using these components to reconstruct a longer data series.

These methods offer a more sophisticated approach to risk assessment and are a significant improvement upon older approaches that are restricted to the available data. However, they are still limited to some extent by the data, because they can only take account of extremes, cycles and trends that lie within the record, not of what happened beyond the record.

Alternative approaches to extending the record are being actively explored. Tree ring analysis has been used to identify wetter and drier periods and even used as surrogates for riverflow (e.g. Jones *et al*, 1984). Written reports in newspapers or diaries have been used to reconstruct climatic events at a qualitative level. The British Hydrological Society's online 'Chronology of British Hydrological Events' is an excellent example of what might be achieved. More quantitative information may be obtained from ships logs. Although these do not relate to the land, they can indicate general weather patterns (Küttel *et al*, 2009; McNally *et al*, 2008). These synoptic patterns may be classified into 'airflow' pattern types with their associated average precipitation and evaporation characteristics, as in the Lamb types in Britain (Lamb, 1950). However, there are two caveats with this approach: the weather associated with each airflow type is very 'average', even when split into seasons, and it assumes that each class of airflow brought the same weather in the past as now, which may not be the case, for example, if polar maritime air is coming from a more ice-covered northern sea, or tropical maritime air crossing a cooler North Atlantic. Mountain and Jones (2006) trialled a more quantitative approach using a methodology developed for estimating the impacts of future climate change on riverflows in a 'hindcasting' mode, reconstructing past events. This uses meteorological data on temperature, pressure and wind fields at a number of levels in the atmosphere, which are available back to the late nineteenth century in the UK, as input into a randomized computer weather generator and feeding the output into a hydrological simulation model. One of the sad facts of data storage is that many hydrometric records from the early twentieth century in Britain have been destroyed and the information kept only in summary form. New methods of reconstruction may help correct this loss. Another problem is that prior to the digital age, data were largely graphical, on charts, and digitizing them has been a slow process often hindered and halted by costs.

Mathematical hydrologists are also actively developing alternative approaches. The methodology of Artificial Neural Networks (ANN) has been applied to such varied topics as pattern recognition and noise reduction. It is now being applied to predicting precipitation and riverflow (e.g. Kung *et al*, 1996; Govindaraju and Rao, 2000). In essence, records are used to 'train' the statistical model, emulating the learning processes of the human brain to recognize a pattern in the data and use this to predict what will happen next. It can be applied in real time, processing incoming observational data in order to forecast. It can also be used to hindcast.

'The multifractal approach, which is still in the developmental stage, is a very promising domain. This approach may provide hydrology with useful concepts that could tomorrow turn into essential tools for water resource management.'

Pierre Hubert, Secretary General
International Association of Hydrological Sciences,
2001

Pierre Hubert is championing multifractals. Multifractals are another sophisticated concept developed by Benoit Mandelbrot (1977) that is slowly but surely permeating through science and technology. One of the more popular applications is in generating images in computer games from a small set of instructions and data. They are based on the principle of 'self-same' reproduction – the notion that the building blocks of an image are 'scale invariant': bigger pictures are merely multiples of similar smaller elements. Although Mandelbrot worked with hydrologists decades ago, the application to water resources is still in the developmental stage, but recent trials are encouraging. Hubert (2001) summarizes some of his work on rainfall and riverflow applications. One possible application of multifractals is to aid distributed rainfall-runoff models by disaggregating rainfall fields from larger scale forecasts into smaller time-space elements that could be used as input to the hydrological models.

Perhaps the most crucial finding is that the long tail of low frequency, high magnitude events in probability distributions (or 'the asymptotic behaviour of extreme values') may be algebraic rather than exponential, as generally assumed in standard approaches to risk assessment. This could fundamentally alter most currently held conceptions, with major consequences.

Revolutionary economic implications ensue. Rainfall and river discharges with a given return period, say once in 50 or 100 years, will be more extreme if the pattern is algebraic rather than exponential: another potential broadside for the risk assessment principles in standard use, which most water resource and flood protection systems are designed on.

There is another, entirely different approach. All the statistical and mathematical approaches are essentially 'black box' methods, that is, they ignore the real physical processes. It is possible to use simulation models to generate riverflows from rainfall records, which are generally longer. The models would be calibrated on present-day data and then extended back: i.e. hindcasting.

They can be used in a 'what if' experiment, whereby notional extreme rainfall is fed into the model. This could be repeated for a range of rainfall events and states of soil wetness within the basin. It could even incorporate the effects of possible changes in land use or in soil properties caused by direct human interference or climate change, so as to predict the impact of these changes on the risk of extreme events – a role that is increasingly necessary with urbanization, deforestation, desertification and global warming threatening to alter river regimes.

The Nile: the longest riverflow record in the world
Mike Marshall

Nile flood levels have been recorded since the beginning of Egyptian civilization because of their critical role in agriculture and society. These records represent the longest written records for any hydrological phenomenon. The Palermo stone records over 60 flood levels starting during dynasty I c.3050 BC, until the middle of dynasty V c.2480 BC. It is difficult to make comparisons between these early measurements and more recent data because the zero point of the scale is generally unknown. This is further complicated by the siltation of the river bed and floodplain throughout historic times, causing the zero point to rise annually, although this can be estimated and corrected (Bell, 1970). The records indicate that the floods near Cairo averaged 0.7 m higher during Dynasty I than

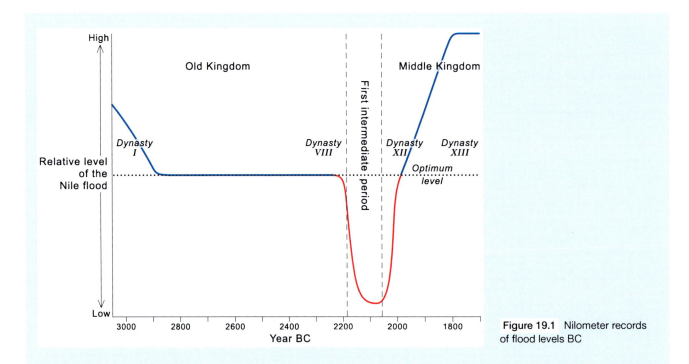

Figure 19.1 Nilometer records of flood levels BC

in subsequent centuries (Bell, 1970; Figure 19.1). More qualitative records are available from ancient texts which make reference to low river levels, famines and social unrest, which in part forced the collapse of the Old Kingdom at the end of dynasty VIII c.2160 BC (Bell, 1971). The next records come during dynasty XII with flood marks at Semna, 50 km above Wadi Halfa near the second cataract, which indicate exceptionally high levels between 1840 and 1770 BC, with floods around 8 m higher than they are today (Bell, 1975).

Although numerous Nilometers were located along the Nile, the records are of a qualitative nature, and many of them have been moved from their original location making quantification and comparison with modern river data problematic. However, a uniquely long and quantitative record originates from the Roda Island Nilometer in Cairo. The series records the annual maximum and minimum river levels between 622 and 1922 AD (compiled by Toussoun, 1925), beginning with the Arab conquest of Egypt. Popper (1951) corrected these compiled data to account for changes in the unit of length used (the cubit), the rise of the bed of the Nile through siltation, and the differences in lunar and solar calendars. Although there are several prolonged gaps in the record after 1471 AD, Kondrashov et al (2005) have applied an iterative gap-filling algorithm (using Singular-Spectrum Analysis) to estimate the missing data (Figure 19.2). Records of Nile discharge as measured monthly at Aswan begin in

1872 AD and continue to the present day. A comparison of annual Nile flood level as measured at Roda with annual peak discharge (July–October) at Aswan, over the common period 1872–1921 AD, shows a high degree of correlation ($r^2=0.88$) between the two. This is encouraging when considering the accuracy of the data derived from the Roda Nilometer. In order to eliminate the bias introduced by water extraction for irrigation in Sudan and by the construction of the Aswan High Dam in Egypt and hence study the natural variability in Nile flow, Eltahir and Wang (1999) naturalized the discharge data measured at the Aswan and Dongola (600 km upstream from Aswan) stations (Figure 19.2).

Climate scientists have demonstrated the significant association between rainfall in the headwaters, the flow of the Nile, the Indian monsoon, and the El Niño Southern Oscillation (ENSO) (Quinn, 1992). Eltahir (1996) showed that 25 per cent of the natural variability in the annual flow of the Nile is associated with El Niño oscillations and went on to suggest a procedure for using the correlation to improve the predictability of the Nile flood. After filling the data gaps, Kondrashov et al (2005) used further advanced spectral methods to reveal, among other periodicities, a quasi-quadriennial (4.2-year) mode and a quasi-biennial (2.2-year) mode, which support the established connection between Nile discharge and ENSO.

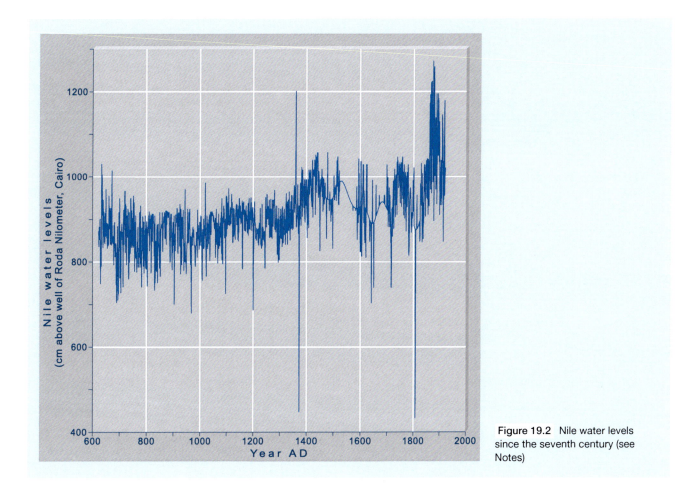

Figure 19.2 Nile water levels since the seventh century (see Notes)

Oscillations and teleconnections

Atmospheric scientists are recognizing a seemingly expanding number of cyclical shifts in pressure patterns and circulation in various parts of the globe that affect weather and water resources, often over great distances. Most of these have only been identified in the last half-century. Some affect water resources very directly, others less so. In many cases, research suggests that weather patterns can best be explained by a combination of atmospheric oscillations. Some of the impacts are more firmly established scientifically than others. Most are as yet purely statistical. Some have clearer physical explanations. Sometimes, adding the phases of sunspot cycles improves the statistical significance of the correlations, with as yet no clear physical explanation.

Hydrologists and water resources managers are gradually coming to accept the relevance of these findings. Much remains tentative. Yet the emerging correlations and physical understanding of these cyclical processes could well prove invaluable for improving hydrological prediction. One hope is that hydrologists in the foreseeable future might be able to improve upon the static approach of current extreme event analysis by placing the period of available records within a wider, maybe even global, framework of spatial and temporal patterns in climatic variation (Jones, 1997). This might enable a more informed judgement to be made as to how representative the observational record may be, whether it is likely to have been taken during unusually wet or dry years, and perhaps open the way to adjustments in average discharges and the predicted probabilities of extreme events.

Oscillations and cycles

Oscillations occur at a variety of timescales in atmospheric and oceanic circulation and their spatial effects range from being confined to a few zones of latitude to covering more than a hemisphere. Nor are they confined to the lowest layer of the atmosphere, the troposphere:

even oscillations in stropospheric circulation seem to have effects. Here is a brief list of some of the main oscillations identified so far and their possible effects on water resources.

El Niño/Southern Oscillation (ENSO) and similar Pacific oscillations

This is now regarded as one of the most important oscillations on the planet and it is especially important because it is the prime example of linkage between quasi-regular fluctuations in the ocean (El Niño) and the atmosphere (Southern Oscillation). The Southern Oscillation Index (SOI) measures the atmosphere shift based on the pressure difference between Tahiti and Darwin, Australia. ENSO events tend to last between six and 18 months. Its effects extend to precipitation and riverflows in the USA and even Europe. Yet only a few decades ago it was generally considered to be of relevance only to the tropics. It has been fully described in the section on *The ocean* in Chapter 10. Part of the onset of El Niño is also related to an oscillation in the movement of the low pressure in the eastern Indian Ocean, the Madden-Julian Oscillation, which varies from 30 to 60 days and may or may not be strong enough to precipitate full migration. We are here concerned about its wider implications for water-related events.

In the Pacific Rim, the immediate effects of El Niño are floods and landslides in the Americas and drought and wildfires in Indonesia and Australia, extending across to India. In the extreme event of 1982–3 it was also associated with drought in East Africa, which may be an indication of the extremity of the event. These are clearly related to the migration in the warm water pool. When the warm pool is in its normal position, it increases evaporation and creates a convective plume which feeds thunderstorms, tropical storms and typhoons. These tend to be less in evidence during El Niño when higher pressure descends over the region. Conversely, the low pressure over the eastern Pacific feeds the rainstorms in the Americas, but does not tend to spawn intense cyclones because the tropical inversion layer in the atmosphere – a layer of warm air caused by air descending and heating as it is sucked into the Intertropical Convergence Zone that forms a barrier to convection – is lower on this side of the ocean.

More distant connections affect the North Atlantic. Caribbean hurricanes tend to be less active during El Niño, as appears to have been the case in 2009–10. There is also a significant effect in the mid-latitude Westerly winds.

A number of other oscillations have been recognized in the Pacific in recent years with generally longer phases than ENSO. The Pacific Decadal Oscillation (PDO) was first recognized in 1996 and has its strongest effects in the Northeast Pacific north of 20°N bringing colder or warmer water to the west coast of North America in phases of 20–30 years. Only two complete cycles seem to have occurred during the last century. It is related to the older North Pacific Oscillation (NPO), which involves air pressure changes in the extratropical North Pacific. The Interdecadal Pacific Oscillation (IPO), identified in 1999, affects the Pacific more equally in both hemispheres with a 15–30 year cycle. The PDO may actually be a part of it. The Quasi-Decadal Oscillation (QDO) has cycles of 8–12 years in the equatorial Pacific, occupying a similar region to ENSO. In fact, these oscillations can interact with or modulate the effects of ENSO. Australian scientists have been particularly concerned about the way the IPO seems to have a strong effect on rainfall patterns related to ENSO.

Pacific/North American (PNA) effect

Following an ENSO event, the Rossby waves in the mid-latitude Westerly jetstream increase in amplitude between the Central Pacific and the Western Atlantic. This Pacific/North American (PNA) effect causes sea level pressure anomalies in the eastern North Atlantic (Hamilton, 1988). The British Met Office has used the cycle in long-range forecasts, although it ceased publishing them after the embarrassment of predicting a 'barbeque summer' in 2009 and a mild winter following that turned out to be among the wettest summers and most severe winters for decades.

Quasi-Biennial Oscillation (QBO)

The Quasi-Biennial Oscillation (QBO) in the middle and upper equatorial stratosphere involves a shift in the position of the transition from westerly flow in the winter hemisphere to easterly flow in the summer hemisphere across the equator and back again. This is linked to fluctuations in atmospheric pressure in the upper troposphere. It has an average periodicity of 2.2 years. Van Loon and Labitzke (1988) found that under the west phase of the QBO westerly winds are stronger across the North Atlantic during sunspot maxima than during sunspot minima, but weaker under the east phase. Tinsley

(1988) found a link between the QBO and the latitude of winter storm tracks crossing the North Atlantic north of 50° N, which shifts by 6° over the cycle, and also a correlation with the solar flux in the 107 mm band, connected with sunspot activity. Russian scientists have found both two-year and 11-year cycles in riverflows that could possibly be related to both QBO and sunspot cycles.

North Atlantic Oscillation (NAO)

The NAO ranks with the Southern Oscillation/Walker circulation as one of oldest recognized fluctuations. Identified by Walker in 1924, its phases are defined by sea level pressure anomalies between the Azores and Iceland, whereby stronger Westerlies cross the North Atlantic in years when temperatures are above average in Greenland and below average in Northern Europe, and *vice versa*. Rainfall in Western Europe is sensitive to the NAO. Daultrey (1996) found that winter rainfall in Ireland is influenced by a combination of the NAO, QBO and ENSO.

North American Monsoon System (NAMS)

This involves the periodic northerly movement of sub-tropical moisture from the Gulf of Mexico or the Pacific into the North American continent. The periodicity lies between those of ENSO and the PDO. It brings unusually wet weather to the south west USA and Mexico, severe thunderstorms and flash floods. It is also one of the least understood cycles. NOAA has been actively tracking monsoon-generated storms in recent years with the aim of improving early warning systems.

Arctic Oscillation

First identified in 2001, the Arctic Oscillation is a close relative of the NAO and occurs in the circumpolar vortex, a ring of strong westerly winds blowing anti-clockwise around the North Pole. It has been in its 'positive' phase for most of the time since the 1970s, blocking the southward spread of cold air. This has meant warmer temperatures, milder winters and wetter weather in northern Eurasia and America. In its negative phase the barrier weakens and the meanders of the Rossby waves within the jetstream increase in amplitude and extend further south. It turned negative in the winter of 2009–10, allowing cold arctic air to penetrate into northern Europe and eastern America, creating the coldest winter in the UK for 31 years, and encouraging the expansion

of sea ice. To what extent the positive phase during the 1980s and 1990s may have contributed to the warming trend in average land surface temperatures is a moot point.

Smaller scale cyclical shifts also occur within the same westerly airstream. The so-called Index Cycle typically lasts just three to eight weeks. It is most noted for creating 'blocking anticyclones' when the jetstream meanders become stuck in a high amplitude, north-south extension. These blocking highs are responsible for persistent droughts in western Europe, especially during summer. They break up when the Westerlies regain strength and re-establish the more usual mild meandering and strong 'zonal' airflow. This frequently results in storms and heavy rains that break the drought.

Antarctic Oscillation

The Antarctic Oscillation, also known as the Southern Annular Mode (SAM) or Southern Hemisphere Annular Mode (rather unfortunately, SHAM), is another low frequency fluctuation mirroring the Arctic Oscillation in the southern hemisphere.

Extraterrestrial cycles

When asked by the *Strand Magazine* in January 1901 to predict what major developments he expected to see during the coming century, the astronomer Sir Norman Lockyer, founder of the respected scientific journal *Nature*, replied as follows:

> 'The first … will … enable us by means of the spectra of sunspots to forecast famines in India and droughts in Australia, as well as other important changes, a long time in advance … rendering it possible to take timely precautions.'

The notion was widely ridiculed for most of the subsequent century. The determinist geographer Ellsworth Huntington was similarly pilloried nearly 50 years later for linking sunspots, agricultural production and the New York stock market – then heavily dependent on agricultural products (Huntington, 1945).

Sir Norman's prediction has yet to be fully realized, but ridicule is no longer on the agenda. An 11-year cycle has been found in rainfall and riverflows in numerous regions of the world, apparently similar to the mean periodicity of the basic sunspot cycle. Jones (1997) illustrates a close match in a 25-year rainfall record from mid-Wales. He also presents a similar length of rainfall record from Lahore, Pakistan, that matches for the first two cycles, then reverses in the third. And this is a common problem that has been cited to discredit the sunspot hypothesis. However, a negative correlation is still a correlation, and one might hypothetically find plausible physical explanations for such shifts, such as changes in average storm tracks predicated upon small changes in solar output crossing some threshold. The sunspot peak associated with a dip rather than a peak in the Lahore rainfall record was the smallest of the three peaks in the data. This is no proof: it is merely to point out the possibility that even where there is no long-term statistical correlation, this may be due to positive and negative relationships cancelling out, and that deeper investigation based on possible physical causes might actually reveal connections.

In fact, sunspot cycles also have cycles within cycles, which modulate sunspot peaks. The double sunspot or Hale cycle is 22 years long, with higher peaks followed by lower ones. The incidence of drought in the USA over the last 300 years has been correlated with the Hale cycle. In the 80–90-year Gleissberg cycle, the amplitude of the sunspot cycles seems to gradually increase and then rapidly fall off. There appears to be a 180-year cycle in solar activity that might be linked to the tidal effects on the sun caused by the alignment of the larger planets, such as Jupiter and Venus (Gribbin and Plagemann, 1977). Links have also been made between climatic patterns and the 18- to 19-year lunar cycle, whose tidal effects also affect atmospheric pressure (Currie and O'Brien, 1992).

If Sir Norman were around today and asked the same question, he could well predict that more detailed exploration of solar mechanisms from satellites like Nasa's Solar Dynamics Observatory, launched in 2010, will lead to better prediction of solar cycles, and that advances in upper atmosphere physics will reveal mechanisms by which these affect the weather.

Teleconnections

We have already alluded to a number of possible links between regional hydrology and large-scale oscillations in the atmosphere and oceans. Teleconnections research is concerned with correlations between events over long distances. At its crudest, this might simply be synchroneity between events such as droughts and floods and be purely statistical. Most teleconnection research, however, now stands at the level of correlating weather events with known atmospheric and oceanic oscillations. As understanding of these oscillations grows, so proper physical explanations are being sought. Considerable progress is now being made in this respect, especially linking the connections with the behaviour of planetary waves within the atmosphere. Physical understanding is now sufficiently advanced for ENSO to be incorporated into GCMs like the Hadley Centre models.

Most research has focused on precipitation and riverflow, but work is also beginning to study the effects on groundwater. What generally emerges is that the hydrological patterns can best be explained by not one but a combination of oscillation patterns, some of which are inter-related and some apparently more independent. In one of the rare cases of groundwater teleconnections, Gurdak *et al* (2007) link fluctuations in water tables beneath the American High Plains to a number of climatic oscillations, especially the 10- to 25-year cycle of the Pacific Decadal Oscillation (PDO), a super-long-term element of the PDO (50–70 years), the North American Monsoon System (NAMS), ENSO (2–6 years), and to a lesser extent the Atlantic Multi-decadal Oscillation (50–80 years). Their analysis of the spectral pattern of groundwater levels indicated that the super-PDO has the greatest influence (explaining 23–94 per cent of the variations), followed by NAMS (13–32 per cent), the PDO (10–25 per cent) and ENSO (2–13 per cent).

The research team at the University of Melbourne, established by Professor Tom McMahon, is a leading player in the work on surface hydrology (see article in box below). Their research demonstrates the global range of some of these oscillations in terms of hydro-climatic effects that are 'statistically significant'. It also demonstrates the weaknesses in these connections, as shown by the low coefficients of determination. There remain large elements of the variability in precipitation and riverflows that are not explained by these teleconnections. It is, of course, entirely logical that influences closer at hand have a greater effect, especially on shorter timescales. Nevertheless, the long distance element is intriguing and potentially of use for longer-term climatic prediction, especially if physical explanations can be found.

Teleconnections between seasonal precipitation and streamflow and large-scale atmospheric/oceanic circulation patterns

Murray C. Peel and Thomas A. McMahon

The search for a link between precipitation and stream-flow and large-scale atmospheric/oceanic circulation processes to improve forecasts has a long history. In the 1920s, while conducting research into forecasting the Indian monsoon, Sir Gilbert Walker identified three large-scale surface level pressure oscillations, the Southern Oscillation (now linked to the oceanic oscillation in ENSO), the North Atlantic Oscillation (NAO) and the North Pacific Oscillation (NPO) (Katz, 2002). These three oscillations have formed the basis of continuing research into fore-casting and understanding the inter-annual variability of precipitation and streamflow around the world.

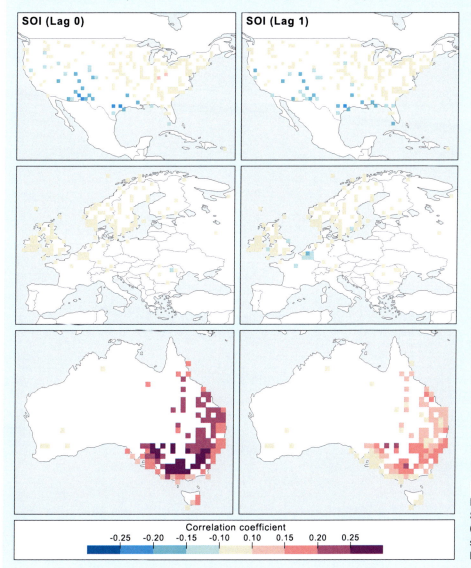

Figure 19.3 Correlations between precipitation and the Southern Oscillation Index. Lag 0; same season, Lag 1; next season. For data sources see Notes

377

Here we calculate direct (no lag) and lag correlations between precipitation and streamflow and a set of circulation indices: the Southern Oscillation Index (SOI), North Atlantic Oscillation (NAO) and the Pacific Decadal Oscillation (PDO) using seasonal data. The PDO is a sea surface temperature-based representation of the surface pressure-based NPO identified by Walker (Mantua *et al*, 1997) and is similar to the Inter-decadal Pacific Oscillation (IPO) of Power *et al* (1999). The lag correlations investigated are one, two and four (one-year) seasons.

Seasonal data for 628 precipitation stations from Version 2 of the Global Historical Climatology Network (GHCN) database (Peterson and Vose, 1997) and 95 streamflow stations from Peel *et al* (2004) were used in the analysis. Seasons are defined as DJF (December, January and Feb-

ruary), MAM, JJA and SON. In order to facilitate comparison between correlations at different stations, all precipitation data have the same record length (90 years, 360 seasons) and cover the same period of time (12/1900–11/1990). Likewise, the streamflow stations have the same record length (50 years, 200 seasons) and cover the same period of time (12/1930–11/1979). This concurrent period restriction limits the spatial distribution of precipitation stations available for analysis to the USA, UK, Scandinavia, Australia and South Africa. The spatial distribution of streamflow stations available for analysis is even further restricted to the USA, Scandinavia and Australia.

The influence of ENSO has been observed in hydroclimate variables all around the world (Ropelewski and Halpert, 1986, 1987; Chiew and McMahon, 2002) and

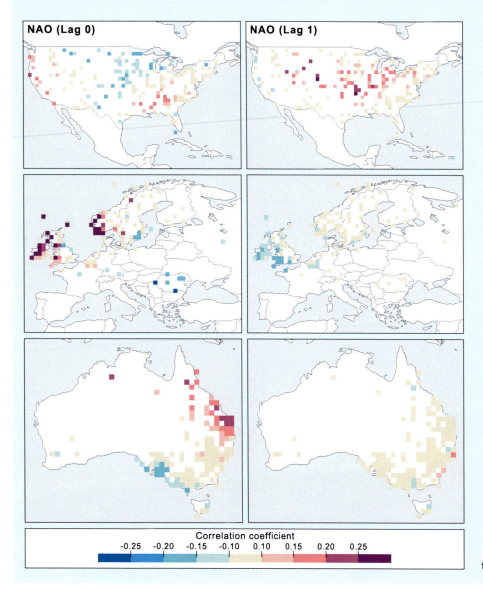

Figure 19.4 Correlations between the NAO and precipitation

is regarded as the largest source of climate variation in the world on an annual time scale (Katz, 2002). Peel *et al* (2002) found that inter-annual variability of precipitation at ENSO influenced locations is generally between five and 25 per cent higher than at non-ENSO influenced locations. Figure 19.3 shows the correlation between seasonal precipitation and the SOI at each station averaged into a 1° × 1° grid, with significant positive correlation in eastern Australia and South Africa and significant negative correlations in South-Western USA and Argentina. The strength of the correlation between precipitation and SOI generally decreases with a one-season lag (Figure 19.3). However, this correlation can, and is, used for seasonal forecasting of precipitation. The number of significant correlations decreases as the lag increases. The correlation between

seasonal streamflow and SOI is largely similar to that for precipitation (not shown).

Figure 19.4 shows the correlation between seasonal precipitation and the NAO, with regions of significant positive correlation occurring in the UK, Scandinavia, North-Eastern Australia and significant negative correlation occurring in Romania and central USA. This pattern of positive and negative correlation is largely reversed when the correlation is lagged by one season (Figure 19.4) and reverses back in the four-season (one-year) lag case (not shown). Few significant correlations between seasonal streamflow and NAO were found; the spatial distribution of the significant ones is similar to the precipitation.

The direct correlation between seasonal precipitation and PDO (Figure 19.5) is very similar to the same correla-

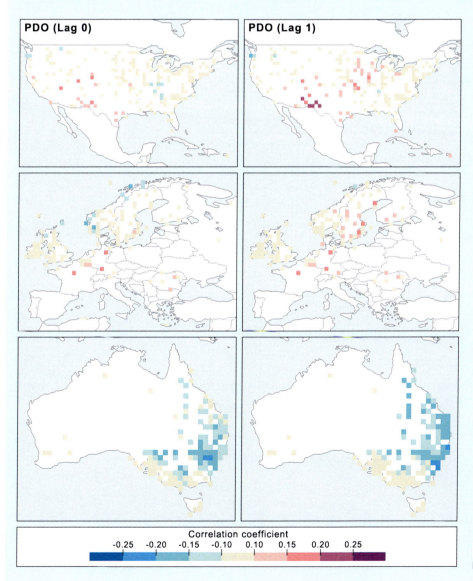

Figure 19.5 Correlations between precipitation and the PDO

tion for IPO (not shown, data from Scott Power, Australian Bureau of Meteorology), which is to be expected since the PDO and IPO are highly correlated (Power *et al*, 1999). Regions of significant positive correlation of precipitation with PDO are in South-Central USA and significant negative correlations are in Eastern Australia. This pattern of positive and negative correlation is largely maintained when the correlation is lagged by one season (Figure 19.5). The number of significant correlations decreases as the lag increases. Few significant correlations between seasonal streamflow and PDO were found; again, the

spatial distribution of the significant ones is similar to the precipitation.

Although many locations display statistically significant correlations between seasonal precipitation, streamflow and circulation indices, it should be remembered that the square of the correlation coefficient is an estimate of the total variance of seasonal precipitation and streamflow explained by the circulation indices. In most cases the correlation coefficients are less than 0.3, which translates into less than nine per cent of the total variance.

Conclusions

Methods of prediction are being continually refined and improved. Real-time forecasts and flood warning systems have been significantly improved by better computer models and better data collection. However, the number of river basins where these are operational is very limited even in advanced economies, and almost non-existent elsewhere. This is an area where vast improvements could and should be made.

Risk assessment in general is hampered by lack of records as well as climate change and climatic variability. Standard methods of assessment still provide a valuable basis for design and decision making, but their limitations are becoming more apparent and there are active moves to improve them. One emerging area is in understanding

the patterns and causes of climatic cycles, which promises to enable limited observational records to be placed in context and maybe extended.

Discussion topics

- Investigate the possible links between climatic cycles and water resources in your part of the world.
- Find out what methods are used for warning of floods and low flows in your area.
- Consider the issues involved in improving risk assessment for extreme hydrological events.

Further reading

A brief introduction to standard extreme event analysis and general principles of modelling can be found in:

Jones (1997) *Global hydrology: processes, resources and environmental management*. Harlow, UK, Addison Wesley Longman, 399pp, Chapters 4 and 6 respectively.

See WMO Guidelines for fuller coverage of extreme value analysis.

A brief introduction to one of the most highly developed simulation models is given in:

Abbott, M.B., Bathurst, J.C., Cunge, J.A., O'Connell, P.E. and Rasmussen, J. 1986. 'An Introduction to the European

System: Système Hydrologique Européen (SHE)'. *Journal of Hydrology* 87, 61–77.

Detailed coverage of simulation models is provided in:

Singh, V.P. and Frevert, D.K. (eds) 2002. *Mathematical Models of Small Watershed Hydrology and Applications*. Water Resources Publications, Highlands Ranch, Colorado.

Singh, V.P. and Frevert, D.K. (eds) 2002. *Mathematical Models of Large Watershed Hydrology*. Water Resources Publications, Highlands Ranch, Colorado.

20 Improving management and justice

Sustainability is 'a set of activities that ensures that the value of the services provided by a given water resource system will satisfy present objectives of society without compromising the ability of the system to satisfy the objectives of future generations.'

M.M. Hufschmidt and K.G. Tejwani
Integrated water resource management – meeting the sustainability challenge.
UNESCO, 1993

We begin this chapter by perhaps belatedly looking at the definition of sustainability. The reason for delaying this issue is that it lies at the heart of improving water management and justice. And, most importantly, the standard definition adopted by UN agencies has had tremendous influence on the way new projects have been assessed, yet some question its reliability.

Defining water sustainability

The definition that Hufschmidt and Tejwani produced for UNESCO in 1993 (above) appears to be very anthropocentric and has been criticized by environmentalists. Despite clarification that the 'services' referred to should be taken to include the protection of ecosystems, many blame the definition for supporting an over-commercial view of water resources. Should water be treated as an economic good? Is 'value' to be measured in monetary terms? Is it only the value we put on it? And ultimately, can we predict what objectives and values future generations might have?

The UN World Commission on Environment and Development (1987) produced a less controversial definition of general sustainability, which maintained that a sustainable society as one that satisfies its needs without compromising the ability of future generations to meet theirs. Nevertheless, there is still an overriding anthropocentric bent here, as also even in Camp's (1994) Environmental Science textbook definition: sustainability 'means that human activity would not degrade the planet's carrying capacity forever for other humans'.

It is possible to think of three different levels of sustainability for water resources: 1) the narrow viewpoint of maintaining the physical water resources, 2) the broader aim of maintaining basin ecosystems or 3) the all-encompassing aim outlined by Dixon and Fallon (1989) of sustaining a balance between social and physical components, between economic returns, social equity, and ecological and hydrological needs. Emphasizing the environmental viewpoint, Jones (1999) proposed two key dicta for water management:

1   That *water* should be managed in such a way as *to minimize the interference with nature and to maximize the benefits for nature*. This means managing water use and manipulation in such a way as to preserve and even enhance the water needed by wildlife and the environment, and

2   That *the environment* should be managed in such a way as *to minimize adverse impacts on and maximize benefits for water resources or flood hazard*.

An economic good or a source for nature and nurture?

The concept of water as an 'economic good' – that can be assigned a monetary value – was promoted by the Dublin water conference and the Rio de Janeiro Earth Summit in 1992. It was subsequently incorporated into a range of principles guiding international policies. The concept helps in the application of cost-benefit analysis to water resources development. But it has also been accused of distorting and misdirecting water management.

There are two areas in which economic values are very difficult if not impossible to assign: wildlife and human lives. There is also the philosophical and religious viewpoint, particularly held in Islam, that water is a natural or god-given gift that should be free to all. In practice, of course, any provision of water facilities beyond the smallest of scales requires money to be spent and therefore implies that water has a cost.

The argument expounded by Professor Malin Falkenmark, of the Stockholm International Water Institute and founder of the annual Stockholm Water Conferences, is twofold. First, that the concept has played a part in the undervaluing by water engineers of water that does not enter the main water bodies, in other words ecological or 'green' water (see *Blue, green, grey, black or gold?* in Chapter 6). The second is the resultant lack of consideration for social justice, for safeguarding access for the poor, in water pricing and cost recovery. The latter problem has been compounded by privatization and commercialization of the water industry. Indeed, the concept of water as an economic good may be seen as a founding principle in the doctrine of privatization that has been supported by UN agencies (see *Privatizing water* in Chapter 4). Social justice is probably better protected when water services are controlled by local or national governments elected by the people.

Falkenmark and Rockström (2004) argue that water is fundamentally different from other economic goods because it is indivisible as it flows through the water cycle and metamorphoses from rain to riverflow, soil moisture, groundwater, and so on. So, any use of water in one form affects some other form. It is also vital for life and it has no substitute – it is a fundamental need. (But is it a right? See below.) Because there is no substitute, the only choice is in finding the most equitable allocation and most efficient use.

A right or a need?

The question of whether water is a human right or simply a need has been the subject of intense debate. It is not trivial. It affects the whole question of whether water should be treated as a commercial good and subject to profit making or not.

If it is a right, then governments have a duty to provide it with equal access for all their citizens. Non-profit organizations, like charities and other NGOs, may also help out, and clean water should be a key target for international aid. But if it is only a need, then private companies can be involved, exploit it for profit and, in the worst-case scenario, select to serve only those who can provide the greatest profit.

The UN decided water is a human right in 2002 under the International Covenant on Economic, Social and Cultural Rights. It is supported by Amnesty International and the WHO. Yet the point is more honoured in the breach than in the practice. Why? Because governments are manifestly failing to deliver. The widespread lack of access to clean water and sanitation tell the tale.

The important question is why have governments failed. Why do nationalized water industries so often fail to maintain infrastructure or invest in new and more efficient systems? The former communist states of Eastern Europe are an illustration (e.g. Stanko and Mahriková, 2009). It is a question of investment priorities and for governments there is often little political capital in water. It is also the huge cost, especially after years of neglect. In a global context, the Global Water Challenge points out that governments spend $3 billion a year on water worldwide, but in order to meet the Millennium Goals in water and sanitation $180 billion needs to be spent. The main argument in favour of privatization in England was that companies would be able to raise money from banks and the stock market to invest more and to afford to employ managers fully trained in commerce. To an extent this has been the case, but recall the details in the section on *Privatizing water*.

In the strictest sense, if water is a right it should be free. We do not normally pay for a right. But what happens when it is free? There is a greater tendency to waste water. In the townships of Kwazulu-Natal people using the standpipes left the tap running when they had finished. The water company found the solution was to create the post of a standpipe manager to police the taps – people are not charged for the water and the managers are not paid, but they have a concession to sell goods to the captive market. Water is saved and the system is popular – a very subtle solution.

There is a more general argument that people do not value what they do not pay for. Conversely, there are cases of real hardship where people cannot pay and there is the public health argument that the poor are in most need of safe, clean water, and if they were forced to economize on its use it would adversely affect their health. The solution must be to price water in a sensitive way, so that those who can afford it pay, without disadvantaging those who cannot. The British solution of a government watchdog, Ofwat, and the independent pressure group the Consumer Council for Water holds profiteering in check. The ban on water cut-offs is another result.

But the industry is not all greedy for profit and there are new pressures and incentives to become 'greener'.

A new business ethos?

Many companies are getting involved in social water-related enterprises. Some are doing it for purely altruistic reasons. Some are finding themselves forced to engage with local communities to maintain their trading position. More are increasingly finding that investing in community projects is profitable. Some are donating to charities, others are doing it themselves (see section headed *Bottling it* in Chapter 8). The old dogma of the economist Milton Friedman – which drove the policies of the British and American governments for most of the 1980s – that the only social responsibility of business is to make profits, is fading, at least in the West. But business opportunities are opening up.

General Electric is buying up water filtration companies and selling water purification units, many to India, that can produce clean water from sewage capable of serving 500 people for ten years, at just $3500 per unit. More than 5000 units have been sold.

Coca-Cola has responded to worldwide criticisms of its work in India (see *Privatizing water*) and elsewhere by funding hygiene education, latrines and wells at schools in Kenya through its Replenish Africa Initiative (RAIN) and providing piped water to schools on Pacific islands. Coca-Cola is establishing regional Foundations to oversee philanthropic water-related projects like RAIN, which was established as a public-private partnership (PPP) in 2009 with $30 million from Coca-Cola to support sustainable clean water, hygiene and sanitation throughout Africa. The Coca-Cola India Foundation was established in 2007 and is already involved in improving water services in Rajasthan. Sceptics have pointed out that Coca-Cola needs to invest in clean water to support its own processing, but its involvement now goes well beyond that.

Dow Water Solutions, an offshoot of The Dow Chemical Company, is not only developing new filtration technologies, but also involved in developing drought-resistant crops, producing higher yields with less water, and collaborating with NGOs and working with the UN CEO Water Mandate, a PPP, to explore strategies for combating the global water crisis. New membranes improve reserve osmosis and nanofiltration for water purification. It aims to reach annual sales of $1 billion by 2015. Dow also part-owns World Health International: one of their projects helped finance 2000 small water systems to serve over ten million people in rural India in 2007. Dow supports the Michigan NGO International Aid, supplying water filters that have provided clean water for over two million people. And Dow Water works with Global Water Challenge, another PPP involving 24 organizations, including Coca-Cola, Proctor & Gamble, Unicef and Water Aid, and supports publicity events like the Blue Planet Run.

Companies have been finding for some time that reducing water usage is a significant aid in cutting manufacturing costs, but pressures have increased this century. Dow, for one, reduced its consumption by ten per cent between 2004 and 2007.

The answer then is part new ethos, part old profit making. But there is little doubt that, human right or no, commercial interests play a vital role in delivering improved services and water savings.

Integrated management

Integrated Water Resources Management

One of the key recommendations that came out of the Dublin conference on Water and the Environment, held as a precursor to the first Earth Summit in Rio de Janeiro in 1992, was the need for more integration in water management. The fragmentation within the water industry, with different bodies controlling public water supply, sewage disposal and flood protection, was seen as an obstacle to efficient management. Integrated Water Resources Management (IWRM) under one overarching management structure was subsequently promoted as the solution by the various UN and UN-related bodies involved with water. International agencies have frequently required its adoption as a prerequisite for the funding of development plans, especially in developing countries.

At the time, this was a major step forward. The basic aim is to avoid developments in one field that might have important detrimental impacts upon another. As Hufschmidt and Tejwani (1993) observed, problems caused by piecemeal developments and fragmented institutional responsibilities have been all too common, especially in developing countries.

Yet in its subsequent implementation it has tended to be too narrow, with less regard for nature and land use than some of its original proponents, including Hufschmidt and Tejwani, envisaged. The emphasis has tended to be on human requirements and on water as an economic good, with cost recovery as a major aim in pricing. Less attention has been given to protection of the environment (Global Water Partnership, 2000). This has been due in large measure to the way it has been promoted in Developing countries, where the most urgent need is seen as providing safe water and sanitation for the people rather than protecting the environment. The chief engineer of the project to tame the Brahmaputra in Bangladesh, M.K. Siddiqi, was not alone when he avowed that protecting the environment is a 'sort of fashion now'. Falkenmark and Rockström (2004) also point to the misapplication of the concept of economic good, which has emphasized human needs over those of the environment.

Integrated River Basin Management

The concept of Integrated River Basin Management (IRBM) grew out of the realization that IWRM alone is insufficient for adequate protection of the environment and also from improved understanding of the important feedbacks between the natural environment, human use of the river basin and water resources. Whereas IWRM focuses more on *integration* activities, the Global Water Partnership (2000) propose that IRBM combines this with *sustainability* and *social justice* in order to 'maximize the resulting economic and social welfare in an equitable manner without compromising the sustainability of vital ecosystems'. The statement has much in common with the guiding principles proposed by Jones (1999).

The concept is gaining adherents, but perhaps not as rapidly as IWRM was spread through the championship of funding agencies. There are two major issues hindering its adoption. One is the deep-rooted belief in many less-developed parts of the world that consideration for the environment is a luxury, when people are dying through lack of food and water and the spread of water-related disease. The other is inherent in the very nature of IRBM: the volume and range of information required, from hydrometric to econometric, the complexity of dealing with it, involving hydrological, ecological and socioeconomic models, and the need to present the analytical results in an uncluttered form to decision makers.

In her article below, Dr Mariele Evers summarizes some of the work undertaken as part of the EU-Flows project and the EU INTERREG 3B Programme on the design of Decision Support Systems that will help in the application and acceptance of IRBM.

Even so, IRBM is not without limitations. It shares some fundamental problems with its ancestor, as proposed in the 1970 UN report on 'Integrated River Basin Development': that water resources systems are not always defined by the natural river basin. Groundwater systems often cross the boundaries of surface drainage basins. Moreover, interbasin transfers mean that many river 'basins' are effectively manmade. Reorganizations of the water industry like those in England in 1973, in Spain in 1985 and others have largely followed the guiding principle of organization based on the river basin as the fundamental administrative unit, together with the other UN (1970) principle of integration from source to reclamation. And organization that tries to follow the natural boundaries of water catchments is more logical for water management than the organization according to political units that it replaced. Yet there are still counter-arguments in favour of political units and communities having control of their own resources, even though this may create transboundary issues. The 1973 reorganization in England and Wales stopped short of full basin unit organization in Wales for political reasons – the initial authority was called the Welsh National Water Development Authority. Its private, but uniquely not-for-profit successor, Welsh Water, inherits the same boundaries.

Ironically, the other great principle advocated by some governments and UN agencies in recent decades – privatization – has worked somewhat against the principle of organization around the river basin unit. Figure 20.1 shows the fragmentation of private water company administrations in England and Wales. Nor is vertical integration source to reclamation totally complete in the UK even now, as shown by Thames Water buying up smaller companies that still survived within its area during the 2009–2010 recession.

Improving finance

As the prime international funding agency and the financial arm of the UN, the World Bank has tremendous influence. It withheld support for Turkey's Southeast

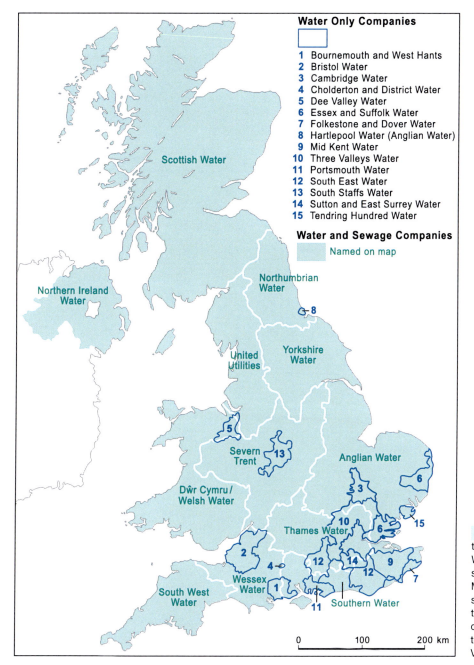

Water Only Companies

1 Bournemouth and West Hants
2 Bristol Water
3 Cambridge Water
4 Cholderton and District Water
5 Dee Valley Water
6 Essex and Suffolk Water
7 Folkestone and Dover Water
8 Hartlepool Water (Anglian Water)
9 Mid Kent Water
10 Three Valleys Water
11 Portsmouth Water
12 South East Water
13 South Staffs Water
14 Sutton and East Surrey Water
15 Tendring Hundred Water

Water and Sewage Companies

Named on map

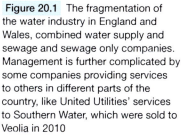

Figure 20.1 The fragmentation of the water industry in England and Wales, combined water supply and sewage and sewage only companies. Management is further complicated by some companies providing services to others in different parts of the country, like United Utilities' services to Southern Water, which were sold to Veolia in 2010

Anatolia Project (GAP) on grounds that it would imperil users downstream in Syria and Iraq, and it refused to fund an Indian-sponsored hydropower scheme in Nepal in the 1990s, in order to put pressure on India to reach an agreement with Bangladesh over the amount of flow it releases across the border in the Ganges.

Over recent decades, the World Bank has progressively moved towards greater emphasis on restricting funding to projects with sound environmental and social foundations. Many other regional development banks, like the Asian Development Bank, the European Bank for Reconstruction and Development and the Council of Europe Development Bank, have followed suit.

In 2006, the World Bank announced a new approach to funding that integrates water resources, agriculture, rural development, and the environment in its publication 'Re-engaging in agricultural water management'. Under these guidelines, the Bank supports projects that involve:

1 increasing water productivity in both irrigated and rainfed agriculture

385

Decision support systems for integrated river basin management

Mariele Evers, EU-Flows Project

A sound understanding of the interrelationships between natural and societal elements is crucial for effective Integrated River Basin Management (IRBM). New Decision Support Systems (DSS) are needed for complex issues such as sustainable management of water resources. These tools must model the natural and the societal systems adequately so that they can be applied in planning and decision making to enable appropriate strategies and cooperative action programmes to be developed. Moreover, the sharing of models and analytical methods, and the mutual exchange of information and ideas within an interdisciplinary team concerned with water, land, ecology and social issues can in themselves improve cooperation and harmonization in research and analysis.

In the context of water sustainability, DSSs can be regarded as socio-technical instruments for analysing, visualizing and collaboration to provide a better understanding and handling of complex systems. A DSS should contain three main components: a user-friendly interface; technical tools, including databases, models and Geographic Information Systems; and a social component developed by an interdisciplinary team that works out DSS requirements in consultation with stakeholders and future users.

Our study of the tools and structures required for a comprehensive planning approach began with the three pillars of IRBM: water quantity and quality, flood risk management, and management of flood prone areas. The study used a prototype DSS developed for the River Elbe basin as a test case (Evers, 2008).

There are severe problems in all aspects of river basin management in Europe. Water quality in surface and groundwater is a severe problem in almost all European river basins. Reduced groundwater quantities are a critical issue for southern and some central European countries. The retention of water within European river basins has been progressively reduced by the lost of 50 per cent of wetlands over the last century and by urban drainage and development on floodplains. Our research identified many gaps in knowledge and data, for example, on the quality of wetlands. The quantities of groundwater are very different when evaluated according to the Water Framework Directive (WFD), which evaluates only the current situation, compared with an historical viewpoint of how levels may have decreased in the recent past, affecting the ecological value of floodplains.

Currently, management and policy concepts are poorly integrated in legal frameworks, for example, with water quality and wetlands being considered as separate items. Flood risk management is not a direct objective of the WFD. Flood risk management and floodplain management are often dealt with separately, with policies sometimes contradicting each other. The new EU Floods Directive will help to rectify this and include a coordinated approach to spatial planning and nature conservation.

The main obstacles hampering the implementation of DSSs are the large information requirements, the complexity of modelling and linking the models of systems, and developing the transparency of the DSS for the users. Each DSS will combine general principles with aspects that are specific to individual basins. The fact that DSSs cannot be simply lifted off the shelf and transported from one basin to another slows progress and increases costs. The most challenging aspects for DSS development are: coordination of planning for agriculture, nature conservation and economic development; stakeholder information and participation; and improving methods for cost-benefit analysis.

2 creating effective and sustainable institutional arrangements, including establishing and strengthening water users' associations, promoting the involvement of the private sector, and ensuring that irrigation systems are financially viable

3 improving links to input supply and marketing chains

4 helping farmers adapt to climate change

5 factoring poverty and gender concerns systematically into agricultural water resources.

The Bank administers a portfolio of $3.7 billion for irrigation and drainage work. Much of this is directed towards sub-Saharan Africa, where the Bank judges that it can make the biggest difference, and it has more than quadrupled its lending to the region in recent years. The Bank also offers advice and analytical support for:

1 private-public partnerships in agricultural development and management
2 the impact of agricultural water operations on poverty
3 river basin management
4 monitoring and evaluation of water management projects in agriculture.

There is also room for smaller-scale funding for individuals and communities, especially in the developing world, to finance their own water projects, such as micro-finance and Islamic banking.

Micro-finance

Dr Muhammad Yunus, founder of the Grammen Bank, has been promoting micro-finance to help small enterprises, especially for women and poor people, to help them out of poverty and set up business. He provides low-interest loans to customers who would not qualify for a loan from mainstream banks and aims to make the facility available to all households in Bangladesh by 2012.

Islamic finance

This is also applicable to large-scale projects. Charging interest is forbidden under Islam. It is based on total transparency, full analysis of risks and debt can only be traded on a par. The IMF believes that it is better at absorbing global financial shocks like the 2009 recession. Unfortunately, the 56 Muslim countries have failed to agree a common strategy for the sector and it currently accounts for less than one per cent of global finance. But the UN is interested and Islamic banking is spreading west from its origins in the Gulf and Asia.

Improving the law

Towards solving conflicts over transboundary water resources

Any solutions to transboundary conflicts must be sought in the twin realms of international political agreements and international law. The UN has had a long-term commitment to brokering international treaties, yet success has been very limited. Similarly, little success has been achieved with either establishing an international legal framework for resolving disputes or enforcing any legal judgements.

The lack of universally recognized international law and the lack of mechanisms for enforcement have severely hampered progress. The old American Harmon Doctrine, also known as the principle of absolute territorial sovereignty, maintains that countries have unlimited rights to use the resources within their territory. This provides an argument that favours upstream users. In contrast, the principle of absolute territorial integrity states that no country may use its resources to the detriment of a downstream state. This is supported and extended by the principle of condominium or common jurisdiction. This maintains that the rights of a state are strictly limited and prior consent from other interested states is needed before water resource developments can take place. This is fundamental to Integrated River Basin Management.

However, the most commonly invoked principle falls slightly short of requiring mutual agreement. This is the principle of equitable utilization or limited territorial sovereignty, whereby developments are permissible if they do not harm the resources of a neighbour. It was established by the International Law Commission in 1966 and forms the core of the Helsinki rules on the use of water in international rivers. It supports a reasonable and equitable sharing of resources and is the most commonly invoked principle, but it falls slightly short of requiring mutual agreement.

Unfortunately, the Helsinki Rules fail to provide an effective mechanism for resolving the conflicts that can arise, and the other principles fall rather short of being mutually supportive.

Nonetheless, matters do show some signs of improvement. The UN Development Programme's persistence led to the Mekong treaty of 1994, but sadly the main owner of the headwaters, China, is not a signatory and is proceeding with its own plans for damming and diversion of waters into the Yangtze Three Gorges system. Even some signatories are now building dams that could soon threaten the UNESCO World Heritage phenomenon of the seasonal Tonle Sap lake, that expands from 2700 to 16,000 km^2 following the monsoon season and provides local inhabitants with 75 per cent of their annual intake

387

of protein from its fish stocks. Even worse could be in prospect for Vietnam, the ultimate downstream user, where the Mekong delta region provides 60 per cent of the country's agricultural production, predominantly from flood irrigation.

A better example might be the cooperative agreement on water resources signed by the 12-nation Southern African Development Community in 1995. This was essentially a home-grown agreement sponsored by the Mandela government of South Africa in cooperation with the other 11 nations and aimed at mitigating the problems of recurrent drought in the region. The accord operates through a Joint Water Administration, and follows the Principle of Equitable Utilisation of Water Resources, which was laid down as the basis for managing international waters by the International Law Commission in 1966 and has since been dubbed the 'Helsinki Rules'. The accord is providing a framework for the amicable resolution of conflicts of interest as Lesotho undertakes major development of its resources and the Administration develops plans for a 'water grid' linking the Zambezi River to other rivers with fewer resources. The Joint Administration has also signed an agreement with DR Congo (Zaire) to buy hydropower from the Congo River dams. The arrangement is allowing the water-poor nations of southern Africa to share the benefits of the two great rivers, the Zambezi and Congo/Zaire that between them hold more water resources than the whole of the Republic of South Africa, in a well-regulated agreement (Jones and van der Walt, 2004).

More recently, a major step forward in international law was taken in 2004 with the establishment of the 'Berlin Rules', which build on the 40 years of experience with the advantages and limitations of the old Helsinki Rules, and include the wider perspective on the water environment and links between water resources and the surrounding environment that has developed in the intervening period (see the article below).

The international legal dimension of sustainable water management

Joseph W. Dellapenna

Water is an ambient resource that largely ignores human boundaries, giving rise to the risk of conflict among neighbouring communities and nations. Cordial and cooperative neighbouring states have had difficulty establishing acceptable arrangements for transboundary surface waters even in relatively humid regions (Teclaff, 1967; Zacklin and Caflisch, 1981). No wonder English derives the word 'rival' from the Latin word *'rivalis'*, meaning persons living on opposite banks of a river. Yet cooperative solutions to water scarcity problems are more likely than prolonged conflict (Dellapenna, 1997; Utton, 1996; Wolf, 1998). The challenges of creating cooperative rather than conflicted arrangements over water have generated a great deal of legal activity at the local, national and international levels. International law (particularly customary international law) cannot solve this problem, yet international law is an essential element of any solution. The relevant international legal developments have only recently been compiled in a single source for lawyers or water managers to find and use. On 21 August 2004, the International Law Association, meeting in Berlin, Germany, approved the *Berlin Rules on Water Resources* as a summary of the customary international law applicable to water resources (ILA, 2004).

The evolution of the customary international law

Customary international law consists of the practices of states undertaken out of a sense of legal obligation – a sense that law requires the practice (Wolfke, 1993). Despite obvious difficulties in determining the precise content of customary international law, the system has been remarkably successful. No form of international life could exist without shared norms that are largely self-effectuating. Focusing on a relatively few highly dramatic instances of international legal failure creates an impression of ineffectiveness. Focusing on the inevitable similar failures in national legal systems would lead to a similar conclusion.

A rich body of customary law regarding internationally shared fresh water has emerged, largely in the last century or so (Dellapenna, 2001; McCaffrey, 2001). *Helsinki Rules on the Use of Waters of International Rivers,* approved by the International Law Association in 1966, crystallized the rules of international law applicable to the transboundary use and development of waters (ILA, 1966). Within a remarkably short time, international lawyers, jurists, governments and scholars accepted the Helsinki Rules as a correct summary of the law it covered. Yet since 1966, the emergence of international law addressing significant environmental problems and the protection of basic human rights has broadened and deepened the customary international law applicable to the world's waters. While the Association approved supplemental rules from time to time to respond to some of these new realities, they were extremely limited, often simply indicating that the basic *Helsinki Rules* applied to the particular issues of the supplemental rules.

The General Assembly of the United Nations approved the *United Nations Convention on the Law of the Non-Navigational Uses of International Watercourses* in May 1997, by a vote of 104 to 3 (UN, 1997). The Convention largely tracked the *Helsinki Rules* and their supplements. While the *Convention* will not come into effect until it receives ratifications from 35 countries, it already has been accepted by the International Court of Justice as a summary of the customary international law (International Court of Justice, 1997, 78, 85).

At about the same time, the International Law Association concluded that the changes in customary international law were so profound, and the attempts to keep abreast of these developments (including the *UN Convention*) were so limited, that it was necessary to undertake a comprehensive review and revision of the *Helsinki Rules* and their supplements. Three reasons lay behind this decision. First, none of the most disputed internationally shared fresh waters are covered by an agreement among all the interested States, suggesting that, despite the growing number of international agreements regarding internationally shared waters, there still is a need for an adequate summary of the relevant customary international law. Second, ratifications of the *UN Convention* have been slow in coming. Third, much of the emerging customary international law applies to waters within a State ('national waters') as well as waters that cross or form international boundaries ('international waters'). Neither the *Helsinki Rules* and their supplements, nor the *UN Convention,* include any provisions directed at national waters.

The *Berlin Rules* set forth a clear, cogent and coherent summary of the relevant customary international law, incorporating the experience of the nearly four decades since the *Helsinki Rules* were adopted. The *Berlin Rules* take into account the development of important bodies of international environmental law, international human rights law, and the humanitarian law relating to war and armed conflict, as well as the adoption by the General Assembly of the *UN Convention.* The *Berlin Rules* include within their scope both national and international waters to the extent that customary international law speaks to those waters. Indeed, some of the rules go beyond speaking strictly about waters and address the surrounding environment that relates to waters (the 'aquatic environment') and the obligation to integrate the management of waters with the surrounding environment. The major changes in the *Berlin Rules* relate to the rules of customary international law applicable to all waters – national as well as international, although there are certain refinements in the rules relating strictly to international waters.

The contemporary customary international law of water resources

Under customary law, riparian states – states across which, or along which, a river flows – have a legal right to use the water of a surface water source (UN, 1997, arts. 2(c), 4). Riparian states in turn are bound by the rule of 'equitable utilization' (*Permanent Court of International Justice, River Oder Case,* 1929; ILA, 2004, art. 12; UN, 1997, art. 5). Equitable utilization requires each state to use water in such a way as not to injure unreasonably other riparian states. Reliance on customary international law, however, to allocate surface or subsurface waters among states is too cumbersome and uncertain to enable the satisfactorily resolution of disputes over international sources of water and too primitive to solve the continuing management problems in a timely fashion (Benvenisti, 1996; Dellapenna, 2001; ILA, 2004, arts. 56–67). Cooperative management has taken many forms around the world, ranging from continuing consultations, to active cooperative management that remains in the hands of the participating states, to the creation of regional institutions capable of making and enforcing their decisions directly (Dellapenna, 2001; Kliot *et al*, 1998).

In addition to cooperative management, contemporary customary international law recognizes important rights for persons, including a right of access to water, a right to a voice in decisions affecting them, and a right to special protections if they are members of especially vulnerable communities (ILA, 2004, arts. 17–21). Contemporary international law also recognizes a duty on states to undertake the conjunctive and integrated management of waters with a view to achieving sustainability in their development and use and to minimizing environmental harms (ILA, 2004, arts. 5–8, 22–35). Further aspects of contemporary customary international law address questions relating to navigation and rights and duties during wars or other armed conflicts (ILA, 2004, arts. 43–55).

Groundwater internationally

In contrast to the considerable state practice regarding the sharing of surface water sources, remarkably little state practice exists regarding the sharing of ground water. Before the spread of vertical turbine pumps after the Second World War, groundwater was a strictly local resource that could not be pumped in large enough volumes to affect users at any considerable distance away. With newer technologies and with the exponential growth in the demand for water, groundwater has emerged as a critical transnational resource that has increasingly become the focus of disputes between nations yet for which no consistent body of state practice has emerged.

Courts and legal scholars have concluded groundwater must be subject to the same rule of equitable utilization as applies to surface sources (*Donauversinkung Case*; Hayton and Utton, 1989). Ground water and surface water are not merely similar, they are the same thing simply moving in differing stages of the hydrologic cycle. As the hydrologic, economic and engineering variables are the same for surface and subsurface waters, the law must also be the same for both. The *UN Convention* did not, however, include groundwater unless it is tributary to an international watercourse (UN, 1997, art. 1). The *Berlin Rules,* however, included an entire chapter on the special concerns relating to groundwater, making it the first comprehensive exploration of the customary international law beginning to emerge on that topic (ILA, 2004, arts. 36–42).

Foremost among the problems in applying equitable utilization to an aquifer is the lack of firm knowledge of the characteristics of the resource, in contrast with surface waters. Investment in acquiring the necessary information, expensive though if might be, is essential if any of the legal duties pertaining to groundwater are to be fulfilled (ILA, 2004, arts. 38, 39). Furthermore, its special vulnerabilities require that groundwater be given special protections beyond those mandated for surface water sources if groundwater is to be managed sustainably (ILA, 2004, arts. 40, 41). Finally, the same sort of transboundary problems arise with regard to groundwater as for surface water sources (ILA, 2004, art. 42).

National law, the public commons and private markets

Laws relating to the right to ownership of water vary enormously between countries. In parts of America it is possible to lay claim to the water in the air above a property. In other countries it may not be possible even to claim the water beneath your property. It may be possible in South Africa to claim that you can do what you like with streams crossing your property. In the UK, the Environment Agency and the Scottish Environmental Protection Agency oversee every intrusion into the river environment.

In his article 'The national legal dimension of sustainable water management' (below), Professor Joseph Dellapenna analyses the conflict between private property and the indivisibility of water resources, and considers the idea of water markets. Water markets have been set up by farmers in California and elsewhere to trade surplus water. Under the market scheme, owners of water rights may sell and transfer their surpluses to anyone willing to pay, be they cities, industries, farmers, fisheries or managers of wildlife habitats. Advocates claim that this can result in water saving and a more equitable use. Dellapenna disagrees. He is not alone: 22 of the 58 counties in California have adopted ordinances restricting water transfers and requiring an environmental review before a sale is sanctioned for groundwater or surface water that is replaced by pumping up extra groundwater. The high cost of these reviews and adverse public opinion have put some sellers off.

Concern has focused on groundwater because it is not protected under California's Water Code as surface water is. Groundwater transfers boomed during the 1990s drought. Now fears of uncontrolled groundwater mining are being increased by evidence from the GRACE satellite that the overdraft is far worse than officially estimated (see *Satellites* in Chapter 18). But there is another, more curious issue: land fallowing. Farmers can leave cropland idle in order to save water to sell. The state set up contracts to encourage this during the 1991 drought. But there are knock-on effects in terms of jobs and the local economy. It caused fierce arguments in the Sacramento Valley in 1991 and injured third parties have no protection under state or federal law. Some recent transfer plans have included compensation for affected communities, but water marketeers fear it could encourage excessive claims if it becomes officially part of the law (Public Policy Institute of California, 2003). The issue of devising sustainable rules for land fallowing remains unresolved.

The system began in the 1970s. By 1990, California's water markets accounted for three per cent of the state's water usage, three-quarters of sales being in the farming hub of Central Valley. The state government has been a major participant, organizing water banks during droughts and buying water for environmental projects. City agencies are major purchasers of long-term contracts, accounting for a fifth of sales, and recent legislation requiring local government to prove there is adequate water to support any new developments is adding to this.

Despite federal and state promotion of water markets, Californian sales have actually fallen since 1996: by 2003 they had fallen 14 per cent and out-of-county sales by 19 per cent. Yet groundwater levels have not recovered and California is heading for serious water crisis, perhaps even sooner than 2020.

The national legal dimension of sustainable water management

Joseph W. Dellapenna

Burgeoning populations and increasing *per capita* demand are putting pressure on resources worldwide, and there is a clear need for greater regulatory control on the consumption and use of water (Dellapenna, 1997). Existing legal regimes already struggle to respond to the increasing and changing demands for water without unduly destabilizing existing expectations expressed in investments in water-use facilities. Global climate change can only increase the stress on existing legal regimes. Too much legal response can produce as much social turmoil as inadequate legal response certainly will.

Challenges to the national dimension of water law

Traditionally, different legal regimes have dealt differently with defined surface water bodies, diffused surface water, groundwater and atmospheric resources. Most societies also have special rules for pollution, navigation and eco-logical or environmental needs. Only the integrated management of water resources can deal adequately with the competing uses and needs for water. This is a long-term trend in legal reform for all levels of water management.

Economists argue that private property and markets would be automatic and nearly painless means for resolving problems of water allocation, distribution and preservation (Anderson and Snyder, 1997; Wolfrum, 1996). Markets, they claim, would introduce the necessary flexibility while allowing the integration of water quality and water quantity into a single managerial model. The results are to be accorded the strong presumption of validity that capitalist societies give to market-based allocations. Still, actual markets for bulk water have always been rare (McCormick, 1994). Markets for water have allowed the transfer of water among small-scale, similar users (Dellapenna, 2000), but water markets have seldom accomplished significant changes in water usage. So-called markets that were used to bring about major changes in water usage functioned only through the rather heavy-handed state intervention. Market supporters seldom address why markets have so seldom been used except to denigrate market critics as holding cultural, religious, even mystical prejudices about water. Water, however, is not like other resources.

The public nature of water

Water is the quintessential 'public good'. A 'public good' shares two qualities: indivisibility and publicness (Kaul *et al*, 1999). Because a public good is indivisible, one cannot simply divide it up and buy as much as one wants, and because it is public, it is impossible to keep others from accessing and enjoying the good so long as it is accessible to anyone. Public goods generally are free goods, i.e. they operate without markets, because consumers cannot be excluded from enjoying the good. Consider a view of the blue sky over another's property. Costs associated with public goods are the costs of capture, transportation and delivery, not a cost for the good itself. This creates an important problem: If one invests in a public good, others who pay nothing will enjoy the benefits of the investment. Such 'free riders' seriously inhibit investment unless some institution can ensure that all pay for the benefits they receive.

Moreover, even the strongest market advocates consider water the paradigm model public good. They use water metaphors to discuss public goods generally: 'common pool resource', 'spill-over effects', etc. In fact, water is not indivisible and public in the strictest sense. Few things are strictly indivisible and public. What is treated as a public good is determined not just by its physical characteristics, but also by its social and economic characteristics. When the costs to exclude others would be so high that it is impractical to exclude others from access to the good, or when there are reasons for a society not excluding some from access to the good, the good is treated as a public good.

Most often something is treated as a public good because transaction costs are so high that no market can function with even minimal effectiveness (Coase, 1960; Howe *et al*, 1990; Shelanski and Klein, 1995). Water is such a commodity. This is most obvious in the case of the protection of instream flows. Less obvious is the public nature of water when withdrawn for private use. While it is easy enough for someone to own and manage water unilaterally in small amounts, e.g. bottled water, a river or an aquifer can never be fully controlled. Large-scale

changes to water use necessarily affects many others, making it impossible to obtain the contractual assent of all significantly affected persons. Transaction costs on all but the smallest water bodies quickly become prohibitive. This reality underlies the tradition of water as a free good – a good available to all at no cost for the water itself, with payment only for the cost of capturing, transporting and disposing of the water.

Economic incentives including fees, taxes and 'water banks', should be introduced for those who use water so they will more realistically evaluate the social consequences of their conduct (Wolfrum, 1996). Resort to economic incentives should not obscure the fact that water remains the prime example of a public good for which prices cannot be set in a marketplace. Ultimately, true markets must remain marginal to the management of large quantities of water for numerous diverse users (Bauer, 2004).

Property in water

In thinking about 'property' in water, one might think of entitlements to water defined in clear and certain terms and protected from change except through market transactions. Examples include the American law of appropriative rights and similar systems in other countries (Dellapenna, 2001, ch. 8). On the other hand, rules allowing anyone with lawful access to use water so long as the use is 'reasonable' (with courts sorting out conflicting uses) is a rule of common property, rather than private property. Examples include the American law of riparian rights and the Roman law of flowing water (Dellapenna, 2001, ch. 7). A third possibility is active public management of water as a public resource. Examples include the American system of regulated riparianism and similar regimes in a growing number of other countries (Dellapenna, 2001, ch. 9). Actual legal regimes might mix aspects of these systems of water law.

The correspondence of these systems to theoretical models enables us to predict whether the model can adapt to changing circumstances, or whether an entirely new system must be substituted. Treating water as common property leads to tragic over-exploitation as soon as water begins to be scarce (Hardin, 1968; Rose, 1990). Yet because of the protection of third-party rights, small-scale transfers of water rights among farmers or ranchers making roughly similar uses at similar locations are the only ones that regularly occur without heavy state intervention (Dellapenna, 2000). Treating water as private property therefore freezes patterns of use rather than create a market. Because markets fail, private property systems experience increasing stress as demands surge and unclaimed water becomes rare. What works best, albeit imperfectly, is treating water as public property managed by public agencies.

One could simply decree that third party rights be ignored, enabling water markets to operate. The Chilean Water Code arguably is an example. The Chilean Water Code has had much less effect than its champions claim because water users resist it (Bauer, 2004). Moreover, to the extent it has been implemented, it caused wealth to be transferred from small (and relatively poor) users to

large (and relative rich) users – a normal effect of such policies (Dellapenna, 2000). Today, nations the world over and states in the United States increasingly turn to active public management for water management (Dellapenna, 2001, ch. 9; Dellapenna, 2003). Such public management might well fall short of its goals and could undoubtedly be improved by using economic incentives (Cummings and Nercissiantz, 1992). One should not confuse this conclusion with the notion that markets will work. One simply cannot have much confidence in a private property/market system, given the scarcity of actual empirical evidence of such a system and the transaction costs and externalities as barriers to the successful operation of a water market.

Social justice

Gender equality

Women and children bear a large proportion of the burden of agricultural labour in Africa and southern Asia. In many of these countries, more than half the women are involved in agriculture (Figure 20.2). Much of their work is concerned with water. Yet women in the developing world generally have little say in the management of water resources beyond collecting water and

tending the crops. A curious feature of the world map is that the burden is very different between many individual countries that are very similar in terms of climate, social structure and religion. This is true of many Islamic countries in the Middle East and Asia-Pacific.

In her article (below), Marcia Brewster reviews the work of the UN-Water's Interagency Task Force on Gender and Water (GWTF), which was established in 2003 in support of the International 'Water for Life' Decade, 2005–2015.

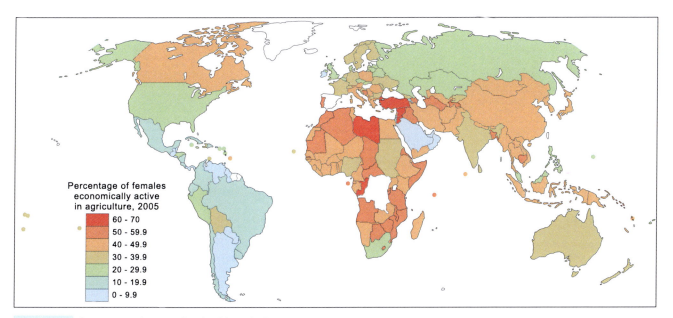

Figure 20.2 Percentage of women involved in agriculture

Gender, water and sanitation

Marcia Brewster, Task Manager, Interagency Gender and Water Task Force
United Nations Division for Sustainable Development

WATER FOR LIFE
2005–2015

The International Decade for Action, 'Water for Life' 2005–2015, coincides with the period set for achieving the Millennium Development Goals in water and sanitation set by Heads of State in 2000 and at the World Summit for Sustainable Development in Johannesburg in 2002. The Interagency Gender and Water Task Force has been entrusted with implementing the gender component of the Water for Life Decade. Countries and their partners in civil society will have to scale up efforts to meet the needs of women in Africa, Asia and other poor parts of the world who lack access to safe drinking water and basic sanitation. These amenities that we in developed nations take for granted are a matter of survival, personal dignity and security for those women.

Control over water is a source of power and economic strength, and it can be a root cause of socio-political stress. Involving both women and men in water management is at the heart of conserving our finite water resources, safeguarding health through proper sanitation and hygiene, and enabling people to rise out of poverty.

In most societies, women have primary responsibility for water supply, sanitation and health at the household level. Water is necessary not only for drinking but also for food preparation, care of domestic animals, crop irrigation, personal hygiene, care of the sick, cleaning, washing and waste disposal – all activities that are largely the responsibility of women. Access to water and sanitation provides greater self esteem, reduced exposure to the threat of violence and health hazards, and increased time available for education, childcare, growing food and income generation from small enterprises.

Some of the major factors that need to be addressed to implement a gender approach to water resources management and sanitation are discussed below.

Providing accessible, clean water is essential for achieving gender equality. For example, in Morocco the Rural Water Supply and Sanitation Project of the World Bank aimed to reduce the burden of girls 'who were traditionally involved in fetching water' in order to improve their school attendance. The project showed that girls' school attendance increased by 20 per cent in four years, attributed in part to the fact that girls spent less time fetching water. At the same time, convenient access to safe water reduced the time spent collecting water by women and young girls by between 50 and 90 per cent (World Bank, 2003).

In the Water for African Cities (WAC) programme, supported by UN-Habitat, an effort is being made to mainstream gender in the provision of adequate water and sanitation in poor urban communities. UN-Habitat, in partnership with the Gender and Water Alliance, is training managers and staff in municipal utilities and water boards to recognize the crucial importance of involving both women and men in water and sanitation management.

Women's access to land ownership, and thus access to water supply, has been limited by the discriminatory legal rights and customs of many countries. Equitable access to water and land for productive use can empower women and address the root causes of poverty and gender inequality.

Each year, more than 2.2 million people in developing countries die from diseases associated with lack of access to safe drinking water, inadequate sanitation and poor hygiene. Simple measures, such as providing schools with latrines and promoting hygiene education in the classroom, can reduce health-related risks for all, and improve attendance of girls and women in schools. In Mozambique, where 80 per cent of all primary schools had no toilets or hand-washing facilities, UNICEF supported the construction of latrines and hand-washing facilities for boys, girls and teachers. Combined with hygiene education, the latrines provided a safer, healthier learning environment, and encouraged girls' education (UNICEF, 2003). Whereas older girls used to drop out of school for lack of privacy, they are now completing their basic schooling.

Targeting women for training and capacity building is critical to the sustainability of water and sanitation initiatives, particularly in technical and managerial roles to

Figure 20.3 Girls carry plastic containers of water they have collected from a new water point installed with UNICEF assistance in the village of Rubungo in Rwanda, outside the capital Kigali. Children can spend many hours collecting water in developing countries, depriving them of any opportunities for education

Figure 20.4 A woman with a baby on her back smiles as she fills a clay pot with water from a tap, as other children and women wait their turn, in Hsai Khao village, Shan state, Myanmar (Burma). UNICEF supplied pipes and fittings for the spring-fed, gravity-flow water supply system. UNICEF has been promoting increased access to safe water and sanitation throughout the country, along with immunization programmes, improved education and assistance to maternal and child health centres, especially in remote areas

ensure their presence in the decision-making process. As of mid-2005, there were about 40 women ministers of water or environment throughout the world. This is a concrete illustration of gender mainstreaming; these ministers constitute the critical mass needed to get gender integrated into water and sanitation policies and programmes.

In South Africa, Lesotho and Uganda, all of which have women as ministers of water, affirmative action programmes have been introduced in the water sector to train women for water and sanitation related careers, including science and engineering. At the local level, women have found their voices and have now been trained to locate water sources in the village, to decide on the location of facilities and to repair pumps. Since these changes were made, the incidence of pump breakdown has decreased

considerably. Our UN Interagency Task Force on Gender and Water has been working with some of the dynamic women ministers to devise ways to secure access to

Figure 20.6 A woman fills jerry cans with water at a well set up with UNICEF support in the village of Manyewa, 18 km from the town of Plumtree, Zimbabwe. A drought emergency was declared in the district in 2002. Almost 13 million people – half of whom were children – required immediate assistance in Zimbabwe, Lesotho, Malawi, Mozambique Swaziland, and Zambia

Figure 20.5 Girls collect water from a Mark II handpump installed by UNICEF outside a house in the village of Awdalok, near the town of Koya in Kurdistan, northern Iraq

water and sanitation for women and men during the Water for Life Decade.

A study of community water and sanitation projects in 88 communities in 15 countries found that projects designed and run with the full participation of women are more sustainable and effective than those that are not. This supports an earlier World Bank study that found that women's participation was strongly associated with water and sanitation project effectiveness.

The volume of external financial assistance is not likely to grow fast enough to meet water and sanitation needs around the world. Formal and informal women's organizations and networks can play important and stimulating roles in mobilizing resources for sustainable and equitable water and land management projects.

A controversial issue is the privatization of water services versus the right to water. Problems arise when corporate profit motives supersede attention to human needs and rights. As water collectors, it is women and girls who often pay more dearly in these cases. Consultation with water users, both women and men, is needed before community or municipal systems are earmarked for private sector management.

Women's lives around the world are closely connected to water. The Water for Life Decade recognizes the central role that women play in providing, managing and safeguarding water and as the main role models within the family for sanitation and hygiene. During the Decade it is crucial to ensure the full participation and equal involvement of women in water-related development efforts and to approach water and sanitation issues from a gender perspective.

Conclusions

Whether access to clean water is a human right or just a need, whether water should be private property, common property or controlled by public agencies, and whether women should be afforded a greater role in decision making over water privatization and community management in developing countries are all crucial issues. Sustainability depends on the answers the world gives. It also depends on how countries apply and police these solutions – it depends on sensitive and sound frameworks for national and international law. Enforcement at the international level is a major current weakness.

Discussion topics

- Consider the practical implications of the definition of water sustainability.
- Consider the implications of the right to water.
- Explore the issues involved in devising enforceable international law.
- Explore issues of gender equality in developing countries.

Further reading

Details of national and international law can be followed up in the publications of Joseph Dellapenna, for example:

Dellapenna, J. 1997. Population and water in the Middle East: The challenge and opportunity for law. *International Journal of Environment and Pollution* 7, 72–111.

Read the full policy brief developed by UN-Water's Interagency Task Force on Gender and Water (GWTF) and other publications:

Gender, water and sanitation – a policy brief. Produced by the Interagency Task Force on Gender and Water Sub-programme of UN-Water and Interagency Network on Women and Gender Equality (IANWGE), UN Water Policy Brief 2, 13pp. At: www.unwater.org/downloads/.

Issues of ecologically sustainable management are discussed in depth in:

Falkenmark, M. and Rockström, J. 2004. *Balancing Water for Humans and Nature: the new approach to ecohydrology.* London, Earthscan, 247pp.

Hardin's famous paper is a classic:

Hardin, G. 1968. The Tragedy of the Commons. *Science* **162**, 1243–50.

The arguments for and against water markets are covered in:

Public Policy Institute of California, 2003. Managing California's water market: issues and prospects. Available at www.ppic.org/.

21 Aid for the developing world

'Trade, unlike aid, confers dignity and self-respect to both sides of the exchange.'

Senator George Mitchell,
former Democratic leader, US Senate

To aid or not to aid?

The quotation from Senator Mitchell encapsulates a common argument advanced by many on both sides of the gift – donors and receivers. Few would doubt that this is a worthy long-term aim, but many, including the United Nations, would realistically say that aid is essential in the shorter term, and particularly so for impoverished nations, failing economies and disaster situations.

The UN encourages developed nations to contribute 0.7% of GDP in aid to poor nations. It is the key element in the drive to 'Make poverty history'. However, in practice, the level has been more like 0.3–0.4% since 2000. Most aid comes from the USA, with Germany, the UK and France each contributing about half as much as the USA.

Nevertheless, economists are strongly divided on the efficacy of aid. Jeffrey Sachs argues that more aid is needed for Africa. Another former World Bank economist, Dambisa Moyo, herself African, argues in her book *Dead Aid* (2009) for a phasing out of aid within five years. The timescale is clearly too short, but she makes the point that more than $1 trillion has been given in aid to Africa over the last 60 years yet people are worse off, the numbers of poor are rising. On its own, this does not prove that aid has been unsuccessful, given the burgeoning population – how many more poor might there have been without international aid? But Moyo argues that over-reliance on aid leads to a vicious cycle of dependency, corruption, and distortions in the markets, which undermines the prosperity of local farmers and increases poverty. She points to South Africa and Botswana as countries that successfully rely on their own regulated free markets rather than aid. And she finds merit in China's approach to Africa, more pro-business, less pro-

charity. But is that the full story of China's involvement? Much governmental aid in the past has been given with an eye to political or commercial advantage. China is also operating out of self-interest (see *The changing geography of wealth and influence* in Chapter 4).

Corruption is one of the key problems which aid has undoubtedly nourished (see Chapter 4 Governance and Finance). It is a major pillar of Dutch reporter Linda Polman's argument in her book *War Games: the Story of Aid and War in Modern Times* (2010) based on her experience with aid agencies in Africa. Corruption can be as obtuse as delaying or refusing permits for aid agencies that are not going to provide money for jobs and largess that officials can use to their advantage. Aid is big business now, she says. She reports cases of aid agencies forcefully competing with each other for contracts to serve emergencies – so many now answer humanitarian calls that local officials can choose the ones that offer the most to them, including lucrative roles as advisors or supervisors. Agencies need to invest considerable time and money to set up operations and so trying to maintain those activities for a reasonable period of time can also make them vulnerable to official demands.

The argument continues, but a balanced judgement would probably be that aid is necessary for now, even if it is inefficient. Few would disagree that ultimately the best way to combat poverty is job creation. This should be a prime aim, to bring money and dignity to young Africans. In an ideal world, international aid should perhaps be phased out within a foreseeable timeframe, but what that timeframe should be is currently illusive. There seems no prospect in the foreseeable future of developing countries being able to cope with some of the extreme disasters experienced in recent years, and these are likely to be magnified by climate change. The

397

words of the 1980 Brandt Report from the Independent Commission on International Issues are even truer today: that many of the poorest countries 'are afflicted by droughts, floods, soil erosion and creeping deserts, which reduce the long-term fertility of the land. Disasters, such as droughts, intensify the malnutrition and ill-health of their people and they are all affected by endemic diseases which undermine their vitality.'

International aid under the UN banner

Individual governments and the EU give aid directly to developing countries, but they also contribute through their support of UN operations. The activities of agencies within or affiliated to the United Nations are as vital as they are diverse and innumerable, ranging from physical supplies to the social, economic and legal framework. Many are directly or indirectly concerned with water and sanitation. Meeting the Millennium Development Goals (MDGs) and coordination by the UN Office to Support the International Decade for Action 'Water for Life' 2005–2015 is resulting in greater collaboration. The annual World Water Days coordinated by the UN Environment Programme (UNEP) demonstrate progress and alert the wider public to the water crisis, with input from FAO Water, UNESCO, the World Health Organization (WHO), the UN Development Programme (UNDP), UN–Habitat, and the UN Economic Council for Europe (UNECE). Similarly, the UN World Water Development Reports coordinated by Unesco publicize the work of over 20 UN agencies. The work crosses the boundary between 'pure' science and measurement to projects aimed at relieving hunger, thirst, poverty, injustice and disease. Aid may be financial and practical, or through

training and knowledge transfer, like that available through the UNESCO–IHE and the UN University.

Here is just a very brief and partial summary of the roles of some of the main agencies involved in water-related activities. UN–Habitat is concerned with improving the lives of the urban poor and aims to help 100 million slum dwellers by 2020. The International Fund for Agricultural Development (IFAD) is also concerned with environmental sustainability and improving drinking water supplies. UNEP is concerned to improve knowledge transfer for sustainability and to promote PPPs. FAO is particularly concerned to develop a fairer trading system in collaboration with the World Trade Organization (WTO). This is supported by the UN Conference on Trade and Development (UNCTAD), which aims to ensure secure markets for people in poverty. UNIFEM (the Development Fund for Women) is concerned with women and poverty, and the International Labour Organization (ILO) aims to improve access to decent employment.

The MDGs relating to alleviating poverty and eradicating extreme poverty, reducing disease and child mortality, and promoting gender equality and environmental sustainability are especially high on the agendas of the UN Children's Agency (Unicef), WHO, FAO and Unesco. Aid in emergencies is provided, for example, by the World Food Programme (WFP), the UN High Commission for Refugees (UNHCR), Unicef and the WHO.

The recent assessment of progress towards the MDGs of improving access to safe drinking water and improved sanitation from WHO and Unicef (2010) is a little more encouraging than hitherto. It shows that while sanitation is definitely off-track, the drinking water goal could be on-track (Table 21.1).

Table 21.1 Projected progress towards the Millennium Goals in drinking water and sanitation

	Improved sanitation		Safe drinking water	
	Number unserved	Percentage	Number unserved	Percentage
1990	2.4 billion	46%	1.7 billion	23%
2015	2.7 billion	36%	672 million	9%
Result	Undershooting target by 1 billion		Out performing target of 12%	
Current status				
Access/collecting		61% served		87% served
		Off track		On track
In household		61% served		57%
		Off track		Off track

Source: WHO/Unicef (2010)

The role of charities

The fact that in June 2009 Britain offered £5 million of much-needed aid for Zimbabwe, in response to Prime Minister Morgan Tsvangirai's plea for help, *not* to the government but to charities working in the country, illustrates some of the more important qualities of charities: that they are apolitical and relatively free of corruption. Where the Burmese military junta refused foreign navies and airforces access to deliver disaster relief to victims following the 2004 tsunami and again in the Cyclone Nargis disaster in 2008, a few non-aligned charities gained entry. Thousands of refugees suffered and died in the camps of war-torn Darfur when Sudanese President al-Bashir forced 16 international charities, including Oxfam and Save the Children, to leave the country following his indictment for war crimes by the International Criminal Court in 2009. The charities were providing food, water and medical aid, which no foreign country was allowed to do. Among the charity workers expelled were the managers of water projects in several of the refugee camps. Save the Children USA and Save the Children UK were supporting over a million children in West Darfur camps. The UK arm had been operating there for over 50 years, providing long-term aid to the poor and water-scarce region, and the US arm joined in 1984 in response to the developing military conflict in the region.

The scale of the humanitarian effort undertaken by international charities is indicated by WaterAid's claim to have helped more than 13 million people gain access to clean water in over 17 countries in Africa and Asia. Charities provide some of the most flexible and wide-ranging aid programmes, covering everything from disaster relief to long-term infrastructure and education projects. The British Red Cross runs long-term water and sanitation programmes throughout Africa and Asia, including programmes for 130,000 people in Ethiopia, and poor communities in Zambia, Zimbabwe and Cambodia. This includes sanitation kits and health education. In Zimbabwe the Red Cross responded to the cholera epidemic that claimed over 4200 lives, helped drill 70 new boreholes and repair 400 latrines and 130 hand pumps.

The charities get most of their funds from private and corporate donations, but Save the Children's link up with an 'ethical' bottled water company in 2008 marks a new departure in fund-raising (see the section headed *Bottling it* in Chapter 8). Similarly, the carbon-offset charity Climate Care, which invests in green energy solutions, such as treadle pumps in India and hydropower in Tajikistan, is now part funded by the Co-operative Bank through a levy on new mortgages.

In disasters, charities can be the first to provide safe bottled water, water purification units or tablets, and oral rehydration supplies. The International Federation of Red Cross and Red Crescent Societies operates a variety of Emergency Response Units (ERUs), three of which are dedicated to water and sanitation – others include field hospitals and basic medical aid. These units comprise expert personnel and equipment ready for deployment at short notice to help local Societies following a disaster, normally for a maximum period of up to four months. Water and sanitation ERUs are on standby in Germany, Sweden, Austria, Spain and France: German water purification equipment was among the first to reach Burma after the 2004 tsunami. Water and Sanitation Module 15 is designed to provide up to 225,000 litres of freshwater a day to support 15,000 people and basic sanitation for up to 5000. This unit requires a local source of water that can be treated. Module 20 is aimed at sanitation and control of disease vectors, including hygiene education, for up to 20,000 people. Module 40 can provide water treatment and distribution using local water sources for up to 40,000 people and 600,000 litres a day. The ERUs include tents and equipment for refugee camps and support for telecommunications. The British Red Cross set up a mass sanitation unit in 2007 which was immediately deployed in the severe floods in Pakistan that year, and subsequently in 2008 in the devastating Chinese earthquake and in the cholera epidemic in Zimbabwe. When Cyclone Aila hit south-west Bangladesh and eastern India in May 2009, affecting ten million people, killing 300 and contaminating drinking water with saltwater, Save the Children deployed nine health teams to isolated communities using a water ambulance and distributed more than 1.2 million litres of drinking water, 15,000 water purification tablets and thousands of doses of oral rehydration supplies by vehicle and boat to treat and prevent water-borne disease and diarrhoea.

Whereas the Red Cross and Red Crescent Societies have divorced themselves from their earlier religious connections, in order to operate in potentially hostile regions, there are still a large number of religiously based charities. All, however, operate on purely humanitarian grounds, irrespective of religious belief, race, nationality or gender. In the floods that devastated large areas of central Africa from Zambia to Mozambique and Zimbabwe to Angola early in 2008, rapid response teams organized by Churches Together International provided water treatment supplies, latrines, mosquito nets and soap. When Cyclone Sidr, the biggest cyclone in 16 years, hit Bangladesh in November 2007, killing more than 3300 people, Christian Aid and its local partners helped evacuate people from the floods, and distributed rehydration salts and water purification tablets to over 100,000 people.

This illustrates a third significant advantage that international charities have: that they can call upon local knowledge and assistance from fellow national charities operating within the afflicted country, while offering reciprocal help in the form of international expertise and funds. Caritas International is one of largest humanitarian networks in the world, operating in 162 countries. The Catholic charity CAFOD is part of that network, offering relief in conflicts and natural disasters. Among its many longer-term projects, CAFOD has helped provide clean water for drinking, washing and irrigation in Zambia aimed specifically to free children from the burden of six-hour walks a day to collect water and so allow them to go to school. CAFOD promotes cheap and simple domestic water filters that consist of a plastic barrel filled with sand, charcoal and gravel filters and a storage drum for the clean water.

Charities have also been prominent in running health and hygiene campaigns. Save the Children pioneered the Stop Polio Campaign. Many charities have participated in the WASH campaign to contain the spread of cholera and diarrhoea.

Nor is the contribution of charities limited to the developing world, although that is where most aid is needed. The charity Loaves and Fishes began providing water, food and washing facilities to the tent cities that started to grow with unemployed and dispossessed migrants around Sacramento, California, during the credit crunch in 2009.

In a very recent development, some bottled water companies have also started charities aimed solely at humanitarian aid rather than commercial advantage (see the section headed *Bottling it* in Chapter 8).

Unfortunately, not all charities are entirely immune from political problems or corruption. In their book *The Charitable Crescent – Politics of Aid in the Muslim World* Jonathan Benthall and Jérôme Bellion-Jourdan relate a number of problems besetting some organizations. In the USA, The Holy Land Foundation charity was found guilty of funding Palestinian charities allegedly linked to Hamas, and in the UK two banks, Lloyds and Barclays, refused to act as banks for two British Muslim charities mounting relief for Pakistan because of suspected links to extremists. But these are still part of a very small minority of cases in a largely blameless and very effective group of organizations that operate impartially across ethnic and religious boundaries, including the flagship organizations, the Red Crescent and Red Cross.

A few of these organizations are listed in Table 21.2.

Table 21.2 Selected international charities and non-profit NGOs involved with water

Organization	Aims and roles
Aquaid	This is actually a commercial company that sells water coolers and dispensers, but it also contributes large sums to charity for water-related schemes, in partnership with Christian Aid and Pump Aid. It has donated over £4.8 million.
British Red Cross	Deals with thousands of emergencies each year, alone and in collaboration with emergency services, followed by long-term aid to recover from disease and conflict.
Business and Professional Women (IFBPW) Taskforce Women for Water	Focuses on bridging gaps between principles and practice in sustainable water management and in particular on the role of women in this process. The key question is not should women play a major role in improving water management, but how gender issues can be incorporated into better water management schemes.
CARE International	Aims to serve individuals and families in the poorest communities in the world. CARE's programmes include Water and Sanitation, and Environment.
Children's Water Fund	A project of Children's Hunger Relief Fund. The aim of this fund is to provide clean, disease-free water systems to families otherwise forced to drink disease-infested water in Africa.
Churches Action on Relief and Development (CARD)	International work based in Malawi offering disaster relief, e.g. earthquakes in Haiti and Chile, floods in India, and supporting projects in Iraq and throughout Africa.
Climate Care	A British carbon-offset charity using contributions from air passengers and others to promote green energy solutions such as treadle pumps for irrigation and hydropower.
Eau Vive	French charity focusing on development in the Sahel rural settlements. Through concrete actions, it aims at creating capacity building in these areas.

Organization	Aims and roles
FogQuest	An innovative, international, non-governmental, non-profit organization, which implements and promotes the environmentally appropriate, socially beneficial and economically viable use of fog, rain and dew as sustainable water resources for people in arid regions of developing countries.
Global Nature Fund	A non-profit NGO working for environmental protection. In 1998 GNF launched the global network Living Lakes.
Global Water	An international non-profit, non-governmental organization. By emphasizing volunteer help, it serves as a vehicle for caring individuals to get involved in the world-wide effort to provide clean drinking water for developing countries.
Green Cross International	Works to prevent conflicts in water-stressed regions. It promotes the need for international mediation to prevent and resolve water related conflicts, the need for an international fund for water, to be used particularly in times of emergency, and the recognition that a basic entitlement to safe water is a universal human right.
Hydraulics without borders	Aims to make the experience and capability of water specialists available for those who need it but lack financial means to pay for related services.
Hydroaid	An Italian association concerned to train personnel in order to build capacity for sustainable water management in developing and emerging economies (see Rossella Monti's article 'Hydroaid', below).
International Development Enterprises	A UK body designing and developing low-cost systems for irrigation, linking with local manufacturers, e.g. working through Oxfam. Supported by BT, Marks & Spencer and Arup.
Islamic Aid	Helps install wells, tube wells, hand pumps and gives training to communities. Particular current concern for Gaza's people and wrecked infrastructure.
Islamic Relief	Fighting poverty, restoring wells, installing water purification plants and supply systems, e.g. in West Java, using bamboo pipes, which are cheaper and do not melt in heat. Also promotes Islamic microfinance – interest-free loans.
Lifewater International	A non-profit organization of Christian water resource specialists based in the USA. It has over 150 serving volunteers, including well drillers, geologists, engineers, health care professionals, scientists and businessmen. Volunteers train nationals in developing countries with technical skills to improve their drinking water supplies.
International Federation of Red Cross and Red Crescent Societies	The umbrella organization for numerous national societies, especially dealing with emergencies in war zones and natural disasters. Red Cross (Christian) and Red Crescent (Muslim) work together and now operate as secular organizations, partially funded by governments.
Oxfam	Originally a British charity now operating worldwide, especially focusing on drought, malnutrition, climate change and the world food crisis, with long-term projects, emergency relief and international campaigns.
Pump Aid	Promotes clean water and sanitation, especially through elephant pumps and toilets, and sustainable use of water for irrigating 'nutrition gardens' for communities.
Samaritan's Purse International	Provides disaster relief worldwide and promotes safe water and sanitation in Africa. Over 4000 water filters installed. Operates 'Turn on the tap' campaign. Based in Essex.
Save the Children	Worldwide operation from national branches in the UK, USA and Canada. Particularly focusing on the needs of children, including long-term water and sanitation projects and emergency relief. Pioneered the Stop Polio Campaign. Promotes education and supports the handwashing campaign to contain diarrhoea and cholera.
Tearfund	A Christian organization currently with a ten-year mission to release 50 million people from poverty through its network of 100,000 local churches worldwide, irrespective of race or creed. Focuses on helping communities adapt to climate change, providing safe water, sanitation and hygiene education. Operates in over 50 countries.
The Center for Sustainability, Environment, Equity and Partnership	Dedicated to issues in water quality. Devotes its time, tools, techniques and resources to address water issues at local, state, national and international levels.
Water for children in Africa	A charitable non-profit organization dedicated to providing safe water for children living in rural villages on the continent of Africa.
Water For People	A non-profit, charitable organization in the United States and Canada that helps people in developing countries obtain safe drinking water. It works with local partner organizations to provide financial and technical assistance to communities, depending on their needs.
WaterAid	An independent charity working through partner organizations to help poor people in developing countries achieve sustainable improvements in their quality of life by improved domestic water supply, sanitation and associated hygiene practices.
WaterLife Foundation	A global organization which helps small, under-served communities in the developing world create and maintain their own safe and sustainable drinking water supplies, leading to better health and higher standards of living.
WaterPartners International	A non-profit organization that addresses the water supply and sanitation needs in developing countries, promoting community water projects.

401

Hydroaid – The Water for Development Management Institute

Rossella Monti, Director

Hydroaid is a non-profit association operating from Italy. Its main goals and objectives are assistance in the field of training and capacity building in the sector of management and protection of water resources and related topics. It was born out of an initiative from public, financial and academic authorities in Piedmont and was founded in Turin in 2001. It holds courses every year.

Hydroaid can count on more than 600 specialized officials and technicians working in 60 developing and emerging countries affected by critical water problems through its network of former participants, more than 25 per cent of whom are women.

The mission

Hydroaid offers training, know-how transfer and capacity building. This is fundamental in fostering local self-sufficiency and the ability to manage local resources through modern systems aimed at sustainable development.

The approach

Many past experiences have proved that infrastructure development without adequate training support can give poor results in terms of the survival, maintenance and long-term management of the infrastructure. The results of appropriate programmes of capacity building are not immediate, but they are fundamental in fostering a better management of public water resources, with positive effects on many related aspects.

The main training courses

Institutional Course: A postgraduate course in the Management of Water Resources and Services for engineers and managers coming from emerging and developing countries has been held in Turin at the International Labour Organization's International Training Centre (ITC ILO), every year since 2002.

Masters Degree: Hydroaid has run a course on 'Integrated Management of Environmental Sanitation' in Brazil in collaboration with Brasilia University since 2005. It takes students from Brazil, Latin America and the Portuguese-speaking countries of Africa.

E-Learning modules: Hydroaid recently started the design and delivery of e-learning training courses on waste and water-related environmental issues.

Conclusions

International aid is still essential, despite doubts in many quarters. Doubts are largely founded on four issues: the need to empower developing nations to support themselves, the failure of many philanthropic schemes designed and funded by western countries due to lack of understanding of local circumstances, corruption in the Third World, and aid given with political or commercial advantages in mind.

In many ways, charities and NGOs are better able to address these issues by having on-the-ground understanding of the problems and links with local organizations with similar aims. But ultimately, charities have less financial muscle than western governments.

International agencies like members of the UNO are increasingly focusing on 'capacity building' and 'technology transfer', training personnel from developing countries to manage and improve their own resources in light of their own, particular problems and cultural backgrounds.

Discussion topics

- Explore the issue of aid.
- Learn more about the work of international aid agencies under the wing of the UN Organization.
- Visit the websites of some of the charities mentioned in this chapter and follow some of their projects.

Further reading

The question of international aid is given detailed treatment in:

Moyo, D. 2009. *Dead Aid – why aid is not working and how there is a better way for Africa*. Penguin Books and Farrar, Straus and Giroux.

Polman, L. 2010. *War Games: the story of aid and war in modern times*. Viking.

The UN and charity websites detail the latest humanitarian campaigns and projects.

Conclusions

22 Is sustainability achievable?

We will intensify our efforts to reach internationally agreed upon goals ... to improve access to safe and clean water, sanitation, hygiene and healthy ecosystems in the shortest possible time ... support scientific research, education, development and adoption of new technologies ... (for) sustainable use and management of water resources ... and promote cooperation on ... transboundary water resources.

Ministerial Statement, World Water Forum, Istanbul, 2009

The first decade of the twenty-first century saw a significant rise in the political profile of water. Sadly, it still lags behind climate change as an issue for governments and the international community at large. Yet in many ways it is a more immediate problem. It is a more visible problem. When it comes to a shortage of water, there can be no arguments like those as to whether an apparent increase in hurricane intensity is due to global warming. There is no need to wait to prove a trend or to diagnose a cause. It is also potentially a more tractable problem. Future climate change will alter conditions, but most of the problems are visible in the here and now. And predicting the future is above all a question of population dynamics. Foretelling the general trend for water shortages does not require the sophisticated computer models needed in the case of climate change, although they will undoubtedly help in policy development, especially at local and regional levels.

It is at the local and regional levels that most of the problems will be solved, yet there are issues that do require broader and even global policy making. The various arms of the UN Organization and the international scientific, economic and legal organizations need to continue and in some cases increase their involvement with water crises.

What is now very apparent is that it is misleading to talk of a single global water crisis. There are multitudes of crises facing individual regions and countries, each with their own sets of causes and their own best solutions. Conservation, reduced waste and more efficient use are indeed fairly universal needs, but the water issue is not the same as global warming. It cannot be solved by a Kyoto-style agreement. Saving water in water-rich countries will not help the water-poor developing countries in the way that reducing carbon emissions may. At least, not in any direct way. But we are in a globalized world, and trade is its lifeblood, which means that virtual water trading is an important issue that needs to be studied and quantified. Does it make sense for Kenya to be the largest outside supplier of fresh flowers to Western Europe, when the country is suffering the results of a multiyear drought that brought four million people to the brink of starvation in 2009? It may make economic sense, but does it make water sense and moral sense?

The newly formed Engineering the Future alliance in the UK took this point up in their publication *Global Water Security – an engineering perspective* (2010). The alliance comprises the respected Royal Academy of Engineering, the Institution of Civil Engineers (which includes the British Hydrological Society) and the Chartered Institution of Water and Environmental Management. They claim that the increasing demand for food and goods by developed countries is putting severe pressure on developing countries that are already short of water, and they encourage the West to do more to help these countries solve their water problems. The alliance also suggests that if the water crisis becomes critical in these exporting countries, it could become a serious threat to the future economy of the UK itself.

While these arguments are cogent, the element that received most publicity is that importing two-thirds of the UK's virtual water is unsustainable. This does not have to be the case. Indeed, it would be impossible for the UK, or any other major developed country – Japan has a similar level of import dependency – to become anywhere near self-sufficient. Nor is it desirable from the viewpoint of the exporters who are making money to support their own country's development. By all means improve self-sufficiency and reduce dependency on outside suppliers, but trade does make the world go round. What the alliance's statement does do is to focus attention on the fact that greater care is needed in sourcing goods. The Fair Trade movement, organic certification, and the Sustainable Forest Management and Forest Stewardship Council marks systems offer role models. This means tracking the source of goods and selecting those that are produced in an environmentally sustainable way and best benefit the society at source.

It is therefore apposite that in 2010 the UN warned big business to expect legislation on water footprints. The Carbon Disclosure Project has switched focus from climate change to water. It aims to produce the first water footprint report on the world's top businesses based on research by Global Compact, the UN green policy group. The Project is evaluating methods for assessment, and using pilot studies in a few businesses in order to design an international standard. Reverting to its original aims, it will also analyse the amount of CO_2 emissions produced during the transportation, treatment and use of water.

The task will not be easy. It will need a lot of data that is not readily available, and some that is not currently collected at all. Accurate water meter data is just one. Also, impacts vary. Just as the most recent virtual water trading calculations have embraced the fact that the same product can require very different amounts of water to produce it in different climatic, social, economic and technological environments, so also the environmental impacts may differ according to the timing and location of water abstraction – for example, whether in the wet season or the dry season, during a drought, from groundwater or surface water, from a large or a small source that is rapidly replenished or not. The same applies to the subsequent release of wastewater, at which point it becomes a question of water quality as well as quantity.

The current Project will focus on large multinationals, but ultimately it must trickle down to all businesses, just as carbon footprints have done. The same hurdles will have to be navigated, especially how to price footprints and what if any free permits might be provided by governments. The very cheapness of water has been a factor in delaying its emergence on corporate agendas. Is it possible to have a global pricing system when economies vary so much? If separate pricing systems were produced by individual nations or trading blocs, would they lead to businesses migrating to cheaper countries? This is the normal result of differences in taxes, regulation and other costs. The problems are immense. Would China, for example, accede to outside regulation and inspection? This was the main stumbling block that prevented agreement at the 2009 Copenhagen climate summit. And there was so much more international pressure for an agreement on climate change than can yet be mustered for water. In reality, the clamour over carbon emissions may have been diverting attention from the water crisis for some time.

It was more than unfortunate that water did not figure in the final draft of the UN climate summit in Copenhagen 2009: it was reprehensible. The Global Water Partnership, the Global Public Policy Network on Water Management, the Stockholm International Water Institute and the Stakeholder Forum were united in telling the conference they were making a dangerous mistake. The session on 'Bridging the Water and Climate Change Agendas' seemingly had no effect, despite strong arguments from participants, including the International Union for Conservation of Nature. Even more oddly, water issues appeared and disappeared many times in the series of drafts from preliminary meetings, the last one written in Barcelona, the city facing one of the most critical water shortages in Europe and less than a day trip away from Zaragoza, the city that hosted the most innovative international expo to date on 'Water and Sustainable Development' in 2008, billed as an expo with no expiry date! Surely this cannot be blamed on oversight? It looks like a deliberate move by vested interests. There is a danger with any system of taxes, fines and regulations that governments and commercial interests see an opportunity for making money. Carbon-offset scheme scams are a case where apparently good ideas have been perverted by some for dishonest gain. The climate change agenda itself was delayed for decades by oil industry lobbying in America.

Roots of the crisis

The burgeoning world population is the largest single cause of the water crisis, but there are many other significant driving factors. Some of these have been with us for a while, like wasteful practices, especially in irrigated agriculture or with ageing and poorly maintained infrastructure. But many are new or are taking on greater importance in the twenty-first century.

Climate change appears to be increasing the frequency of extremes. Recent climatic trends may be partly or wholly natural, but there is a new set of threats that are totally man-made. These range from terrorism to the apparently innocuous change of ownership of water resources. Somewhere in between are some critical socio-economic trends that have been intensifying during the early years of this century: the geography of wealth is changing. More people now live in cities than in the countryside. Last century, people in cities tended to consume and waste considerably more water than the rural population: typically between four and six times as much. But most of the new urban population is in the Less Economically Developed Countries, where a considerable proportion live in makeshift housing and do not have mains services. This may reduce water consumption, but it also increases the risks from water-related diseases.

Finally, security means more than having the resources. It means having control of them as well. The commercialization, privatization and globalization of water resources began with the exploitation of water for profit, with water companies being driven by the need to produce a dividend for investors ahead of providing a service to customers. Since the turn of the millennium, it has developed into international takeovers by large multinational corporations with interests as diverse as electricity and hotels. The latter development is perhaps the most worrying, especially for developing countries, as they can lose control of their national resources.

The whole question of funding for water resources projects needs fundamental re-evaluation to ensure water security. This ranges from the long-held policies of the World Bank and the IMF to encourage privatization and public-private partnerships through to the recent highly leveraged forays of private equity into water companies, and, with the collapse of the easy credit market in 2009, the beginning of a switch to 'sovereign wealth funds' (SWFs) to finance takeovers. SWFs can be highly political like the China State Investment Corporation, adding an element of foreign control that is even more difficult to shake off than the profit-oriented private companies. China's expanding investment in Africa, building dams and offering loans for water projects in return for access to minerals, is repeating much of the

self-interest that marred colonialism. Combined with its policy of non-interference with local politics, however, it is bolstering some corrupt regimes to the potential detriment of the people.

The prospects

The international community and countless individual nations are finally realizing the perilous state of many water resources. Access to adequate supplies of clean water and sanitation is a fundamental human right and the starting point for wellbeing and a civilized life. The UN Millennium Development Goals aim of halving the number of people without access to safe water by 2015 is laudable and essential, but we are already halfway there in time and way short on the numbers. International aid to developing countries has fallen well short of the promises. And the prospects seem bleaker following the credit crunch of 2008–9.

At the G8 summit in Evian in 2003, world leaders produced a Water Action Plan, which included improving access to financial resources; to give high priority in official development aid to environmentally and socially sound proposals as a catalyst to mobilize other monetary sources; to encourage investment from international financial institutions like the World Bank and IMF; and to promote public-private partnerships (PPPs). We have heard less on water from the G8 since then, and at the time of writing (2010) the financial climate does not bode well. During the IMF's multiple bail out of countries like Iceland, Hungary, Serbia and the Ukraine in 2008, the British government expressed the fear that even the IMF's wealth may not be enough to support all the needs of the faltering global economy: small hope for many water projects.

President Obama's determination to redesign the architecture of the global financial system to avoid a repeat of the financial crash, and similar support from countries like the UK and France, is encouraging. If it can be achieved, if risky investment banking can be separated from retail banks offering loans, and if foreign arms of multinational banks can be separated from their parent banks so that contagion does not spread so rapidly around the world, then real progress will have been made.

Another encouraging move in recent years has been the decision by the World Bank and many regional development banks to include environmental and social impacts in considering proposals for new water projects along the lines proposed at Evian. As too was the first Asia Pacific Water Summit held in Beppu, Japan in 2007, which drew attention to the plight of water supply and sanitation in a region that accounts for 57 per cent of world water withdrawals and 70 per cent of consumption. The Asia Development Bank emphasized that effective water management is the key to safeguarding the continued development of Asian economies. Given the potential for conflict if water governance remains weak, the Bank proposed to double its investment in Asian water projects to $2 billion p.a. from 2007.

However, the G8 and World Bank policy of encouraging project financing through PPPs has been far from an unmitigated success, with companies often taking the profits and leaving the debts for the public purse. The credit crunch and global recession have reduced the amount of money available for new projects for years to come. Even established water firms may find it difficult when loans come up for refinancing. One month before the crash in the financial markets in September 2008, Water UK announced a proposed £27 billion spend to replace century-old pipework and other infrastructure between 2010 and 2015. It remains to be seen whether such plans, or the more vital plans for Third World developments, will come to fruition in the foreseeable future. And all the while there are more mouths to feed – perhaps 50 per cent more by 2050.

Climate change is generally viewed as aggravating the problems. It is certainly true that most of the current regions of water stress will suffer worse hardships in terms of the amount, reliability and variability of resources. But the warmer ocean will create more rainfall and increase *global* water resources by perhaps 5–6 per cent from 2050. It is the distribution of that rainfall that is the problem. However, the changes in physical resources are nothing compared with a 50 per cent reduction in *per capita* global water resources due to population growth.

New technologies and new approaches are developing that will help alleviate many of the factors that trigger and aggravate water stress and scarcity. Greater use of salt and brackish water will undoubtedly be made. World trade will grow and with it virtual water transfers will supplement local water resources. Notwithstanding its potentially negative effect on some exporting nations with water shortages of their own, in the Middle East and North Africa importing virtual water will be a vital survival strategy.

The Developed World now has more options at its disposal than ever before, ranging from geo-engineering to international legislation. Inventiveness and environmental knowledge are at an all-time high. Political attention and international collaboration is greater than ever. And peak water is not quite like peak oil: water returns. Ultimately, however, whether water stress is overcome or not depends on human management, on the dissemination of knowledge to all levels of society and government, on the eradication of greed and corruption, on amicable sharing of resources and above all on what happens to population growth. Perhaps populations will stabilize towards the end of the century in the emerging economies as affluence increases, as it has in many western countries. But there seems little prospect at present of eradicating the poverty that is a major force behind population growth and dwindling resources in Africa and many other parts of the world. The problems are regional. Many solutions are global, but in the final analysis the solutions must be rooted in the local physical and socio-economic environment if the misapplication of western approaches in the past is to be avoided in the future.

Notes

Chapter 1

Data on available water resources and water demand are taken from the Food and Agriculture Organization's AQUASTAT data files.

Chapter 2

Table 2.1. The effect of dams on regional water resources: The original data (black) is from Lvovich (1970). The 1992 values (orange) are from Jones (1997), who used dam capacity data from Gleick (1993) and the International Water Power and Dam Construction (1992). The 2003 values (red) are derived from dam capacities in Gleick (2003) and the Foundation for Water Research (2005), together with 2003 population data from *The Times* (2005).

There are many reasons for treating these data with caution. Lvovich's original estimates of the amounts of stable and unstable riverflows have been kept for consistency rather than because they are necessarily the best. Similarly, Chao *et al* (2008) found numerous errors in official archives of dam capacities, and their data are the best available to date, extending to the end of the 2000s in the table accompanying their paper online. Unfortunately, it is not very user-friendly: a pdf of around 30,000 entries listed by country, dam name and date from around 1900. Also, not all reservoirs are listed. The data collated by Chao *et al* only relate to large dams. They note that in China alone, the Chinese committee of the International Commission on Large Dams listed 83,000 small to moderate dams, but Chao and colleagues assume that the volume held in these is 'presumably insignificant' and omit them, as they do for similar dams elsewhere. Perhaps more significantly, many dams are not operated primarily for water supply purposes. There is also insufficient information on operating practices for dams around the world: if reservoirs are maintained full in the interests, for example, of leisure or hydropower when a flood occurs, they will not trap the floodwaters.

Chapter 3

Figure 3.3 Sahel rainfall: The graph is based on the Terrestrial Precipitation 1900–2008 Gridded Monthly Time Series (V2.01) available at http://climate.geog.udel.edu/~climate/html_pages/Global2_Ts_2009/README.global_p_ts_2009.html. The Sahel is defined slightly differently from the image shown in Figure 3.2 – 10–20°N – to include areas north of 8°N in the Horn of Africa. The Sahel now starts much further north than this in West Africa, but it dips down to the latitude of 8° in the Horn of Africa. Since the data are based on a grid, this makes the boundaries of the Sahel rather fuzzy.

Figure 3.4 Water prices: Data from UN Economic and Social Commission for Asia and the Pacific, Human Settlements, Urban Poverty Alleviation, at www.unescap.org/huset/hangzhou/urban_poverty.htm.

Chapter 5

Figures 5.1, 5.3, 5.6, 5.9, 5.17 and 5.21: data on water quality are taken from the GEMS/Water archive and Eurostat.

Mackenzie Basin footnote: September 2010, Canadian Environment Minister Jim Prentice appoints an independent panel of scientists to investigate pollution of the Athabasca River following photos of deformed fish and a report showing high levels of lead and mercury. He claims water management is a federal responsibility and rejects participation from the Alberta government. The report's author, Dr David Schindler, criticises RAMP as inadequate and visiting Avatar film director James Cameron says the oil sands development is unfettered and environmentally appalling and publicly lobbies governments and the UN Permanent Forum on Indigenous Issues.

Figures 5.22–5.26 maps of diseases: based on WHO data.

Figure 5.27 onchocerciasis: based on Maps 48–50 in World Health Organization 1989. *Geographical distribution of arthropod-borne diseases and their principal vectors.* Geneva, Vector Biology and Control Division. Unpublished document WHO/VBC/89.967, chapter 10, at: www.ciesin.org/docs/001-613/001-613.html.

Chapter 6

Figures 6.6, 6.7 and 6.15: based on GEMS/Water data.

Figures 6.11 and 6.13: based on data from the International Union for Conservation of Nature.

Chapter 8

Dambusting: Germany planned retaliation on dams near Sheffield. Attacks on dams were banned under Protocol 1 article 56 (1977).

Chapter 11

Map projections and data sources
The Interrupted Goode's Homolosine projection was selected for these global maps because it offers the best combined preservation of true area on the globe and of continental shapes. The projection is therefore best for representing areal data like rainfall, which is measured in millimetres per square millimetre of surface. Elsewhere in the book, the standard projection for global maps is Robinson, which is a good compromise projection for use where data are not measured in units per unit area. Robinson is commonly used by the UN Organization.

Climatic data are derived from the archives of the Climatic Research Unit, University of East Anglia, averaged over the period 1960–1990. This period is considered the most recent stable period prior to the marked warming phase in the 1990s. It is therefore the baseline climatology used in modelling studies of future anthropogenic climate change.

Riverflow data were provided by the Global Runoff Data Centre in Koblenz, and groundwater information by the WHYMAP programme (The World-wide Hydrogeological Mapping and Assessment) at the Federal Institute for Geosciences and Natural Resources (BGR), Hanover.

Seasonality index (Figure 11.11)
The rainfall seasonality index expresses the relative seasonal contrasts through the year by accumulating the differences between the rainfall in each month and the mean expectation if the annual precipitation were to be evenly distributed through the year. It is presented as a percentage according to Ayoade (1970):

$$\text{SI\%} = \left\{ \frac{100}{22\bar{R}_y} \sum_1^{12} \left| 12\,\bar{R}_m - \bar{R}_y \right| \right\}$$

where $\bar{R}_m$ is the mean monthly precipitation for each month, 1 to 12, and $\bar{R}_y$ is the mean annual precipitation. A climate in which all precipitation falls in just one or two months has a seasonality index approaching 100 per cent. A very equitable climate has an index of under 20 per cent. An index of zero indicates a perfectly even distribution of precipitation throughout the year.

Chapter 18

Improved capacity to develop DEMs: In October 2010 the European Space Agency's TanDEM-X radar satellite links operationally with its identical twin, TerraSAR-X, lauched in 2007, to provide high accuracy maps for the first time in many areas. The satellites will map the whole world in a 12 metre resolution grid with an altitudinal accuracy of 2 metres, providing valuable input to flood-warning models.

Figures 18.3 and 18.4: The mean annual number of days with snow cover was computed from the NOAA monthly snow cover dataset maintained by Dr David Robinson at Rutgers University. This dataset includes corrections to the NOAA land/sea mask and uses the Rutgers weighting scheme to partition each chart into the appropriate month (Robinson *et al*, 1993). The average was computed over the 1972 to 1998 period when the analysis method was relatively stable. The native resolution of the NOAA weekly dataset is relatively coarse – a 190.5 km polar stereographic grid centred on the North Pole. Snowcover duration isolines were created using a bilinear interpolation algorithm. Original maps and data supplied by Ross Brown, Meteorological Service of Canada.

Chapter 19

Figures 19.1 and 19.2: Time-continuous, digital River Nile gauge data were provided by D. Kondrashov, based on the gap filling of Kondrashov *et al* (2005). The original, gappy data were largely based on the compilations of Toussoun (1925), Ghaleb (1951), and Popper (1951), further revised and edited by Th. De Putter and D. Percival (see De Putter *et al*, 1998). River Nile discharge data were provided by E.A. Eltahir, based on the naturalization of Eltahir and Wang (1999).

Data sources for the teleconnections analysis by Peel and McMahon:

1 The data used in the SOI analysis were provided by the Australian Bureau of Meteorology, and calculations were made using the method of Troup (1965).

2 The NAO data were obtained from work at the Climate Research Unit, University of East Anglia, by Jones *et al* (1997).

3 The PDO data are from the Joint Institute for the Study of the Atmosphere and Ocean, University of Washington (Mantua *et al*, 1997).

Bibliography

Abbott, M.B., Bathurst, J.C., Cunge, J.A., O'Connell, P.E. and Rasmussen, J. 1986. An introduction to the European System: Système Hydrologique Européen (SHE). *Journal of Hydrology* 87, 61–77.

Acharyya, S.K. 2002. Arsenic contamination in groundwater affecting major parts of southern West Bengal and parts of western Chhattisgarh: source and mobilization process. *Current Science* 82(6), 740–744.

Alcamo, J., Döll, P., Henrichs, T. Kaspar, F., Lehner, B., Rösch, T. and Siebert, S. 2003. Development and testing of the WaterGAP 2 global model of water use and availability. *Hydrological Sciences* 48(3), 317–337.

Alcamo, J., Henrichs, T. and Rösch, T. 2000. *World water in 2025 – global modeling scenarios for the World Commission on Water for the 21st Century.* World Water Series Report 2, Center for Environmental Systems Research, University of Kassel, Germany, p.48.

Ali, M.A., Badruzzaman, A.B.M, Jalil, M.A. and Hossain, M.D. 2001. *Development of low-cost technologies for removal of arsenic from tubewell water.* Final Report to United Nations University, Tokyo, Japan.

Allan, J.A. 1998. Virtual water: a strategic resource: global solutions to regional deficits. *Groundwater* 36(4), 546.

Allan, J.A. 2001. *The Middle East water question: hydropolitics and the global economy.* New York, I.B. Tauris, 382pp.

Allan, J.A. 2003. Virtual water eliminates water wars? A case study from the Middle East. In: Hoekstra, J.A. (ed.) *Virtual water trade: Proceedings of the International Expert Meeting on Virtual Water Trade,* Value of Water Research Report Series No. 12, chapter 9, 137–145.

Alloway, B.J. (ed.) 1995. *Heavy metals in soils.* London, Blackie Academic, 368pp.

Alloway, B.J. and Ayres, D.C. 1993. *Chemical principles of environmental pollution.* Glasgow, Blackie, 291pp.

Almisnid, A. 2005. *Climate change and water use for irrigation: a case study in the Gassim area of Saudi Arabia.* Unpub. PhD thesis, University of East Anglia, UK, 245pp.

Almond, P. 2006. Beware the new Goths are coming. *The Sunday Times,* 11 June.

Altmann, P., Cunningham, J., Dhanesha, U., Ballard, M., Thompson, J. and Marsh, F. 1999. Disturbance of cerebral function in people exposed to drinking water contaminated with aluminium sulphate: retrospective study of the Camelford water incident. *British Medical Journal* 319, 807–811.

Anderson, T. and Snyder, P. 1997. *Water Markets: Priming the Invisible Pump.* Washington, DC, Cato Institute.

Annez, P.C. 2006. Urban Infrastructure Finance From Private Operators: What Have We Learned From Recent Experience? *World Bank Policy Research Working Paper* 4045.

Anon. 2009. Sigma Xi's year of water 2008. In: Critical Issues in Science, *American Scientist* 97(2), 175.

Apadula, F., Stella, G., D'Agata, C., Diolaiuti, G. and Smiraglia, C. 2001. Database and website of Italian glaciers: a contribution to the knowledge and the divulgation of a fundamental climatic indicator. *Proceedings of CNR Global Change, Censimento Ricerche Italiane, Roma,* 27–29.11.2000, 322–323. (In Italian.)

Arlosoroff, S. and Jones, J.A.A. 2009. Working group II: the threat from armed conflict and terrorism. In: Jones, J.A.A., Vardanian, T. and Hakopian, C. (eds) *Threats to global water security,* NATO Science for Peace and Security Series – C, Springer, Dordrecht, 123–128.

Arnell, N.W. 1992. Impacts of climatic change on river flow regimes in the UK. *Journal of Institution of Water and Environmental Management* 6(4), 432–442.

Arnell, N.W. 1998. Climate change and water resources in Britain. *Climatic Change* 39, 83–110.

Arnell, N.W. and King, R. 1997. The impact of climate change on water resources. In: *Climate change and its impacts: a global perspective,* Department of the Environment, Transport and the Regions and The Met Office, UK, 10–11.

Arnell, N.W. and Reynard, N. 1993. *Impact of climate change on river flow regimes in the United Kingdom.* Wallingford, Institute of Hydrology, 129pp.

Arnell, N.W., Jenkins, A. and George, D.G. 1994. *The implications of climate change for the National Rivers Authority*. Bristol, National Rivers Authority, R. & D. Report 12, 94pp.

Arnold, G. 2006. *Africa: a modern history*. Atlantic, 1028pp.

Avro, S.R. 2009. IBM to build the world's most powerful supercomputer for energy dept. *Consumer Energy Reports* at: www.consumerenergyreports/2009/02/09/ ibm-to-build-the-worlds-most-powerful-super computer-for-energy-dept/.

Ayoade, J.O. 1970. The seasonal incidence of rainfall. *Weather* 25, 414–418.

Bamber, J.L., Riccardo, E.M.R., Vermeersen, B.L.A. and LeBroq, A.M. 2009. Reassessment of the potential sea-level rise from a collapse of the West Antarctic Ice Sheet. *Science* 324(5929), 901. doi: 10.1126/ science.1169335.

Bandyopandhyay, J. and Perveen, S. 2004. The doubtful science of interlinking. *India Together*. Available at: www.indiatogether.org/2004/feb/env-badsci-p1.htm.

Bao, H., Fairchild, I.J., Wynn, P.M. and Spötl, C. 2009. Stretching the envelope of past surface environments: Neoproterozoic times. *Science* 323, 119–122.

Barlow, M. 2001. The last frontier. *The Ecologist* 31(1), 38–42.

Barlow, M. and Clarke, T. 2002. *Blue gold: the fight to stop corporate theft of the world's water*. Toronto, Stoddart / New York, The New Press.

Barlow, M. and Clarke, T. 2003. *Blue gold: the battle against corporate theft of the world's water*. London, Earthscan.

Barrett, J.H., Parslow, R.C., McKinney, P.A., Law, G.R. and Forman, D. 1998. Nitrate in drinking water and the incidence of gastric, esophageal, and brain cancer in Yorkshire, England. *Cancer Causes and Control* 9(2), 155–159. doi: 10.1023/A:1008878126535.

Barry, R.G. and Chorley, R.J. 2010. Atmosphere, weather and climate. 9th edition, London and New York, Routledge, 516pp.

Bass, B., Akkur, N., Russo, J. and Zack, J. 1996. Modelling the biospheric aspects of the hydrological cycle: upscaling processes and downscaling weather data. In: Jones, J.A.A, Liu, C., Woo, M-K. and Kung, H-T. (eds) *Regional hydrological response to climate change*, Dordrecht, Kluwer Academic Publishers, 39–62.

Bauer, C. 2004. *Siren Song: Chilean Water Law as a model for international reform*. Washington, DC, Resources for the Future Press.

Bell, B. 1970. The oldest record of the Nile Floods. *The Geographical Journal* 136, 569–573.

Bell, B. 1971. The Dark Ages in ancient history: The first dark age in Egypt. *American Journal of Archaeology* 75, 1–25.

Bell, B. 1975. Climate and the history of Egypt: The Middle Kingdom. *American Journal of Archaeology* 79, 223–269.

Beltman, B., Van den Broek, T., Barendregt, A., Bootsma, M.C. and Grootjans, A.P. 2001. Rehabilitation of acidified and eutrophied fens in The Netherlands: Effects of hydrologic manipulation and liming. *Ecological Engineering* 17(1), 21–31. doi: 10.1016/ S0925-8574(00)00128-2.

Benthall, J. and Bellion-Jourdan, J. 2009. *The charitable crescent – politics of aid in the Muslim world*. London, I.B. Tauris.

Benvenisti, E. 1996. Collective Action in the Utilization of Shared Freshwater: The Challenges of International Water Resources Law. *American Journal of International Law* 90, 384–415.

Benvenisti, E. 1996. Collective Action in the Utilization of Shared Freshwater: The Challenges of International Water Resources Law. *American Journal of International Law* 90, 384–415.

Bertell, R. 1999. Conflict of interest between IAEA and WHO. *WISE News Communique*, November 19. At: www.antenna-wise.nl.

Betton, C., Webb, B.W. and Walling, D.E. 1991. Recent trends in NO_3-N concentration and loads in British rivers. In: *Sediment and stream water quality in a changing environment: trends and explanation*, International Association of Hydrological Sciences Pub. No. 203, 169–180.

Biancotti, A. and Motta, L. 2002. Recent and actual evolution of Italian glaciers. *Bollettino Geofisico* 23(3–4), 27–35. (In Italian.)

Bleier, R. 2009. War on Sudan: another war for Israel – will Nile water go to Israel? *Islamic Intelligence*, 11 April. At: http://islamic-intelligence.blogspot. com/2009/war-on-sudan-another-war-for-israel. html.

Block, S. 2001. The growing threat of biological weapons. *American Scientist* 89, 1–3.

Boeye, D., van Straaten, D. and Verheyen, R.F. 1995. A recent transformation from poor to rich fen caused by artificial groundwater storage. *Journal of Hydrology* 169(1–4), 111–129.

Bonacci, O., Trninić, D. and Roje-Bonacci, T. 2008. Analysis of the water temperature regime of the

Danube and its tributaries in Croatia. *Hydrological Processes* **22**(7), 1014–1021. doi: 10.1002/hyp.6975.

Boon, P.J., Calow, P. and Petts, G.E. (eds) 1992. *River conservation and management.* Chichester, Wiley, 470pp.

Brink, E., McLain, S. and Rothert, S. 2004. *Beyond Dams – options and alternatives.* Report by American Rivers and International Rivers Network. At: www.internationalrivers.org/en/the-way-forward/water-energy-solutions/beyond-dams-options-alternatives.

Broecker, W.S. 1989. Greenhouse surprises. In: Abrahamson, D.E. (ed.) *The Challenge of global warming,* Washington, DC, Island Press, 196–208.

Bruntland, G. (ed.) 1987. *Our common future: The World Commission on Environment and Development.* Oxford, Oxford University Press, 374pp.

Bryden, H.L., Longworth, H.R. and Cuningham, S.A. 2005. Slowing of the Atlantic meridional overturning circulation at 25°N. *Nature* **438**, 655–657. doi: 10.1038/nature04385.

Bryden, H.L., Roemmich, D. and Church, J.A. 1991. Ocean heat transport across 24°N in the Pacific. *Deep Sea Research* **38**(3A), 297–324.

Bryson, R.A. and Murray, T.J. 1977. *Climates of hunger.* Madison, University of Wisconsin Press.

Bull, K.B. 1991. The critical loads/levels approach to gaseous pollutant emissions control. *Environmental Pollution* **69**, 105–123.

Bultot, F., Coppens, A., Dupriez, G.L., Gellens, D. and Meulenberghs, F. 1988. Repercussions of a CO_2 – doubling on the water cycle and on the water balance: a case study from Belgium. *Journal of Hydrology* **99**, 319–347.

Burnett, W.C., Aggarwal, P.K., Aureli, A., Bokuniewicz, H., Cable, J.E., Charette, M.A., Kontar, E., Krupa, S., Kulkarni, K.M., Loveless, A., Moore, W.S., Oberdorfer, J.A., Oliveira, J., Ozyurt, N., Povinec, P., Privitera, A.M.G., Rajar, R., Ramessur, R.T., Scholten, J., Stieglitz, T., Taniguchi, M. and Turner, J.V. 2006. Quantifying submarine groundwater discharge in the coastal zone via multiple methods. *Science of the Total Environment* **367**, 498–543.

Burt, T.P. and Haycock, N.E. 1992. Catchment planning and the nitrate issue: a UK perspective. *Progress in Physical Geography* **74**(4), 358–400.

Busby, C. and Cato, M.S. 2000. Increases in leukemia in infants in Wales and Scotland following Chernobyl: evidence for errors in statutory risk estimates. *Energy and Environment* **11**(2), 127–139.

Butler, D. and Dixon, A. 2002. Financial viability of in-building grey water recycling. *Proceedings of the International Conference on Wastewater Management and Technologies for Highly Urbanized Cities,* Hong Kong, June.

Caldeira, K. and Wood, L. 2008. Global and Arctic climate engineering: numerical model studies. *Philosophical Transactions of the Royal Society A* **366**. doi 10.1098/rsta.2008.0132.

Calow, P. and Petts, G.E. (eds) 1992. *The river handbook: hydrological and ecological principles.* vol. 1, Oxford, Blackwell, 526pp.

Camp, W.G. 1994. *Environmental Science for agriculture and the life sciences.* Albany, New York, Delmar, 439pp.

Campins, H., Hargrove, K., Pinilla-Alonso, N., Howell, E.S., Kelley, M.S., Licandro, J., Mothé-Diniz, T., Fernández, Y. and Ziffer, J. 2010. Water ice and organics on the surface of the asteroid 24 Themis. *Nature* **464**, 1320–1321. doi: 10.1038/nature09029.

Carson, R. 1962. *The silent spring.* London, Hamish Hamilton, 304pp.

Cassardo, C., Badino, G., Mercalli, L., Acordon, V., Cat Berro, D. and Di Napoli, G. 2006. Cambiamenti climatici in Valle d'Aosta: opportunità e strategie di risposta (parts I and II). *Società Meteorologica Subalpina,* ISBN 88-900099-9-3, 149pp. (Climate changes in the Aosta Valley: opportunities and strategies of adaptation.) (In Italian.)

Centre for Public Policy for the Regions (CPPR) 2010. *Scottish Water – threats and opportunities.* Scottish Government Budget Options, Briefing Series No. 2, authors: Armstrong, J., McLaren, J. and Harris, R., University of Glasgow, 18pp. At: www.cppr.ac.uk/media/media_146434_en.pdf.

Chahine, M.T. 1992. The hydrological cycle and its influence on climate. *Nature* **359**, 373–380.

Chao, B.F., Wu, Y.H. and Li, Y.S. 2008. Impact of artificial reservoir water impoundment on global sea level. *Science* **320**(5873), 212–214. doi: 10.1126/science.1154580.

Chapagain, A.K. 2006. *Globalisation of water: opportunities and threats of virtual water trade.* Taylor and Francis, London. Download at: http://re[pository.tudeft.nl/view/ihe/ (12MB).

Chapagain, A.K. and Orr, S. 2009. An improved water footprint methodology linking global consumption to local water resources: A case of Spanish tomatoes. *Journal of Environmental Management* **90**, 1219–1228.

Chapagain, A.K. and Hoekstra, A.Y. 2004. *Water*

footprints of nations. Value of Water Research Report Series No. 16, UNESCO-IHE, Delft, the Netherlands. Download at: www.unesco-ihe.org/ (2.5MB).

Chapagain, A.K., Hoekstra, A.Y., Savenije, H.H.G. and Guatam, R. 2006. The water footprint of cotton consumption: an assessment of the impact of the world wide consumption of cotton products on water resources in the cotton producing countries. *Ecological Economics* **60**(1), 186–203.

Charbonneau, D., Berta, Z.K., Irwin, J., Burke, C.J., Nutzman, P., Buchhave, L.A., Lovis, C., Bonfils, X., Latham, D.W., Udry, S., Murray-Clay, R.A., Holman, M.J., Falco, E.E., Winn, J.N., Queloz, D., Pepe, F., Mayor, M., Delfosse, X. and Forveille, T. 2009. A super-Earth transiting a nearby low-mass star. *Nature* **462**, 891–894. doi: 10.1038/nature08679 Letter.

Chen, J. and Ohmura, A. 1990. On the influence of Alpine glaciers on runoff. International Association of Hydrological Sciences Pub. No. 193, 117–125.

Chen, S. and Ravallion, M. 2004. How have the world's poorest fared since the early 1980s? *World Bank Policy Research Working Paper* 3341.

Chichilnisky, G. and Heal, G. 1998. Economic returns from the biosphere. *Nature* **391**(12), 629–630.

Chiew, F.H.S. and McMahon, T.A. 2002. Global ENSO-streamflow teleconnection, streamflow forecasting and interannual variability. *Hydrological Sciences Journal* **47**, 505–522.

Christner, B.C., Morris, C.E., Foreman, C.M., Rongman C. and Sands, D.C. 2008. Ubiquity of biological ice nucleators in snowfall. *Science* **319**, 1214.

Citterio, M., Diolaiuti, G., Smiraglia, C., D'agata, C., Carnielli, T., Stella, G. and Siletto, G.B. 2007. The recent fluctuations of Italian glaciers during the last century: a contribution to knowledge about Alpine glacier changes, *Geografiska Annaler A*, **89**(3), 164–182.

Clark, R.M. and Deininger, R.A. 2000. Protecting the nation's critical infrastructure: the vulnerability of US water supply systems. *Journal of Contingencies and Crisis Management* **8**(2), 73–80.

Cliff, A., Haggett, P. and Smallman-Raynor, M.R. 2004. *World atlas of epidemic diseases*. London, Arnold, 212pp.

Climatic Research Unit (CRU) 2000. *Information sheet 9 – Climate change scenarios*. Norwich, Climatic Research Unit, University of East Anglia.

Cluckie, I.D. and Rico-Ramirez, M.A. 2004. Weather Radar Technology and Future Developments. In: Chen, Y., Takara, K., Cluckie, I.D. and Hilaire De Smedt, F. (eds) 2004. *GIS and Remote Sensing in Hydrology, Water Resources and Environment*. International Association of Hydrological Sciences Pub. 289, 11–20.

Coase, R. 1960. The problem of social cost. *Journal of Law & Economics* 3, 1–44.

Coetzee, H., Winde, F. and Wade, P. 2006. An assessment of sources, pathways, mechanisms and risks of current and potential future pollution of water and sediments in gold mining areas of the Wonderfonteinspruit catchment (Gauteng/North West Province, South Africa). Pretoria, WRC report no. 1214/1/06, ISBN 1-77005-419-7.

Cogley, J.G. 2005. Mass and energy balances of glaciers and ice sheets. In: Anderson, M.G. (ed.) *Encyclopaedia of Hydrological Sciences*, volume 4, New York, Wiley, 2555–2573.

Cogley, J.G. 2009. Geodetic and direct mass-balance measurements: comparison and joint analysis. *Annals of Glaciology* **50**(50), 96–100.

Cogley, J.G. and Adams, W.P. 1998. Mass balance of glaciers other than the ice sheets. *Journal of Glaciology* **44**(147), 315–325.

Collins, D.N. 1989. Influence of glacierization on the response of runoff from Alpine basins to climate variability. In: *Conference on Climate and Water*, Helsinki, Academy of Finland, **1**, 319–328.

Colombo, C., Monhemius, A.J. and Plant, J.A. 2008. Platinum, palladium and rhodium release from vehicle exhaust catalysts and road dust exposed to simulated lung fluids. *Ecotoxicology and Environmental Safety* **71**(3), 722–730. doi: 10.1016/j.ecoenv.2007.11.011.

Condon, M. 2006. High and Dry. *Courier Mail*, 15–16 July, Qweekend, 19–22.

Cooley, H., Gleick, P.H. and Wolff, G. 2006. *Desalination: with a grain of salt. A California perspective*. A Report of the Pacific Institute for Studies in Development, Environment, and Security, Oakland, California. Available at: www.pacinst.org.

Copeland, C. and Cody, B. 2005. *Terrorism and security issues facing the water infrastructure sector*. Federation of American Scientists CRS Report for Congress at: www.fas.org/irp/crs/RL32189.pdf.

Cosgrove, D. and Petts, G.E. (eds) 1990. *Water, Engineering, and Landscape: Water Control and Landscape Transformation in the Modern Period*. London, Belhaven Press, 192pp.

Cox, P.M., Harris, P.P., Huntingford, C., Betts, R.A., Collins, M., Jones, C.D., Jupp, T.E., Marengo, J.A. and Nobre, C.A. 2008. Increasing risk of Amazonian drought due to decreasing aerosol pollution. *Nature* **453**, Letters. doi: 10.1038/nature06960.

Crawford, M.D. 1972. Hardness of drinking-water and cardiovascular disease. *Proceedings of the Nutrition Society* **37**, 347–353.

Crowther, J., Kay, D. and Wyer, M.D. 2002a. Relationship between microbial water quality and environmental conditions in coastal recreational waters: the Fylde coast, UK. *Water Research* **35**, 4029–4038.

Crowther, J., Kay, D. and Wyer, M.D. 2002b. Faecal indicator concentrations in waters draining lowland catchments in the UK: relationships with land use and farming practices. *Water Research* **36**, 1725–1734.

Crowther, J., Wyer, M.D., Bradford, M., Kay, D. and Francis, C.A. 2003. Modelling faecal indicator concentrations in large rural catchments using land use and topographic data. *Journal of Applied Microbiology* **94**, 962–973.

Cummings, R. and Nercissiantz, V. 1992. The use of water pricing as a means for enhancing water use efficiency in irrigation: case studies in Mexico and the United States. *Natural Resources Journal* **32**, 731–755.

Cunningham, S.A., Kanzow, T., Rayner, D., Baringer, M.O., Johns, W.E., Marotzke, J., Longworth, H.R., Grant, E.M., Hirschi, J.J-M., Beal, L.M., Meinen, C.S. and Bryden, H.L. 2007. Temporal Variability of the Atlantic Meridional Overturning Circulation at 26.5°N. *Science* **317**(5840), 935–938. doi: 10.1126/science.1141304.

Currie, R.G. and O'Brien, D.P. 1992. Deterministic signals in Southeast United States precipitation data. *Journal of Environmental Science and Health*, Part A **27**(3), 827–841. doi: 10.1080/10934529209375763.

Dabbagh, T., Sadler, P., al-Saqabi, A. and Sadeqi, M. 1993. *Desalination: the neglected option*. Water in the Arab World Symposium, Harvard University, Kuwait Fund for Arab Economic Development, 44pp.

Damstra, T., Barlow, S., Bergman, A., Kavlock, R. and Van Der Kraak, G. (eds) 2002. *Global assessment of the state-of-the-science of endocrine disruptors: An assessment prepared by an expert group on behalf of the World Health Organization, the International Labour Organisation, and the United Nations Environment Programme*. International Programme on Chemical Safety. Downloadable at: www.who.int/ipcs/publications/new_issues/endocrine_disruptors/en/index.html.

Daultrey, S. 1996. The influences of the North Atlantic Oscillation, the El Niño/Southern Oscillation and the Quasi-Biennial Oscillation on winter precipitation in Ireland. In: Jones, J.A.A, Liu, C., Woo, M-K. and Kung, H-T. (eds) *Regional hydrological response to climate change*, Dordrecht, Kluwer Academic Publishers, 213–236.

Dawdy, D.R. and O'Donnell, T. 1965. Mathematical models of catchment behavior. *Proceedings of American Society of Civil Engineers*, HY4, **91**, 123–137.

De Putter, T., Loutre, M.F. and Wansard, G. 1998. Decadal periodicities, Nile River historical discharge (AD 622–1470) and climatic implications. *Geophysical Research Letters* **25**, 3193–3196.

Defra (Department for Environment, Food and Rural Affairs) 2009. *Safeguarding our soils: a strategy for England*. London, Defra, 42pp.

Dellapenna, J. (ed.) 2003. *The regulated riparian model water code*. Washington, DC, American Society of Civil Engineers, ASCE Standard 40–03.

Dellapenna, J. 1997. Population and water in the Middle East: the challenge and opportunity for law. *International Journal of Environment and Pollution* **7**, 72–111.

Dellapenna, J. 2000. The importance of getting names right: the myth of markets for water. *William and Mary Environmental Law and Policy Review* **25**, 317–377.

Dellapenna, J. 2001. Riparianism. In: Beck, R. (ed.) *Waters and Water Rights*, Charlottesville, VA, The Michie Co., volume 1, chapters 6–9.

Dellapenna, J. 2001. The customary international law of transboundary fresh waters. *International Journal of Global Environmental Issues* **1**, 264–305.

Delta Commission 2008. *Working together with water*. Delta Commissie Report, at: www.deltacommissie.com.

Department of Health and Health Protection Agency 2008. *Health effects of climate change in the UK*. Crown, 124pp. Downloadable at: www.dh.gov.uk/en/Publicationsandstatistics.

DETR (Department of the Environment, Transport and the Regions) 1999. *A Better Quality of Life: A strategy for sustainable development for the UK*. London, DETR.

Dettinger, M.D., Cayan, D.R., Meyer, M.K. and Jeton, A.E. 2004. Simulated hydrologic responses to climate variations and change in the Merced, Carson, and American River basins, Sierra Nevada, California, 1900–2099. *Climatic Change* **62**, 283–317.

Devidze, M. 2009. Emergency response and water security of the BTC pipeline in ecologically sensitive areas of Georgia. In: Jones, J.A.A., Vardanian, T.G. and

Hakopian, C. (eds) *Threats to global water security*, Nato Science for Peace and Security Series, Dordrecht, Springer, 333–336.

DfID (Department for International Development) 2009. *Credit crunch 'tsunami' to hit world's poor as 90 million forced into poverty*. Reporting speech by Douglas Alexander. At: http://webarchive.nationalarchives.gov.uk/+/http://www.dfid.gov.uk/Media-Room/Press-release/2009/Credit-crunch-'tsunami'-to-hit-world's-poor-as-90-million-forced-into-poverty.

Diaz, H.F. and Murnane, R.J. (eds) 2008. *Climate extremes and society*. Cambridge, Cambridge University Press, 356pp.

Dibb, S. 1995. Swimming in a sea of oestrogens – chemical hormone disrupters. *The Ecologist* 25(1), 27–31.

Diolaiuti, G. 2001. *Mass balance and frontal variations of Alpine glaciers: a contribution to the modeling and the knowledge of the spatial and temporal variability of the glaciological signal*. PhD. thesis, XIV cycle, Department of Earth Sciences, University of Milan, Italy. (In Italian.)

Diolaiuti, G., D'Agata, C., Smiraglia, C., Apadula, F. and Stella, G. 1999. The Italian glaciers. Database and recent variations of a precious hydric resource. *Nimbus*, **23–24**, 57–59. (In Italian.)

Dixon, A., Butler, D. and Fewkes, A. 1999. Guidelines for greywater reuse – health issues. *Journal of Chartered Institution of Water & Environmental Management* 13(5), 322–326.

Dixon, J.A. and Fallon, L.A. 1989. The concept of sustainability: origins, extensions, and usefulness for policy. *Society and Natural Resources* 2(2), 73–84.

DoE (Department of Environment) 1986. *Nitrate in water*. Nitrate Coordination Group, Pollution Papers No. 26, London, HMSO.

Doglioni, A., Giustolisi, O., Savic, D.A. and Webb, B.W. 2008. An investigation on stream temperature analysis based on evolutionary computing. *Hydrological Processes* 22(3), 315–326.

Döll, P., Kaspar, F. and Lehner, B. 2003. A global hydrological model for deriving water availability indicators: model tuning and validation. *Journal of Hydrology* 270, 105–134.

Dragoni, W. and Sukija, B.S. 2008. *Climate change and groundwater: a short review*. Geological Society, London, Special Publications 288, 1–12. doi: 10.1144/SP288.1.

Dudley, N. 1990. *Nitrates: the threat to food and water*. London, Green Print.

Dukhovny, V.A., Tuchin, A.I., Sorokin, A.G., Ruziev, I. and Stulina, G.V. 2009. Future of the Aral Sea and the Aral Sea Coast. In: Jones, J.A.A., Vardanian, T.G. and Hakopian, C. (eds) *Threats to global water security*, Nato Science for Peace and Security Series, Dordrecht, Springer, 377–380.

Dunne, T. and Leopold, L.B. 1978. *Water in environmental planning*. San Francisco, Freeman, 818pp.

DWAF (South African Department of Water Affairs and Forestry) 1986. *Management of water resources of the Republic of South Africa*. Department of Water Affairs, Pretoria.

DWAF 1993. *South African Water Quality Guidelines. Volume 1: Domestic Water Use*. 1st edition, Pretoria, DWAF.

DWAF 1996a. *South African Water Quality Guidelines. Volume 1: Domestic Water Use*. Pretoria, DWAF.

DWAF 1996b. *South African Water Quality Guidelines. Volume 4: Agricultural use: Irrigation*. Pretoria, DWAF.

Economist 2005. Are you being served? *The Economist*, 23–29 April, 76–78.

Edwards, R.W., Gee, A.S. and Stoner, J.H. 1990. *Acid waters in Wales*. Dordrecht, Kluwer, 337pp.

Edworthy, K.J. and Downing, R.A. 1979. Artificial groundwater recharge and its relevance in Britain. *Journal of the Institution of Water Engineers and Scientists* 33, 151–172.

Eltahir, E.A. 1996. El Niño and the natural variability in the flow of the Nile river. *Water Resources* 32, 131–137.

Eltahir, E.A. and Wang, G. 1999. Nilometers, El Niño, and climate variability. *Geophysical Research Letters* 26, 489–492.

Engineering the Future 2010. *Global water security – an engineering perspective*. The Royal Academy of Engineering, London, 40pp. Online at: www.raeng.org.uk/gws.

Environment Agency 2008. *Water resources in England and Wales – current state and future pressures*. Bristol, Environment Agency, 22pp.

Environment Agency 2009. General quality assessment. At: www.environment-agency.gov.uk/.

Environment Agency 2010. Opportunity and environmental sensitivity mapping for hydropower. Available at: www.environment-agency.gov.uk/shell/hydropowerswf.html.

EPA (Environmental Protection Agency) 2000. Low impact development: a literature review. EPA-841-B-00-005, Washington, DC, Office of Water. At: www.epa.gov/nps/lid.pdf, 41pp.

EPA (Environmental Protection Agency) 2001. *Final rule for arsenic in drinking water*. Technical Fact Sheet EPA 815-F-00-016. www.epa.gov/safewater/ars/ars_rule_techfactsheet.html.

Espizua, L.E. and Maldonado, G.I. 2007. Glacier variations in the Central Andes of Mendoza Province, Argentina, from 1986 to 2005. In: Scarpati, O.E. and Jones, J.A.A. (eds) *Environmental change and rational water use*, Buenos Aires, Orientacion, 353–366.

Essenwanger. O.M. 2001. Classification of climates. In: *World Survey of Climatology 1C, General Climatology*, Elsevier, Amsterdam, 102pp.

European Environment Agency 2009. *Water resources across Europe – confronting water scarcity and drought*. EEA Report No 2/2009, European Environment Agency and Office of Official Publications of the European Communities.

European Environment Agency undated. Golf courses and washing machines: obstacles and opportunities for sustainable water management. At: www.eea.europa.eu/article/golf-courses-and-washing-machines-obstacles-and-opportunities-for-sustainable-water-management/.

Evers, M. 2008. *Decision Support Systems in Integrated River Basin Management – requirements for appropriate tools and structures for a comprehensive planning approach*. Aachen, Shaker Verlag, 322pp. plus appendices.

Falkenmark, M. and Rockström, J. 2004. *Balancing water for humans and nature: the new approach to ecohydrology*. London, Earthscan, 247pp.

Falkenmark, M. and Rockström, J. 2006. The new blue and green water paradigm: breaking new ground for water resources planning and management. Editorial, *Journal of Water Resources Planning and Management* 132(3), 129–132.

Faulkner, H. 2006. Piping hazard in collapsible and dispersible soils in Europe. In: Boardman, J. and Poesen, J. (eds) *Soil erosion in Europe*, Wiley, London, 537–562.

Fenton, I., Geissler, P.E. and Haberle, R.M. 2007. Global warming and climate forcing by recent albedo changes on Mars. *Nature* 446, 646–649. doi 10.1038/nature05718.

Fewtrell, L. and Kay, D. (eds) 2008. *A guide to the health impact assessment of sustainable water management*. London, International Water Association, 320pp.

Flanner, M.G., Zender, C.S., Randerson, J.T. and Rasch, P.J. 2007. Present-day climate forcing and response from black carbon in snow. *Journal of Geophysical Research* 112, D11202. doi: 10.1029/2006JD008003.

Foodwatch, e.V. 2008. Uran-Belastung von Trinkwasser in Deutschland (Messung FAL-PB August bis November 2006, Stand 27.3.2008), Analytical results, 10pp. At: www.foodwatch.de/.

Foundation for Water Research 2005 (updated 2010). *World water: resources, usage and the role of man-made reservoirs*. FR/R0012, Marlow, UK, Foundation for Water Research. At: www.fwr.org/wwtrstrg.pdf.

Foxon, T.J., Butler, D., Dawes, J.K., Hutchinson, D., Leach, M.A., Pearson, P.J.G. and Rose, D. 2000. An assessment of water demand management options from a systems approach. *Journal of the Chartered Institution of Water and Environmental Management* 14(3), 171–178.

Frijters, I.D. and Leentvaar, J. 2003. *Rhine case study*. UNESCO-IHP, PCCP Series Publication, 33pp. http://www.unesco.org/water/wwap/pccp/case_studies.shtml.

Fu, G., Butler, D. and Khu, S-T. 2009. The impact of new developments on river water quality from an integrated system modelling perspective. *Science of the Total Environment* 407, 1257–1267.

Fu, G., Khu, S-T. and Butler, D. 2010. Optimal distribution and control of storage tank to mitigate the impact of new developments on receiving water quality. *ASCE Journal of Environmental Engineering* 136(3), 335–342.

Fukuyama, F. 1992. *The end of history and the last man*. New York, Free Press.

Fuller, J.F. (ed.) 1990. *Thor's Legions: Weather Support to the U.S. Air Force and Army, 1937–1987*. Boston, MA, American Meteorological Society.

Galatchi, L-D. 2009. Environmental management of intentional and accidental environmental threats to water security in the Danube Delta. In: Jones, J.A.A., Vardanian, T.G. and Hakopian, C. (eds) *Threats to global water security*, Nato Science for Peace and Security Series, Dordrecht, Springer, 305–315.

Gasnier, C., Dumont, C., Benachour, N., Clair, E., Chagnon, M-C. and Séralini, G-E. 2009. Glyphosate-based herbicides are toxic and endocrine disruptors in human cell lines. *Toxicology* 262(3), 184–191. doi: 10.1016/j.tox.2009.06.006.

Gauzer, B. 1993. *Effect of air temperature change on the Danube flow regime*. VITUKI, Budapest. (In Hungarian.)

Gething, P.W., Smith, D.L., Patil, A.P., Tatem, A.J., Snow, R.W. and Hay, S.I. 2010. Climate change and the global malaria recession. *Nature* 465(7296), 342–345.

Ghaleb, K.O. 1951. Le Mikyas ou Nilomètre de l'île de Rodah. *Memoire a L'Institut D'Egypte* **54**, 182.

Giannini, A., Saravanan, R. and Chang, P. 2003. Oceanic forcing of Sahel rainfall on interannual to interdecadal time scales. *Science* **302**, 1027–1030.

Gilbertson, D.D., Hunt, C.O., Fieller, N.R.J. and Barker, G.W.W. 1994. The environmental consequences and context of ancient floodwater farming in the Tripolitanian pre-desert. In: Millington, A.C. and Pye, K. (eds) *Environmental change in drylands: biogeographical and geomorphological perspectives*, Chichester, Wiley, 229–251.

Giles, K.A., Laxon, S.W. and Ridout, A.L. 2008. Circumpolar thinning of Arctic sea ice following the 2007 record ice extent minimum. *Geophysical Research Letters* **35**, L22502. doi: 10.1029/2008GL035710.

GIWA, 2006. Global International Water Assessment Report. UNEP, 20pp. Available at: www.unep.org/dewa/giwa/.

Glantz, M. 1994. The West African Sahel, In: Glantz, M. (ed.) *Drought follows the plow*, Cambridge University Press, 33–43.

Glanz, M.H. 1982. Consequences and responsibilities in drought forecasting: the case of Yakima, 1977. *Water Resources Research* **18**(1), 3–13.

Gleick, P.H. (ed.) 1993. *Water in crisis: a guide to the world's fresh water resources*. New York and Oxford, Oxford University Press, 473pp.

Gleick, P.H. 2003. Global freshwater resources: soft path solutions for the 21st century. *Science* **302**(5650), 1524–1528. doi 10.1126/science.1089967.

Gleick, P.H. 2009. Stealing water from the future: California's massive groundwater overdraft newly revealed. *AlterNet* at: www.alternet.org/story/144676.

Goklany, I.M. 2007. Death and death rates due to extreme weather events. Civil Society Coalition on Climate Change. At: www.cscc.info/report/report_23.pdf.

Goldsworthy, A.K. 2002. *Roman warfare*. Toronto, McArthur and Co. Also pb London, Phoenix, 2007.

Gordon, C., Cooper, C., Senior, C.A., Banks, H., Gregory, J.M., Johns, T.C., Mitchell, J.F.B. and Wood, R.A. 2000. The simulation of SST, sea ice extents and ocean heat transports in a version of the Hadley Centre coupled model without flux adjustments. *Climate Dynamics* **16**, 147–168.

Govindaraju, R.S. and Rao, A.R. (eds) 2000. *Artificial neural networks in hydrology*. Dordrecht, Kluwer, 292pp.

Gray, W.M., Frank, W.M., Myron L., Corrin, M.L. and Stokes, C.A. 1976. Weather-modification by carbon dust absorption of solar energy. *Journal of Applied Meteorology* **15**(4), 355–386.

Greenspan, A. 2007. *The age of turbulence: adventures in a new world*. New York, Penguin, 505pp.

Gregory, J.M., Dixon, K.W., Stouffer, R.J., Weaver, A.J., Driesschaert, E., Eby, M., Fichefet, T., Hasumi, H., Hu, A., Jungclaus, J.H., Kamenkovich, I.V., Levermann, A., Montoya, M., Murakami, S., Nawrath, S., Oka, A., Sokolov, A.P. and Thorpe, R.B. 2005. A model intercomparison of changes in the Atlantic thermohaline circulation in response to increasing atmospheric CO_2 concentration. *Geophysical Research Letters* **32**, L12703. doi: 10.1029/2005GL023209.

Gribbin, J. and Plagemann, S. 1977. *The Jupiter effect*. Glasgow, Fontana/Collins, 178pp.

Groves, B. 2009. The bottle boom – why buy bottled water? Exposing dietary misinformation, at: www.second-opinions.co.uk/bottle1.html.

Gurdak, J.J., Hanson, R.T., McMahon, P.B., Bruce, B.W., McCray, J.E., Thyne, G.D. and Reedy, R.C. 2007. Climate variability controls on unsaturated water and chemical movement, High Plains Aquifer, USA. Published in special section: groundwater resources assessment under the pressures of humanity and climate change, *Vadose Zone Journal* **6**, 533–547. doi: 10.2136/vzj2006.0087.

GWP (Global Water Partnership) 2000. *Towards water security: a framework for action*. Stockholm, Global Water Partnership, 18pp.

Haag, A.L. 2007. The even darker side of brown clouds: atmospheric aerosols compete with carbon dioxide as an agent of warming. *Nature Reports – Climate Change*. doi: 10.1038/climate.2007.41.

Haeberli, W., Bosh, H., Scherler, K., Ostrem, G. and Wallen, C.C. (eds) 1989. *World Glacier Inventory: Status 1988*, International Association of Hydrological Sciences, Nairobi.

Hall, D. 2008. *Public-Private Partnerships (PPPs): Summary paper*. London, Public Services International Research Unit, University of Greenwich, 26pp. At: www.psiru.org/reports/2008-11-PPPs-summ.pdf.

Halliday, S. 2004. *Water: a turbulent history*. Sutton Publishing, Stroud, UK, 246pp.

Hand, E. 2009. Water on the Moon? *Nature News* (18 September). doi: 10.1038/news.2009.931.

Hanna, E., Irvine-Fynn, T., Wise, S., Huybrechts, P., Steffen, K., Huff, R., Cappelen, J., Shuman, C. and Griffiths, M. 2008. Increased runoff from melt from the Greenland ice sheet: a response to global warming. *Journal of Climate*, **21**(2), 331–341.

Hannah, D.M., Webb, B.W. and Nobilis, F. (eds) 2008. River and stream temperature: dynamics, processes, models and implications. *Hydrological Processes*, special issue, **22**(7).

Hansen, J., Sato, M., Ruedy, R., Nazarenko, L., Lacis, A., Schmidt, G.A., Russell, G., Aleinov, I., Bauer, M., Bauer, S., Bell, N., Cairns, B., Canuto, V., Chandler, M., Cheng, Y., Del Genio, A., Faluvegi, G., Fleming, E., Friend, A., Hall, T., Jackman, C., Kelley, M., Kiang, N., Koch, D., Lean, J., Lerner, J., Lo, K., Menon, S., Miller, R., Minnis, P., Novakov, T., Oinas, V., Perlwitz, Ja., Perlwitz, Ju., Rind, D., Romanou, A., Shindell, D., Stone, P., Sun, S., Tausnev, N., Thresher, D., Wielicki, B., Wong, T., Yao, M. and Zhang, S. 2005. Efficacy of climate forcings. *Journal of Geophysical Research* **110**, D18104, doi: 10.1029/2005JD005776.

Hardin, G. 1968. The tragedy of the commons. *Science* **162**, 1243–1250.

Hardy, J. and Gucinski, H. 1989. Stratospheric ozone depletion: implications for marine ecosystems. *Oceanography* **2**(2), 18–21.

Hargrove, T. 2001. China announces seawater irrigation of GM crops. At: www.greenbio.checkbiotech. org/news/china_announces_seawater_irrigation_gm_ crops.

Harper, D.M. and Ferguson, A.J.D. 1995. *The ecological basis for river management*. Chichester, Wiley, 614pp.

Harries-Rees, K. 2005. Desert beetle provides model for fog-free nanocoating. *Chemistry World News* (Royal Society of Chemistry). http://www.rsc.org/ chemistryworld/News/2005/August/31080502.asp.

Harris, P.P., Huntingford, C. and Cox, P.M. 2008. Amazon basin climate under global warming: the role of sea surface temperature. *Philosophical Transactions of the Royal Society of London B* **363**(1498), 1753–1759. doi: 10.1098/rstb.2007.0037.

Hart-Davis, A. 2004. *What the past did for us*. London, BBC Books, 224pp.

Hasell, N. 2009. Greenspan says financial crisis will reoccur. *The Times*, 9 September.

Hay, J.E. and Mimura, N. 2006. Sea-level rise: Implications for water resources management. *Mitigation and Adaptation Strategies for Global Change* **10**, 717–737.

Hayton, R. and Utton, A. 1989. Transboundary groundwaters: The Bellagio draft treaty. *Natural Resources Journal* **29**, 663–722.

Hermon-Taylor, J., Bull, T.J., Sheridan, J.M., Cheng, J., Stellakis, M.L. and Sumar, N. 2000. Causation of Crohn's disease by *Mycobacterium avium*

subspecies *paratuberculosis*. *Canadian Journal of Gastroenterology* **14**(6), 521–539.

Hickman, D.C. 1999 (updated 2002). *A chemical and biological warfare threat: USAF water systems at risk*. The Counterproliferation Papers Future Warfare Series No. 3, USAF Counterproliferation Center, Air War College, Air University, Maxwell Air Force Base, Alabama.

Hillel, D. 1992. *Civilization and the life in the soil*. London, Anrom Books, 321pp.

Hiscock, K. 2005. *Hydrogeology. principles and practice*. Oxford, Blackwell, 389pp.

Hoekstra, A.Y. (ed.) 2003. *Virtual water trade: Proceedings of the International Expert Meeting on Virtual Water Trade*. Value of Water Research Report Series, No. 12, Unesco Institute for Water Education, 244pp. At: www.unesco-ihe.org/Value-of-Water-Research-Report-Series/.

Hoekstra, A.Y. 2009. Water security of nations: how international trade affects national water scarcity and dependency. In: Jones, J.A.A., Vardanian, T.G. and Hakopian, C. (eds) *Threats to global water security*. NATO Science for Peace and Security Series – C, Dordrecht, Springer, 27–36.

Hoekstra, A.Y. and Hung, P.Q. 2005. Globalisation of water resources: International virtual water flows in relation to crop trade. *Global Environmental Change* **15**(1), 45–56.

Hoekstra, A.Y. and Chapagain, A.K. 2008. *Globalization of water: sharing the planet's freshwater resources*. Oxford, Blackwell.

Holland, M.M., Bitz, C.M. and Tremblay, B. 2006. Future abrupt reductions in the summer Arctic sea ice. *Geophysical Research Letters* **33**, L23503. doi: 10.1029/2006GL028024.

Hollis, G.E. (ed.) 1979. *Man's influence on the hydrological cycle in the United Kingdom*. Norwich, Geo-Books, 278pp.

Holt, C.P. and Jones, J.A.A. 1996. Equilibrium and transient global warming scenario implications for water resources in Wales. *Water Resources Bulletin* **32**(4), 711–721.

Holt, T. 1983. Soviet river diversions – a social need, a climatic hazard? *Climate Monitor* **12**(3), 91–95.

Hoover, J.E. 1941. Water supply facilities and national defense. *Journal of American Water Works Association* **33**(11), 1861.

Hornig, J.F. (ed.) 1999. *Social and environmental impacts of the James Bay Hydroelectric Project*. Montreal, McGill-Queens Press, 169pp.

Houghton, Sir J. 2009. *Global warming: the complete briefing.* 4th edition, Cambridge, Cambridge University Press, 438pp.

House of Commons Transport Committee 2008. *The London Underground and the Public-Private Partnership Agreements: government response to the Committee's Second Report of Session 2007–08.* London, The Stationery Office, 15pp.

House, T.J., Near, J.B. Jnr, Shields, W.B., Celentano, R.J., Husband, D.M., Mercer, A.E. and Pugh, J.E. 1996. *Weather as a Force Multiplier: Owning the Weather in 2025.* A Research Paper presented to Air Force 2025.

Howat, I.M., Joughin, I. and Scambos, T.A. 2007. Rapid changes in ice discharge from Greenland outlet glaciers. *Science,* **315**(5818), 1559–1561. doi: 10.1126/science.1138478.

Howe, C., Boggs, C. and Butler, P. 1990. Transaction costs as a determinant of water transfers. *University of Colorado Law Review* **61**, 393–402.

Howell, W.E. 1972. Impact of snowpack management on snow and ice hydrology. In: *The role of snow and ice in hydrology,* International Association of Hydrological Sciences Pub. No. 107, 1464–1472.

Hoyle, F. and Wickramasinghe, N.C. 1993. *Our place in the cosmos.* London, Phoenix.

Hubert, P. 2001. Multifractals as a tool to overcome scale problems in hydrology. In: Can Science and Society Avert the World Water Crisis in the 21st Century? Special Issue, *Hydrological Sciences Journal* **46**(6), 897–905.

Hufschmidt, M.M. and Tejwani, K.G. 1993. *Integrated water resource management – meeting the sustainability challenge.* Paris, UNESCO, International Hydrological Programme Humid Tropics Programme Series No. 5, 37pp.

Hughes, C.E., Cendrón, D.I., Johansen, M.P. and Meredith, K.T. in press. Climate change and groundwater. In: Jones, J.A.A. (ed.) *Sustaining groundwater resources,* Dordrecht, Springer Legacy Series.

Hulme, M. 2001. Climatic perspectives on Sahelian desiccation: 1973–1998. *Global Environmental Change* **11,** 19–29.

Hulme, M. and Jenkins, G.J. 1998. *Climate Change Scenarios for the United Kingdom: Scientific Report.* UKCIP Technical Report No. 1, Norwich, Climatic Research Unit, 80pp.

Huntington, E. 1945. *Mainsprings of civilization.* New York, Wiley, London, Chapman and Hall, 660pp.

Huntsman-Mapila, P., Mapila, T., Letshwenyo, M., Wolski, P. and Hemond, C. 2006. Characterization of arsenic occurrence in the water and sediments of the Okavango Delta, NW Botswana. *Applied Geochemistry* **21**(2), 1376–1391.

Huntsman-Mapila, P. *et al.* in press. Arsenic distribution and geochemistry in groundwater of a recharge wetland in NW Botswana. In: Jones, J.A.A. (ed.) *Sustaining groundwater resources,* Dordrecht, Springer.

ICE (Institution of Civil Engineers) 1960. *Floods in relation to reservoir practice.* London, Institution of Civil Engineers, 66pp.

ICOLD (International Commission on Large Dams) 1999. *Benefits and concerns about dams.* At: www. icold-cigb.org/PDF/BandC.PDF.

IELRC (International Environmental Law Research Centre) 1995. *Further report of the five member group on certain issues relating to the Sardar Sarovar Project.* Geneva, IELRC. Available online at: www. ielrec.org/.

ILA (International Law Association) 1966. Helsinki Rules on the use of waters of international rivers. In: *Report of the Fifty-Second Conference,* London, International Law Association, 484–524.

ILA (International Law Association) 2004. The Berlin Rules on water resources. In: *Report of the Seventy-First Conference,* London, International Law Association, 337–411.

IMAGE Modeling Team 2001. *The IMAGE 2.2 implementation of the SRES scenarios. A comprehensive analysis of emissions, climate change and impacts in the 21st century,* Main disc, RIVM CD-ROM publication 481508018, Bilthoven, The Netherlands, National Institute for Public Health and the Environment.

International Court of Justice 1997. *Danube River Case (Hungary v. Slovakia),* 1997 ICJ No. 92. At: www.internationalwaterlaw.org/.../ IJGEI/06ijgenvl2001v1n34fuyane.pdf.

IPCC (International Panel on Climate Change) 2000. *Special report on emission scenarios.* Intergovernmental Panel on Climate Change, Cambridge, Cambridge University Press, 599pp.

IPCC (International Panel on Climate Change) 2007. *Climate Change 2007: The Physical Science Basis. Contribution of Working Group I to the Fourth Assessment Report of the Intergovernmental Panel on Climate Change.* Solomon, S., Qin, D., Manning, M., Chen, Z., Marquis, M., Averyt, K.B., Tignor, M.

and Miller, H.L. (eds), Cambridge and New York, Cambridge University Press.

ISARM (Internationally Shared Aquifer Resources Management) 2009. *Atlas of transboundary aquifers*. Paris, UNESCO.

ISRIC (International Soil Reference and Information Centre) 2007. *The spark has jumped the gap: Green Water Credits proof of concept*. Green Water Credits Report 7, ISRIC – World Soil Information, Wageningen, The Netherlands.

IUCN/UNEP 1991. *Island Directory*. Available at: www.islands.unep.ch/Isldir.htm.

IUCN/UNEP/WWF 1991. Caring for the Earth: a strategy for sustainable living. Gland, Switzerland, IUCN/UNEP/WWF.

Jacques, M. 2009. *When China rules the world: the rise of the Middle Kingdom and the end of the Western World*. Allen Lane, 550pp.

Jamieson, D.G. and Wilkinson, J.C. 1972. River Dee research programme. *Water Resources Research* **8**(4), 899–920.

Jenkins, A. and Holland, D. 2007. Melting of floating ice and sea level rise. *Geophysical Research Letters* **34**(16), L16609. doi: 10.1029/2007GL030784.

Jobling, S., Williams, R., Johnson, A., Taylor, A., Gross-Sorokin, M., Nolan, M., Tyler, C.R., van Aerle, R., Santos, E. and Brighty, G. 2006. Predicted exposures to steroid estrogens in U.K. rivers correlate with widespread sexual disruption in wild fish populations. *Environmental Health Perspectives* **114**(1), 32–39. doi: 10.1289/ehp.8050.

Johnson, R.W. and Walker, T. 2005. Scandal of officials who devour African aid. *The Sunday Times* World News, 13 March, p.31.

Jones, D.N. 2008. Debate: a grand design, or the best we can expect. *Public Money and Management* **28**(3), 136–138.

Jones, J.A.A. 1969. The growth and significance of white ice at Knob Lake, Quebec. *Canadian Geographer* **8**, 354–372.

Jones, J.A.A. 1996. Current evidence on the likely impact of global warming on hydrological regimes in Europe. In: Jones, J.A.A., Liu, C.M., Woo, M-K. and Kung, H-T. (eds) *Regional hydrological response to climate change*, Dordrecht, Kluwer Academic Publishers, 87–131.

Jones, J.A.A. 1997. *Global hydrology: processes, resources and environmental management*. Harlow, UK, Addison Wesley Longman, 399pp.

Jones, J.A.A. 1999. Climate change and sustainable water resources: placing the threat of global warming in perspective. *Hydrological Sciences Journal* **44**(4), 541–557.

Jones, J.A.A. 2000. The physical causes and characteristics of floods. In: Parker, D.J. (ed.) *Floods*, volume II, London and New York, Routledge, 91–112.

Jones, J.A.A. 2004. Civilization – the basis for water science. In: Rodda, J.C. and Ubertini, L. (eds) *The basis of civilization – water science?* International Association of Hydrological Sciences Publication No. 286, 277–283.

Jones, J.A.A. 2010. Soil piping and catchment response. *Hydrological Processes* **24**, 1548–1566.

Jones, J.A.A. and Mountain, N.C. 2000. Hindcasting and forecasting flows for water resource planning using an airflow-index-based weather generator and a hydrological simulation model: Elan and Wye valleys, Wales. *British Hydrological Society Occasional Paper* No. 11, 85–92.

Jones, J.A.A. and van der Walt, I.J. (eds) 2004. Barriers and solutions to sustainable water resources in Africa. Special Issue, *GeoJournal* **61**(2), 105–214.

Jones, J.A.A., Mountain, N.C., Pilling, C.G. and Holt, C.P. 2007. Implications of climate change for river regimes in Wales – a comparison of scenarios and models. In: Lobo Ferreira, J-P. and Vieira, J.M.P. (eds) *Water in Celtic countries: quantity, quality and climatic variability*, International Association of Hydrological Sciences Publication 310, 71–77.

Jones, P.D., Briffa, K.R. and Picher, J.R. 1984. Riverflow reconstruction from tree rings in southern Britain. *Journal of Climatology* **4**, 461–472.

Jones, P.D., Jonsson, T. and Wheeler, D. 1997. Extension to the North Atlantic Oscillation using early instrumental pressure observations from Gibraltar and South-West Iceland. *International Journal of Climatology* **17**, 1433–1450.

Jurgieliwicz, L. 1996. *Global environmental change and international law: prospects for progress in the legal order*. Lanham, MD, University Press of America.

Kabat, P., Claussen, M., Dirmeyer, P.A., Gash, J.H.C., de Guenni, L.B., Meybeck, M., Vorosmarty, C.J., Hutjes, R.W.A. and Lütkemeier, S. (eds) 2004. *Vegetation, water, humans and the climate: a new perspective on an interactive system*. The IGBP Series, Springer Verlag, 566pp.

Kaletsky, A. 2009. Now is the time for a revolution in economic thought. *The Times*, 9 February, p.37.

Kamalov, Y. 2009. The Aral Sea: a matter of mutual trust. In: Jones, J.A.A., Vardanian, T.G. and Hakopian, C.

(eds) *Threats to global water security*. NATO Science for Peace and Security Series – C, Dordrecht, Springer, 367–369.

Kamrin, M.A. and Rodgers, P.W. (eds) 1985. *Dioxins in the Environment*. Washington, DC, Hemisphere Publishing Co.

Kasprzyk-Holdern, B., Drusdale, R.M. and Guwy, A.J. 2008. The occurrence of pharmaceuticals, personal care products, endocrine disruptors and illicit drugs in surface water in South Wales, UK. *Water Research* 42(13), 3498–3518.

Katz, R.W. 2002. Sir Gilbert Walker and a connection between El Niño and statistics. *Statistical Science* 17(1), 97–112.

Kaul, I., Grunberg, I. and Stern, M. (eds) 1999. *Global Public Goods: International Cooperation in the 21st Century*. Amsterdam, Elsevier Science B.V.

Kay, D., Anthony, S., Crowther, J., Chambers, B.J., Nicholson, F.A., Chadwick, D., Stapleton, C.M. and Wyer, M. 2010. Microbial water pollution: a screening tool for initial catchment-scale assessment and source apportionment. *Science of the Total Environment* online. doi: 10.1016/j.scitotenv.2009.07.033.

Kay, D., Francis, C., Edwards, A., Kay, C., McDonald, A., Lowe, N., Stapleton, C.M., Watkins, J. and Wyer, M. 2005a. *The efficacy of natural wastewater treatment systems in removing faecal indicator bacteria*. London, UK Water Industry Research (UKWIR) Report Number 07/WW/21/5 Water UK, ISBN 1 84057 385 6.

Kay, D., Wyer, M.D., Crowther, J., Stapleton, C.M., Bradford, M., McDonald, A.T., Greaves, J., Francis, C. and Watkins, J. 2005b. Predicting faecal indicator fluxes using digital land use data in the UK's sentinel Water Framework Directive catchment: the Ribble study. *Water Research* 39, 655–667.

Kay, D., Wyer, M.D., Crowther, J., Wilkinson, J., Stapleton, C. and Glass, P. 2005c. Sustainable reduction in the flux of microbial compliance parameters to coastal bathing waters by a wetland ecosystem produced by a marine flood defence structure. *Water Research* 39, 3320–3332.

Kaye, P.H., Hirst, E., Greenaway, R.S., Ulanowski, Z., Hesse, E., DeMott, P.J., Saunders, C. and Connolly, P. 2008. Classifying atmospheric ice crystals by spatial light scattering. *Optics Letters*, 33(13), 1545–1547. doi: 10.1364/OL.33.001545.

Khaiter, P.A., Nikanorov, A.M., Yereschukova, M.G., Prach, K., Vadineanu, A., Oldfield, J. and Petts, G.E. 2000. River conservation in Central and Eastern Europe (incorporating the European parts of the Russian Federation). In: Boon, P.J., Davis, B.R. and Petts, G.E. (eds) *Global perspectives on river conservation: science, policy and practice*, Chichester, Wiley, 105–126.

Khan, A. 2007. Environmental and socio-cultural impacts of a water resources development project: a case study of the Ghazi-Barotha hydropower project, Pakistan. In: Robinson, P.J., Jones, J.A.A. and Woo, M-K. (eds) *Managing water resources in a changing physical and social environment*, Rome, International Geographical Union, 115–126.

Khaydarov, R.A., Khaydarov, R.R. and Cho, S.Y. 2009. Natural disaster: prevention of drinking water scarcity. In: Jones, J.A.A., Vardanian, T.G. and Hakopian, C. (eds) *Threats to global water security*. NATO Science for Peace and Security Series – C, Dordrecht, Springer, 381–383.

King, R.D., Rowland, J., Oliver, S.G., Young, M., Aubrey, W., Byrne, E., Liakata, M., Markham, M., Pir, P., Soldatova, L.N., Sparkes, A., Whelan, K.E. and Clare, A. 2009. The automation of science. *Science* 324(5923), 85–89. doi: 10.1126/science.1165620.

King, S.D. 2010. *Losing control: the emerging threats to western prosperity*. Yale University Press, 302pp.

Kirkby, J. 2008. Cosmic rays and climate. *Surveys in Geophysics* 28, 333–375.

Kirkby, J. 2009. Cosmic rays and climate. Presentation, CERN Colloquium, 4 June 2009 at http://indico.cern.ch/.

Kliot, N., Shmueli, D. and Shamir, U. 1998. *Institutional frameworks for the management of transboundary water resources*. Haifa, Water Research Institute, Technion.

Knight, C. 1979. Urbanization and natural stream channel morphology: the case of two English new towns. In: Hollis, G.E. (ed.) *Man's influence on the hydrological cycle in the United Kingdom*, Norwich, GeoBooks, 181–198.

Kondrashov, D., Feliks, Y. and Ghil, M. 2005. Oscillatory modes extended Nile River records (A.D. 622–1922). *Geophysical Research Letters* 32, L10702.

Kondratev, S.I., Dolotov, V.V., Moiseev, Y.G. and Shchetinin, Y.T. 1999. Submarine springs of fresh water in the region from Cape Feolent to Cape Sarych. *Physical Oceanography* 10(3), 257–272.

Kondratjeva, L. and Fhisher, N. 2009. Estimation of ecological risk of transboundary pollution of the Amur River. In: Jones, J.A.A., Vardanian, T.G. and Hakopian, C. (eds) *Threats to global water security*. NATO Science for Peace and Security Series – C, Dordrecht, Springer, 385–388.

Köppen, W. 1936. Das geographisca System der Klimate. In: Köppen, W. and Geiger, G. (eds) *Handbuch der Klimatologie*, 1. C. Gebr, Borntraeger, 1–44.

Koppenjan, J., Charles, B.M. and Ryan, N. 2008. Editorial: managing competing public values in public infrastructure projects. *Public Money and Management* **28**(3), 131–134. doi: 10.1111/j.1467-9302.2008.00632.x.

Kothari, A. and Ram, R.N. 1994. Environmental impacts of the Sardar Sarovar Project. Online at the Friends of Narmada website, www.narmada.org/ENV/index.html.

Kotikov, O. 2003. Toxic overload in Tomsk: Siberian region faces variety of environmental hazards. *Give and Take* **6**(2), 11–14.

Kottek, M., Grieser, J., Beck, C., Rudolf, B. and Rubel, F. 2006. World map of the Köppen-Geiger climate classification updated. *Meteorologische Zeitschrift* **15**(3), 259–263.

Krachler, M. and Shotyk, W. 2008. Trace and ultratrace metals in bottled waters: Survey of sources worldwide and comparison with refillable metal bottles. *Science of the Total Environment* **407**, 1089–1096.

Krenke, A.N. and Kravchenko, G.N. 1996. Impact of future climate change on glacier runoff and the possibilities for artificially increasing melt water runoff in the Aral ASea basin. In: Jones, J.A.A, Liu, C., Woo, M-K. and Kung, H-T. (eds) *Regional hydrological response to climate change*, Dordrecht, Kluwer Academic Publishers, 259–267.

Kundzewicz, Z.W. and Takeuchi, K. 1999. Flood protection and management: quo vadimus. *Hydrological Sciences Journal* **44**(3), 417–432.

Kung, H-T., Lin, L.Yu. and Malasri, S. 1996. Use of artificial neural networks in precipitation forecasting. In: Jones, J.A.A, Liu, C., Woo, M-K. and Kung, H-T. (eds) *Regional hydrological response to climate change*, Dordrecht, Kluwer Academic Publishers, 173–179.

Küttel, M., Xoplaki, E., Gallego, D., Luterbacher, J., García-Herrera, R., Allan, R., Barriendos, M., Jones, P.D., Wheeler, D. and Wanner, H. 2009. The importance of ship log data: reconstructing North Atlantic, European and Mediterranean sea level pressure fields back to 1750. *Climate Dynamics* online. doi: 10.1007/s00382-009-0577-9.

Kwadijk, J.C.J. 1993. The impact of climate change on the discharge of the River Rhine. *Netherlands Geographical Studies* No. 171, 299pp.

Lafferty, K.D. 2009. The ecology of climate change and infectious diseases. *Ecology* **90**(4), 888–900. doi: 10.1890/08.0079.1.

Lamb, H.H. 1950. Types and spells of weather around the year in the British Isles: annual trends, seasonal structure of the year, singularities. *Quarterly Journal of the Royal Meteorological Society* **76**, 393–438.

Lamb, P.J. and Peppler, R.A. 1992. Further case studies of tropical Atlantic surface atmospheric and oceanic patterns associated with subsaharan drought. *Journal of Climate* **5**, 476–488.

Lartigue, C., Vashee, S., Algire, M.A., Chuang, R-Y., Benders, G.A., Ma, L., Noskov, V.N., Denisova, E.A., Gibson, D.G., Assad-Garcia, N., Alperovich, N., Thomas, D.W., Merryman, C., Hutchison, C.A., III, Smith, H.O., Craig Venter, J. and Glass, J.I. 2009. Creating bacterial strains from genomes that have been cloned and engineered in yeast. *Science* **325**(5948), 1693–1696. doi: 10.1126/science.1173759.

Latham, J. 2002. Amelioration of global warming by controlled enhancement of the albedo and longevity of low-level maritime cloud. *Atmospheric Research Letters* **3**(2–4), 52–58. doi: 10.1006/asle.2002.0099.

Lauterpacht, E. and McNair, A.D. (eds) 1931. The Donauversinkung Case (Württemberg & Prussia vs. Baden), 116 RGZ 1 (SGH 1927). *Annual Digest Public International Law Cases* 128.

Lawler, A. 2005. Reviving Iraq's wetlands. *Science* **307**(5713), 1186–1189.

Lebel, T., Redelsperger, J.L. and Thorncroft, C. 2003. African Monsoon Multidisciplinary Analysis (AMMA): An international research project and field campaign. *CLIVAR Exchanges* **8**, 52–54.

Ledger, M.E. and Hildrew, A.G. 2005. The ecology of acidification and recovery: changes in herbivore-algal food web linkages across a pH gradient in streams. *Environmental Pollution* **137**, 103–118.

Lemke, P., Ren, J., Alley, R.B., Allison, I., Carrasco, J., Flato, G., Fujii, Y., Kaser, G., Mote, P., Thomas, R.H. and Zhang, T. 2007. Observations: changes in snow, ice and frozen ground. In: Solomon, S., Qin, D., Manning, M., Chen, Z., Marquis, M., Averyt, K.B., Tignor, M. and Miller, H.L. (eds) *Climate Change 2007: The Physical Science Basis*, New York, Cambridge University Press, 337–383.

Lin, D-F. and Weng, C-H. 2001. Use of sewage sludge ash as brick material. *Journal of Environmental Engineering* **127**(10), 6pp. At: www.aseanenvironment.info/Abstract/41013456.pdf.

Lipkowski, J. 2006. Hydrophobic hydration – ecological aspects. *Journal of Thermal Analysis and Calorimetry* 83(3), 525–531.

Liu, C. and Zuo, D. 1983. Impact of south-to-north water transfer upon the natural environment. In: Biswas, A.K., Zuo, D., Nickum, J. and Liu, C-M. (eds) *Long-Distance Water Transfer – A Chinese Case Study and international experience*, United Nations University Press, Japan, chapter 12. Available at: www.unu.edu/unupress/unupbooks/.

Lomborg, B. 2007. *Cool it: the sceptical environmentalist's guide to global warming*. London, Random House, 272pp.

Lvovich, M.I. 1970. *Water resources of the world and their future*. International Association of Scientific Hydrology Pub. No. 92, 317–322.

Lvovich, M.I. 1977. World water resources present and future. *Ambio* 6(1), 13–21.

Lvovich, M.I. 1979. *World water resources and their future*. Washington, D.C., American Geophysical Union, 415pp.

Lvovich, M.I. and Chernishov, E.P. 1977. Experimental studies of changes in the water balance of an urban area. In: *Effects of urbanization and industrialization on the hydrological regime and on water quality*, International Association of Hydrological Sciences Pub. No. 123, 63–67.

Lvovich, M.I., Karasik, G.Ya., Bratseva, N.L., Medvedeva, G.P. and Maleshko, A.V. 1991. *Contemporary intensity of the world land intracontinental erosion*. Moscow, USSR Academy of Sciences.

MacKenzie, C. in press. *International environmental law and forestry policy*. Cambridge, Cambridge University Press.

MacKenzie, J. undated. Saskatchewan's uranium bonanza. Online research documents, Saskatchewan Western Development Museum. At: www.wdm.ca/skteacherguide/WDMResearch/Saskatchewan's%20Uranium%20Bonanza%20by%20Janet%20Mackenzie.

Maclean, N. (ed.) 2010. *Silent summer: the state of wildlife in Britain and Ireland*. Cambridge, Cambridge University Press, 600pp.

Mandelbrot, B. and Hudson, R.L. 2004. *The (mis)behaviour of markets: a fractal view of financial turbulence*. New York, Basic Books.

Mandelbrot, B. and Taleb, N.N. 2005. How the financial gurus get risk all wrong. *Brainstorm Retirement Guide*.

Mandelbrot, B. and Wallis, J.R. 1968. Noah, Joseph, and operational hydrology. *Water Resources Research* 4, 909–918.

Mandelbrot, B.B. 1977. *Fractals: form, chance and dimension*. San Francisco, Freeman.

Mantua, N.J., Hare, S.R., Zhang, Y., Wallace, J.M. and Francis, R.C. 1997. A Pacific interdecadal climate oscillation with impacts on salmon production, *Bulletin of the American Meteorological Society* 78(6), 1069–1079.

Marier, J.R. 1978. Cardio-protective contributions of hard waters to magnesium intake. *Reviews of Canadian Biology* 37(2), 115–125. Download at: www.mgwater.com/marier.html.

Marr, B. 2008. *Managing and delivering performance: how government, public sector and not-for-profit organisations can measure and manage what really matters*. Oxford, Butterworth-Heinemann, 312pp.

Marsalek, J. 1977. Runoff control in urbanizing catchments. In: *Effects of urbanization and industrialization on the hydrological regime and on water quality*, International Association of Hydrological Sciences, Pub. No. 123, 153–161.

Mason, B.J. 1957. *The Physics of clouds*. Oxford, Oxford University Press.

Mason, B.J. 1962. *Clouds, rain and rainmaking*. Cambridge, Cambridge University Press.

McCaffrey, S. 2001. *The law of international watercourses: non-navigational uses*. Oxford, Oxford University Press.

McCormick, Z. 1994. Institutional Barriers to Water Marketing in the West. *Water Resources Bulletin* 30, 953–962.

McCully, P. 1996. *Silenced rivers: the ecology and politics of large dams*. Zed Books, London, 350pp.

McGuffie, K. and Henderson-Sellers, A. 2005. *A climate modeling primer*. 3rd edition, New York, Wiley, 280pp.

McNally, L.K., III, Maasch, K.A. and Zuill, K.J. 2008. The use of ships' protests for reconstruction of synoptic-scale weather and tropical storm identification in the late eighteenth century. *Weather* 63(7), 208–213.

McSweegan, E. 1996. The infectious diseases impact statement: a mechanism for addressing emerging diseases. *Emerging Infectious Diseases* 2(2), 103–108.

Mercado, L.M., Bellouin, N., Sitch, S., Boucher, O., Huntingford, C., Wild, M. and Cox, P.M. 2009. Impact of changes in diffuse radiation on global land carbon sink. *Nature* 458, 1014–1017. doi: 10.1038/nature07949.

Merli, F., Pavan, M., Rossi, G.C., Smiraglia, C., Tamburini, A. and Ubiali, G. 2001. Depth and volume variations of the tongue of Ghiacciaio dei Forni (central Alps, Ortles-Cevedale chain) in the XX century. Results and methodology comparison. *Suppl. Geog. Phys. Dynam. Quat.* **V**, 121–128. (In Italian.)

Met Office 2009. The decline in arctic summer sea ice. Press release, 15th October, at: www.metoffice.gov.uk/corporate/pressoffice/2009/pr20091015b.html.

Mileta, M., Beysens, D., Nikolayev, V., Milimouk, I., Clus, O. and Muselli, M. 2006. Fog and dew collection projects in Croatia. *Proceedings International Conference on 'Water Observation and Information System for Decision Support' (BALWOIS 2006), Ohrid, Republic of Macedonia*, 1. Available at: www.balwois.com/balwois/administration/full_paper/ffp-587.pdf.

Miller, S., Muir-Wood, R. and Boissonnade, A. 2008. An exploration of trends in normalized weather-related catastrophe losses. In: Diaz, H.F. and Murnane, R.J. (eds) *Climate extremes and society*, Cambridge, Cambridge University Press, 356pp.

Minschwaner, K. and Dessler, A.E. 2004. Water vapor feedback in the tropical upper troposphere: model results and observations. *Journal of Climate* **17**, 1272–1282.

Mitchell, J.F.B. 1991. The equilibrium response to doubling atmospheric CO_2. In: Schlesinger, M.E. (ed.) *Greenhouse-gas-induced climatic change: a critical appraisal of simulations and observations*, Amsterdam, Elsevier, 49–61.

Mitchell, N.T., Camplin, W.C. and Leonard, D.R.P. 1986. The Chernobyl reactor accident and the aquatic environment of the UK: a fisheries viewpoint. *Journal of the Society for Radiological Protection* **6**, 167–172. doi: 10.1088/0260-2814/6/4/002.

Mitrovica, J.X., Gomez, N. and Clark, P.U. 2009. The sea-level fingerprint of West Antarctic collapse. *Science* **323**(5915), 753. doi: 10.1126/science.1166510.

Moatar, F. and Gailhard, J. 2006. Water temperature behaviour in the River Loire since 1976 and 1881. *Comptes Rendus Geoscience* **338**, 319–328.

Monosowski, E. 1985. Environmental impact of dams. *International Water Power and Dam Construction* **37**(4), 48–50.

Morely, D. (ed.) 2000. *NGOs and water perspectives on freshwater, issues and recommendations of NGOs*. London, 2000 UN Environment and Development Forum.

Moriarty, F. 1988. *Ecotoxicology: the study of pollutants in ecosystems*. London, Academic Press, 289pp.

Morimoto, R. and Hope, C. 2004. Applying a cost-benefit model to the Three Gorges project in China. *Impact Assessment and Project Appraisal* **22**(3), 205–220.

Morris, G.L., Annandale, G. and Hotchkiss, R. 2007. Reservoir sedimentation. In: Garcia, M.H. (ed.) *Sedimentation engineering: processes, measurements, modeling, and practice*, ASCE Manuals and Reports on Engineering Practice No. 110, Reston, VA, American Society of Civil Engineers, 578–612.

Mountain, N.C. and Jones, J.A.A. 2002. Estimating low flow frequencies in the mid to late 21st century for two basins in central Wales. In: García-Ruiz, J.M., Jones, J.A.A. and Arnáez, J. (eds) *Environmental Change and Water Sustainability*, Zaragoza, Consejo Superior de Investigationes Científicas (CSIC), 111–127.

Mountain, N.C. and Jones, J.A.A. 2006. Reconstructing extreme flows using an airflow index-based stochastic weather generator and a hydrological simulation model. *Catena* **66**, 120–134.

Moyo, D. 2009. *Dead Aid: why aid is not working and how there is a better way for Africa*. New York, Farrar, Straus and Giroux, and London, Penguin, 188pp.

Muro, A.I. and Raybould, J.N. 1990. Population decline of Simulium woodi and reduced onchocerciasis transmission at Amani, Tanzania, in relation to deforestation. *Acta Leiden* **59**(1–2), 153–159.

Mysore, N. 2009. Feasibility study of the Ken Betwa Project Super Reservoir Hydrological Model. Duke Space at: http://hdl.handle.net/10161/1374.

Nabulo, G., Oryem-Origa, H. and Diamond, M. 2006. Assessment of lead, cadmium and zinc contamination of roadside soils, surface films and vegetables in Kampala City, Uganda. *Environmental Research* **101**, 42–52.

Nagel, J.F. 1956. Fog precipitation on Table Mountain. *Quarterly Journal of Royal Meteorological Society* **82**, 452–60.

Nakicenovic, N. and Swart, R. (eds) 2000. *IPCC Special report on emissions scenarios*. Available online at: www.grida.no/publications/other/ipcc_srs/?src=/climate/ipcc/emission/.

National Geographic 2005. *Visual history of the world*. Berlin, Peter Delius Verlag, 656pp.

National Infrastructure Protection Center 2002. *Terrorist Interest in Water Supply and SCADA Systems*. Information Bulletin 02-001, 30 January.

National Research Council 2004. A review of the EPA

427

water security research and technical support action plan. National Academies Press, Washington DC, 121pp.

National Rivers Authority (NRA) 1994. *Water – nature's precious resource. An environmentally sustainable water resources development strategy for England and Wales.* London, HMSO, 93pp.

Navruzov, S. 2009. On the development of a strategy for the optimal use of the upstream water resources of the Amudary basin in the national interests of the Tajik Republic. In: Jones, J.A.A., Vardanian, T.G. and Hakopian, C. (eds) *Threats to global water security*, NATO Science for Peace and Security Series, Dordrecht, Springer, 389–393.

Nicholson, S.E. 2000. Land surface processes and Sahel climate. *Reviews of Geophysics* 38, 117–139.

Nikiforuk, A. 2010. *Tar Sands: dirty oil and the future of a continent.* Revised edition, Vancouver and Toronto, Greystone Books, 240pp. (Original version 2008.)

Noerdlinger, P.D. and Brower, K.R. 2007. The melting of floating ice raises the ocean level. *Geophysical Journal International* 170(1), 145–150. doi: 10.1111/j.1365-246X.2007.03472.

Normatov, I.Sh. and Petrov, G.N. 2009. Integrated management strategy for transboundary water resources in Central Asia. In: Jones, J.A.A., Vardanian, T.G. and Hakopian, C. (eds) *Threats to global water security*, NATO Science for Peace and Security Series, Dordrecht, Springer, 395–400.

Nyberg, J., Malmgren, B.A., Winter, A., Jury, M.R., Kilbourne, K.H. and Quinn, T.M. 2007. Low Atlantic hurricane activity in the 1970s and 1980s compared to the past 270 years. *Nature* 447, 698–701.

O'Neill, J. 2009. Where do the BRICs stand? Presentation to the Canadian International Council, Vancouver Branch, May 2009. Video and summary available at: www.onlinecic.org/resources/multimedia/wheredothe.

Obled, Ch. and Harder, H. 1979. A review of snow melt in the mountain environment. In: Colbeck, S.C. and Ray, M. (eds) *Modeling of snow cover runoff*, Hanover, New Hampshire, US Army Corps of Engineers, Cold Regions Research and Engineering Laboratory, 179–204.

Office of Deputy Prime Minister (2002). *Projections of households in England 2021.* http://www.housing.odpm.gov.uk/research/project/01.htm [29 July 2002].

Office of Deputy Prime Minister 2002. Projections of households in England 2021. At: www.housing.odpm.gov.uk/research/project/01.htm [29 July 2002].

Olajire, A.A. and Ayodele, E.T. 1997. Contamination of roadside soil and grass with heavy metals. *Environment International* 23, 91–101.

Olivier, J. 2004. Fog Harvesting: an alternative source of water on the west coast of South Africa. *GeoJournal* 61, 203–214.

Olivier, J. and van Heerden, J. 1999. *The South African Fog Water Collection Project.* Water Research Commission Report No. 671/1/99. 159pp.

Olivier, J. and van Heerden, J. 2002. *Implementation of an operational prototype fog water collection system.* Water Research Commission Report No. 902/1/02, 91.

Omran, J. 2004. Declining water tables in Alasa, Saudi Arabia. *Asharq Alawsat.* http://www.asharqalawsat.com/view/ksa-local/ksa-local.html#2004,11,07,264363.

Orange, R. 2008. Relentless rise of the endless city. In: *Sanitation: a global crisis*, The Times Focus Report, 22 March, pp.4–5.

Orr, S. and Chapagain, A.K. 2006. Virtual water: a case study of green beans and flowers exported to the UK from Africa. *Fresh Insights* No. 3, published as part of 'Small-scale producers and standards in agrifood supply chains 2005–2008', funded by DfID and the Natural Resources Institute; London, International Institute for Environment and Development, 24pp.

Otterpohl, R., Grottker, M. and Lange, J. 1997. Sustainable water and waste management in urban areas. *Water Science and Technology* 35(9), 121–133.

Pagli, C. and Sigmundsson, F. 2008. Will present day glacier retreat increase volcanic activity? Stress induced by recent glacier retreat and its effect on magmatism at the Vatnajokull ice cap, Iceland. *Geophysical Research Letters* 35, L09304, 9.

Palaniappan, M. 2008. *Peak water.* Pacific Institute, 11pp. At: www.pacinst.org.

Palaniappan, M. and Gleick, P.H. 2009. Peak water. In: Gleick, P.H. (ed.) *The World's Water 2008–2009*, Pacific Institute, 1–16.

Palmer, M.A., Reidy Liermann, C., Nilsson, C., Flörke, M., Alcamo, J., Lake, P. and Bond, N. 2008. Climate change and the world's river basins: anticipating management options. *Frontiers of Ecology and the Environment* 6, 81–89. doi: 10.1890/060148.

Parker, A.R. and Lawrence, C.R. 2001. Water capture by a desert beetle. *Nature* 414, 33–34. doi: 10.1038/35102108.

Parrott, J.L., McMaster, M.E. and Tetreault, G.R. 2009. Effects of municipal wastewater effluents and pharmaceuticals in fish. *Abstracts of the 52nd Annual Conference on Great Lakes Research*, 18–22 May, Toledo, Ohio.

Parry, T. 2009. Credit crunch hits the poorest hardest. *The Mirror*, 2 February. At: www.mirror.co.uk/developing-world-stories/2009/02/credit-crunch-hits-the-poorest-hardest.html.

Pearce, F. 2006/2007. *When the rivers run dry: what happens when our water runs out?* London, Eden Project Books (2006); London, Transworld Publishers, paperback (2007), 368pp.

Peel, M.C. and McMahon, T.A. 2006. Recent frequency component changes in interannual climate variability. *Geophysical Research Letters* **33**, L16810, doi: 10.1029/2006GL025670.

Peel, M.C., Finlayson, B.L. and McMahon, T.A. 2007. Updated world map of the Köppen-Geiger climate classification. Hydrology and Earth System Sciences Discussions **4**, 439–473.

Peel, M.C., McMahon, T.A. and Finlayson, B.L. 2002. Variability of annual precipitation and its relationship to the El Niño-Southern Oscillation. *Journal of Climate* **15**, 545–551.

Peel, M.C., McMahon, T.A. and Finlayson, B.L. 2004. Continental differences in the variability of annual runoff – update and reassessment, *Journal of Hydrology* **295**, 185–197.

Permanent Court of International Justice online. Case relating to the territorial jurisdiction of the International Commission of the River Oder *(Germany v. Poland)*, 1929. 1929 P.C.I.J., ser. A, no. 23.

Peterson, T.C. and Vose, R.S. 1997. An overview of the Global Historical Climatology Network temperature database, *Bulletin of the American Meteorological Society* 78(12), 2837–2849.

Petts, G.E. 1990. Water, engineering and landscape: development, protection and restoration. In: Cosgrove, D. and Petts, G.E. (eds) *Water, engineering and landscape*, London, Belhaven Press, 188–205.

Pew Center for Global Climate Change undated (accessed March 2010). Global warming basics – facts and figures. The Pew Center, Arlington, Virginia, part of the Strategies for the Global Environment. At: www.pewclimate.org/global-warming-basics/facts_and_figures/.

Pickup, R.W., Rhodes, G., Bull, T.J., Arnott, S., Sidi-Boumedine, K., Hurley, M. and Hermon-Taylor, J. 2006. *Mycobacterium avium* subsp. *paratuberculosis* in lake catchments, in river water abstracted for domestic use, and in effluent from domestic sewage treat-

ment works: diverse opportunities for environmental cycling and human exposure. *Applied Environmental Microbiology* 72(6), 4067–4077.

Pilling, C. and Jones, J.A.A. 1999. High resolution equilibrium and transient climate change scenario implications for British runoff. *Hydrological Processes* **13**(17), 2877–2895.

Pilling, C. and Jones, J.A.A. 2002. The impact of future climate change on seasonal discharge, hydrological processes and extreme flows in the Upper Wye experimental catchment, mid-Wales. *Hydrological Processes* **16**(6), 1201–1213.

Pilling, C., Wilby, R.L. and Jones, J.A.A. 1998. Downscaling of catchment hydrometeorology from GCM scenarios using airflow indices. In: Wheater, H.S. and Kirby, C. (eds) *Hydrology in a Changing Environment*, volume 1, John Wiley/British Hydrological Society, 191–208.

Pinna, M. 1996. *The recent history of climate*. Milan, Angeli Ed. (In Italian.)

Polman, L. 2010. *War games: the story of aid and war in modern times*. London, Viking, Penguin, 218pp.

Pope, V., Gallani, M., Rowntree, P. and Stratton, R. 2000. The impact of new physical parameterizations in the Hadley Centre climate model – HadCM3. *Climate Dynamics* **16**, 123–146.

Popper, W. 1951. *The Cairo Nilometer*. Berkeley, University of California Press, 269pp.

Postel, S.L. and Wolf, A.T. 2001. Dehydrating conflict. *Foreign Policy* **126**, 60–67.

Power, S., Casey, T., Folland, C., Colman, A. and Mehta, V. 1999. Inter-decadal modulation of the impact of ENSO on Australia, *Climate Dynamics* **15**, 319–324.

Pradas, M., Imeson, A. and van Mulligen, E. 1994. The infiltration and runoff characteristics of burnt soils in N.E. Catalonia and the implications for erosion. In: Sala, M. and Rubio, J.L. (eds) *Soil erosion and degradation as a consequence of forest fires*, Geoforma Ediciones, 229–240.

President's Critical Infrastructure Assurance Office 1998. *Preliminary Research and Development Roadmap for Protecting and Assuring Critical National Infrastructures*, July 1998, E-3 û E-4; online, 26 January 1999. Available from http://ciao.gov/roadmap-e.pdf.

Pretorius, S.J. 2005. *Geohydrology: Groundwater dependency. Module 5.6: Environmental Risk Assessment and Management, 3–4*. Potchefstroom, South Africa, Centre for Environmental Management, University of the North-West, August 2005.

Pretor-Pinney, G. 2006. *The cloudspotter's guide.* London, Sceptre, 360pp.

Public Policy Institute of California 2003. *Managing California's water market: issues and prospects.* Issue 74. At: www.ppic.org.

Quinn, W.H. 1992. A study of Southern Oscillation-related climatic activity for AD 622–1990 incorporating Nile river flood data. In: Diaz, H.F. and Markgraf, V. (eds) *El Niño: historical and paleoclimatic aspects of the Southern Oscillation,* Cambridge, Cambridge University Press, 119–149.

Radić, V. and Hock, R. 2010. Regional and global volumes of glaciers derived from statistical upscaling of glacier inventory data. *Journal of Geophysical Research* **115**, F01010, 10pp. doi: 10.1029/2009JF001373.

Rahman, R. 1995. The law of international uses of international watercourses: dilemma for lower riparians. *Fordham International Law Journal* **19**, 9–24.

Rajvanshi, A.K. 1981. Large scale dew collection as a source of fresh water supply. *Desalination* 3(3), 299–306.

Ramaker, T.A.B., Meuleman, A.F.M., Bernhardi, L. and Cirkel, G. 2005. Climate change and drinking water production in the Netherlands. *Water, Science and Technology* **51**(5), 37–44.

Ramanathan, V., Ramana, M.V., Roberts, G., Kim, D., Corrigan, C., Chung, C. and Winker, D. 2007. Warming trends in Asia amplified by brown cloud solar absorption. *Nature* **448**, 575–578. doi: 10.1038/nature06019.

Rang, M.C. and Schouten, C.J.J. 1988. Major obstacles to water quality management, Part 2: hydro-inertia. *Proceedings of International Association of Theoretical and Applied Limnology* **23**, 1482–1487.

Rang, M.C., Kleijn, C.E. and Schouten, C.J.J. 1986. Historic changes in the enrichment of fluvial deposits with heavy metals. In: *Monitoring to detect changes in water quality series,* International Association of Hydrological Sciences Pub. No. 157, 47–59.

Ranjan, P., Kazama, S., Sawamoto, M. and Sana, A. 2009. Global scale evaluation of coastal fresh groundwater resources. *Ocean and Coastal Management* **52**(3–4), 197–206.

Rasch, P.J., Tilmes, S., Turco, R.P., Robock, A., Oman, L., Chan, C.-C., Stenchikov, G.L. and Garcia, R.R. 2008. An overview of geoengineering of climate using stratospheric sulphate aerosols. *Philosophical Transactions of the Royal Society A* **366**(1882), 4007–4037. doi 10.1098/rsta.2008.0131.

Raymo, M.E. and Ruddiman, W.F. 1992. Tectonic forcing of late Cenozoic climate. *Nature* **359**, 117–122.

Raymond-Wish, S., Mayer, L.P., O'Neal, T., Martinez, A., Sellers, M.A., Christian, P.J., Marion, S.L., Begay, C., Propper, C.R., Hoyer, P.B. and Dyer, A.C. 2007. Drinking water with uranium below the U.S: EPA water standard causes estrogen receptor-dependent responses in female mice. *Environmental Health Perspectives* **115**(12), 1711–1716.

Rees, Lord M. 2003/2004. *Our Final Century?: Will the human race survive the twenty-first century?* Heinemann (2003), Arrow pb (2004). Published in USA as *Our Final Hour: A scientist's warning: How terror, error, and environmental disaster threaten humankind's future in this century – on Earth and beyond.* Basic Books (2003), pb (2004).

Reichel, F., Hans-Reinhard, V., Kramer, S., Cluckie, I.D. and Rico-Ramirez, M.A. 2009. Radar-based flood forecasting for river catchments. *Proceedings of the Institution of Civil Engineer, Water Management,* **162**, 159–168. doi: 10.1680/wama.2009.162.2.159.

Reisner, M. 1993. *Cadillac desert: the American west and its disappearing water.* New York, Penguin, 582pp.

Richardson, C.J., Reiss, P., Hussain, N.A., Alwash, A.J. and Pool, D.J. 2005. The restoration potential of the Mesopotamian marshes of Iraq. *Science* **307**(5713), 1307–1311.

Rignot, E. and Jacobs, S.S. 2002. Rapid bottom melting widespread near Antarctic Ice Sheet grounding lines. *Science* **296**(5575), 2020–2023.

Rignot, E., Bamber, J.L., van den Broeke, M.R., Davis, C., Yonghong, Li, van de Berg, W.J. and van Meijgaard, E. 2008. Recent Antarctic ice mass loss from radar interferometry and regional climate modelling. *Nature Geoscience* **1**, 106–110. doi: 10.1038/ngeo102.

Rijssenbeek, W. 2007. *Jatropha global position.* At: www.europa.eu/research/agriculture/pdf/events/1jatropha_eu.pdf.

Roberts, G. and Marsh, T. 1987. The effects of agricultural practices on the nitrate concentrations in the surface water domestic supply sources of Western Europe. In: Rodda, J.C. and Matalas, N.C. (eds) *Water for the future: hydrology in perspective,* International Association of Hydrological Sciences Pub. No. 164, 365–80.

Roberts, J.M. 1985. *The triumph of the West.* London, BBC, and Boston, Little Brown, 464pp.

Roberts, J.M. 2004. *Ancient history: from the first civilizations to the Renaissance.* London, Duncan Baird Publishers, 911pp.

Roberts, W.O. and Lansford, H. 1979. *The climate mandate*. San Francisco, W.H. Freeman, 197pp.

Robinson, D.A., Dewey, K.F. and Heim, R.R. 1993. Global snow cover monitoring: an update. *Bulletin of the American Meteorological Society* **74**, 1689–1696.

Roche, P-A., Valiron, F., Coulomb, R. and Villessot, D. 2001. Infrastructure integration issues. In: Maksimovic, C. and Tejada-Guibert, J.A. (eds) *Frontiers in urban water management. Deadlock or hope*, London, International Water Association Publishing, chapter 4.

Rodda, J.C. 1970. Rainfall excesses in the United Kingdom. *Transactions of the Institute of British Geographers* **49**, 49–60.

Rodda, J.C. 2006. Sustaining water resources in South East England. *Atmospheric Science Letters* **7**(3), 75–77.

Rodda, J.C. and Ubertini, L. (eds) 2004. *The basis of civilization – water science?* International Association of Hydrological Sciences Publication No. 286.

Rodda, J.C., Pieyns, S.A., Sehmi, N.S. and Matthews, G.R. 1993. Towards a world hydrological cycle observing system. *Hydrological Sciences Journal* **38**, 373–378.

Rogers, D.J. and Randolph, S.E. 2000. The global spread of malaria in a future, warmer world. *Science* **289**, 1763–1766. doi 10.1126/science.289.5485.1763.

Ropelewski, C.F. and Halpert, M.S. 1986. North American precipitation and temperature patterns associated with the El Niño/Southern Oscillation (ENSO). *Monthly Weather Review* **114**, 2352–2362.

Ropelewski, C.F. and Halpert, M.S. 1987. Global and regional scale precipitation patterns associated with the El Niño/Southern Oscillation. *Monthly Weather Review* **115**, 1606–1626.

Rose, C. 1990. Energy and efficiency in the realignment of common-law water rights. *Journal of Legal Studies* **19**, 261–296.

Rowntree, P.R. 1990. Estimates of future climatic changes over Britain. Part 2: results. *Weather* **45**(3), 79–89.

Roy, A. 1999. *The cost of living: the greater common good and the end of imagination*. London, Flamingo (HarperCollins).

Royal Society 2009. *Geoengineering the climate: science, governance and uncertainty*. Panel chair: John Shepherd, London, The Royal Society, 83pp.

Royal Society for the Protection of Birds (RSPB) 2007. *The Uplands – time to change?* Download at: www.rspb.org.uk/ourwork/conservation/projects/uplands/index.asp.

Sachs, J.D. 2005. *The End of Poverty: how we can make it happen in our lifetime*. London, Penguin, 396pp.

Sagan, D. 1990. *Biospheres: metamorphosis of Planet Earth*. Arkana, Penguin Books, London, 198pp.

Sagoff, M. 2002. On the value of natural ecosystems: The Catskills Parable. *Philosophical and Public Policy Quarterly* **22**(1/2), 10–16.

Sala, M. and Rubio, J.L. (eds) 1994. *Soil erosion as a consequence of forest fires*. Geoforma Ediciones, Logroño, 275pp.

Salem, O. and Pallas, P. 2001. The Nubian Sandstone Aquifer System (NSAS). In: Puri, S. (ed.) *Internationally shared (transboundary) aquatic resources management – a framework document*, IHP-VI, Non-serial Documents in Hydrology, International Hydrological Programme, UNESCO, Paris, 41–44.

Salter, S., Sortino, G. and Latham, J. 2008. Sea-going hardware for the cloud albedo method of reversing global warming. *Philosophical Transactions of the Royal Society of London A* **366**. doi: 10.1098/rsta.2008.0136.

Sancho Marco, T.A. 2006. Social acceptability of dams: facts and arguments. Public awareness concerning dams. In: Berga, L., Buil, J.M., Bofill, E., De Cea, J.C., Garcia Perez, J.A., Mañueco, G., Polimon, J., Soriano, A. and Yagüe, J. (eds) *Dams and Reservoirs, Societies and Environment in the 21st century*, volume 2. Leiden, Taylor and Francis/Balkema, 1327–1332.

Sauer, J. 2009. No-plumbing disease. At: www.huffingtonpost.com/john-sauer/no-plumbing-disease_b_158567.html.

Saunders, M.A. and Lea, A.S. 2008. Large contribution of sea surface warming to recent increase in Atlantic hurricane activity. *Nature* **451**, 557–560.

Scanlon, B.R., Keese, K.E., Flint, A.L., Flint, L.E., Gaye, C.B., Edmunds, W.M. and Simmers, I. 2006. Global synthesis of groundwater recharge in semiarid and arid regions. *Hydrological Processes* **20**, 3335–3370.

Schijven, J.F. and de Roda Husman, A.M. 2005. Effects of climate change on waterborne disease in The Netherlands. *Water, Science and technology* **51**(5), 79–87.

Schnug, E., Steckel, H. and Haneklaus, S. 2005. Contribution of uranium in drinking waters to the daily uranium intake of humans – a case study from Northern Germany. *Landbauforsch Volkenrode* **55**, 227–236.

Seager, A. 2009. World's poor suffering most in the credit crunch. *The Guardian*, 5 March.

431

Seely, M. 1987. *The Namib: Natural history of an ancient desert.* Shell Oil South West Africa, Windhoek, 104pp.

Shahdeganova, M., Hagg, W., Hassell, D. and Stokes, C.R. 2009. Climate change, glacier retreat, and water availability in the Caucasus region. In: Jones, J.A.A., Vardanian, T.G. and Hakopian, C. (eds) *Threats to global water security*, NATO Science for Peace and Security Series, Dordrecht, Springer, 131–143.

Sharpe, R. and Skakkebaek, N. 1993. Are oestrogens involved in falling sperm counts and disorders of the male reproductive tract? *The Lancet* 341, 1392–1395.

Shea, D.A. 2003. *Critical infrastructure control systems and the terrorist threat.* Report to Congress, Order Code RL315334, 14pp.

Shelanski, H. and Klein, P. 1995. Empirical research in transaction cost economics: a review and assessment. *Journal of Law, Economics and Organization* 11, 335–361.

Shiklomanov, A.I. and Lammers, R.B. 2009. Record Russian river discharge in 2007 and the limits of analysis. *Environmental Research Letters* 4, 045015. doi: 10.1088/1748-9326/4/4/045015.

Singer, F.S. 1973. Is growth obsolete? Comment. In: Moss, M. (ed.) *The measurement of economic and social performance*, National Bureau of Economic Research, 532–536.

Singh, V.P. and Frevert, D.K. (eds) 2002a. *Mathematical models of small watershed hydrology and applications.* Water Resources Publications, Highlands Ranch, Colorado, 950pp.

Singh, V.P. and Frevert, D.K. (eds) 2002b. *Mathematical models of large watershed hydrology.* Water Resources Publications, Highlands Ranch, Colorado, 928pp.

Smith, R.P. 2009. The blue baby syndrome. *American Scientist* 97(2), 94–96.

Soil Association 2009. *Soil carbon and organic farming: a review of the evidence of agriculture's potential to combat climate change.* Soil Association, Bristol, 212pp. At: www.soilassociation.org/.

Solomon, S., Qin, D., Manning, M., Chen, Z., Marquis, M., Averyt, K.B., Tigor, M. and Miller, H.L. (eds) 2007. *Climate Change 2007: The physical science basis. Contribution of Working Group I to the Fourth assessment Report of the Intergovernmental Panel on Climate Change*, Cambridge, Cambridge University Press.

Speidel, D.H. and Agnew, A.F. 1988. World water budget. In: Speidel, D.H., Ruedisili, L.C. and Agnew, A.F. (eds) *Perspectives on water uses and abuses*, Oxford, Oxford University Press.

Srivastava, B.C., Singh, R.B. and Mohanty, R.K. 2004. Integrated farming approach for run-off recycling schemes in humid plateaus areas of Eastern India. *Agricultural Water Management* 64(3), 197–222.

Stanko, Š. and Mahriková, I. 2009. Sewer system condition: type of sewers and their impacts. In: Jones, J.A.A., Vardanian, T.G. and Hakopian, C. (eds) *Threats to Global Water Security*, Nato Science for Peace and Security Series, Dordrecht, Springer, 359–364.

Starodub, N.F. 2009. Biosensors in a system of instrumental tools to prevent effects of bioterrorism and automotive control of water process purification. In: Jones, J.A.A., Vardanian, T.G. and Hakopian, C. (eds) *Threats to Global Water Security*, Nato Science for Peace and Security Series, Dordrecht, Springer, 59–71.

Stavrev, E., Cihák, M. and Harjes, T. 2009. Euro Area monetary policy in uncharted waters. *International Monetary Fund Working Paper* No. 09/185.

Steinberg, C.E.W. and Wright, R.F. (eds) 1994. *Acidification of freshwater ecosystems – implications for the future.* Chichester, Wiley, 404pp.

Steppuhn, H. 1981. Snow and agriculture. In: Gray, D.M. and Male, D.H. (eds) *Handbook of snow: principles, processes, management and use.* Oxford, Pergamon, 60–125.

Steppuhn, H., Falk, K.C. and Stumborg, M.A. 2008. Producing biodiesel oil from saline lands. In: *Proceedings of the 45th Annual Alberta Soil Science Workshop*, 19–21 February, 2008, 25–33. Downloadable at: www.soilworkshop.ab.ca/2008/ASSW_Proceedings_Volume_2008.pdf.

Stetson, T.M. 1980. Let's look at some legal implications of streamflow forecasting. In: *Improved hydrologic forecasting: how and why*, Pacific Grove, California, American Society of Civil Engineers and American Meteorological Society, 312–319.

Stott, P.A., Gillett, N.P. and Hegerl, G.C. 2010. Detection and attribution of climate change: a regional perspective. *Wiley Interdisciplinary Reviews: Climate Change*, doi: 10.1002/wcc.34.

Street-Perrott, F.A., Holmes, J.A., Waller, M.P., Allen, M.J., Barber, N.G.H., Fothergill, P.A., Harkness, D.D., Ivanovich, M., Kroon, D. and Perrott, R.A. 2000. Drought and dust deposition in the West African Sahel: a 5500-year record from Kajemarum Oasis, northeastern Nigeria. *The Holocene* 10, 293–302.

Struckmeier, W.F., Gilbrich, W.H., Richts, A. and Zaepke, M. (eds) 2008. WHYMAP and the Groundwater Resources Map of the World at the scale of

1:50 000 000. Hannover, BGR and Paris, UNESCO. Accessible at: http://www.whymap.org.

Struckmeier, W.F., Rubin, Y. and Jones, J.A.A. 2005. *Groundwater – reservoir for a thirsty planet*. Leiden, Netherlands, International Union of Geological Sciences, Earth Sciences for Society Foundation, 14pp.

Stuart, T. 2009. *Waste: uncovering the global food scandal*. London, Penguin, 496pp.

Stubenrauch, C.J. and Schumann, U. 2005. Impact of air traffic on cirrus coverage. *Geophysical Research Letters* 32(14), L14813.1-L14813.4.

STWR (Share The World's Resources) 2010. Water super profits in a time of crisis: who controls the world's water? Sustainable economics to end global poverty. At: www.stwr.org/land-energy-water/water-super-profits-in-a-time-of-crisis-who-controls-the-worlds-water/.

Svensmark, H., Bondo, T. and Svensmark, J. 2009. Cosmic ray decreases affect atmospheric aerosols and clouds. *Geophysical Research Letters* 36, L15101. doi: 10.1029/2009GL 038429.

Svensmark, H., Pedersen, J.O.P., Marsh, N., Enghoff, M. and Uggerhøj, U. 2007. Experimental evidence for the role of ions in particle nucleation under atmospheric conditions. *Proceedings of the Royal Society A*, 463(2078), 385–396. doi: 10.1098/rspa.2006.1773.

Swan, S.H., Main, K.M., Liu, F., Stewart, S.L., Kruse, R.L., Calafat, A.M., Mao, C.S., Redmon, J.B., Ternand, C.L., Sullivan, S. and Teague, J.L. 2005. Decrease in anogenital distance among male infants with prenatal phthalate exposure. *Environmental Health Perspectives*. 113(8), 1056–61. PMID 16079079. PMC 1280349. http://ehpnet1.niehs.nih.gov/members/2005/8100/8100.html.

Takeuchi, K. 2002a. Flood management in Japan – from rivers to basins. In: Simonovic, S.P. (ed.) *Non-structural flood protection and sustainability*, Proceedings of the International Workshop, London, Ontario, Canada, October 2001, 37–44.

Takeuchi, K. 2002b. Flood management in Japan – from rivers to basins. *Water International* 27, 20–26.

Taleb, N.N. 2007. *The Black Swan: the impact of the highly improbable*. London, Allen Lane.

Teclaff, L. 1967. *The river basin in history and law*. Buffalo, W.S. Hein.

Tennant, D.L. 1976. Instream flow regimens for fish, wildlife, recreation and related environmental resources. *Fisheries* 1(4), 6–10.

Theroux, P. 2006. Bono aid is making Africa sick. *The Sunday Times* News Review, 1 January, p.7.

Thomas, L., Jenkins, D.B., Wareing, D.P., Vaughan, G. and Farrington, M. 1987. Lidar observations of stratospheric aerosols associated with the El Chichón eruption. *Annales Geophysicae* 5A, 47–56.

Thompson, L.G., Brecher, H.H., Mosley-Thompson, E., Hardy, D.R. and Mark, B.G. 2009. Glacier loss on Kilimanjaro continues unabated. *Proceedings of the National Academy of Sciences of the United States of America*, 106, 19770–19775. doi 10.1073/pnas.0906029106.

Tickell, Sir C. 1977. *Climate change and world affairs*. Cambridge, Mass., Harvard Studies in International Affairs No. 37, 76pp.

Tickell, Sir C. 1991. The human species: a suicidal success? *Geographical Journal*, 159(2), 219–226.

Tiisetso, T. 2009. Is South Africa heading for a recession? South Africa's economic outlook (2008–2009). At: http://international-financial-affairs.suite101.com/article.cfm/is_South_Africa_heading_for_a_recession.

Tinsley, B.A. 1988. The solar cycle and the QBO influences on the latitude of storm tracks in the North Atlantic. *Geophysical Research Letters* 15, 409–412.

Tooley, M.J. 1991. Future sea level rise – a working paper. *Proceedings of the Institution of Civil Engineers* 90(5, part 1), 1101–1105.

Topping, A.B. 1996. Dai Qing, voice of the Yangtze River Gorges. *The Earth News* At: http://weber.cusd.edu/dmccubbi/chinadaiqingjan11_97.htm.

Toussoun, O. 1925. Mémoire sur l'histoire du Nil. *Memoire a L'Institut D'Egypte* 18, 366–404.

Troup, A.J. 1965. The Southern Oscillation. *Quarterly Journal of the Royal Meteorological Society* 91, 490–506.

Tutu, Archbishop D. 2008. World's poor will die from credit crunch. *The Sun*, 20 August. At: www.thesun.co.uk/sol/homepage/news/money/article 1580756.ece.

Uluocha, N. and Okeke, I. 2004. Implications of wetlands degradation for water resources management: Lessons from Nigeria. *GeoJournal*, special issue 61(2), 151–154.

UN 1970. *Integrated river basin development: report of a panel of experts*. New York, UN Department of Economic and Social Affairs, 60pp.

UN 1997. *Convention on the Law of Non-Navigational Uses of International Watercourses,* approved May 21, UN Doc. No. A/51/869, reprinted in *International Legal Materials* 36, 700–20.

UNECE 2004. Forest and wetland: suppliers of water and first line of defence against floods. At: www.unece.org/press/pr2004/04env_p22e.htm.

UNEP 2007. *Global environment outlook.* 4th report (GEO-4). Downloadable at: www.unep.org/geo/geo4.

UNEP GEMS/Water 2007. *Water quality for ecosystem and human health.* At: http://www.gemswater.org/freshwater_assessments/index-e.html.

UNEP/FAO/UNESCO 2009. *Blue Carbon – the role of healthy oceans in binding carbon.* At: http://www.grida.no/publications/rr/blue-carbon/.

UNESCO 2006. *Water: a shared responsibility. The United Nations World Water Developmernt Report 2.* UNESCO, Paris and Berghahn Books, New York, 584pp.

UN-Habitat Water and Sanitation Programme 2006. Framework for Gender Mainstreaming, Water and Sanitation for Cities. Gender Mainstreaming Strategy Initiative, WAC II.

UN Human Settlements Programme, Nairobi, Kenya, 31pp. At: www.unhabitat.org/downloads/docs/GenderMainstreamingReport.pdf.

UNICEF 2003. At a glance: Mozambique. At: http://www.unicef.org/infobycountry/mozambique_2231.html.

UN-Water Interagency Task Force on Gender and Water (GWTF). *Gender, water and sanitation – a policy brief.* Interagency Task Force on Gender and Water Sub-programme of UN-Water and Interagency Network on Women and Gender Equality (IANWGE), UN Water Policy Brief 2, 13pp. At: www.unwater.org/downloads/.

Utton, A. 1996. Regional cooperation: the example of international waters systems in the twentieth century. *Natural Resources Journal* 36, 151–154.

van den Broeke, M., Bamber, J., Ettema, J., Rignot, E., Schrama, E., van de Berg, W.J., van Meijgaard, E., Velicogna, I. and Wouters, B. 2009. Partitioning recent Greenland mass loss. *Science* 326(5955), 984. doi: 10.1126/science.1178176.

van der Gun in press. Hydrogeological zoning: from local to global scales. In: Jones, J.A.A. (ed.) *Sustaining groundwater resources*, Dordrecht, Springer Legacy Series.

Van Loon, H. and Labitzke, K. 1988. Association between the 11-year solar cycle, the QBO, and the atmosphere. Part II: surface and 700 mb in the northern hemisphere in winter. *Journal of Climatology* 1, 905–920.

Vardanian, T.G. and Robinson, P.J. 2007. Lake Sevan: the loss and restoration of the 'Pearl of Armenia'. In: Robinson, P.J., Jones, J.A.A. and Woo, M-K. (eds) *Managing water resources in a changing physical and social environment*, Rome, International Geographical Union, 67–77.

Vecchi, G.A. and Soden, B.J. 2007. Effects of remote sea surface temperature change on tropical cyclone potential intensity. *Nature* 450, 1066–1071.

Vehlainen, B. and Lohvansuu, J. 1991. The effects of climate change on discharges and snow cover in Finland. *Hydrological Sciences Journal* 36(2), 109–121.

Velicogna, I. 2009. Increasing rates of mass loss from the Greenland and Antarctic ice sheets revealed by GRACE, *Geophysical Research Letters*, 36, L19503. doi: 10.1029/2009GL040222.

Verbiscer, A.J., Skrutskie, M.F. and Hamilton, D.P. 2009. Saturn's largest ring. *Nature* 461, 1098–1100. doi: 10.1038/nature08515, PMID: 19812546.

Vermeer, M. and Rahmstorf, S. 2009. Global sea level linked to global temperature. *Proceedings of the National Academy of Sciences (USA)*, 106, 21527–21532. www.pnas.org/cgi/doi/10.1073/pnas.0907765106.

Vermeulen, A. 2005. Presentation at the 2nd Annual Water Services Convention, Midrand, June 2005. *Water Wheel* 4(4), 6.

Vidal, J., 2006. How your supermarket flowers empty Kenya's rivers. *The Guardian*, at www.guardian.co.uk/kenya/story/0,,1928004.html.

Vigneswaran, S. and Sundaravadivel, M. 2004. Recycle and reuse of domestic wastewater. In: Saravanamuthu, V. (ed.) *Encyclopedia of life support systems*, Oxford, EOLSS Publishers, under auspices of Unesco, ebooks at www.eolss.net/ebooks/.

Vonnegut, B. 1947. The nucleation of ice formation by silver iodide. *Journal of Applied Physics* 18, 593–595. doi: 10.1063/1.1697813.

Walling, D.E. 2006. Human impact on land–ocean sediment transfer by the world's rivers. *Geomorphology* 79, 192–216.

Walling, D.E., Quine, T.A. and Rowan, J.S. 1992. Fluvial transport and redistribution of Chernobyl fallout radionuclides. *Hydrobiologia* 235–236(1), 231–246.

Walling, D.E. and Webb, B.W. 1983. Patterns of sediment yield. In: Gregory, K.J. (ed.) *Background to palaeohydrology*, Chichester, Wiley, 69–100.

Walling, D.E. and Webb, B.W. 1996. Erosion and sediment yield: a global overview. In: *Erosion and Sediment Yield: Global and Regional Perspectives*, Proceedings of the Exeter Symposium, IAHS Publ.no. 236, 3–19.

Wallingford, H.R. 1993. *Environmental changes*

downstream of Sardar Sarovar. Report EX 2750. Available online at the Friends of Narmada website, www.narmada.org/ENV/index.html.

Wang, M. and Overland, J.E. 2009. A sea ice free summer Arctic within 30 years? *Geophysical Research Letters* 36(7), 2–6. doi: 10.1029/2009GL037820.

Ward, D., Holmes, N. and José, P. 1994. *The new rivers and wildlife handbook.* Reprinted 2001, Sandy, Bedfordshire, Royal Society for the Protection of Birds, NRA, and Wildlife Trusts, 426pp.

Warner, J. 2003. Virtual water – virtual benefits? Scarcity, distribution and conflict reconsidered. In: Hoekstra, J.A. (ed.) *Virtual water trade: Proceedings of the International Expert Meeting on Virtual Water Trade.* Value of Water Research Report Series No. 12, chapter 8, 125–133.

Watts, G. 2008. How clean is your water? *British Medical Journal* 337, 80–82.

Webb, B.W. 1992. *Climate change and the thermal regime of rivers.* Unpublished report to the Department of the Environment, University of Exeter, Department of Geography, 79pp.

Webb, B.W. and Nobilis, F. 1995. Long term water temperature trends in Austrian rivers. *Hydrological Sciences Journal* 40, 83–96.

Webb, B.W., David, M., Hannah, D.M., Moore, R.D., Brown, L.E. and Nobilis, F. 2008. Recent advances in stream and river temperature research. *Hydrological Processes* 22(7), 902–915. doi: 10.1002/hyp.6994.

Webb, B.W., Zhang, Y. and Nobilis, F. 1995. Scale-related water temperature behavior. In: *Effects of Scale on Interpretation and Management of Sediment and Water Quality,* IAHS Publ. No. 226, 231–239.

Weeks, W.F. and Campbell, W.J. 1973. Icebergs as a freshwater source: an appraisal. *Journal of Glaciology* 12(65), 207–233.

Weston, D.P., Amweg, E.L., Mekebri, A., Ogle, R.S. and Lydy, M.J. 2006. Aquatic effects of aerial spraying for mosquito control over an urban area. *Environmental Science Technology* 40(18), 5817–5822. doi: 10.1021/es0601540.

Wetz, M.S. and Paerl, H.W. 2008. Estuarine Phytoplankton Responses to Hurricanes and Tropical Storms with Different Characteristics. *Estuaries and coasts* 31(2), 419–429.

WHO 1989. Geographical distribution of arthropod-borne diseases and their principal vectors. Geneva, WHO Vector Biology and Control Division.

WHO 1993. Guidelines for drinking-water quality. 2nd edition, volume 1, Geneva, World Health Organization.

WHO 1999. Agreement between the International Atomic Energy Agency (IAEA) and the WHO, ratified by the 12th WHO-General Assembly, 28 May 1959, Resolution WHA 1240, Organisation mondiale de la santé, 'Documents fondamentaux', 42nd edition, Geneva, World Health Organization.

WHO 2003. *Hardness in drinking-water. Background document for development of WHO Guidelines for drinking-water.* WHO/SDE/WSH/03.04/06 (Original 1996).

WHO 2005. *Water Safety Plans: Managing drinking-water quality from catchment to consumer.* 3rd edition, prepared by: Davison, A., Howard, G., Stevens, M., Callan, P., Fewtrell, L., Deere, D. and Bartram, J., WHO, Geneva, 235pp. Download at: www.who.int/water_sanitation_health/dwq/wsp0506/en/index.html.

WHO 2008. *Guidelines for drinking-water quality.* 3rd edition, incorporating 1st and 2nd addenda, WHO, Geneva. Available at: www.who.int/water_sanitation_health/dwq/gdwq3rev/en/.

WHO 2008. *Safer water, better health.* Prepared by: Prüss-Üstün, A., Bos, R., Gore, F., Bartram, J. WHO, Geneva, 53pp. Download at: www.who.int/quantifying_ehimpacts/publications/saferwater/en/index.html.

WHO 2009. *Flooding and communicable diseases fact sheet.* Download at: www.who.int/hac/techguidance/ems/flood_cds/en/.

WHO Commission on Social Determinants of Health, 2008. *Closing the Gap in a Generation: Health Equity through Action on the Social Determinants of Health.* World Health Organization.

Wilby, R.L., Conway, D. and Jones, P.D. 2002. Prospects for downscaling seasonal precipitation variability using conditioned weather generator parameters. *Hydrological Processes* 16, 1215–1234.

Wilcock, A.A. 1968. Köppen after fifty years. *Annals of the Association of American Geographers* 58(1), 12–28.

Winde, F. 2008. *Development of a map ranking sites with known radioactive pollution in the Wonderfonteinspruit catchment according to the urgency of required intervention ('Intervention site map') – Underlying methodology and results.* Confidential report to Joint Coordinating Committee of the Department of Water Affairs and Forestry (DWAF) and the National Nuclear Regulator (NNR), DWAF Pretoria, unpublished, 17pp.

Winde, F. 2009. *Urankontamination von Fliessgewässern – Prozessdynamik, Mechanismen und Steuerfaktoren. Untersuchungen zum Transport von gelöstem Uran in bergbaulich gestörten Landschaften unterschiedlicher Klimate.* (Uranium contamination of streams – process dynamics, mechanisms and governing factors. Investigations into the transfer of dissolved uranium in mining-impacted landscapes of different climatic conditions), ISBN 978-3-86866-087-6, Taunusstein, Driesen Verlag, 657pp. (In German.)

Winde, F. 2010a. Uranium pollution of the Wonderfonteinspruit, 1997–2008. Part I: U-toxicity, regional background and mining-related sources of U-pollution. *Water SA* 36(3), 239–256. At: www.wrc.org.za/.

Winde, F. 2010b. Uranium pollution of the Wonderfonteinspruit: 1997–2008. Part II: U in water – concentrations, loads and associated risks. *Water SA* 36(3), 257–278.

Winter, T.C. 2000. The vulnerability of wetlands to climate change: a hydrologic landscape perspective. *Journal of American Water Resources Association* 36, 305–311.

Wisner, B., Blaikie, P., Cannon, T. and Davis, I. 2004. *At risk: natural hazards, people's vulnerability and disasters.* 2nd edition, London and New York, Routledge, 471pp.

Wither, A., Greaves, J., Dunhill, I., Wyer, M., Stapleton, C., Kay, D., Humphrey, N., Watkins, J., Francis, C., McDonald, A. and Crowther, J. 2005. Estimation of diffuse and point source microbial pollution in the Ribble catchment discharging to bathing waters in the north west of England. *Water Science and Technology* 51(3–4), 191–198.

WMO in press. *Guide to hydrological practices.* 6th edition, Geneva, WMO.

Wolf, A.T. 1998. Conflict and cooperation along international waterways. *Water Policy* 1(2), 251–265.

Wolfke, K. 1993. *Custom in present international law.* 2nd revised edition, Dordrecht, Martinus Nijhoff Publishers.

Wolfrum, R. (ed.) 1996. *Enforcing Environmental Standards: Economic Mechanisms as Viable Means?* London, Kluwer Law International.

Woo, M.K. and Xia, Z.J. 1996. Effects of hydrology on the thermal conditions of the active layer. *Nordic Hydrology* 27, 129–142.

Woo, M.K. and Young, K.L. 2003. Hydrogeomorphology of patchy wetlands in the High Arctic, polar desert environment. *Wetlands* 23, 291–309.

World Bank 2003. Report No. 25917. Available at: http://www-wds.worldbank.org/servlet/WDSContentServer/WDSP/IB/2003/06/17/000090341_20030617084733/Rendered/PDF/259171MA1Rural1ly010Sanitation01ICR.pdf.

World Commission on Dams (WCD) 2000. *Dams and Development: a new framework for decision-making.* World Commission on Dams. At: www.unep.org/damsWCD/.

World Meteorological Organization (WMO) 1994. *Guide to hydrological practices.* 5th edition, WMO Pub. No. 168, Geneva, WMO, 735pp.

World Nuclear Association 2009. *World uranium mining.* At: http://www.world-nuclear.org/info/inf23.html.

World Resources Institute 2008. *Watching water: a guide to evaluating corporate risks in a thirsty world.* A report by J.P. Morgan. Available at: www.wri.org/publication/watching-water.

World Water Assessment Programme (WWAP) 2009. *The United Nations World Water Development Report 3: water in a changing world.* Unesco, Paris, and Earthscan, London.

World Water Council (WWC) 2004. *E-conference synthesis: virtual water trading – conscious choices.* World Water Council Publication No. 2, 31pp. Available at: www.worldwatercouncil.org/fileadmin/wwc/Library/Publications_and_reports/ virtual_water_final_synthesis.pdf.

World Wildlife Fund (WWF) online. Dams and infrastructure. At: http://www.wwf.org.uk/what_we_do/safeguarding_the_natural_world/rivers_and_lakes/dams_and_infastructure.

World Wildlife Fund (WWF) online. The value of wetlands – a source of life for people and wildlife. At: wwf.panda.org/about_our_earth/about_freshwater/intro/value.

WRc (Water Research Council) 2003. *Sustainable Water Management. A Future View.* Sustainable Water Management Associate Foresight Programme. WRc plc, June.

Xia, J. 2007. Water problems and sustainability in North China. In: Robinson, P.J., Jones, J.A.A. and Woo, M-K. (eds) *Managing water resources in a changing physical and social environment*, Rome, International Geographical Union, 79–87.

Yperlaan, G.J. 1977. Statistical evidence of the influence of urbanization on precipitation in the Rijmond area. In: *Effects of urbanization and industrialization on the hydrological regime and on water quality*, International Association of Hydrological Sciences, Pub. No. 123, 20–30.

Zacklin, R. and Caflisch, L. (eds) 1981. *The legal regime of international rivers and lakes*. The Hague, Martinus Nijhoff.

Zahran, N.A. 2010. Impending regional firestorm over Nile River. *Ethiopian Review*, 20 May. At: www.ethiopianreview.com/article/32107.

Zenith International 2009. *Global bottled water report*. At: www.zenithinternational.com.

Zherelina, I. 2003. In a turn to the past, Moscow proposes to reverse Siberia's rivers. *Give and Take* 6(2), 10–11.

Zuccato, E. and Castiglioni, S. 2009. Illicit drugs in the environment. *Philosophical Transactions of the Royal Society A* 367(1904), 3965–3978. doi: 10.1098/rsta.2009.0107.

Zuccato, E., Castiglioni, S., Bagnati, R., Chiabrando, C., Grassi, P. and Fanelli, R. 2008. Illicit drugs, a novel group of environmental contaminants. *Water Research*, 42(4–5), 961–968.

Index